Springer Monographs in Mathematics

Editorial Board
S. Axler
K.A. Ribet

For further volumes:
www.springer.com/series/3733

Asen L. Dontchev · R. Tyrrell Rockafellar

Implicit Functions and Solution Mappings

A View from Variational Analysis

With 12 Illustrations

Asen L. Dontchev
Mathematical Reviews
416 Fourth Street
Ann Arbor, MI 48107-8604
USA
ald@ams.org

R. Tyrrell Rockafellar
University of Washington
Department of Mathematics
PO Box 354350
Seattle, WA 98195-4350
USA
rtr@math.washington.edu

ISSN 1439-7382   e-ISSN
ISBN 978-0-387-87820-1   e-ISBN 978-0-387-87821-8
DOI 10.1007/978-0-387-87821-8
Springer Dordrecht Heidelberg London New York

Library of Congress Control Number: 2009926485

Mathematics Subject Classification (2000): 26B10, 47J07, 58C15, 49J53, 49K40, 90C31, 93C70

©Springer Science+Business Media, LLC 2009
All rights reserved. This work may not be translated or copied in whole or in part without the written permission of the publisher (Springer Science+Business Media, LLC, 233 Spring Street, New York, NY 10013, USA), except for brief excerpts in connection with reviews or scholarly analysis. Use in connection with any form of information storage and retrieval, electronic adaptation, computer software, or by similar or dissimilar methodology now known or hereafter developed is forbidden.
The use in this publication of trade names, trademarks, service marks, and similar terms, even if they are not identified as such, is not to be taken as an expression of opinion as to whether or not they are subject to proprietary rights.

Printed on acid-free paper

Springer is part of Springer Science+Business Media (www.springer.com)

Preface

Setting up equations and solving them has long been so important that, in popular imagination, it has virtually come to describe what mathematical analysis and its applications are all about. A central issue in the subject is whether the solution to an equation involving parameters may be viewed as a function of those parameters, and if so, what properties that function might have. This is addressed by the classical theory of implicit functions, which began with single real variables and progressed through multiple variables to equations in infinite dimensions, such as equations associated with integral and differential operators.

A major aim of the book is to lay out that celebrated theory in a broader way than usual, bringing to light many of its lesser known variants, for instance where standard assumptions of differentiability are relaxed. However, another major aim is to explain how the same constellation of ideas, when articulated in a suitably expanded framework, can deal successfully with many other problems than just solving equations.

These days, forms of modeling have evolved beyond equations, in terms, for example, of problems of minimizing or maximizing functions subject to constraints which may include systems of inequalities. The question comes up of whether the solution to such a problem may be expressed as a function of the problem's parameters, but differentiability no longer reigns. A function implicitly obtainable this manner may only have one-sided derivatives of some sort, or merely exhibit Lipschitz continuity or something weaker. Mathematical models resting on equations are replaced by "variational inequality" models, which are further subsumed by "generalized equation" models.

The key concept for working at this level of generality, but with advantages even in the context of equations, is that of the set-valued *solution mapping* which assigns to each instance of the parameter element in the model *all* the corresponding solutions, if any. The central question is whether a solution mapping can be localized graphically in order to achieve single-valuedness and in that sense produce a function, the desired *implicit function*.

In modern variational analysis, set-valued mappings are an accepted workhorse in problem formulation and analysis, and many tools have been developed for handling them. There are helpful extensions of continuity, differentiability, and regularity of several types, together with powerful results about how they can be applied. A corresponding further aim of this book is to bring such ideas to wider attention by demonstrating their aptness for the fundamental topic at hand.

In line with classical themes, we concentrate primarily on local properties of solution mappings that can be captured metrically, rather than on results derived from topological considerations or involving exotic spaces. In particular, we only briefly discuss the Nash–Moser inverse function theorem. We keep to finite dimensions in Chapters 1 to 4, but in Chapters 5 and 6 provide bridges to infinite dimensions. Global implicit function theorems, including the classical Hadamard theorem, are not discussed in the book.

In Chapter 1 we consider the implicit function paradigm in the classical case of the solution mapping associated with a parameterized equation. We give two proofs of the classical inverse function theorem and then derive two equivalent forms of it: the implicit function theorem and the correction function theorem. Then we gradually relax the differentiability assumption in various ways and even completely exit from it, relying instead on the Lipschitz continuity. We also discuss situations in which an implicit function fails to exist as a graphical localization of the solution mapping, but there nevertheless exists a function with desirable properties serving locally as a selection of the set-valued solution mapping. This chapter does not demand of the reader more than calculus and some linear algebra, and it could therefore be used by both teachers and students in analysis courses.

Motivated by optimization problems and models of competitive equilibrium, Chapter 2 moves into wider territory. The questions are essentially the same as in the first chapter, namely, when a solution mapping can be localized to a function with some continuity properties. But it is no longer an equation that is being solved. Instead it is a condition called a generalized equation which captures a more complicated dependence and covers, as a special case, variational inequality conditions formulated in terms of the set-valued normal cone mapping associated with a convex set. Although our prime focus here is variational models, the presentation is self-contained and again could be handled by students and others without special background. It provides an introduction to a subject of great applicability which is hardly known to the mathematical community familiar with classical implicit functions, perhaps because of inadequate accessibility.

In Chapter 3 we depart from insisting on localizations that yield implicit functions and approach solution mappings from the angle of a "varying set." We identify continuity properties which support the paradigm of the implicit function theorem in a set-valued sense. This chapter may be read independently from the first two. Chapter 4 continues to view solution mappings from this angle but investigates substitutes for classical differentiability. By utilizing concepts of generalized derivatives, we are able to get implicit mapping theorems that reach far beyond the classical scope.

Chapter 5 takes a different direction. It presents extensions of the Banach open mapping theorem which are shown to fit infinite-dimensionally into the paradigm of the theory developed finite-dimensionally in Chapter 3. Some background in basic functional analysis is required. Chapter 6 goes further down that road and illustrates how some of the implicit function/mapping theorems from earlier in the book can be used in the study of problems in numerical analysis.

This book is targeted at a broad audience of researchers, teachers and graduate students, along with practitioners in mathematical sciences, engineering, economics and beyond. In summary, it concerns one of the chief topics in all of analysis, historically and now, an aid not only in theoretical developments but also in methods for solving specific problems. It crosses through several disciplines such as real and functional analysis, variational analysis, optimization, and numerical analysis, and can be used in part as a graduate text as well as a reference. It starts with elementary results and with each chapter, step by step, opens wider horizons by increasing the complexity of the problems and concepts that generate implicit function phenomena.

Many exercises are included, most of them supplied with detailed guides. These exercises complement and enrich the main results. The facts they encompass are sometimes invoked in the subsequent sections.

Each chapter ends with a short commentary which indicates sources in the literature for the results presented (but is not a survey of all the related literature). The commentaries to some of the chapters additionally provide historical overviews of past developments.

Whidbey Island, Washington *Asen L. Dontchev*
August, 2008 *R. Tyrrell Rockafellar*

Acknowledgements

Special thanks are owed to our readers Marius Durea, Shu Lu, Yoshiyuki Sekiguchi and Hristo Sendov, who gave us valuable feedback on the entire manuscript, and to Francisco J. Aragón Artacho, who besides reviewing most of the book helped us masterfully with all the figures. During various stages of the writing we also benefited from discussions with Aris Daniilidis, Darinka Dentcheva, Hélène Frankowska, Michel Geoffroy, Alexander Ioffe, Stephen Robinson, Vladimir Veliov, and Constantin Zălinescu. We are also grateful to Mary Anglin for her help with the final copy-editing of the book.

The authors

Contents

Preface   v

Acknowledgements   ix

Chapter 1. Functions defined implicitly by equations   1

   1A. The classical inverse function theorem   9
   1B. The classical implicit function theorem   17
   1C. Calmness   21
   1D. Lipschitz continuity   26
   1E. Lipschitz invertibility from approximations   35
   1F. Selections of multi-valued inverses   47
   1G. Selections from nonstrict differentiability   51

Chapter 2. Implicit function theorems for variational problems   61

   2A. Generalized equations and variational problems   62
   2B. Implicit function theorems for generalized equations   74
   2C. Ample parameterization and parametric robustness   83
   2D. Semidifferentiable functions   88
   2E. Variational inequalities with polyhedral convexity   95
   2F. Variational inequalities with monotonicity   106
   2G. Consequences for optimization   112

Chapter 3. Regularity properties of set-valued solution mappings   131

   3A. Set convergence   134
   3B. Continuity of set-valued mappings   142
   3C. Lipschitz continuity of set-valued mappings   148
   3D. Outer Lipschitz continuity   154

3E. Aubin property, metric regularity and linear openness ... 159
3F. Implicit mapping theorems with metric regularity ... 169
3G. Strong metric regularity ... 178
3H. Calmness and metric subregularity ... 182
3I. Strong metric subregularity ... 186

Chapter 4. Regularity properties through generalized derivatives ... 197

4A. Graphical differentiation ... 198
4B. Derivative criteria for the Aubin property ... 205
4C. Characterization of strong metric subregularity ... 217
4D. Applications to parameterized constraint systems ... 221
4E. Isolated calmness for variational inequalities ... 224
4F. Single-valued localizations for variational inequalities ... 228
4G. Special nonsmooth inverse function theorems ... 237
4H. Results utilizing coderivatives ... 245

Chapter 5. Regularity in infinite dimensions ... 251

5A. Openness and positively homogeneous mappings ... 253
5B. Mappings with closed and convex graphs ... 259
5C. Sublinear mappings ... 265
5D. The theorems of Lyusternik and Graves ... 274
5E. Metric regularity in metric spaces ... 280
5F. Strong metric regularity and implicit function theorems ... 292
5G. The Bartle–Graves theorem and extensions ... 297

Chapter 6. Applications in numerical variational analysis ... 311

6A. Radius theorems and conditioning ... 312
6B. Constraints and feasibility ... 320
6C. Iterative processes for generalized equations ... 326
6D. An implicit function theorem for Newton's iteration ... 336
6E. Galerkin's method for quadratic minimization ... 348
6F. Approximations in optimal control ... 352

References ... 363

Notation ... 371

Index ... 373

Chapter 1
Functions Defined Implicitly by Equations

The idea of solving an equation $f(p,x) = 0$ for x as a function of p, say $x = s(p)$, plays a huge role in classical analysis and its applications. The function obtained in this way is said to be defined *implicitly* by the equation. The closely related idea of solving an equation $f(x) = y$ for x as a function of y concerns the *inversion* of f. The circumstances in which an implicit function or an inverse function exists and has properties like differentiability have long been studied. Still, there are features which are not widely appreciated and variants which are essential to seeing how the subject might be extended beyond solving only equations. For one thing, properties other than differentiability, such as Lipschitz continuity, can come in. But fundamental expansions in concept, away from thinking just about functions, can serve in interesting ways as well.

As a starter, consider for real variables x and y the extent to which the equation $x^2 = y$ can be solved for x as a function of y. This concerns the inversion of the function $f(x) = x^2$ in Figure 1.1 below, as depicted through the reflection that interchanges the x and y axes. The reflection of the graph is not the graph of a function, but some parts of it may have that character. For instance, a function is obtained from a neighborhood of the point B, but not from one of the point A, no matter how small.

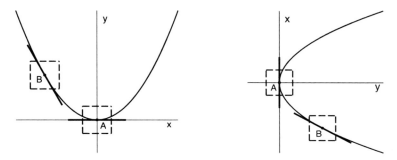

Fig. 1.1 Graphical localizations of the function $y = x^2$ and its inverse.

A.L. Dontchev and R.T. Rockafellar, *Implicit Functions and Solution Mappings: A View from Variational Analysis*, Springer Monographs in Mathematics, DOI 10.1007/978-0-387-87821-8_1, © Springer Science+Business Media, LLC 2009

Although the reflected graph in this figure is not, as a whole, the graph of a function, it can be regarded as the graph of something more general, a "set-valued mapping" in terminology which will be formalized shortly. The question revolves then around the extent to which a "graphical localization" of a set-valued mapping might be a function, and if so, what properties that function would possess. In the case at hand, the reflected graph assigns two different x's to y when $y > 0$, but no x when $y < 0$, and just $x = 0$ when $y = 0$.

To formalize that framework for the general purposes of this chapter, we focus on set-valued mappings F from $\mathbb{R}^n$ and $\mathbb{R}^m$, signaled by the notation

$$F : \mathbb{R}^n \rightrightarrows \mathbb{R}^m,$$

by which we mean correspondences which assign to each $x \in \mathbb{R}^n$ one or more elements of $\mathbb{R}^m$, or possibly none. The set of elements $y \in \mathbb{R}^m$ assigned by F to x is denoted by $F(x)$. However, instead of regarding F as going from $\mathbb{R}^n$ to a space of subsets of $\mathbb{R}^m$ we identify as the *graph* of F the set

$$\text{gph } F = \big\{ (x,y) \in \mathbb{R}^n \times \mathbb{R}^m \,\big|\, y \in F(x) \big\}.$$

Every subset of $\mathbb{R}^n \times \mathbb{R}^m$ serves as gph F for a uniquely determined $F : \mathbb{R}^n \rightrightarrows \mathbb{R}^m$, so this concept is very broad indeed, but it opens up many possibilities.

When F assigns more than one element to x we say it is *multi-valued* at x, and when it assigns no element at all, it is *empty-valued* at x. When it assigns exactly one element y to x, it is *single-valued* at x, in which case we allow ourselves to write $F(x) = y$ instead of $F(x) = \{y\}$ and thereby build a bridge to handling functions as special cases of set-valued mappings.

Domains and ranges get flexible treatment in this way. For $F : \mathbb{R}^n \rightrightarrows \mathbb{R}^m$ the *domain* is the set

$$\text{dom } F = \big\{ x \,\big|\, F(x) \neq \emptyset \big\},$$

while the *range* is

$$\text{rge } F = \big\{ y \,\big|\, y \in F(x) \text{ for some } x \big\},$$

so that dom F and rge F are the projections of gph F on $\mathbb{R}^n$ and $\mathbb{R}^m$ respectively. Any subset of gph F can freely be regarded then as itself the graph of a set-valued submapping which likewise projects to some domain in $\mathbb{R}^n$ and range in $\mathbb{R}^m$.

The *functions* from $\mathbb{R}^n$ to $\mathbb{R}^m$ are identified in this context with the set-valued mappings $F : \mathbb{R}^n \rightrightarrows \mathbb{R}^m$ such that F is single-valued at every point of dom F. When F is a function, we can emphasize this by writing $F : \mathbb{R}^n \to \mathbb{R}^m$, but the notation $F : \mathbb{R}^n \rightrightarrows \mathbb{R}^m$ doesn't preclude F from actually being a function. Usually, though, we use lower case letters for functions: $f : \mathbb{R}^n \to \mathbb{R}^m$. Note that in this notation f can still be empty-valued in places; it's single-valued only on the subset dom f of $\mathbb{R}^n$. Note also that, although we employ "mapping" in a sense allowing for potential multi-valuedness (as in a "set-valued mapping"), no multi-valuedness is ever involved when we speak of a "function."

A clear advantage of the framework of set-valued mappings over that of only functions is that *every set-valued mapping $F : \mathbb{R}^n \rightrightarrows \mathbb{R}^m$ has an inverse*, namely the

1 Functions Defined Implicitly by Equations

set-valued mapping $F^{-1}: \mathbb{R}^m \rightrightarrows \mathbb{R}^n$ defined by

$$F^{-1}(y) = \{x \mid y \in F(x)\}.$$

The graph of F^{-1} is generated from the graph of F simply by reversing (x,y) to (y,x), which in the case of $m = n = 1$ corresponds to the reflection in Figure 1.1. In this manner a function f always has an inverse f^{-1} as a *set-valued mapping*. The question of an inverse *function* comes down then to passing to some piece of the graph of f^{-1}. For that, the notion of "localization" must come into play, as we are about to explain after a bit more background. Traditionally, a function $f : \mathbb{R}^n \to \mathbb{R}^m$ is *surjective* when rge $f = \mathbb{R}^m$ and *injective* when dom $f = \mathbb{R}^n$ and f^{-1} is a function; full *invertibility* of f corresponds to the juxtaposition of these two properties.

In working with $\mathbb{R}^n$ we will, for now, keep to the Euclidean norm $|x|$ associated with the canonical inner product

$$\langle x, x' \rangle = \sum_{j=1}^n x_j x'_j \text{ for } x = (x_1, \ldots, x_n) \text{ and } x' = (x'_1, \ldots, x'_n),$$

namely

$$|x| = \sqrt{\langle x, x \rangle} = \left[\sum_{j=1}^n x_j^2\right]^{1/2}.$$

The closed ball around $\bar{x}$ with radius r is then

$$\mathbb{B}_r(\bar{x}) = \{x \mid |x - \bar{x}| \leq r\}.$$

We denote the closed unit ball $\mathbb{B}_1(0)$ by $\mathbb{B}$. A *neighborhood* of $\bar{x}$ is any set U such that $\mathbb{B}_r(\bar{x}) \subset U$ for some $r > 0$. We recall for future needs that the interior of a set $C \subset \mathbb{R}^n$ consists of all points x such that C is a neighborhood of x, whereas the closure of C consists of all points x such that the *complement* of C is *not* a neighborhood of x; C is *open* if it coincides with its interior and *closed* if it coincides with its closure. The interior and closure of C will be denoted by int C and cl C.

Graphical localization. For $F : \mathbb{R}^n \rightrightarrows \mathbb{R}^m$ and a pair $(\bar{x}, \bar{y}) \in \text{gph } F$, a graphical localization of F at $\bar{x}$ for $\bar{y}$ is a set-valued mapping $\tilde{F}$ such that

$$\text{gph } \tilde{F} = (U \times V) \cap \text{gph } F \text{ for some neighborhoods } U \text{ of } \bar{x} \text{ and } V \text{ of } \bar{y},$$

so that

$$\tilde{F} : x \mapsto \begin{cases} F(x) \cap V & \text{when } x \in U, \\ \emptyset & \text{otherwise.} \end{cases}$$

The inverse of $\tilde{F}$ then has

$$\tilde{F}^{-1}(y) = \begin{cases} F^{-1}(y) \cap U & \text{when } y \in V, \\ \emptyset & \text{otherwise,} \end{cases}$$

and is thus a graphical localization of the set-valued mapping F^{-1} at $\bar{y}$ for $\bar{x}$.

Often the neighborhoods U and V can conveniently be taken to be closed balls $\mathbb{B}_a(\bar{x})$ and $\mathbb{B}_b(\bar{y})$. Observe, however, that the domain of a graphical localization $\tilde{F}$ of F with respect to U and V may differ from $U \cap \operatorname{dom} F$ and may well depend on the choice of V.

Single-valuedness in localizations. By a single-valued localization of F at $\bar{x}$ for $\bar{y}$ will be meant a graphical localization that is a function, its domain not necessarily being a neighborhood of $\bar{x}$. The case where the domain is indeed a neighborhood of $\bar{x}$ will be indicated by referring to a single-valued localization of F around $\bar{x}$ for $\bar{y}$ instead of just at $\bar{x}$ for $\bar{y}$.

For the function $f(x) = x^2$ from $\mathbb{R}$ to $\mathbb{R}$ we started with, the set-valued inverse mapping f^{-1}, which is single-valued only at 0 with $f^{-1}(0) = 0$, fails to have a single-valued localization at 0 for 0. But as observed in Figure 1.1, it has a single-valued localization around $\bar{y} = 1$ for $\bar{x} = -1$.

In passing from inverse functions to implicit functions more generally, we need to pass from an equation $f(x) = y$ to one of the form

$$(1) \qquad f(p,x) = 0 \text{ for a function } f : \mathbb{R}^d \times \mathbb{R}^n \to \mathbb{R}^m$$

in which p acts as a parameter. The question is no longer that of inverting f, but the framework of set-valuedness is valuable nonetheless because it allows us to immediately introduce the *solution mapping*

$$(2) \qquad S : \mathbb{R}^d \rightrightarrows \mathbb{R}^n \text{ with } S(p) = \{x \mid f(p,x) = 0\}.$$

We can then look at pairs $(\bar{p},\bar{x})$ in $\operatorname{gph} S$ and ask whether S has a single-valued localization s around $\bar{p}$ for $\bar{x}$. Such a localization is exactly what constitutes an implicit function coming out of the equation. The classical implicit function theorem deduces the existence from certain assumptions on f. A review of the form of this theorem will help in setting the stage for later developments because of the pattern it provides. Again, some basic background needs to be recalled, and this is also an opportunity to fix some additional notation and terminology for subsequent use.

A function $f : \mathbb{R}^n \to \mathbb{R}$ is *upper semicontinuous* at a point $\bar{x}$ when $\bar{x} \in \operatorname{int} \operatorname{dom} f$ and for every $\varepsilon > 0$ there exists $\delta > 0$ for which

$$f(x) - f(\bar{x}) < \varepsilon \quad \text{whenever } x \in \operatorname{dom} f \text{ with } |x - \bar{x}| < \delta.$$

If instead we have

$$-\varepsilon < f(x) - f(\bar{x}) \quad \text{whenever } x \in \operatorname{dom} f \text{ with } |x - \bar{x}| < \delta,$$

then f is said to be *lower semicontinuous* at $\bar{x}$. Such upper and lower semicontinuity combine to *continuity*, meaning the existence for every $\varepsilon > 0$ of a $\delta > 0$ for which

$$|f(x) - f(\bar{x})| < \varepsilon \quad \text{whenever } x \in \operatorname{dom} f \text{ with } |x - \bar{x}| < \delta.$$

1 Functions Defined Implicitly by Equations

This condition, in our norm notation, carries over to defining the continuity of a vector-valued function $f : \mathbb{R}^n \to \mathbb{R}^m$ at a point $\bar{x} \in \operatorname{int dom} f$. However, we also speak more generally then of f being *continuous at $\bar{x}$ relative to a set D* when $\bar{x} \in D \subset \operatorname{dom} f$ and this last estimate holds for $x \in D$; in that case $\bar{x}$ need not belong to $\operatorname{int dom} f$. When f is continuous relative to D at every point of D, we say it is continuous on D. The graph $\operatorname{gph} f$ of a function $f : \mathbb{R}^n \to \mathbb{R}^m$ with closed domain $\operatorname{dom} f$ that is continuous on $D = \operatorname{dom} f$ is a closed set in $\mathbb{R}^n \times \mathbb{R}^m$.

A function $f : \mathbb{R}^n \to \mathbb{R}^m$ is *Lipschitz continuous* relative to a set D, or on a set D, if $D \subset \operatorname{dom} f$ and there is a constant $\kappa \geq 0$ such that

$$|f(x') - f(x)| \leq \kappa |x' - x| \quad \text{for all } x', x \in D.$$

If f is Lipschitz continuous relative to a neighborhood of a point $\bar{x} \in \operatorname{int dom} f$, f is said to be Lipschitz continuous *around* $\bar{x}$. A function $f : \mathbb{R}^d \times \mathbb{R}^n \to \mathbb{R}^m$ is Lipschitz continuous with respect to x uniformly in p near $(\bar{p}, \bar{x}) \in \operatorname{int dom} f$ if there is a constant $\kappa \geq 0$ along with neighborhoods U of $\bar{x}$ and Q of $\bar{p}$ such that

$$|f(p, x') - f(p, x)| \leq \kappa |x' - x| \quad \text{for all } x', x \in U \text{ and } p \in Q.$$

Differentiability entails consideration of linear mappings. Although we generally allow for multi-valuedness and even empty-valuedness when speaking of "mappings," single-valuedness everywhere is required of a linear mapping, for which we typically use a letter like A. A linear mapping from $\mathbb{R}^n$ to $\mathbb{R}^m$ is thus a function $A : \mathbb{R}^n \to \mathbb{R}^m$ with $\operatorname{dom} A = \mathbb{R}^n$ which obeys the usual rule for linearity:

$$A(\alpha x + \beta y) = \alpha A x + \beta A y \text{ for all } x, y \in \mathbb{R}^n \text{ and all scalars } \alpha, \beta \in \mathbb{R}.$$

The *kernel* of A is

$$\ker A = \{x \,|\, Ax = 0\}.$$

In the finite-dimensional setting, we carefully distinguish between a linear mapping and its matrix, but often use the same notation for both. A linear mapping $A : \mathbb{R}^n \to \mathbb{R}^m$ is represented then by a matrix A with m rows, n columns, and components $a_{i,j}$:

$$A = \begin{pmatrix} a_{11} & a_{12} & \cdots & a_{1n} \\ a_{21} & a_{22} & \cdots & a_{2n} \\ \vdots & & & \vdots \\ a_{m1} & a_{m2} & \cdots & a_{mn} \end{pmatrix}.$$

The inverse A^{-1} of a linear mapping $A : \mathbb{R}^n \to \mathbb{R}^m$ always exists in the set-valued sense, but it isn't a linear mapping unless it is actually a function with all of $\mathbb{R}^m$ as its domain, in which case A is said to be *invertible*. From linear algebra, of course, that requires $m = n$ and corresponds to the matrix A being nonsingular. More generally, if $m \leq n$ and the rows of the matrix A are linearly independent, then the rank of the matrix A is m and the mapping A is surjective. In terms of the transpose of A, denoted by A^T, the matrix AA^T is in this case nonsingular. On the other hand, if $m \geq n$ and the columns of A are linearly independent then $A^\mathsf{T}A$ is nonsingular.

Both the identity mapping and its matrix will be denoted by I, regardless of dimensionality. By default, $|A|$ is the operator norm of A induced by the Euclidean norm,

$$|A| = \max_{|x| \leq 1} |Ax|.$$

A function $f : \mathbb{R}^n \to \mathbb{R}^m$ is *differentiable* at a point $\bar{x}$ when $\bar{x} \in \operatorname{int} \operatorname{dom} f$ and there is a linear mapping $A : \mathbb{R}^n \to \mathbb{R}^m$ with the property that for every $\varepsilon > 0$ there exists $\delta > 0$ with

$$|f(\bar{x}+h) - f(\bar{x}) - Ah| \leq \varepsilon |h| \quad \text{for every } h \in \mathbb{R}^n \text{ with } |h| < \delta.$$

If such a mapping A exists at all, it is unique; it is denoted by $Df(\bar{x})$ and called the *derivative* of f at $\bar{x}$. A function $f : \mathbb{R}^n \to \mathbb{R}^m$ is said to be *twice differentiable* at a point $\bar{x} \in \operatorname{int} \operatorname{dom} f$ when there is a bilinear mapping $N : \mathbb{R}^n \times \mathbb{R}^n \to \mathbb{R}^m$ with the property that for every $\varepsilon > 0$ there exists $\delta > 0$ with

$$|f(\bar{x}+h) - f(\bar{x}) - Df(\bar{x})h - N(h,h)| \leq \varepsilon |h|^2 \quad \text{for every } h \in \mathbb{R}^n \text{ with } |h| < \delta.$$

If such a mapping N exists it is unique and is called the *second derivative* of f at $\bar{x}$, denoted by $D^2 f(\bar{x})$. Higher-order derivatives can be defined accordingly.

The $m \times n$ matrix that represents the derivative $Df(\bar{x})$ is called the *Jacobian* of f at $\bar{x}$ and is denoted by $\nabla f(\bar{x})$. In the notation $x = (x_1, \ldots, x_n)$ and $f = (f_1, \ldots, f_m)$, the components of $\nabla f(\bar{x})$ are the partial derivatives of the component functions f_i:

$$\nabla f(\bar{x}) = \left(\frac{\partial f_i}{\partial x_j}(\bar{x}) \right)_{i,j=1}^{m,n}.$$

In distinguishing between $Df(\bar{x})$ as a linear mapping and $\nabla f(\bar{x})$ as its matrix, we can guard better against ambiguities which may arise in some situations. When the Jacobian $\nabla f(x)$ exists and is continuous (with respect to the matrix norms associated with the Euclidean norm) on a set $D \subset \mathbb{R}^n$, then we say that the function f is *continuously differentiable* on D; we also call such a function *smooth* or $\mathscr{C}^1$ on D. Accordingly, we define k times continuously differentiable ($\mathscr{C}^k$) functions.

For a function $f : \mathbb{R}^d \times \mathbb{R}^n \to \mathbb{R}^m$ and a pair $(\bar{p}, \bar{x}) \in \operatorname{int} \operatorname{dom} f$, the *partial derivative* mapping $D_x f(\bar{p}, \bar{x})$ of f with respect to x at $(\bar{p}, \bar{x})$ is the derivative of the function $g(x) = f(\bar{p}, x)$ at $\bar{x}$. If the partial derivative mapping is continuous as a function of the pair (p, x) in a neighborhood of $(\bar{p}, \bar{x})$, then f is said to be continuously differentiable with respect to x around $(\bar{p}, \bar{x})$. The partial derivative $D_x f(\bar{p}, \bar{x})$ is represented by an $m \times n$ matrix, denoted $\nabla_x f(\bar{p}, \bar{x})$ and called the partial Jacobian. Respectively, $D_p f(\bar{p}, \bar{x})$ is represented by the $m \times d$ partial Jacobian $\nabla_p f(\bar{p}, \bar{x})$. It's a standard fact from calculus that if f is differentiable with respect to both p and x around $(\bar{p}, \bar{x})$ and the partial Jacobians $\nabla_x f(p, x)$ and $\nabla_p f(p, x)$ depend continuously on p and x, then f is continuously differentiable around $(\bar{p}, \bar{x})$.

Fig. 1.2 The front page of Dini's manuscript from 1877/78.

With this notation and terminology in hand, let us return to the setting of implicit functions in equation (1), as traditionally addressed with tools of differentiability. Most calculus books present a result going back to Dini[1], who formulated and proved it in lecture notes of 1877/78; the cover of Dini's manuscript is displayed above. The version typically seen in advanced texts is what we will refer to as the *classical implicit function theorem* or *Dini's theorem*. In those texts the set-valued solution mapping S in (2) never enters the picture directly, but a brief statement in that mode will help to show where we are headed in this book.

---

[1] Ulisse Dini (1845–1918). Many thanks to Danielle Ritelli from the University of Bologna for a copy of Dini's manuscript.

The statement centers on a pair $(\bar{p},\bar{x})$ satisfying the equation (1), or equivalently such that $\bar{x} \in S(\bar{p})$. It makes two assumptions: f is continuously differentiable around $(\bar{p},\bar{x})$ and the partial Jacobian $\nabla_x f(\bar{p},\bar{x})$ is nonsingular (requiring of course that $m = n$). The conclusion then is that a single-valued localization s of S exists around $\bar{p}$ for $\bar{x}$ which moreover is continuously differentiable around $\bar{p}$ with Jacobian given by the formula

$$\nabla s(p) = -\nabla_x f(p,s(p))^{-1} \nabla_p f(p,s(p)).$$

The Dini classical implicit function theorem and its variants will be taken up in detail in Section 1B after the development in Section 1A of an equivalent inverse function theorem. Later in Chapter 1 we gradually depart from the assumption of continuously differentiability of f to obtain far-reaching extensions of this classical theorem. It will be illuminating, for instance, to reformulate the assumption about the Jacobian $\nabla_x f(\bar{p},\bar{x})$ as an assumption about the function

$$h(x) = f(\bar{p},\bar{x}) + \nabla_x f(\bar{p},\bar{x})(x - \bar{x})$$

giving the partial linearization of f at $(\bar{p},\bar{x})$ with respect to x and having $h(\bar{x}) = 0$. The condition corresponding to the invertibility of $\nabla_x f(\bar{p},\bar{x})$ can be turned into the condition that the inverse mapping h^{-1}, with $\bar{x} \in h^{-1}(0)$, has a single-valued localization around 0 for $\bar{x}$. In this way the theme of single-valued localizations can be carried forward even into realms where f might not be differentiable and h could be some other kind of "local approximation" of f. We will be able to operate with a broad *implicit function paradigm*, extending in later chapters to much more than solving equations. It will deal with single-valued localizations s of solution mappings S to "generalized equations." These localizations s, if not differentiable, will at least have other key properties.

1A. The Classical Inverse Function Theorem

In this section of the book, we state and prove the classical inverse function theorem in two ways. In these proofs, and also later in the chapter, we will make use of the following two observations from calculus.

Fact 1 (estimates for differentiable functions). *If a function $f : \mathbb{R}^n \to \mathbb{R}^m$ is differentiable at every point in a neighborhood of $\bar{x}$ and the Jacobian mapping $x \mapsto \nabla f(x)$ is continuous at $\bar{x}$, then for every $\varepsilon > 0$ there exists $\delta > 0$ such that*

(a) $|f(x') - f(x) - \nabla f(x)(x' - x)| \leq \varepsilon |x' - x|$ *for every* $x', x \in \mathbb{B}_\delta(\bar{x})$.

Equivalently, for every $\varepsilon > 0$ there exists $\delta > 0$ such that

(b) $|f(x') - f(x) - \nabla f(\bar{x})(x' - x)| \leq \varepsilon |x' - x|$ *for every* $x', x \in \mathbb{B}_\delta(\bar{x})$.

Proof. For a vector $h \in \mathbb{R}^m$ with $|h| = 1$ and points x, x', $x \neq x'$, in an open neighborhood of $\bar{x}$ where f is differentiable, define the function $\varphi : \mathbb{R} \to \mathbb{R}$ as $\varphi(t) = \langle h, f(x+t(x'-x))\rangle$. Then φ is continuous on $[0, 1]$ and differentiable in $(0, 1)$ and also $\varphi'(t) = \langle h, Df(x+t(x'-x))(x'-x)\rangle$. A basic result in calculus, the *mean value theorem*, says that when a function $\psi : \mathbb{R} \to \mathbb{R}$ is continuous on an interval $[a, b]$ with $a < b$ and differentiable in (a, b), then there exists a point $c \in (a, b)$ such that $\psi(b) - \psi(a) = \psi'(c)(b - a)$; see, e.g., Bartle and Sherbert [1992], p. 197. Applying the mean value theorem to the function φ we obtain that there exists $\bar{t} \in (0, 1)$ such that

$$\langle h, f(x')\rangle - \langle h, f(x)\rangle = \langle h, Df(x+\bar{t}(x'-x))(x'-x)\rangle.$$

Then the triangle inequality and the assumed continuity of Df at $\bar{x}$ give us (a). The equivalence of (a) and (b) follows from the continuity of Df at $\bar{x}$.   □

Fact 2 (stability of matrix nonsingularity). *Suppose A is a matrix-valued function from $\mathbb{R}^n$ to the space $\mathbb{R}^{m \times m}$ of all $m \times m$ real matrices, such that the determinant of $A(x)$, as well as those of its minors, depends continuously on x around $\bar{x}$ and the matrix $A(\bar{x})$ is nonsingular. Then there is a neighborhood U of $\bar{x}$ such that $A(x)$ is nonsingular for every $x \in U$ and, moreover, the function $x \mapsto A(x)^{-1}$ is continuous in U.*

Proof. Since the nonsingularity of $A(x)$ corresponds to its determinant being nonzero, it is sufficient to observe that the determinant of $A(x)$ (along with its minors) depends continuously on x.   □

The classical inverse function theorem which parallels the classical implicit function theorem described in the introduction to this chapter reads as follows.

Theorem 1A.1 (classical inverse function theorem). *Let $f : \mathbb{R}^n \to \mathbb{R}^n$ be continuously differentiable in a neighborhood of a point $\bar{x}$ and let $\bar{y} := f(\bar{x})$. If $\nabla f(\bar{x})$ is nonsingular, then f^{-1} has a single-valued localization s around $\bar{y}$ for $\bar{x}$. Moreover, the function s is continuously differentiable in a neighborhood V of $\bar{y}$, and its Jacobian satisfies*

(1) $$\nabla s(y) = \nabla f(s(y))^{-1} \quad \textit{for every } y \in V.$$

Examples.
1) For the function $f(x) = x^2$ considered in the introduction, the inverse f^{-1} is a set-valued mapping whose domain is $[0, \infty)$. It has two single-valued localizations around any $\bar{y} > 0$ for $\bar{x} \neq 0$, represented by either $x(y) = \sqrt{y}$ if $\bar{x} > 0$ or $x(y) = -\sqrt{y}$ if $\bar{x} < 0$. The inverse f^{-1} has no single-valued localization around $\bar{y} = 0$ for $\bar{x} = 0$.

2) The inverse f^{-1} of the function $f(x) = x^3$ is single-valued everywhere; it is the function $x(y) = \sqrt[3]{y}$. The inverse $f^{-1} = \sqrt[3]{y}$ is not differentiable at 0, which fits with the observation that $f'(0) = 0$.

3) For a higher-dimensional illustration, we look at diagonal real matrices

$$A = \begin{pmatrix} \lambda_1 & 0 \\ 0 & \lambda_2 \end{pmatrix}$$

and the function $f : \mathbb{R}^2 \to \mathbb{R}^2$ which assigns to (λ_1, λ_2) the trace $y_1 = \lambda_1 + \lambda_2$ of A and the determinant $y_2 = \lambda_1 \lambda_2$ of A,

$$f(\lambda_1, \lambda_2) = \begin{pmatrix} \lambda_1 + \lambda_2 \\ \lambda_1 \lambda_2 \end{pmatrix}.$$

What can be said about the inverse of f? The range of f consists of all $y = (y_1, y_2)$ such that $4y_2 \leq y_1^2$. The Jacobian

$$\nabla f(\lambda_1, \lambda_2) = \begin{pmatrix} 1 & 1 \\ \lambda_2 & \lambda_1 \end{pmatrix}$$

has determinant $\lambda_1 - \lambda_2$, so it is nonsingular except along the line where $\lambda_1 = \lambda_2$, which corresponds to $4y_2 = y_1^2$. Therefore, f^{-1} has a smooth single-valued localization around $y = (y_1, y_2)$ for (λ_1, λ_2) as long as $4y_2 < y_1^2$, in fact two such. But it doesn't have such a localization around other (y_1, y_2).

It will be illuminating to look at two proofs[2] of the classical inverse function theorem. The one we lay out first requires no more background than the facts listed at the beginning of this section, and it has the advantage of actually "calculating" a single-valued localization of f^{-1} by a procedure which is well known in numerical

---

[2] These two proofs are not really different, if we take into account that the contraction mapping principle is proved by using a somewhat similar iterative procedure, see Section 5E.

1 Functions Defined Implicitly by Equations

analysis, namely *Newton's iterative method*[3] for solving nonlinear equations. The second, which we include for the sake of connections with later developments, utilizes a nonconstructive, but very broad, fixed-point argument.

Proof I of Theorem 1A.1. First we introduce some constants. Let $a > 0$ be a scalar so small that, by appeal to Fact 2 in the beginning of this section, the Jacobian matrix $\nabla f(x)$ is nonsingular for every x in $\mathbb{B}_a(\bar{x})$ and the function $x \mapsto \nabla f(x)^{-1}$ is continuous in $\mathbb{B}_a(\bar{x})$. Set

$$c = \max_{x \in \mathbb{B}_a(\bar{x})} |\nabla f(x)^{-1}|.$$

Take $a > 0$ smaller if necessary to obtain, on the basis of the estimate (a) in Fact 1, that

(2) $\quad |f(x') - f(x) - \nabla f(x)(x' - x)| \leq \dfrac{1}{2c}|x' - x| \quad$ for every $x', x \in \mathbb{B}_a(\bar{x})$.

Let $b = a/(16c)$. Let s be the localization of f^{-1} with respect to the neighborhoods $\mathbb{B}_b(\bar{y})$ and $\mathbb{B}_a(\bar{x})$:

(3) $\quad \text{gph } s = \left[\mathbb{B}_b(\bar{y}) \times \mathbb{B}_a(\bar{x})\right] \cap \text{gph } f^{-1}.$

We will show that s has the properties claimed. The argument is divided into three steps.

STEP 1: The localization s is nonempty-valued on $\mathbb{B}_b(\bar{y})$ with $\bar{x} \in s(\bar{y})$, in particular.

The fact that $\bar{x} \in s(\bar{y})$ is immediate of course from (3), inasmuch as $\bar{x} \in f^{-1}(\bar{y})$. Pick any $y \in \mathbb{B}_b(\bar{y})$ and any $x^0 \in \mathbb{B}_{a/8}(\bar{x})$. We will demonstrate that the iterative procedure

(4) $\quad x^{k+1} = x^k - \nabla f(x^k)^{-1}(f(x^k) - y), \quad k = 0, 1, \ldots$

produces a sequence of vectors $x^1, x^2, \ldots$ which is convergent to a point $x \in f^{-1}(y) \cap \mathbb{B}_a(\bar{x})$. The procedure (4) is the celebrated Newton's iterative method for solving the equation $f(x) = y$ with a starting point x^0. By using induction we will show that this procedure generates an infinite sequence $\{x^k\}$ satisfying for $k = 1, 2, \ldots$ the following two conditions:

(5a) $\quad\quad\quad\quad\quad\quad\quad\quad\quad x^k \in \mathbb{B}_a(\bar{x})$

and

---

[3] Isaac Newton (1643–1727). In 1669 Newton wrote his paper *De Analysi per Equationes Numero Terminorum Infinitas*, where, among other things, he describes an iterative procedure for approximating real roots of a polynomial equation of third degree. In 1690 Joseph Raphson proposed a similar iterative procedure for solving more general polynomial equations and attributed it to Newton. It was Thomas Simpson who in 1740 stated the method in today's form (using Newton's fluxions) for an equation not necessarily polynomial, without making connections to the works of Newton and Raphson; he also noted that the method can be used for solving optimization problems by setting the gradient to zero.

(5b) $$|x^k - x^{k-1}| \leq \frac{a}{2^{k+1}}.$$

To initialize the induction, we establish (5a) and (5b) for $k = 1$. Since $x^0 \in \mathbb{B}_{a/8}(\bar{x})$, the matrix $\nabla f(x^0)$ is indeed invertible, and (4) gives us x^1. The equality in (4) for $k = 0$ can also be written as

$$x^1 = -\nabla f(x^0)^{-1}(f(x^0) - y - \nabla f(x^0)x^0),$$

which we subtract from the obvious equality

$$\bar{x} = -\nabla f(x^0)^{-1}(f(\bar{x}) - \bar{y} - \nabla f(x^0)\bar{x}),$$

obtaining

$$\bar{x} - x^1 = -\nabla f(x^0)^{-1}(f(\bar{x}) - f(x^0) - \bar{y} + y - \nabla f(x^0)(\bar{x} - x^0)).$$

Taking norms on both sides and utilizing (2) with $x' = \bar{x}$ and $x = x^0$ we get

$$|x^1 - \bar{x}| \leq |\nabla f(x^0)^{-1}|(|f(\bar{x}) - f(x^0) - \nabla f(x^0)(\bar{x} - x^0)| + |y - \bar{y}|) \leq \frac{c}{2c}|x^0 - \bar{x}| + cb.$$

Inasmuch as $|x^0 - \bar{x}| \leq a/8$, this yields

$$|x^1 - \bar{x}| \leq \frac{a}{16} + cb = \frac{a}{8} \leq a.$$

Hence (5a) holds for $k = 1$. Moreover, by the triangle inequality,

$$|x^1 - x^0| \leq |x^1 - \bar{x}| + |\bar{x} - x^0| \leq \frac{a}{8} + \frac{a}{8} = \frac{a}{4},$$

which is (5b) for $k = 1$.

Assume now that (5a) and (5b) hold for $k = 1, 2, \ldots, j$. Then the matrix $\nabla f(x^k)$ is nonsingular for all such k and the iteration (4) gives us for $k = j$ the point x^{j+1}:

(6) $$x^{j+1} = x^j - \nabla f(x^j)^{-1}(f(x^j) - y).$$

Through the preceding iteration, for $k = j - 1$, we have

$$y = f(x^{j-1}) + \nabla f(x^{j-1})(x^j - x^{j-1}).$$

Substituting this expression for y into (6), we obtain

$$x^{j+1} - x^j = -\nabla f(x^j)^{-1}(f(x^j) - f(x^{j-1}) - \nabla f(x^{j-1})(x^j - x^{j-1})).$$

Taking norms, we get from (2) that

$$|x^{j+1} - x^j| \leq c|f(x^j) - f(x^{j-1}) - \nabla f(x^{j-1})(x^j - x^{j-1})| \leq \frac{1}{2}|x^j - x^{j-1}|.$$

1 Functions Defined Implicitly by Equations

The induction hypothesis on (5b) for $k = j$ then yields

$$|x^{j+1} - x^j| \leq \frac{1}{2}|x^j - x^{j-1}| \leq \frac{1}{2}\left(\frac{a}{2^{j+1}}\right) = \frac{a}{2^{j+2}}.$$

Hence, (5b) holds for $k = j+1$. Further,

$$|x^{j+1} - \bar{x}| \leq \sum_{i=1}^{j+1} |x^i - x^{i-1}| + |x^0 - \bar{x}| \leq \sum_{i=1}^{j+1} \frac{a}{2^{i+1}} + \frac{a}{8}$$

$$\leq \frac{a}{4} \sum_{i=0}^{\infty} \frac{1}{2^i} + \frac{a}{8} = \frac{a}{2} + \frac{a}{8} = \frac{5a}{8} \leq a.$$

This gives (5a) for $k = j+1$ and the induction step is complete. Thus, both (5a) and (5b) hold for all $k = 1, 2, \ldots$.

To verify that the sequence $\{x^k\}$ converges, we observe next from (5b) that, for any k and j satisfying $k > j$, we have

$$|x^k - x^j| \leq \sum_{i=j}^{k-1} |x^{i+1} - x^i| \leq \sum_{i=j}^{\infty} \frac{a}{2^{i+2}} \leq \frac{a}{2^{j+1}}.$$

Hence, the sequence $\{x^k\}$ satisfies the Cauchy criterion, which is known to guarantee that it is convergent.

Let x be the limit of this sequence. Clearly, from (5a), we have $x \in \mathbb{B}_a(\bar{x})$. Through passing to the limit in (4), x must satisfy $x = x - \nabla f(x)^{-1}(f(x) - y)$, which is equivalent to $f(x) = y$. Thus, we have proved that for any $y \in \mathbb{B}_b(\bar{y})$ there exists $x \in \mathbb{B}_a(\bar{x})$ such that $x \in f^{-1}(y)$. In other words, the localization s of the inverse f^{-1} at $\bar{y}$ for $\bar{x}$ specified by (3) has nonempty values. In particular, $\mathbb{B}_b(\bar{y}) \subset \operatorname{dom} f^{-1}$.

STEP 2: The localization s is single-valued on $\mathbb{B}_b(\bar{y})$.

Let $y \in \mathbb{B}_b(\bar{y})$ and suppose x and x' belong to $s(y)$. Then $x, x' \in \mathbb{B}_a(\bar{x})$ and also

$$x = -\nabla f(x)^{-1}[f(x) - y - \nabla f(x)x] \quad \text{and} \quad x' = -\nabla f(x)^{-1}[f(x') - y - \nabla f(x)x'].$$

Consequently

$$x' - x = -\nabla f(x)^{-1}[f(x') - f(x) - \nabla f(x)(x' - x)].$$

Taking norms on both sides and invoking (2), we get

$$|x' - x| \leq c|f(x') - f(x) - \nabla f(x)(x' - x)| \leq \frac{1}{2}|x' - x|$$

which can only be true if $x' = x$.

STEP 3: The localization s is continuously differentiable in int $\mathbb{B}_b(\bar{y})$ with $\nabla s(y)$ expressed by (1).

An extension of the argument in Step 2 will provide a needed estimate. Consider any y and y' in $\mathbb{B}_b(\bar{y})$ and let $x = s(y)$ and $x' = s(y')$. These elements satisfy

$$x = -\nabla f(x)^{-1}\left[f(x) - y - \nabla f(x)x\right] \quad \text{and} \quad x' = -\nabla f(x)^{-1}\left[f(x') - y' - \nabla f(x)x'\right],$$

so that

$$x' - x = -\nabla f(x)^{-1}\left[f(x') - f(x) - \nabla f(x)(x' - x) - (y' - y)\right].$$

This implies through (2) that

$$|x' - x| \le c|f(x') - f(x) - \nabla f(x)(x' - x)| + c|y' - y| \le \frac{1}{2}|x' - x| + c|y' - y|,$$

hence $|x' - x| \le 2c|y' - y|$. Thus,

(7) $$\qquad\qquad |s(y') - s(y)| \le 2c|y' - y| \quad \text{for } y, y' \in B_b(\bar{y}).$$

This estimate means that the localization s is Lipschitz continuous on $B_b(\bar{y})$.

Now take any $\varepsilon > 0$. Then, from (a) in Fact 1, there exists $\delta > 0$ such that

(8) $$\qquad |f(x') - f(x) - \nabla f(x)(x' - x)| \le \frac{\varepsilon}{2c^2}|x' - x| \quad \text{for every } x', x \in B_\delta(\bar{x}).$$

Choose $y \in \text{int } B_b(\bar{y})$; then there exists $\tau > 0$ such that $\tau < \delta/(2c)$ and $y + h \in B_b(\bar{y})$ for any $h \in \mathbb{R}^n$ with $|h| \le \tau$. From the estimate (7) we get that

$$|s(y+h) - s(y)| \le 2c|h| \le 2c\tau \le \delta.$$

Writing the equalities $f(s(y+h)) = y + h$ and $f(s(y)) = y$ as

$$s(y+h) = -\nabla f(s(y))^{-1}(f(s(y+h)) - y - h - \nabla f(s(y))s(y+h))$$

and

$$s(y) = -\nabla f(s(y))^{-1}(f(s(y)) - y - \nabla f(s(y))s(y))$$

and subtracting the second from the first, we obtain

$$s(y+h) - s(y) - \nabla f(s(y))^{-1}h$$
$$= -\nabla f(s(y))^{-1}(f(s(y+h)) - f(s(y)) - \nabla f(s(y))(s(y+h) - s(y))).$$

Once again taking norms on both sides, and using (7) and (8), we get

$$|s(y+h) - s(y) - \nabla f(s(y))^{-1}h| \le \frac{c\varepsilon}{2c^2}|s(y+h) - s(y)| \le \varepsilon|h| \quad \text{whenever } h \in B_\tau(0).$$

By definition, this says that the function s is differentiable at y and that its Jacobian equals $\nabla f(s(y))^{-1}$, as claimed in (1). This Jacobian is continuous in $\text{int } B_b(\bar{y})$; this comes from the continuity of ∇f^{-1} in $B_a(\bar{x})$ where are the values of s, and the continuity of s in $\text{int } B_b(\bar{y})$, and also taking into account that a composition of continuous functions is continuous. □

1 Functions Defined Implicitly by Equations

We can make a shortcut through Steps 1 and 2 of the Proof I, arriving at the promised Proof II, if we employ a deeper result of analysis far beyond the framework so far, namely the contraction mapping principle. Although we work here in Euclidean spaces, we state this theorem in the framework of a complete metric space, as is standard in the literature. A more general version of this principle for set-valued mapping will be proved in Section 5E. The reader who wants to stick with Euclidean spaces may assume that X is a closed nonempty subset of $\mathbb{R}^n$ with metric $\rho(x,y) = |x - y|$.

Theorem 1A.2 (contraction mapping principle). *Let X be a complete metric space with metric ρ. Consider a point $\bar{x} \in X$ and a function $\Phi : X \to X$ for which there exist scalars $a > 0$ and $\lambda \in [0,1)$ such that:*
(a) $\rho(\Phi(\bar{x}), \bar{x}) \leq a(1 - \lambda)$;
(b) $\rho(\Phi(x'), \Phi(x)) \leq \lambda \rho(x', x)$ *for every $x', x \in \mathbb{B}_a(\bar{x})$.*
Then there is a unique $x \in \mathbb{B}_a(\bar{x})$ satisfying $x = \Phi(x)$, that is, Φ has a unique fixed point in $\mathbb{B}_a(\bar{x})$.

Most common in the literature is another formulation of the contraction mapping principle which seems more general but is actually equivalent to 1A.2. To distinguish it from 1A.2, we call it *basic*.

Theorem 1A.3 (basic contraction mapping principle). *Let X be a complete metric space with metric ρ and let $\Phi : X \to X$. Suppose that there exists $\lambda \in [0,1)$ such that*

$$\rho(\Phi(x'), \Phi(x)) \leq \lambda \rho(x', x) \quad \text{for every } x', x \in X.$$

Then there is a unique $x \in X$ satisfying $x = \Phi(x)$.

Another equivalent version of the contraction mapping principle involves a parameter.

Theorem 1A.4 (parametric contraction mapping principle). *Let P be a metric space with metric σ and X be a complete metric space with metric ρ, and let $\Phi : P \times X \to X$. Suppose that there exist $\lambda \in [0,1)$ and $\mu \geq 0$ such that*

(9) $\qquad \rho(\Phi(p,x'), \Phi(p,x)) \leq \lambda \rho(x',x) \quad \text{for every } x', x \in X \text{ and } p \in P$

and

(10) $\qquad \rho(\Phi(p',x), \Phi(p,x)) \leq \mu \sigma(p',p) \quad \text{for every } p', p \in P \text{ and } x \in X.$

Then the mapping

(11) $\qquad \psi : p \mapsto \{x \in X \mid x = \Phi(p,x)\} \quad \text{for } p \in P$

is single-valued on P, which is moreover Lipschitz continuous on P with Lipschitz constant $\mu/(1-\lambda)$.

Exercise 1A.5. *Prove that theorems 1A.2, 1A.3 and 1A.4 are equivalent.*

Guide. Let 1A.2 be true and let Φ satisfy the assumptions in 1A.3 with some $\lambda \in [0,1)$. Choose $\bar{x} \in X$; then $\Phi(\bar{x}) \in X$. Let $a > \rho(\bar{x}, \Phi(\bar{x}))/(1-\lambda)$. Then (a) and (b) are satisfied with this a and hence there exists a unique fixed point x of Φ in $\mathbb{B}_a(\bar{x})$. The uniqueness of the fixed point in the whole X follows from the contraction property. To prove the converse implication first use (a)(b) to obtain that Φ maps $\mathbb{B}_a(\bar{x})$ into itself and then use the fact that the closed ball $\mathbb{B}_a(\bar{x})$ equipped with metric ρ is a complete metric space. Another way to have equivalence of 1A.2 and 1A.3 is to reformulate 1A.2 with a being possibly ∞.

Let 1A.3 be true and let Φ satisfy the assumptions (9) and (10) in 1A.4 with corresponding λ and μ. Then, by 1A.3, for every fixed $p \in P$ the set $\{x \in X \mid x = \Phi(p,x)\}$ is a singleton; that is, the mapping ψ in (11) is a function with domain P. To complete the proof, choose $p', p \in P$ and the corresponding $x' = \Phi(p', x')$, $x = \Phi(p, x)$, and use (9), (10) and the triangle inequality to obtain

$$\rho(x', x) = \rho(\Phi(p', x'), \Phi(p, x))$$
$$\leq \rho(\Phi(p', x'), \Phi(p', x)) + \rho(\Phi(p', x), \Phi(p, x)) \leq \lambda \rho(x', x) + \mu \sigma(p', p).$$

Rearranging the terms gives us the desired Lipschitz continuity. $\square$

Proof II of Theorem 1A.1. Denote $A = \nabla f(\bar{x})$ and let $c := |A^{-1}|$. There exists $a > 0$ such that from the estimate (b) in Fact 1 (in the beginning of this section) we have

(12) $\quad |f(x') - f(x) - \nabla f(\bar{x})(x' - x)| \leq \dfrac{1}{2c}|x' - x| \quad$ for every $x', x \in \mathbb{B}_a(\bar{x})$.

Let $b = a/(4c)$. The space $\mathbb{R}^n$ equipped with the Euclidean norm is a complete metric space, so in this case X in Theorem 1A.2 is identified with $\mathbb{R}^n$. Fix $y \in \mathbb{B}_b(\bar{y})$ and consider the function

$$\Phi_y : x \mapsto x - A^{-1}(f(x) - y) \quad \text{for } x \in \mathbb{B}_a(\bar{x}).$$

We have

$$|\Phi_y(\bar{x}) - \bar{x}| = |-A^{-1}(\bar{y} - y)| \leq cb = \frac{ca}{4c} < a\left(1 - \frac{1}{2}\right),$$

hence condition (a) in the contraction mapping principle 1A.2 holds with the so chosen a and $\lambda = 1/2$. Further, for any $x, x' \in \mathbb{B}_a(\bar{x})$, from (12) we obtain that

$$|\Phi_y(x) - \Phi_y(x')| = |x - x' - A^{-1}(f(x) - f(x'))| \leq |A^{-1}||f(x) - f(x') - A(x - x')|$$
$$\leq c\frac{1}{2c}|x - x'| = \frac{1}{2}|x - x'|.$$

Thus condition (b) in 1A.2 is satisfied with the same λ. Hence, there is a unique $x \in \mathbb{B}_a(\bar{x})$ such that $\Phi_y(x) = x$; that is equivalent to $f(x) = y$.

Translated into our terminology, this tells us that f^{-1} has a single-valued localization around $\bar{y}$ for $\bar{x}$ whose graph satisfies (3). The continuous differentiability is argued once more through Step 3 of Proof I. □

Exercise 1A.6. *Prove Theorem 1A.1 by using, instead of iteration (4), the iteration*

$$x^{k+1} = x^k - \nabla f(\bar{x})^{-1}(f(x^k) - y), \quad k = 0, 1, \ldots.$$

Guide. Follow the argument in Proof I with respective adjustments of the constants involved. □

In this and the following chapters we will derive the classical inverse function theorem 1A.1 a number of times and in different ways from more general theorems or utilizing other basic results. For instance, in Section 1F we will show how to obtain 1A.1 from Brouwer's invariance of domain theorem and in Section 4B we will prove 1A.1 again with the help of the Ekeland variational principle.

There are many roads to be taken from here, by relaxing the assumptions in the classical inverse function theorem, that lead to a variety of results. Some of them are paved and easy to follow, others need more advanced techniques, and a few lead to new territories which we will explore later in the book.

1B. The Classical Implicit Function Theorem

In this section we give a proof of the classical implicit function theorem stated by Dini and described in the introduction to this chapter. We consider a function $f : \mathbb{R}^d \times \mathbb{R}^n \to \mathbb{R}^n$ with values $f(p,x)$, where p is the parameter and x is the variable to be determined, and introduce for the equation $f(p,x) = 0$ the associated *solution mapping*

$$(1) \qquad S : p \mapsto \{x \in \mathbb{R}^n \,|\, f(p,x) = 0\} \quad \text{for } p \in \mathbb{R}^d.$$

We restate the result, furnishing it with a label for reference.

Theorem 1B.1 (Dini classical implicit function theorem). *Let $f : \mathbb{R}^d \times \mathbb{R}^n \to \mathbb{R}^n$ be continuously differentiable in a neighborhood of $(\bar{p}, \bar{x})$ and such that $f(\bar{p}, \bar{x}) = 0$, and let the partial Jacobian of f with respect to x at $(\bar{p}, \bar{x})$, namely $\nabla_x f(\bar{p}, \bar{x})$, be nonsingular. Then the solution mapping S defined in (1) has a single-valued localization s around $\bar{p}$ for $\bar{x}$ which is continuously differentiable in a neighborhood Q of $\bar{p}$ with Jacobian satisfying*

$$(2) \qquad \nabla s(p) = -\nabla_x f(p, s(p))^{-1} \nabla_p f(p, s(p)) \quad \text{for every } p \in Q.$$

The classical inverse function theorem is the particular case of the classical implicit function theorem in which $f(p,x) = -p + f(x)$ (with a slight abuse of notation). However, it will also be seen now that the classical implicit function theorem can be obtained from the classical inverse function theorem. For that, we first state an easy-to-prove fact from linear algebra.

Lemma 1B.2. *The square matrix*

$$J = \begin{pmatrix} I & 0 \\ B & A \end{pmatrix},$$

where I is the $d \times d$ identity matrix, 0 is the $d \times n$ zero matrix, B is an $n \times d$ matrix, and A is an $n \times n$ nonsingular matrix, is nonsingular.

Proof. If J is singular, then there exists

$$y = \begin{pmatrix} p \\ x \end{pmatrix} \neq 0 \quad \text{such that} \quad Jy = 0,$$

which reduces to the equation

$$\begin{pmatrix} p \\ Bp + Ax \end{pmatrix} = 0.$$

Hence there exists $x \neq 0$ with $Ax = 0$, which contradicts the nonsingularity of A. □

Proof of Theorem 1B.1. Consider the function

$$\varphi(p,x) = \begin{pmatrix} p \\ f(p,x) \end{pmatrix}$$

acting from $\mathbb{R}^d \times \mathbb{R}^n$ to itself. The inverse of this function is defined by the solutions of the equation

(3) $$\varphi(p,x) = \begin{pmatrix} p \\ f(p,x) \end{pmatrix} = \begin{pmatrix} y_1 \\ y_2 \end{pmatrix},$$

where the vector $(y_1, y_2) \in \mathbb{R}^d \times \mathbb{R}^n$ is now the parameter and (p,x) is the dependent variable. The nonsingularity of the partial Jacobian $\nabla_x f(\bar{p}, \bar{x})$ implies through Lemma 1B.2 that the Jacobian of the function φ in (3) at the point $(\bar{x}, \bar{p})$, namely the matrix

$$J(\bar{p}, \bar{x}) = \begin{pmatrix} I & 0 \\ \nabla_p f(\bar{p}, \bar{x}) & \nabla_x f(\bar{p}, \bar{x}) \end{pmatrix},$$

is nonsingular as well. Then, according to the classical inverse function theorem 1A.1, the inverse φ^{-1} of the function in (3) has a single-valued localization

$$(y_1, y_2) \mapsto (q(y_1, y_2), r(y_1, y_2)) \quad \text{around } (\bar{p}, 0) \text{ for } (\bar{p}, \bar{x})$$

which is continuously differentiable around $(\bar{p}, 0)$. To develop formula (2), we note that
$$\begin{cases} q(y_1, y_2) & = y_1, \\ f(y_1, r(y_1, y_2)) = y_2. \end{cases}$$
Differentiating the second equality with respect to y_1 by using the chain rule, we get

(4) $\quad \nabla_p f(y_1, r(y_1, y_2)) + \nabla_x f(y_1, r(y_1, y_2)) \cdot \nabla_{y_1} r(y_1, y_2) = 0.$

When (y_1, y_2) is close to $(\bar{p}, 0)$, the point $(y_1, r(y_1, y_2))$ is close to $(\bar{p}, \bar{x})$ and then $\nabla_x f(y_1, r(y_1, y_2))$ is nonsingular (Fact 2 in Section 1A). Thus, solving (4) with respect to $\nabla_{y_1} r(y_1, y_2)$ gives
$$\nabla_{y_1} r(y_1, y_2) = -\nabla_x f(y_1, r(y_1, y_2))^{-1} \nabla_p f(y_1, r(y_1, y_2)).$$
In particular, at points $(y_1, y_2) = (p, 0)$ close to $(\bar{p}, 0)$ we have that the mapping $p \mapsto s(p) := r(p, 0)$ is a single-valued localization of the solution mapping S in (1) around $\bar{p}$ for $\bar{x}$ which is continuously differentiable around $\bar{p}$ and its derivative satisfies (2). □

Thus, the classical implicit function theorem, as stated above, is *equivalent* to the classical inverse function theorem as stated in the preceding section. We now look at yet another equivalent result.

Theorem 1B.3 (correction function theorem). *Let $f : \mathbb{R}^n \to \mathbb{R}^n$ be continuously differentiable in a neighborhood of $\bar{x}$. If $\nabla f(\bar{x})$ is nonsingular, then the correction mapping*
$$\Xi : x \mapsto \{ u \in \mathbb{R}^n \,|\, f(x+u) = f(\bar{x}) + \nabla f(\bar{x})(x - \bar{x}) \} \quad \text{for } x \in \mathbb{R}^n$$
has a single-valued localization ξ around $\bar{x}$ for 0. Moreover, ξ is continuously differentiable in a neighborhood U of $\bar{x}$ with $\nabla \xi(\bar{x}) = 0$.

Proof. Consider the function
$$\varphi : (x, u) \mapsto f(x+u) - f(\bar{x}) - \nabla f(\bar{x})(x - \bar{x}) \quad \text{for } (x, u) \in \mathbb{R}^n \times \mathbb{R}^n$$
in a neighborhood of $(\bar{x}, \bar{u})$ for $\bar{u} := 0$. Since $\nabla_u \varphi(\bar{x}, \bar{u}) = \nabla f(\bar{x})$ is nonsingular, we apply the classical implicit function theorem 1B.1 obtaining that the solution mapping
$$\Xi : x \mapsto \{ u \in \mathbb{R}^n \,|\, \varphi(x, u) = 0 \} \quad \text{for } x \in \mathbb{R}^n$$
has a smooth single-valued localization ξ around $\bar{x}$ for 0. The chain rule gives us $\nabla \xi(\bar{x}) = 0$. □

Exercise 1B.4. *Prove that the correction function theorem implies the inverse function theorem.*

Guide. Let $\bar{y} := f(\bar{x})$ and assume $A := \nabla f(\bar{x})$ is nonsingular. In these terms the correction function theorem 1B.3 claims that the mapping

$$\Xi : z \mapsto \{\xi \in \mathbb{R}^n \mid f(z+\xi) = \bar{y} + A(z-\bar{x})\} \text{ for } z \in \mathbb{R}^n$$

has a single-valued localization ξ around $\bar{x}$ for 0 and that ξ is continuously differentiable around $\bar{x}$ and has zero derivative at $\bar{x}$. The affine function $y \mapsto z(y) := \bar{x} + A^{-1}(y-\bar{y})$ is the solution mapping of the linear equation $\bar{y} + A(z-\bar{x}) = y$ having $z(\bar{y}) = \bar{x}$. The composite function $y \mapsto \xi(z(y))$ hence satisfies

$$f(z(y) + \xi(z(y))) = \bar{y} + \nabla f(\bar{x})(z(y) - \bar{x}) = y.$$

The function $s(y) := z(y) + \xi(z(y))$ is a single-valued localization of the inverse f^{-1} around $\bar{y}$ for $\bar{x}$. To show that $\nabla s(y) = \nabla f(s(y))^{-1}$, use the chain rule. □

Inasmuch as the classical inverse function theorem implies the classical implicit function theorem, and the correction function theorem is a corollary of the classical implicit function theorem, all three theorems — the inverse, the implicit and the correction function theorems, stated in 1A.1, 1B.1 and 1B.3 respectively — are equivalent.

Proposition 1B.5 (higher derivatives). *In Theorem 1B.1, if f is k times continuously differentiable around $(\bar{p}, \bar{x})$ then the localization s of the solution mapping S is k times continuously differentiable around $\bar{p}$. Likewise in Theorem 1A.1, if f is k times continuously differentiable around $\bar{x}$, then the localization s of f^{-1} is k times continuously differentiable around $\bar{y}$.*

Proof. For the implicit function theorem 1B.1, this is an immediate consequence of the formula in (2) by way of the chain rule for differentiation. It follows then for the inverse function theorem 1A.1 as a special case. □

If we relax the differentiability assumption for the function f, we obtain a result of a different kind, the origins of which go back to the work of Goursat [1903].

Theorem 1B.6 (Goursat's implicit function theorem). *For the solution mapping S defined in (1), consider a pair $(\bar{p}, \bar{x})$ with $\bar{x} \in S(\bar{p})$. Assume that:*
  (a) *$f(p,x)$ is differentiable with respect to x in a neighborhood of the point $(\bar{p}, \bar{x})$, and both $f(p,x)$ and $\nabla_x f(p,x)$ depend continuously on (p,x) in this neighborhood;*
  (b) *$\nabla_x f(\bar{p}, \bar{x})$ is nonsingular.*
Then S has a single-valued localization around $\bar{p}$ for $\bar{x}$ which is continuous at $\bar{p}$.

We will prove a far reaching generalization of this result in Section 2B, which we supply with a detailed proof. In the following exercise we give a guide for a direct proof.

Exercise 1B.7. *Prove Theorem 1B.6.*

Guide. Mimic the proof of 1A.1 by choosing a and q sufficiently small so that if

$$c = \max_{\substack{x \in B_a(\bar{x}) \\ p \in B_q(\bar{p})}} |\nabla_x f(p,x)^{-1}|,$$

1 Functions Defined Implicitly by Equations

then one has, as in the estimate (a) in Fact 1, that for every $x', x \in \mathbb{B}_a(\bar{x})$ and $p \in \mathbb{B}_q(\bar{p})$

(5) $$|f(p,x') - f(p,x) - \nabla_x f(p,x)(x'-x)| \leq \frac{1}{2c}|x'-x|.$$

Then use the iteration

$$x^{k+1} = x^k - \nabla_x f(\bar{p},\bar{x})^{-1} f(p,x^k)$$

to obtain that S has a nonempty graphical localization s around $\bar{p}$ for $\bar{x}$. As in Step 2 in Proof I of 1A.1, show that s is single-valued. To show continuity at $\bar{p}$, for $x = s(p)$ subtract the equalities

$$x = \nabla_x f(\bar{p},\bar{x})^{-1}(f(p,x) - \nabla_x f(\bar{p},\bar{x})x)$$

and

$$\bar{x} = \nabla_x f(\bar{p},\bar{x})^{-1}(f(\bar{p},\bar{x}) - \nabla_x f(\bar{p},\bar{x})\bar{x}),$$

and after adding and subtracting terms, use (5). □

Exercise 1B.6. *Consider a polynomial of degree* $n > 0$,

$$p(x) = \sum_{i=0}^{n} a_i x^i,$$

where the coefficients $a_0, \ldots, a_n$ *are real numbers. For each coefficient vector* $a = (a_0, \ldots, a_n) \in \mathbb{R}^{n+1}$ *let* $S(a)$ *denote the set of all real zeros of* p, *so that* S *is a mapping from* $\mathbb{R}^{n+1}$ *to* $\mathbb{R}$ *whose domain consists of the vectors* a *such that* p *has at least one real zero. Let* $\bar{a}$ *be a coefficient vector such that* p *has a simple real zero* $\bar{s}$; *thus* $p(\bar{s}) = 0$ *but* $p'(\bar{s}) \neq 0$. *Prove that* S *has a smooth single-valued localization around* $\bar{a}$ *for* $\bar{s}$. *Is such a statement correct when* $\bar{s}$ *is a double zero?*

1C. Calmness

In this section we introduce a continuity property of functions which will play an important role in the book.

Calmness. *A function* $f : \mathbb{R}^n \to \mathbb{R}^m$ *is said to be* calm *at* $\bar{x}$ *relative to a set* D *in* $\mathbb{R}^n$ *if* $\bar{x} \in D \cap \text{dom } f$ *and there exists a constant* $\kappa \geq 0$ *such that*

(1) $$|f(x) - f(\bar{x})| \leq \kappa |x - \bar{x}| \quad \text{for all } x \in D \cap \text{dom } f.$$

The calmness property (1) can alternatively be expressed in the form of the inclusion
$$f(x) \in f(\bar{x}) + \kappa |x - \bar{x}| B \quad \text{for all } x \in D \cap \text{dom } f.$$
That expression connects with the generalization of the definition of calmness to set-valued mappings, which we will discuss at length in Chapter 3.

Note that a function f which is calm at $\bar{x}$ may have empty values at some points x near $\bar{x}$ when $\bar{x}$ is on the boundary of dom f. If $\bar{x}$ is an isolated point of $D \cap \text{dom } f$, then trivially f is calm at $\bar{x}$ relative to D with $\kappa = 0$.

We will mostly use a local version of the calmness property where the set D in the condition (1) is a neighborhood of $\bar{x}$; if such a neighborhood exists we simply say that f is calm at $\bar{x}$. Calmness of this kind can be identified with the finiteness of the *modulus* which we proceed to define next.

Calmness modulus. *For a function $f : \mathbb{R}^n \to \mathbb{R}^m$ and a point $\bar{x} \in \text{dom } f$, the calmness modulus of f at $\bar{x}$, denoted $\text{clm}(f;\bar{x})$, is the infimum of the set of values $\kappa \geq 0$ for which there exists a neighborhood D of $\bar{x}$ such that (1) holds.*

According to this, as long as $\bar{x}$ is *not an isolated point* of dom f, the calmness modulus satisfies
$$\text{clm}(f;\bar{x}) = \limsup_{\substack{x \in \text{dom } f, x \to \bar{x} \\ x \neq \bar{x}}} \frac{|f(x) - f(\bar{x})|}{|x - \bar{x}|}.$$

If $\bar{x}$ is an isolated point we have $\text{clm}(f;\bar{x}) = 0$. When f is not calm at $\bar{x}$, from the definition we get $\text{clm}(f;\bar{x}) = \infty$. In this way,
$$f \text{ is calm at } \bar{x} \iff \text{clm}(f;\bar{x}) < \infty.$$

Examples.
1) The function $f(x) = x$ for $x \geq 0$ is calm at any point of its domain $[0, \infty)$, always with calmness modulus 1.
2) The function $f(x) = \sqrt{|x|}, x \in \mathbb{R}$ is not calm at zero but calm everywhere else.
3) The linear mapping $A : x \mapsto Ax$, where A is an $m \times n$ matrix, is calm at every point $x \in \mathbb{R}^n$ and everywhere has the same modulus $\text{clm}(A;x) = |A|$.

Straight from the definition of the calmness modulus, we observe that
(i) $\text{clm}(f;\bar{x}) \geq 0$ for any $\bar{x} \in \text{dom } f$;
(ii) $\text{clm}(\lambda f;\bar{x}) = |\lambda| \text{clm}(f;\bar{x})$ for any $\lambda \in \mathbb{R}$ and $\bar{x} \in \text{dom } f$;
(iii) $\text{clm}(f+g;\bar{x}) \leq \text{clm}(f;\bar{x}) + \text{clm}(g;\bar{x})$ for any $\bar{x} \in \text{dom } f \cap \text{dom } g$.

These properties of the calmness modulus resemble those of a norm on a space of functions f, but because $\text{clm}(f;\bar{x}) = 0$ does not imply $f = 0$, one could at most contemplate a seminorm. However, even that falls short, since the modulus can take on ∞, as can the functions themselves, which do not form a linear space because they need not even have the same domain.

Exercise 1C.1 (properties of the calmness modulus). *Prove that*
(a) $\text{clm}(f \circ g; \bar{x}) \leq \text{clm}(f; g(\bar{x})) \cdot \text{clm}(g; \bar{x})$ *whenever $\bar{x} \in \text{dom } g$ and $g(\bar{x}) \in \text{dom } f$;*

(b) $\operatorname{clm}(f-g;\bar{x}) = 0 \Rightarrow \operatorname{clm}(f;\bar{x}) = \operatorname{clm}(g;\bar{x})$ whenever $\bar{x} \in \operatorname{int}(\operatorname{dom} f \cap \operatorname{dom} g)$, *but the converse is false.*

With the concept of calmness in hand, we can interpret the differentiability of a function $f : \mathbb{R}^n \to \mathbb{R}^m$ at a point $\bar{x} \in \operatorname{int} \operatorname{dom} f$ as the existence of a linear mapping $A : \mathbb{R}^n \to \mathbb{R}^m$, represented by an $n \times m$ matrix, such that

(2) $\qquad \operatorname{clm}(e;\bar{x}) = 0 \text{ for } e(x) = f(x) - [f(\bar{x}) + A(x - \bar{x})].$

According to property (iii) before 1C.1 there is at most one mapping M satisfying (2). Indeed, if A_1 and A_2 satisfy (2) we have for the corresponding approximation error terms $e_1(x)$ and $e_2(x)$ that

$$|A_1 - A_2| = \operatorname{clm}(e_1 - e_2; \bar{x}) \leq \operatorname{clm}(e_1; \bar{x}) + \operatorname{clm}(e_2; \bar{x}) = 0.$$

Thus, A is unique and the associated matrix has to be the Jacobian $\nabla f(\bar{x})$. We conclude further from property (b) in 1C.1 that

$$\operatorname{clm}(f;\bar{x}) = |\nabla f(\bar{x})|.$$

The following theorem complements Theorem 1A.1. It shows that the invertibility of the derivative is a necessary condition to obtain a calm single-valued localization of the inverse.

Theorem 1C.2 (Jacobian nonsingularity from inverse calmness). *Given $f : \mathbb{R}^n \to \mathbb{R}^n$ and $\bar{x} \in \operatorname{int} \operatorname{dom} f$, let f be differentiable at $\bar{x}$ and let $\bar{y} := f(\bar{x})$. If f^{-1} has a single-valued localization around $\bar{y}$ for $\bar{x}$ which is calm at $\bar{y}$, then the matrix $\nabla f(\bar{x})$ must be nonsingular.*

Proof. The assumption that f^{-1} has a calm single-valued localization s around $\bar{y}$ for $\bar{x}$ means several things: first, s is nonempty-valued around $\bar{y}$, that is, $\operatorname{dom} s$ is a neighborhood of $\bar{y}$; second, s is a function; and third, s is calm at $\bar{y}$. Specifically, there exist positive numbers a, b and κ and a function s with $\operatorname{dom} s \supset \mathbb{B}_b(\bar{y})$ and values $s(y) \in \mathbb{B}_a(\bar{x})$ such that for every $y \in \mathbb{B}_b(\bar{y})$ we have $s(y) = f^{-1}(y) \cap \mathbb{B}_a(\bar{x})$ and s is calm at $\bar{y}$ with constant κ. Taking b smaller if necessary we have

(3) $\qquad |s(y) - \bar{x}| \leq \kappa |y - \bar{y}| \quad \text{for any } y \in \mathbb{B}_b(\bar{y}).$

Choose τ to satisfy $0 < \tau < 1/\kappa$. Then, since $\bar{x} \in \operatorname{int} \operatorname{dom} f$ and f is differentiable at $\bar{x}$, there exists $\delta > 0$ such that

(4) $\qquad |f(x) - f(\bar{x}) - \nabla f(\bar{x})(x - \bar{x})| \leq \tau |x - \bar{x}| \quad \text{for all } x \in \mathbb{B}_\delta(\bar{x}).$

If the matrix $\nabla f(\bar{x})$ were singular, there would exist $d \in \mathbb{R}^n$, $|d| = 1$, such that $\nabla f(\bar{x})d = 0$. Pursuing this possibility, let ε satisfy $0 < \varepsilon < \min\{a, b/\tau, \delta\}$. Then, by applying (4) with $x = \bar{x} + \varepsilon d$, we get $f(\bar{x} + \varepsilon d) \in \mathbb{B}_b(\bar{y})$. In terms of $y_\varepsilon := f(\bar{x} + \varepsilon d)$, we then have $\bar{x} + \varepsilon d \in f^{-1}(y_\varepsilon) \cap \mathbb{B}_a(\bar{x})$, hence $s(y_\varepsilon) = \bar{x} + \varepsilon d$. The calmness condition (3) then yields

$$1 = |d| = \frac{1}{\varepsilon}|\bar{x} + \varepsilon d - \bar{x}| = \frac{1}{\varepsilon}|s(y_\varepsilon) - \bar{x}| \leq \frac{\kappa}{\varepsilon}|y_\varepsilon - \bar{y}| = \frac{\kappa}{\varepsilon}|f(\bar{x} + \varepsilon d) - f(\bar{x})|.$$

Combining this with (4) and taking into account that $\nabla f(\bar{x})d = 0$, we arrive at $1 \leq \kappa\tau|d| < 1$ which is absurd. Hence $\nabla f(\bar{x})$ is nonsingular. □

Note that in the particular case of an affine function $f(x) = Ax + b$, where A is a square matrix and b is a vector, calmness can be dropped from the set of assumptions of Theorem 1C.2; the existence of a single-valued localization of f^{-1} around any point is already equivalent to the nonsingularity of the Jacobian. This is not always true even for polynomials. Indeed, the inverse of $f(x) = x^3$, $x \in \mathbb{R}$, has a single-valued localization around the origin (which is not calm), but $\nabla f(0) = 0$.

The classical inverse function theorem 1A.1 combined with Theorem 1C.2 above gives us

Theorem 1C.3 (symmetric inverse function theorem). *Let $f : \mathbb{R}^n \to \mathbb{R}^n$ be continuously differentiable around $\bar{x}$. Then the following are equivalent:*
  (i) *$\nabla f(\bar{x})$ is nonsingular;*
  (ii) *f^{-1} has a single-valued localization s around $\bar{y} := f(\bar{x})$ for $\bar{x}$ which is continuously differentiable around $\bar{y}$.*

The formula for the Jacobian of the single-valued localization s of the inverse,

$$\nabla s(y) = \nabla f(s(y))^{-1} \quad \text{for } y \text{ around } \bar{y},$$

comes as a byproduct of the statement (ii) by way of the chain rule.

A modification of the proof of Theorem 1C.2 gives us the converse to the correction function theorem.

Theorem 1C.4 (Jacobian nonsingularity from correction differentiability). *Let $f : \mathbb{R}^n \to \mathbb{R}^n$ be differentiable at $\bar{x}$ and suppose that the correction mapping*

$$\Xi : x \mapsto \left\{ u \in \mathbb{R}^n \,\middle|\, f(x + u) = f(\bar{x}) + \nabla f(\bar{x})(x - \bar{x}) \right\} \quad \text{for } x \in \mathbb{R}^n$$

has a single-valued localization ξ around $\bar{x}$ for 0 such that ξ is calm at $\bar{x}$ with clm$(\xi; \bar{x}) = 0$. Then $\nabla f(\bar{x})$ is nonsingular.

Proof. If $\nabla f(\bar{x})$ is singular, there must exist a vector $d \in \mathbb{R}^n$ with $|d| = 1$ such that $\nabla f(\bar{x})d = 0$. Then for all sufficiently small $\varepsilon > 0$ we have

$$f(\bar{x} + \varepsilon d + \xi(\bar{x} + \varepsilon d)) = f(\bar{x}).$$

Thus, $\varepsilon d + \xi(\bar{x} + \varepsilon d) \in \Xi(\bar{x})$ for all small $\varepsilon > 0$. Since Ξ has a single-valued localization around $\bar{x}$ we get $\bar{x} + \varepsilon d + \xi(\bar{x} + \varepsilon d) = \bar{x}$. Then

$$1 = |d| = \frac{1}{\varepsilon}|\xi(\bar{x} + \varepsilon d)|.$$

1 Functions Defined Implicitly by Equations

The right side of this equation goes to zero as $\varepsilon \to 0$, and that produces a contradiction. $\square$

Next, we extend the definition of calmness to its partial counterparts.

Partial calmness. *A function $f : \mathbb{R}^d \times \mathbb{R}^n \to \mathbb{R}^m$ is said to be calm with respect to x at $(\bar{p},\bar{x}) \in \mathrm{dom}\, f$ when the function φ with values $\varphi(x) = f(\bar{p},x)$ is calm at $\bar{x}$. Such calmness is said to be uniform in p at $(\bar{p},\bar{x})$ when there exists a constant $\kappa > 0$ and neighborhoods Q of $\bar{p}$ and U of $\bar{x}$ such that actually*

$$|f(p,x) - f(p,\bar{x})| \leq \kappa |x - \bar{x}| \quad \text{for all } (p,x) \in (Q \times U) \cap \mathrm{dom}\, f.$$

Correspondingly, the partial calmness modulus of f with respect to x at $(\bar{p},\bar{x})$ is denoted as $\mathrm{clm}_x(f;(\bar{p},\bar{x}))$, while the uniform partial calmness modulus is

$$\widehat{\mathrm{clm}}_x(f;(\bar{p},\bar{x})) := \limsup_{\substack{x \to \bar{x},\, p \to \bar{p},\\ (p,x) \in \mathrm{dom}\, f,\, x \neq \bar{x}}} \frac{|f(p,x) - f(p,\bar{x})|}{|x - \bar{x}|}$$

provided that every neighborhood of $(\bar{p},\bar{x})$ contains points $(p,x) \in \mathrm{dom}\, f$ with $x \neq \bar{x}$.

Observe in this context that differentiability of $f(p,x)$ with respect to x at $(\bar{p},\bar{x}) \in \mathrm{int\, dom}\, f$ is equivalent to the existence of a linear mapping $A : \mathbb{R}^n \to \mathbb{R}^m$, the partial derivative of f with respect to x at $(\bar{p},\bar{x})$, which satisfies

$$\mathrm{clm}\,(e;\bar{x}) = 0 \text{ for } e(x) = f(\bar{p},x) - [f(\bar{p},\bar{x}) + A(x - \bar{x})],$$

and then A is the partial derivative $D_x f(\bar{p},\bar{x})$. In contrast, under the stronger condition that

$$\widehat{\mathrm{clm}}_x(e;(\bar{p},\bar{x})) = 0, \text{ for } e(p,x) = f(p,x) - [f(\bar{p},\bar{x}) + A(x - \bar{x})],$$

we say f is *differentiable with respect to x uniformly in p* at $(\bar{p},\bar{x})$. This means that for every $\varepsilon > 0$ there are neighborhoods Q of $\bar{p}$ and U of $\bar{x}$ such that

$$|f(p,x) - f(p,\bar{x}) - D_x f(\bar{p},\bar{x})(x - \bar{x})| \leq \varepsilon |x - \bar{x}| \quad \text{for } p \in Q \text{ and } x \in U.$$

Exercise 1C.5 (joint calmness criterion). *Let $f : \mathbb{R}^d \times \mathbb{R}^n \to \mathbb{R}^m$ be calm in x uniformly in p and calm in p, both at $(\bar{p},\bar{x})$. Show that f is calm at $(\bar{p},\bar{x})$.*

Exercise 1C.6 (nonsingularity characterization). *Let $f : \mathbb{R}^n \to \mathbb{R}^n$ be differentiable at $\bar{x}$, let $\bar{y} = f(\bar{x})$, and suppose that f^{-1} has a single-valued localization s around $\bar{y}$ for $\bar{x}$ which is continuous at $\bar{y}$. Prove in this setting that s is differentiable at $\bar{y}$ if and only if the Jacobian $\nabla f(\bar{x})$ is nonsingular.*

Guide. The "only if" part can be obtained from Theorem 1C.2, using the fact that if s is differentiable at $\bar{x}$, it must be calm at $\bar{x}$. In the other direction, starting from the assumption that $\nabla f(\bar{x})$ is nonsingular, argue in a manner parallel to the first part of Step 3 of Proof I of Theorem 1A.1. □

1D. Lipschitz Continuity

Calmness is a "one-point" version of the well-known "two-point" property of functions named after Rudolf Otto Sigismund Lipschitz (1832–1903). That property has already entered our deliberations in Section 1A in connection with the Proof II of the classical inverse function theorem by way of the contraction mapping principle, but we investigate it now more directly. For convenience we recall the definition:

Lipschitz continuous functions. *A function $f : \mathbb{R}^n \to \mathbb{R}^m$ is said to be Lipschitz continuous relative to a set D, or on a set D, if $D \subset \mathrm{dom}\, f$ and there exists a constant $\kappa \geq 0$ (Lipschitz constant) such that*

$$(1) \qquad |f(x') - f(x)| \leq \kappa |x' - x| \quad \text{for all } x', x \in D.$$

It is said to be Lipschitz continuous around $\bar{x}$ when this holds for some neighborhood D of $\bar{x}$. We say further, in the case of an open set C, that f is locally Lipschitz continuous on C if it is a Lipschitz continuous function around every point x of C.

Lipschitz modulus. *For a function $f : \mathbb{R}^n \to \mathbb{R}^m$ and a point $\bar{x} \in \mathrm{int}\, \mathrm{dom}\, f$, the Lipschitz modulus of f at $\bar{x}$, denoted $\mathrm{lip}(f;\bar{x})$, is the infimum of the set of values of κ for which there exists a neighborhood D of $\bar{x}$ such that (1) holds. Equivalently,*

$$(2) \qquad \mathrm{lip}(f;\bar{x}) := \limsup_{\substack{x',x \to \bar{x} \\ x \neq x'}} \frac{|f(x') - f(x)|}{|x' - x|}.$$

Note that, by this definition, for the Lipschitz modulus we have $\mathrm{lip}(f;\bar{x}) = \infty$ precisely in the case where, for every $\kappa > 0$ and every neighborhood D of $\bar{x}$, there are points $x', x \in D$ violating (1). Thus,

$$f \text{ is Lipschitz continuous around } \bar{x} \iff \mathrm{lip}(f;\bar{x}) < \infty.$$

A function f with $\mathrm{lip}(f;\bar{x}) < \infty$ is also called *strictly continuous* at $\bar{x}$. For an open set C, a function f is *locally Lipschitz continuous* on C exactly when $\mathrm{lip}(f;x) < \infty$ for every $x \in C$. Every continuously differentiable function on an open set C is locally Lipschitz continuous on C.

1 Functions Defined Implicitly by Equations

Examples.
1) The function $x \mapsto |x|$, $x \in \mathbb{R}^n$, is Lipschitz continuous everywhere with $\mathrm{lip}\,(|x|;x) = 1$; it is not differentiable at 0.
2) An affine function $f : x \mapsto Ax + b$, corresponding to a matrix $A \in \mathbb{R}^{m \times n}$ and a vector $b \in \mathbb{R}^m$, has $\mathrm{lip}\,(f;\bar{x}) = |A|$ for every $\bar{x} \in \mathbb{R}^n$.
3) If f is continuously differentiable in a neighborhood of $\bar{x}$, then $\mathrm{lip}\,(f;\bar{x}) = |\nabla f(\bar{x})|$.

Like the calmness modulus, the Lipschitz modulus has the properties of a seminorm, except in allowing for ∞:

(i) $\mathrm{lip}\,(f;\bar{x}) \geq 0$ for any $\bar{x} \in \mathrm{int\,dom}\,f$;
(ii) $\mathrm{lip}\,(\lambda f;\bar{x}) = |\lambda|\,\mathrm{lip}\,(f;\bar{x})$ for any $\lambda \in \mathbb{R}$ and $\bar{x} \in \mathrm{int\,dom}\,f$;
(iii) $\mathrm{lip}\,(f+g;\bar{x}) \leq \mathrm{lip}\,(f;\bar{x}) + \mathrm{lip}\,(g;\bar{x})$ for any $\bar{x} \in \mathrm{int\,dom}\,f \cap \mathrm{int\,dom}\,g$.

Exercise 1D.1 (properties of the Lipschitz modulus). *Prove that*
(a) $\mathrm{lip}\,(f \circ g;\bar{x}) \leq \mathrm{lip}\,(f;g(\bar{x})) \cdot \mathrm{lip}\,(g;\bar{x})$ *when* $\bar{x} \in \mathrm{int\,dom}\,g$ *and* $g(\bar{x}) \in \mathrm{int\,dom}\,f$;
(b) $\mathrm{lip}\,(f-g;\bar{x}) = 0 \Rightarrow \mathrm{lip}\,(f;\bar{x}) = \mathrm{lip}\,(g;\bar{x})$ *when* $\bar{x} \in \mathrm{int\,dom}\,f \cap \mathrm{int\,dom}\,g$;
(c) $\mathrm{lip}\,(f;\cdot)$ *is upper semicontinuous at any* $\bar{x} \in \mathrm{int\,dom}\,f$ *where it is finite*;
(d) *the set* $\{x \in \mathrm{int\,dom}\,f \mid \mathrm{lip}\,(f;x) < \infty\}$ *is open.*

Bounds on the Lipschitz modulus lead to Lipschitz constants relative to sets, as long as convexity is present. First, recall that a set $C \subset \mathbb{R}^n$ is *convex* if

$$(1-\tau)x_0 + \tau x_1 \in C \quad \text{for all } \tau \in (0,1) \text{ when } x_0, x_1 \in C,$$

or in other words, if C contains for any pair of its points the entire line segment that joins them. The most obvious convex set is the ball $\mathbb{B}$ as well as its interior, while the boundary of the ball is of course nonconvex.

Exercise 1D.2 (Lipschitz continuity on convex sets). *Show that if C is a convex subset of* $\mathrm{int\,dom}\,f$ *such that* $\mathrm{lip}\,(f;x) \leq \kappa$ *for all* $x \in C$, *then* f *is Lipschitz continuous relative to C with constant κ.*

Guide. It is enough to demonstrate for an arbitrary choice of points x and x' in C and $\varepsilon > 0$ that $|f(x') - f(x)| \leq (\kappa + \varepsilon)|x' - x|$. Argue that the line segment joining x and x' is a compact subset of $\mathrm{int\,dom}\,f$ which can be covered by finitely many balls on which f is Lipschitz continuous with constant $\kappa + \varepsilon$. Moreover these balls can be chosen in such a way that a finite sequence of points $x_0, x_1, \ldots, x_r$ along the segment, starting with $x_0 = x$ and ending with $x_r = x'$, has each consecutive pair in one of them. Get the Lipschitz inequality for x and x' from the Lipschitz inequalities for these pairs. □

Exercise 1D.3 (Lipschitz continuity from differentiability). *If f is continuously differentiable on an open set O and C is a compact convex subset of O, then f is Lipschitz continuous relative to C with constant* $\kappa = \max_{x \in C} |\nabla f(x)|$.

Convexity also provides an important class of examples of Lipschitz continuous functions from $\mathbb{R}^n$ into itself which are not everywhere differentiable, namely distance and projection mappings; for an illustration see Fig. 1.3.

Distance and projection. *For a point $x \in \mathbb{R}^n$ and a set $C \subset \mathbb{R}^n$, the quantity*

$$(3) \qquad d_C(x) = d(x,C) = \inf_{y \in C} |x - y|$$

is called the distance *from x to C. (Whether the notation $d_C(x)$ or $d(x,C)$ is used is a matter of convenience in a given context.) Any point y of C which is closest to x in the sense of achieving this distance is called a* projection *of x on C. The set of such projections is denoted by $P_C(x)$. Thus,*

$$(4) \qquad P_C(x) = \operatorname*{argmin}_{y \in C} |x - y|.$$

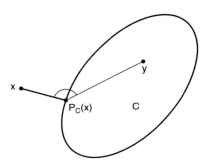

Fig. 1.3 Distance and projection.

In this way, C gives rise to a *distance function d_C* and a *projection mapping P_C*. If C is empty, then trivially $d_C(x) = \infty$ for all x, whereas if C is nonempty, then $d_C(x)$ is finite (and nonnegative) for all x. As for P_C, it is, in general, a set-valued mapping from $\mathbb{R}^n$ into C, but additional properties follow from particular assumptions on C, as we explore next.

Proposition 1D.4 (properties of distance and projection).
(a) *For a nonempty set $C \subset \mathbb{R}^n$, one has $d_C(x) = d_{\operatorname{cl} C}(x)$ for all x. Moreover, C is closed if and only if every x with $d_C(x) = 0$ belongs to C.*
(b) *For a nonempty set $C \subset \mathbb{R}^n$, the distance function d_C is Lipschitz continuous on $\mathbb{R}^n$ with Lipschitz constant $\kappa = 1$. As long as C is closed, one has*

$$\operatorname{lip}(d_C; \bar{x}) = \begin{cases} 0 & \text{if } \bar{x} \in \operatorname{int} C, \\ 1 & \text{otherwise.} \end{cases}$$

1 Functions Defined Implicitly by Equations

(c) *For a nonempty, closed set $C \subset \mathbb{R}^n$, the projection set $P_C(x)$ is nonempty, closed and bounded for every $x \in \mathbb{R}^n$.*

Proof. For (a), we fix any $x \in \mathbb{R}^n$ and note that obviously $d_{\text{cl}\,C}(x) \leq d_C(x)$. This inequality can't be strict because for any $\varepsilon > 0$ we can find $y \in \text{cl}\,C$ making $|x - y| < d_{\text{cl}\,C}(x) + \varepsilon$ but then also find $y' \in C$ with $|y - y'| < \varepsilon$, in which case we have $d_C(x) \leq |x - y'| < d_{\text{cl}\,C}(x) + 2\varepsilon$. In particular, this argument reveals that $d_{\text{cl}\,C}(x) = 0$ if and only if $x \in \text{cl}\,C$. Having demonstrated that $d_C(x) = d_{\text{cl}\,C}(x)$, we may conclude that $C = \{x \mid d_C(x) = 0\}$ if and only if $C = \text{cl}\,C$.

For (b), consider any points x and x' along with any $\varepsilon > 0$. Take any point $y \in C$ such that $|x - y| \leq d_C(x) + \varepsilon$. We have $d_C(x') \leq |x' - y| \leq |x' - x| + |x - y| \leq |x' - x| + d_C(x) + \varepsilon$, and through the arbitrariness of ε therefore $d_C(x') - d_C(x) \leq |x' - x|$. The same thing must hold with the roles of x and x' reversed, so this demonstrates that d_C is Lipschitz continuous with constant 1.

Let C be nonempty and closed. If $\bar{x} \in \text{int}\,C$, we have $d_C(x) = 0$ for all x in a neighborhood of $\bar{x}$ and consequently $\text{lip}\,(d_C;\bar{x}) = 0$. Suppose now that $\bar{x} \notin \text{int}\,C$. We will show that $\text{lip}\,(d_C;\bar{x}) \geq 1$, in which case equality must actually hold because we already know that d_C is Lipschitz continuous on $\mathbb{R}^n$ with constant 1. According to the property of the Lipschitz modulus displayed in Exercise 1D.1(c), it is sufficient to consider $\bar{x} \notin C$. Let $\tilde{x} \in P_C(\bar{x})$. Then on the line segment from $\tilde{x}$ to $\bar{x}$ the distance increases linearly, that is, $d_C(\tilde{x} + \tau(\bar{x} - \tilde{x})) = \tau d(\bar{x}, C)$ for $0 \leq \tau \leq 1$ (prove!). Hence for the two points $x = \tilde{x} + \tau(\bar{x} - \tilde{x})$ and $x' = \tilde{x} + \tau'(\bar{x} - \tilde{x})$ we have $|d_C(x') - d_C(x)| = |\tau' - \tau||\bar{x} - \tilde{x}| = |x' - x|$. Note that $\bar{x}$ can be approached by such pairs of points and hence $\text{lip}\,(d_C;\bar{x}) \geq 1$.

Turning now to (c), we again fix any $x \in \mathbb{R}^n$ and choose a sequence of points $y_k \in C$ such that $|x - y_k| \to d_C(x)$ as $k \to \infty$. This sequence is bounded and therefore has an accumulation point y in C, inasmuch as C is closed. Since $|x - y_k| \geq d_C(x)$ for all k, it follows that $|x - y| = d_C(x)$. Thus, $y \in P_C(x)$, so $P_C(x)$ is not empty. Since by definition $P_C(x)$ is the intersection of C with the closed ball with center x and radius $d_C(x)$, it's clear that $P_C(x)$ is furthermore closed and bounded. □

It has been seen in 1D.4(c) that for any nonempty closed set $C \subset \mathbb{R}^n$ the projection mapping $P_C : \mathbb{R}^n \rightrightarrows C$ is nonempty-compact-valued, but when might it actually be single-valued as well? The convexity of C is the additional property that yields this conclusion, as will be shown in the following proposition[4].

Proposition 1D.5 (Lipschitz continuity of projection mappings). *For a nonempty, closed, convex set $C \subset \mathbb{R}^n$, the projection mapping P_C is single-valued (a function) from $\mathbb{R}^n$ onto C which moreover is Lipschitz continuous with Lipschitz constant $\kappa = 1$. Also,*

(5) $$P_C(\bar{x}) = \bar{y} \iff \langle \bar{x} - \bar{y}, y - \bar{y} \rangle \leq 0 \quad \text{for all } y \in C.$$

---

[4] A set C such that P_C is single-valued is called a Chebyshev set. A nonempty, closed, convex set is always a Chebyshev set, and in $\mathbb{R}^n$ the converse is also true; for proofs of this fact see Borwein and Lewis [2006] and Deutsch [2001]. The question of whether a Chebyshev set in an arbitrary infinite-dimensional Hilbert space must be convex is still open.

Proof. We have $P_C(x) \neq \emptyset$ in view of 1D.4(c). Suppose $\bar{y} \in P_C(\bar{x})$. For any $\tau \in (0,1)$, any $y \in \mathbb{R}^n$ and $y_\tau = (1-\tau)\bar{y} + \tau y$ we have the identity

(6)
$$\begin{aligned}|\bar{x} - y_\tau|^2 - |\bar{x} - \bar{y}|^2 &= |(y_\tau - \bar{y}) - (\bar{x} - \bar{y})|^2 - |\bar{x} - \bar{y}|^2 \\ &= |y_\tau - \bar{y}|^2 - 2\langle \bar{x} - \bar{y}, y_\tau - \bar{y}\rangle \\ &= \tau^2 |y - \bar{y}|^2 - 2\tau \langle \bar{x} - \bar{y}, y - \bar{y}\rangle.\end{aligned}$$

If $y \in C$, we also have $y_\tau \in C$ by convexity, so the left side is nonnegative. This implies that $\tau |y - \bar{y}|^2 \geq 2\langle \bar{x} - \bar{y}, y - \bar{y}\rangle$ for all $\tau \in (0,1)$. Thus, the inequality in (5) holds. On the other hand, let $\langle \bar{x} - \bar{y}, y - \bar{y}\rangle \leq 0$ for all $y \in C$. If $y \in C$ is such that $|\bar{x} - y| \leq |\bar{x} - \bar{y}|$ then for y_τ as in (6) we get $\tau^2 |y - \bar{y}|^2 \leq 2\langle \bar{x} - \bar{y}, y - \bar{y}\rangle \leq 0$ showing that $y = \bar{y}$. Thus (5) is fully confirmed along with the fact that $P_C(\bar{x})$ can't contain any $y \neq \bar{y}$.

Consider now two points x_0 and x_1 and their projections $y_0 = P_C(x_0)$ and $y_1 = P_C(x_1)$. On applying (5), we see that

$$\langle x_0 - y_0, y_1 - y_0\rangle \leq 0 \quad \text{and} \quad \langle x_1 - y_1, y_0 - y_1\rangle \leq 0.$$

When added, these inequalities give us

$$0 \geq \langle x_0 - y_0 - x_1 + y_1, y_1 - y_0\rangle = |y_1 - y_0|^2 - \langle x_1 - x_0, y_1 - y_0\rangle$$

and consequently

$$|y_1 - y_0|^2 \leq \langle x_1 - x_0, y_1 - y_0\rangle \leq |x_1 - x_0||y_1 - y_0|.$$

It follows that
$$|y_1 - y_0| \leq |x_1 - x_0|.$$

Thus, P_C is Lipschitz continuous with Lipschitz constant 1. □

Projection mappings have many uses in numerical analysis and optimization. Note that P_C always fails to be differentiable on the boundary of C. As an example, when C is the set of nonpositive reals $\mathbb{R}_-$ one has

$$P_C(x) = \begin{cases} 0 & \text{for } x \geq 0, \\ x & \text{for } x < 0 \end{cases}$$

and this function is not differentiable at $x = 0$.

It is clear from the definitions of the calmness and Lipschitz moduli that we always have
$$\mathrm{clm}(f;\bar{x}) \leq \mathrm{lip}(f;\bar{x}).$$

This relation is illustrated in Fig. 1.4.

In the preceding section we showed how to characterize differentiability through calmness. Now we introduce a sharper concept of derivative which is tied up with the Lipschitz modulus.

1 Functions Defined Implicitly by Equations

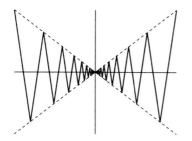 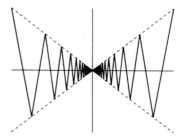

Fig. 1.4 Plots of calm and Lipschitz continuous functions. On the left is the plot of the function $f(x) = (-1)^{n+1}9x + (-1)^n 2^{2n+1}/5^{n-2}$, $|x| \in [x_{n+1}, x_n]$ for $x_n = 4^{n-1}/5^{n-2}$, $n = 1, 2, \ldots$ for which $\mathrm{clm}(f;0) < \mathrm{lip}(f;0) < \infty$. On the right is the plot of the function $f(x) = (-1)^{n+1}(6+n)x + (-1)^n 210(5+n)!/(6+n)!$, $|x| \in [x_{n+1}, x_n]$ for $x_n = 210(4+n)!/(6+n)!$, $n = 1, 2, \ldots$ for which $\mathrm{clm}(f;0) < \mathrm{lip}(f;0) = \infty$.

Strict differentiability. *A function $f : \mathbb{R}^n \to \mathbb{R}^m$ is said to be* strictly differentiable *at a point $\bar{x}$ if there is a linear mapping $A : \mathbb{R}^n \to \mathbb{R}^m$ such that*

$$\mathrm{lip}(e; \bar{x}) = 0 \ \ \textit{for} \ e(x) = f(x) - [f(\bar{x}) + A(x - \bar{x})].$$

In particular, in this case we have that $\mathrm{clm}(e; \bar{x}) = 0$ and hence f is differentiable at $\bar{x}$ with $A = Df(\bar{x})$, but strictness imposes a requirement on the difference

$$e(x) - e(x') = f(x) - [f(x') + Df(\bar{x})(x - x')]$$

also when $x' \neq \bar{x}$. Specifically, it demands the existence for each $\varepsilon > 0$ of a neighborhood U of $\bar{x}$ such that

$$|f(x) - [f(x') + Df(\bar{x})(x - x')]| \leq \varepsilon |x - x'| \ \ \text{for every } x, x' \in U.$$

Exercise 1D.6 (strict differentiability from continuous differentiability). *Prove that any function f that is continuously differentiable in a neighborhood of $\bar{x}$ is strictly differentiable at $\bar{x}$.*

Guide. Adopt formula (b) in Fact 1 in the beginning of Section 1A. □

The converse to the assertion in Exercise 1D.6 is false, however: f can be strictly differentiable at $\bar{x}$ without being continuously differentiable around $\bar{x}$. This is demonstrated in Fig. 1.5 showing the graphs of two functions that are both differentiable at origin but otherwise have different properties. On the left is the graph of the continuous function $f : [-1, 1] \to \mathbb{R}$ which is even, and on $[0, 1]$ has values $f(0) = 0$, $f(1/n) = 1/n^2$, and is linear in the intervals $[1/n, 1/(n+1)]$. This function is strictly differentiable at 0, but in every neighborhood of 0 there are points where differentiability is lacking. On the right is the graph of the function[5]

$$f(x) = \begin{cases} x/2 + x^2 \sin(1/x) & \text{for } x \neq 0, \\ 0 & \text{for } x = 0, \end{cases}$$

which is differentiable at 0 but not strictly differentiable there.

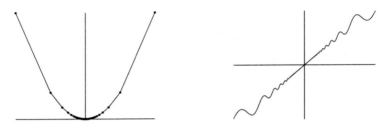

Fig. 1.5 Plots of functions differentiable at the origin. The function on the left is strictly differentiable at the origin but not continuously differentiable. The function on the right is differentiable at the origin but not strictly differentiable there.

The second of these examples has the interesting feature that, even though $f(0) = 0$ and $f'(0) \neq 0$, no single-valued localization of f^{-1} exists around 0 for 0. In contrast, we will see in 1D.9 that strict differentiability would ensure the availability of such a localization.

Exercise 1D.7 (strict differentiability from differentiability). *Consider a function f which is differentiable at every point in a neighborhood of $\bar{x}$. Prove that f is strictly differentiable at $\bar{x}$ if and only if the Jacobian ∇f is continuous at $\bar{x}$.*

Guide. Let f be strictly differentiable at $\bar{x}$ and let $\varepsilon > 0$. Then there exists $\delta_1 > 0$ such that for any $x_1, x_2 \in B_{\delta_1}(\bar{x})$ we have

(7) $$|f(x_2) - f(x_1) - \nabla f(\bar{x})(x_2 - x_1)| \leq \frac{1}{2}\varepsilon|x_1 - x_2|.$$

Fix an $x_1 \in B_{\delta_1/2}(\bar{x})$. For this x_1 there exists $\delta_2 > 0$ such that for any $x' \in B_{\delta_2}(u_1)$,

(8) $$|f(x') - f(x_1) - \nabla f(x_1)(x' - x_1)| \leq \frac{1}{2}\varepsilon|x' - x_1|.$$

Make δ_2 smaller if necessary so that $B_{\delta_2}(x_1) \subset B_{\delta_1}(\bar{x})$. By (7) with x_2 replaced by x' and by (8), we have

$$|\nabla f(x_1)(x' - x_1) - \nabla f(\bar{x})(x' - x_1)| \leq \varepsilon|x' - x_1|.$$

---

[5] These two examples are from Nijenhuis [1974], where the introduction of strict differentiability is attributed to Leach [1961]. By the way, Nijenhuis dedicated his paper to Carl Allendoerfer "for *not* taking the implicit function theorem for granted." In the book we follow this advice.

1 Functions Defined Implicitly by Equations

This implies
$$|\nabla f(x_1) - \nabla f(\bar{x})| \leq \varepsilon.$$
Since x_1 is arbitrarily chosen in $\mathbb{B}_{\delta_1/2}(\bar{x})$, we obtain that the Jacobian is continuous at $\bar{x}$.

For the opposite direction, use Fact 1 in the beginning of Section 1A. □

Exercise 1D.8 (continuous differentiability from strict differentiability). *Prove that a function f is strictly differentiable at every point of an open set O if and only if it is continuously differentiable on O.*

Guide. Apply 1D.7. □

With the help of the strict derivative we can obtain a new version of the classical inverse function theorem 1A.1.

Theorem 1D.9 (symmetric inverse function theorem under strict differentiability). *Let $f : \mathbb{R}^n \to \mathbb{R}^n$ be strictly differentiable at $\bar{x}$. Then the following are equivalent:*

(i) $\nabla f(\bar{x})$ *is nonsingular;*

(ii) f^{-1} *has a single-valued localization s around $\bar{y} := f(\bar{x})$ for $\bar{x}$ which is strictly differentiable at $\bar{y}$. In that case, moreover, $\nabla s(\bar{y}) = \nabla f(\bar{x})^{-1}$.*

Proof. The implication (i) ⇒ (ii) can be accomplished by combining various pieces already present in the proofs of Theorem 1A.1, since strict differentiability of f at $\bar{x}$ gives us by definition the estimate (b) in Fact 1 in Section 1A. Parallel to Proof II of 1A.1 we find positive constants a and b and a single-valued localization of f^{-1} of the form
$$s : y \mapsto f^{-1}(y) \cap \mathbb{B}_a(\bar{x}) \quad \text{for } y \in \mathbb{B}_b(\bar{y}).$$
Next, by using the equation
$$s(y) = -A^{-1}(f(s(y)) - y - As(y)) \quad \text{for } y \in \mathbb{B}_b(\bar{y}),$$
where $A = \nabla f(\bar{x})$, we demonstrate Lipschitz continuity of s around $\bar{y}$ as in the beginning of Step 3 of Proof I of Theorem 1A.1. Finally, to obtain strict differentiability of s at $\bar{y}$, repeat the second part of Step 3 of Proof I with $\nabla f(s(y))$ replaced with A. For the converse implication we invoke Theorem 1C.2 and the fact that strict differentiability entails calmness. □

Working now towards a corresponding version of the implicit function theorem, we look at additional forms of Lipschitz continuity and strict differentiability.

Partial Lipschitz continuity. A function $f : \mathbb{R}^d \times \mathbb{R}^n \to \mathbb{R}^m$ is said to be *Lipschitz continuous with respect to x around* $(\bar{p}, \bar{x}) \in \text{int dom } f$ when the function $x \mapsto f(\bar{p}, x)$ is Lipschitz continuous around $\bar{x}$; the associated Lipschitz modulus of f with respect to x is denoted by $\text{lip}_x(f; (\bar{p}, \bar{x}))$. We say f is *Lipschitz continuous with respect to x*

uniformly in p around $(\bar{p},\bar{x}) \in \operatorname{int} \operatorname{dom} f$ when there are neighborhoods Q of $\bar{p}$ and U of $\bar{x}$ along with a constant κ and such that

$$|f(p,x) - f(p,x')| \leq \kappa |x - x'| \quad \text{for all } x,x' \in U \text{ and } p \in Q.$$

Accordingly, the partial uniform Lipschitz modulus with respect to x has the form

$$\widehat{\operatorname{lip}}_x(f;(\bar{p},\bar{x})) := \limsup_{\substack{x,x' \to \bar{x}, p \to \bar{p},\\ x \neq x'}} \frac{|f(p,x') - f(p,x)|}{|x' - x|}.$$

Exercise 1D.10 (partial uniform Lipschitz modulus with differentiability). *Show that if the function $f : \mathbb{R}^d \times \mathbb{R}^n \to \mathbb{R}^m$ is differentiable with respect to x at all points (p,x) in some neighborhood of $(\bar{p},\bar{x})$, then*

$$\widehat{\operatorname{lip}}_x(f;(\bar{p},\bar{x})) = \limsup_{(p,x) \to (\bar{p},\bar{x})} |\nabla_x f(p,x)|.$$

Strict partial differentiability. A function $f : \mathbb{R}^d \times \mathbb{R}^n \to \mathbb{R}^m$ is said to be *strictly differentiable with respect to x at* $(\bar{p},\bar{x})$ if the function $x \mapsto f(\bar{p},x)$ is strictly differentiable at $\bar{x}$. It is said to be *strictly differentiable with respect to x uniformly in p at* $(\bar{p},\bar{x})$ if

$$\widehat{\operatorname{lip}}_x(e;(\bar{p},\bar{x})) = 0 \quad \text{for } e(p,x) = f(p,x) - [f(\bar{p},\bar{x}) + D_x f(\bar{p},\bar{x})(x - \bar{x})],$$

or in other words, if for every $\varepsilon > 0$ there are neighborhoods Q of $\bar{p}$ and U of $\bar{x}$ such that

$$|f(p,x) - [f(p,x') + D_x f(\bar{p},\bar{x})(x - x')]| \leq \varepsilon |x - x'| \quad \text{for all } x,x' \in U \text{ and } p \in Q.$$

Exercise 1D.11 (joint differentiability criterion). *Let $f : \mathbb{R}^d \times \mathbb{R}^n \to \mathbb{R}^m$ be strictly differentiable with respect to x uniformly in p and be differentiable with respect to p, both at $(\bar{p},\bar{x})$. Prove that f is differentiable at $(\bar{p},\bar{x})$.*

Exercise 1D.12 (joint strict differentiability criterion). *Prove that $f : \mathbb{R}^d \times \mathbb{R}^n \to \mathbb{R}^m$ is strictly differentiable at $(\bar{p},\bar{x})$ if and only if it is strictly differentiable with respect to x uniformly in p and strictly differentiable with respect to p uniformly in x, both at $(\bar{p},\bar{x})$.*

We state next the implicit function counterpart of Theorem 1D.9.

Theorem 1D.13 (implicit functions under strict partial differentiability). *Given $f : \mathbb{R}^d \times \mathbb{R}^n \to \mathbb{R}^n$ and $(\bar{p},\bar{x})$ with $f(\bar{p},\bar{x}) = 0$, suppose that f is strictly differentiable at $(\bar{p},\bar{x})$ and let the partial Jacobian $\nabla_x f(\bar{p},\bar{x})$ be nonsingular. Then the solution mapping*

1 Functions Defined Implicitly by Equations

$$S : p \mapsto \{x \in \mathbb{R}^n \,|\, f(p,x) = 0\}$$

has a single-valued localization s around $\bar{p}$ for $\bar{x}$ which is strictly differentiable at $\bar{p}$ with its Jacobian expressed by

$$\nabla s(\bar{p}) = -\nabla_x f(\bar{p},\bar{x})^{-1} \nabla_p f(\bar{p},\bar{x}).$$

Proof. We apply Theorem 1D.9 in a manner parallel to the way that the classical implicit function theorem 1B.1 was derived from the classical inverse function theorem 1A.1. □

Exercise 1D.14. Let $f : \mathbb{R} \to \mathbb{R}$ be strictly differentiable at 0 and let $f(0) \neq 0$. Consider the following equation in x with a parameter p:

$$pf(x) = \int_0^x f(pt)\,dt.$$

Prove that the solution mapping associated with this equation has a strictly differentiable single-valued localization around 0 for 0.

Guide. The function $g(p,x) = pf(x) - \int_0^x f(pt)\,dt$ satisfies $(\partial g/\partial x)(0,0) = -f(0)$, which is nonzero by assumption. For any $\varepsilon > 0$ there exist open intervals Q and U centered at 0 such that for every $p \in Q$ and $x, x' \in U$ we have

$$\begin{aligned}
&\left| g(p,x) - g(p,x') - \frac{\partial g}{\partial x}(0,0)(x-x') \right| \\
&= \left| p(f(x) - f(x')) - \int_{x'}^x f(pt)\,dt + f(0)(x - x') \right| \\
&= \left| p(f(x) - f(x')) - (f(p\tilde{x}) - f(0))(x - x') \right| \\
&\leq \left| p(f(x) - f(x') - f'(0)(x - x')) \right| + \left| pf'(0)(x - x') \right| \\
&\quad + \left| (f(p\tilde{x}) - f(0))(x - x') \right| \leq \varepsilon |x - x'|,
\end{aligned}$$

where the mean value theorem guarantees that $\int_{x'}^x f(pt)\,dt = (x - x')f(p\tilde{x})$ for some $\tilde{x}$ between x' and x. Hence, g is strictly differentiable with respect to x uniformly in p at $(0,0)$. Prove in a similar way that g is strictly differentiable with respect to p uniformly in x at $(0,0)$. Then apply 1D.12 and 1D.13. □

1E. Lipschitz Invertibility from Approximations

In this section we completely depart from differentiation and develop inverse and implicit function theorems for equations in which the functions are merely continuous. The price to pay is that the single-valued localization of the inverse that is obtained might not be differentiable, but at least it will have a Lipschitz property.

The way to do that is found through notions of how a function f may be "approximated" by another function h around a point $\bar{x}$. Classical theory focuses on f being differentiable at $\bar{x}$ and approximated there by the function h giving its "linearization" at $\bar{x}$, namely $h(x) = f(\bar{x}) + \nabla f(\bar{x})(x - \bar{x})$. Differentiability corresponds to having $f(x) = h(x) + o(|x - \bar{x}|)$ around $\bar{x}$, which is the same as $\text{clm}\,(f - h;\bar{x}) = 0$, whereas strict differentiability corresponds to the stronger requirement that $\text{lip}\,(f - h;\bar{x}) = 0$. The key idea is that conditions like this, and others in a similar vein, can be applied to f and h even when h is not a linearization dependent on the existence of $\nabla f(\bar{x})$. Assumptions on the nonsingularity of $\nabla f(\bar{x})$, corresponding in the classical setting to the invertibility of the linearization, might then be replaced by assumptions on the invertibility of some other approximation h.

First-order approximations of functions. *Consider a function $f : \mathbb{R}^n \to \mathbb{R}^m$ and a point $\bar{x} \in \text{int dom}\,f$. A function $h : \mathbb{R}^n \to \mathbb{R}^m$ with $\bar{x} \in \text{int dom}\,h$ is a first-order approximation to f at $\bar{x}$ if $h(\bar{x}) = f(\bar{x})$ and*

$$\text{clm}\,(e;\bar{x}) = 0 \ \ \text{for } e(x) = f(x) - h(x),$$

which can also be written as $f(x) = h(x) + o(|x - \bar{x}|)$. It is a strict first-order approximation if the stronger condition holds that

$$\text{lip}\,(e;\bar{x}) = 0 \ \ \text{for } e(x) = f(x) - h(x).$$

In other words, h is a first-order approximation to f at $\bar{x}$ when $f(\bar{x}) = h(\bar{x})$ and for every $\varepsilon > 0$ there exists $\delta > 0$ such that

$$|f(x) - h(x)| \leq \varepsilon |x - \bar{x}| \ \ \text{for every } x \in \mathbb{B}_\delta(\bar{x}),$$

and a strict first-order approximation when

$$|[f(x) - h(x)] - [f(x') - h(x')]| \leq \varepsilon |x - x'| \ \ \text{for all } x,x' \in \mathbb{B}_\delta(\bar{x}).$$

Clearly, if h is a (strict) first-order approximation to f, then f is a (strict) first-order approximation to h.

First-order approximations obey calculus rules which follow directly from the corresponding properties of the calmness and Lipschitz moduli:

(i) If q is a (strict) first-order approximation to h at $\bar{x}$ and h is a (strict) first-order approximation to f at $\bar{x}$, then q is a (strict) first-order approximation to f at $\bar{x}$.

(ii) If f_1 and f_2 have (strict) first-order approximations h_1 and h_2, respectively, at $\bar{x}$, then $h_1 + h_2$ is a (strict) first-order approximation of $f_1 + f_2$ at $\bar{x}$.

(iii) If f has a (strict) first-order approximation h at $\bar{x}$, then for any $\lambda \in \mathbb{R}$, λh is a (strict) first-order approximation of λf at $\bar{x}$.

(iv) If h is a first-order approximation of f at $\bar{x}$, then $\text{clm}\,(f;\bar{x}) = \text{clm}\,(h;\bar{x})$. Similarly, if h is a strict first-order approximation of f at $\bar{x}$, then $\text{lip}\,(f;\bar{x}) = \text{lip}\,(h;\bar{x})$.

1 Functions Defined Implicitly by Equations

The next proposition explains how first-order approximations can be chained together.

Proposition 1E.1 (composition of first-order approximations). *Let h be a first-order approximation of f at $\bar{x}$ which is calm at $\bar{x}$. Let v be a first-order approximation of u at $\bar{y}$ for $\bar{y} := f(\bar{x})$ which is Lipschitz continuous around $\bar{y}$. Then $v \circ h$ is a first-order approximation of $u \circ f$ at $\bar{x}$.*

Proof. By the property (iv) of the first-order approximations displayed before the statement, the function f is calm at $\bar{x}$. Choose $\varepsilon > 0$ and let μ and λ be such that $\mathrm{clm}(f;\bar{x}) < \mu$ and $\mathrm{lip}(v;\bar{x}) < \lambda$. Then there exist neighborhoods U of $\bar{x}$ and V of $\bar{y}$ such that $f(x) \in V$ and $h(x) \in V$ for $x \in U$,

$$|f(x) - h(x)| \leq \varepsilon |x - \bar{x}| \text{ and } |f(x) - f(\bar{x})| \leq \mu |x - \bar{x}| \text{ for } x \in U,$$

and moreover,

$$|u(y) - v(y)| \leq \varepsilon |y - \bar{y}| \text{ and } |v(y) - v(y')| \leq \lambda |y - y'| \text{ for } y, y' \in V.$$

Then for $x \in U$, taking into account that $\bar{y} = f(\bar{x}) = h(\bar{x})$, we have

$$\begin{aligned} |(u \circ f)(x) - (v \circ h)(x)| &= |u(f(x)) - v(h(x))| \\ &\leq |u(f(x)) - v(f(x))| + |v(f(x)) - v(h(x))| \\ &\leq \varepsilon |f(x) - f(\bar{x})| + \lambda |f(x) - h(x)| \\ &\leq \varepsilon \mu |x - \bar{x}| + \lambda \varepsilon |x - \bar{x}| \leq \varepsilon (\mu + \lambda) |x - \bar{x}|. \end{aligned}$$

Since ε can be arbitrarily small, the proof is complete. □

Exercise 1E.2 (strict approximations through composition). *Let the function f satisfy $\mathrm{lip}(f;\bar{x}) < \infty$ and let the function g have a strict first-order approximation q at $\bar{y}$, where $\bar{y} := f(\bar{x})$. Then $q \circ f$ is a strict first-order approximation of $g \circ f$ at $\bar{x}$.*

Guide. Mimic the proof of 1E.1. □

For our purposes here, and in later chapters as well, first-order approximations offer an appealing substitute for differentiability, but an even looser notion of approximation will still lead to important conclusions.

Estimators beyond first-order approximations. *Consider a function $f : \mathbb{R}^n \to \mathbb{R}^m$ and a point $\bar{x} \in \mathrm{int\,dom}\, f$. A function $h : \mathbb{R}^n \to \mathbb{R}^m$ with $\bar{x} \in \mathrm{int\,dom}\, h$ is an estimator of f at $\bar{x}$ with constant μ if $h(\bar{x}) = f(\bar{x})$ and*

$$\mathrm{clm}(e;\bar{x}) \leq \mu < \infty \text{ for } e(x) = f(x) - h(x),$$

which can also be written as $|f(x) - h(x)| \leq \mu |x - \bar{x}| + o(|x - \bar{x}|)$. It is a strict estimator if the stronger condition holds that

$$\mathrm{lip}(e;\bar{x}) \leq \mu < \infty \text{ for } e(x) = f(x) - h(x).$$

In this terminology, a first-order approximation is simply an estimator with constant $\mu = 0$. Through that, any result involving estimators can immediately be specialized to a result about first-order approximations.

Estimators can be of interest even when differentiability is present. For instance, in the case of a function f that is strictly differentiable at $\bar{x}$ a strict estimator of f at $\bar{x}$ with constant μ is furnished by $h(x) = f(\bar{x}) + A(x - \bar{x})$ for any matrix A with $|\nabla f(\bar{x}) - A| \leq \mu$. Such relations have a role in certain numerical procedures, as will be seen at the end of this section and later in the book.

Theorem 1E.3 (inverse function theorem beyond differentiability). *Let $f : \mathbb{R}^n \to \mathbb{R}^n$ be a function with $\bar{x} \in \text{int dom } f$, and let $h : \mathbb{R}^n \to \mathbb{R}^n$ be a strict estimator of f at $\bar{x}$ with constant μ. At the point $\bar{y} = f(\bar{x}) = h(\bar{x})$, suppose that h^{-1} has a Lipschitz continuous single-valued localization σ around $\bar{y}$ for $\bar{x}$ with $\text{lip}(\sigma; \bar{y}) \leq \kappa$ for a constant κ such that $\kappa\mu < 1$. Then f^{-1} has a Lipschitz continuous single-valued localization s around $\bar{y}$ for $\bar{x}$ with*

$$\text{lip}(s; \bar{y}) \leq \frac{\kappa}{1 - \kappa\mu}.$$

This result is a particular case of the implicit function theorem 1E.13 presented later in this section, which in turn follows from a more general result proven in Chapter 2. The reader who does not want to wait for a proof until the next chapter is encouraged to do the following exercise which is supplied with a detailed guide.

Exercise 1E.4. *Prove Theorem 1E.3.*

Guide. Below we outline a direct proof by following the steps in Proof I of Theorem 1A.1. One may also prove this theorem in parallel to Proof II of 1A.1, by showing that $\Phi_y(x) = r(y - (f - h)(x))$ has a fixed point.

First, fix any

$$\lambda \in (\kappa, \infty) \text{ and } \nu \in (\mu, \kappa^{-1}) \text{ with } \lambda\nu < 1.$$

Without loss of generality, suppose that $\bar{x} = 0$ and $\bar{y} = 0$ and take a small enough that the mapping

$$y \mapsto h^{-1}(y) \cap a\mathbb{B} \quad \text{for } y \in a\mathbb{B}$$

is a localization of σ that is Lipschitz continuous with constant λ and also the difference $e = f - h$ is Lipschitz continuous on $a\mathbb{B}$ with constant ν. Next we choose α satisfying

$$0 < \alpha < \frac{1}{4}a(1 - \lambda\nu)\min\{1, \lambda\}$$

and let $b := \alpha/(4\lambda)$. Pick any $y \in b\mathbb{B}$ and any $x^0 \in (\alpha/4)\mathbb{B}$; this gives us

$$|y - e(x^0)| \leq |y| + |e(x^0) - e(0)| \leq |y| + \nu|x^0| \leq \frac{\alpha}{4\lambda} + \nu\frac{\alpha}{4} \leq \frac{\alpha}{2\lambda}.$$

1 Functions Defined Implicitly by Equations

In particular $|y - e(x^0)| < a$, so the point $y - e(x^0)$ lies in the region where σ is Lipschitz continuous. Let $x^1 = \sigma(y - e(x^0))$; then

$$|x^1| = |\sigma(y - e(x^0))| = |\sigma(y - e(x^0)) - \sigma(0)| \leq \lambda |y - e(x^0)| \leq \frac{\alpha}{2},$$

so in particular x^1 belongs to the ball $a\mathbb{B}$. Furthermore,

$$|x^1 - x^0| \leq |x^1| + |x^0| \leq \alpha/2 + \alpha/4 = 3\alpha/4.$$

We also have

$$|y - e(x^1)| \leq |y| + v|x^1| \leq \alpha/(4\lambda) + \alpha/(2\lambda) \leq a,$$

so that $y - e(x^1) \in a\mathbb{B}$.

Having started in this pattern, proceed to construct an infinite sequence of points x^k by taking

$$x^{k+1} = \sigma(y - e(x^k))$$

and prove by induction that

$$x^k \in a\mathbb{B}, \quad y - e(x^k) \in a\mathbb{B} \text{ and } |x^k - x^{k-1}| \leq (\lambda v)^{k-1}|x^1 - x^0| \text{ for } k = 2, 3, \ldots.$$

Observe next that, for $k > j$,

$$|x^k - x^j| \leq \sum_{i=j}^{k-1} |x^{i+1} - x^i| \leq \sum_{i=j}^{k-1} (\lambda v)^i a \leq \frac{a}{1 - \lambda v}(\lambda v)^j,$$

hence

$$\lim_{\substack{j,k \to \infty \\ k > j}} |x^k - x^j| = 0.$$

Then the sequence $\{x^k\}$ is Cauchy and hence convergent. Let x be its limit. Since all x^k and all $y - e(x^k)$ are in $a\mathbb{B}$, where both e and σ are continuous, we can pass to the limit in the equation $x^{k+1} = \sigma(y - e(x^k))$ as $k \to \infty$, getting

$$x = \sigma(y - e(x)), \text{ that is, } x \in f^{-1}(y).$$

According to our construction, we have

$$|x| \leq \lambda(|y| + |e(x) - e(0)|) \leq \lambda |y| + \lambda v |x|,$$

so that, since $|y| \leq b$, we obtain

$$|x| \leq \frac{\lambda b}{1 - \lambda v}.$$

Thus, it is established that for every $y \in bI\!\!B$ there exists $x \in f^{-1}(y)$ with $|x| \leq \lambda b/(1-\lambda v)$. In other words, we have shown the nonempty-valuedness of the localization of f^{-1} given by

$$s: y \mapsto f^{-1}(y) \cap \frac{\lambda b}{1-\lambda v} I\!\!B \quad \text{for } y \in bI\!\!B.$$

Next, demonstrate that this localization s is in fact single-valued and Lipschitz continuous. If for some $y \in bI\!\!B$ we have two points $x \neq x'$, both of them in $s(y)$, then subtracting $x = \sigma(y - e(x))$ from $x' = \sigma(y - e(x'))$ gives

$$0 < |x' - x| = |\sigma(y - e(x')) - \sigma(y - e(x))| \leq \lambda |e(x') - e(x)| \leq \lambda v |x' - x| < |x' - x|,$$

which is absurd. Further, considering $y', y \in bI\!\!B$ and recalling that $s(y) = \sigma(y - e(s(y)))$, one gets

$$|s(y') - s(y)| = |\sigma(y' - e(s(y'))) - \sigma(y - e(s(y)))| \leq \lambda(|y' - y| + v|s(y') - s(y)|),$$

and hence s is Lipschitz continuous relative to $bI\!\!B$ with constant $\lambda/(1-\lambda v)$. This expression is continuous and increasing as a function of λ and v, which are greater than κ and μ but can be chosen arbitrarily close to them, hence the Lipschitz modulus of s at $\bar{y}$ satisfies the desired inequality. □

When $\mu = 0$ in Theorem 1E.3, so that h is a strict first-order approximation of f at $\bar{x}$, the conclusion about the localization s of f^{-1} is that $\text{lip}(s; \bar{y}) \leq \kappa$. The strict derivative version 1D.9 of the inverse function theorem corresponds to the case where $h(x) = f(\bar{x}) + Df(\bar{x})(x - \bar{x})$. The assumption on h^{-1} in Theorem 1E.3 is tantamount then to the invertibility of $Df(\bar{x})$, or equivalently the nonsingularity of the Jacobian $\nabla f(\bar{x})$; we have $\text{lip}(\sigma; \bar{y}) = |Df(\bar{x})^{-1}| = |\nabla f(\bar{x})^{-1}|$, and κ can be taken to have this value. Again, though, Theorem 1E.3 does not, in general, insist on h being a first-order approximation of f at $\bar{x}$.

The following example sheds light on the sharpness of the assumptions in 1E.3 about the relative sizes of the Lipschitz modulus of the localization of h^{-1} and the Lipschitz modulus of the "approximation error" $f - h$.

Example 1E.5 (illustration of invertibility without strict differentiability). With $\alpha \in (0, \infty)$ as a parameter, let $f(x) = \alpha x + g(x)$ for the function

$$g(x) = \begin{cases} x^2 \sin(1/x) & \text{for } x \neq 0, \\ 0 & \text{for } x = 0, \end{cases}$$

noting that f and g are differentiable with

$$g'(x) = \begin{cases} 2x \sin(1/x) - \cos(1/x) & \text{for } x \neq 0, \\ 0 & \text{for } x = 0, \end{cases}$$

but f and g are not strictly differentiable at 0, although g is Lipschitz continuous there with $\text{lip}(g;0) = 1$.

Let $h(x) = \alpha x$ and consider applying Theorem 1E.3 to f and h at $\bar{x} = 0$, where $f(0) = h(0) = 0$. Since $f - h = g$, we have for every $\mu > 1$ that $f - h$ is Lipschitz continuous with constant μ on some neighborhood of 0. On the other hand, h^{-1} is Lipschitz continuous with constant $\kappa = 1/\alpha$. Therefore, as long as $\alpha > 1$, the assumptions of Theorem 1E.3 are fulfilled (by taking μ in $(1, \alpha)$ arbitrarily close to 1). We are able to conclude from 1E.3 that f^{-1} has a single-valued localization s around 0 for 0 such that $\text{lip}(s;0) \leq (\alpha - 1)^{-1}$, despite the inapplicability of 1D.9.

When $0 < \alpha < 1$, however, f^{-1} has no single-valued localization around the origin at all. This comes out of the fact that, for such α, the derivative $f'(x)$ has infinitely many changes of sign in any neighborhood of $\bar{x} = 0$, hence infinitely many consecutive local maximum values and minimum values of f in any such neighborhood, with both values tending to 0 as the origin is approached. Let x_1 and x_2, $0 < x_1 < x_2$, be two consecutive points where f' vanishes and f has a local maximum at x_1 and local minimum at x_2. For a value $y > 0$ the equation $f(x) = y$ must have not only a solution in (x_1, x_2), but also one in $(0, x_1)$. Hence, regardless of the size of the neighborhood U of $\bar{x} = 0$, there will be infinitely many y values near $\bar{y} = 0$ for which $U \cap f^{-1}(y)$ is not a singleton. Both cases are illustrated in Fig. 1.6.

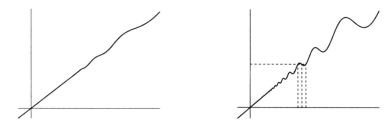

Fig. 1.6 Graphs of the function f in Example 1E.5 when $\alpha = 2$ on the left and $\alpha = 0.5$ on the right.

When the approximation h to f in Theorem 1E.3 is itself strictly differentiable at the point in question, a simpler statement of the result can be made.

Corollary 1E.6 (estimators with strict differentiability). *For $f : \mathbb{R}^n \to \mathbb{R}^n$ with $\bar{x} \in \text{int dom } f$ and $f(\bar{x}) = \bar{y}$, suppose there is a strict estimator $h : \mathbb{R}^n \to \mathbb{R}^n$ of f at $\bar{x}$ with constant μ which is strictly differentiable at $\bar{x}$ with nonsingular Jacobian $\nabla h(\bar{x})$ satisfying $\mu |\nabla h(\bar{x})^{-1}| < 1$. Then f^{-1} has a Lipschitz continuous single-valued localization s around $\bar{y}$ for $\bar{x}$ with*

$$\text{lip}(s;\bar{y}) \leq \frac{|\nabla h(\bar{x})^{-1}|}{1 - \mu |\nabla h(\bar{x})^{-1}|}.$$

Proof. A localization σ of h^{-1} having $\text{lip}(\sigma;\bar{y}) = |\nabla h(\bar{x})^{-1}|$ is obtained by applying Theorem 1D.9 to h. Theorem 1E.3 can be invoked then with $\kappa = |\nabla h(\bar{x})^{-1}|$. □

An even more special application, to the case where both f and h are linear, yields a well-known estimate for matrices.

Corollary 1E.7 (estimation for perturbed matrix inversion). *Let A and B be $n \times n$ matrices such that A is nonsingular and $|B| < |A^{-1}|^{-1}$. Then $A + B$ is nonsingular with*
$$|(A+B)^{-1}| \leq \left(|A^{-1}|^{-1} - |B|\right)^{-1}.$$

Proof. This comes from Corollary 1E.6 by taking $f(x) = (A+B)x$, $h(x) = Ax$, $\bar{x} = 0$, and writing $\left(|A^{-1}|^{-1} - |B|\right)^{-1}$ as $|A|/(1 - |A||B|)$. □

We can state 1E.7 in other two equivalent ways, which we give next as an exercise. More about the estimate in 1E.7 will be said in Chapter 5.

Exercise 1E.8 (equivalent estimation rules for matrices). *Prove that the following two statements are equivalent to Corollary 1E.7:*

(a) *For any $n \times n$ matrix C with $|C| < 1$, the matrix $I + C$ is nonsingular and*

(1) $$|(I+C)^{-1}| \leq \frac{1}{1-|C|}.$$

(b) *For any $n \times n$ matrices A and B with A nonsingular and $|BA^{-1}| < 1$, the matrix $A + B$ is nonsingular and*
$$|(A+B)^{-1}| \leq \frac{|A^{-1}|}{1-|BA^{-1}|}.$$

Guide. Clearly, (b) implies Corollary 1E.7 which in turn implies (a) with $A = I$ and $B = C$. Let (a) hold and let the matrices A and B satisfy $|BA^{-1}| < 1$. Then, by (a) with $C = BA^{-1}$ we obtain that $I + BA^{-1}$ is nonsingular, and hence $A + B$ is nonsingular, too. Using the equality $A + B = (I + BA^{-1})A$ in (1) we have
$$|(A+B)^{-1}| = |A^{-1}(I+BA^{-1})^{-1}| \leq \frac{|A^{-1}|}{1-|BA^{-1}|},$$
that is, (a) implies (b). □

Corollary 1E.7 implies that, given a nonsingular matrix A,

(2) $$\inf\left\{|B| \,\big|\, A+B \text{ is singular}\right\} \geq |A^{-1}|^{-1}.$$

It turns out that this inequality is actually equality, another classical result in matrix perturbation theory.

Theorem 1E.9 (radius theorem for matrix nonsingularity). *For any nonsingular matrix A,*
$$\inf\{\,|B|\,\big|\,A+B \text{ is singular}\,\} = |A^{-1}|^{-1}.$$

Proof. It is sufficient to prove the inequality opposite to (2). Choose $\bar{y} \in \mathbb{R}^n$ with $|\bar{y}| = 1$ and $|A^{-1}\bar{y}| = |A^{-1}|$. For $\bar{x} = A^{-1}\bar{y}$ we have $|\bar{x}| = |A^{-1}|$. The matrix
$$B = -\frac{\bar{y}\bar{x}^{\mathsf{T}}}{|\bar{x}|^2}$$
satisfies
$$|B| = \max_{|x|=1} \frac{|\bar{y}\bar{x}^{\mathsf{T}} x|}{|\bar{x}|^2} \leq \max_{|x|=1} \frac{|\bar{x}^{\mathsf{T}} x|}{|\bar{x}|^2} = \frac{|\bar{x}^{\mathsf{T}}\bar{x}|}{|\bar{x}|^3} = \frac{1}{|\bar{x}|} = |A^{-1}|^{-1}.$$

On the other hand $(A+B)\bar{x} = A\bar{x} - \bar{y} = 0$, and since $\bar{x} \neq 0$, the matrix $A+B$ is singular. Thus the infimum in (2) is not greater than $|A^{-1}|^{-1}$. □

Exercise 1E.10 (radius theorem for function invertibility). *Consider a function $f : \mathbb{R}^n \to \mathbb{R}^n$ and a point $\bar{x} \in \operatorname{int} \operatorname{dom} f$. Call f smoothly locally invertible at $\bar{x}$, for short, when f^{-1} has a Lipschitz continuous single-valued localization around $f(\bar{x})$ for $\bar{x}$ which is strictly differentiable at $f(\bar{x})$. In this terminology, prove in the case of f being strictly differentiable at $\bar{x}$ with its Jacobian $\nabla f(\bar{x})$ nonsingular, that*
$$\inf\{\,|B|\,\big|\,f+B \text{ is not smoothly locally invertible at } \bar{x}\,\} = |A^{-1}|^{-1},$$
where $A = \nabla f(\bar{x})$ and the infimum is taken over all linear mappings $B : \mathbb{R}^n \to \mathbb{R}^n$.

Guide. Combine 1E.9 and 1D.9. □

We will extend the facts in 1E.9 and 1E.10 to the much more general context of set-valued mappings in Chapter 6.

In the case of Theorem 1E.3 with $\mu = 0$, an actual equivalence emerges between the invertibility of f and that of h, as captured by the following statement. The key is the fact that first-order approximation is a *symmetric* relationship between two functions.

Theorem 1E.11 (Lipschitz invertibility with first-order approximations). *Let $h : \mathbb{R}^n \to \mathbb{R}^n$ be a strict first-order approximation to $f : \mathbb{R}^n \to \mathbb{R}^n$ at $\bar{x} \in \operatorname{int} \operatorname{dom} f$, and let $\bar{y}$ denote the common value $f(\bar{x}) = h(\bar{x})$. Then f^{-1} has a Lipschitz continuous single-valued localization s around $\bar{y}$ for $\bar{x}$ if and only if h^{-1} has such a localization σ around $\bar{y}$ for $\bar{x}$, in which case*

(3) $$\operatorname{lip}(s;\bar{y}) = \operatorname{lip}(\sigma;\bar{y}),$$

and moreover σ is then a first-order approximation to s at $\bar{y}$: $s(y) = \sigma(y) + o(|y - \bar{y}|)$.

Proof. Applying Theorem 1E.3 with $\mu = 0$ and $\kappa = \operatorname{lip}(\sigma;\bar{y})$, we see that f^{-1} has a single-valued localization s around $\bar{y}$ for $\bar{x}$ with $\operatorname{lip}(s;\bar{y}) \leq \operatorname{lip}(\sigma;\bar{y})$. In these circumstances, though, the symmetry in the relation of first-order approximation allows us to conclude from the existence of this s that h^{-1} has a single-valued localization σ' around $\bar{y}$ for $\bar{x}$ with $\text{lip}\,(\sigma';\bar{y}) \leq \text{lip}\,(s;\bar{y})$. The two localizations of h have to agree graphically in a neighborhood of $(\bar{y},\bar{x})$, so we can simply speak of σ and conclude the validity of (3).

To argue that σ is a first-order approximation of s, we begin by using the identity $h(\sigma(y)) = y$ to get $f(\sigma(y)) = y + e(\sigma(y))$ for the function $e = f - h$ and then transform that into

(4) $$\sigma(y) = s(y + e(\sigma(y))) \quad \text{for } y \text{ near } \bar{y}.$$

Let $\kappa > \text{lip}\,(s;\bar{y})$. From (3) there exists $b > 0$ such that

(5) $$\max\left\{|\sigma(y) - \sigma(y')|, |s(y) - s(y')|\right\} \leq \kappa|y - y'| \quad \text{for } y, y' \in \mathbb{B}_b(\bar{y}).$$

Because $e(\bar{x}) = 0$ and $\text{lip}\,(e;\bar{x}) = 0$, we know that for any $\varepsilon > 0$ there exists a positive $a > 0$ for which

(6) $$|e(x) - e(x')| \leq \varepsilon|x - x'| \quad \text{for all } x, x' \in \mathbb{B}_a(\bar{x}).$$

Choose
$$0 < \beta \leq \min\left\{\frac{a}{\kappa}, \frac{b}{(1+\varepsilon\kappa)}\right\}.$$

Then, for every $y \in \mathbb{B}_\beta(\bar{y})$ from (5) we have

$$|\sigma(y) - \bar{x}| \leq \kappa\beta \leq a,$$

and
$$|y + e(\sigma(y)) - \bar{y}| \leq |y - \bar{y}| + \varepsilon|\sigma(y) - \bar{x}| \leq \beta + \varepsilon\kappa\beta \leq b.$$

Hence, utilizing (4), (5) and (6), we obtain

$$\begin{aligned}|\sigma(y) - s(y)| &= |s(y + e(\sigma(y))) - s(y)| \\ &\leq \kappa|e(\sigma(y)) - e(\sigma(\bar{y}))| \\ &\leq \kappa\varepsilon|\sigma(y) - \sigma(\bar{y})| \leq \kappa^2\varepsilon|y - \bar{y}|.\end{aligned}$$

Since ε can be arbitrarily small, we arrive at the equality $\text{clm}\,(s - \sigma;\bar{y}) = 0$, and this completes the proof. $\square$

Finally, we observe that these results make it possible to deduce a slightly sharper version of the equivalence in Theorem 1D.9.

Theorem 1E.12 (extended equivalence under strict differentiability). *Let $f : \mathbb{R}^n \to \mathbb{R}^n$ be strictly differentiable at $\bar{x}$ with $f(\bar{x}) = \bar{y}$. Then the following are equivalent:*
  (i) $\nabla f(\bar{x})$ *is nonsingular;*
  (ii) f^{-1} *has a Lipschitz continuous single-valued localization around $\bar{y}$ for $\bar{x}$;*

1 Functions Defined Implicitly by Equations

(iii) f^{-1} has a single-valued localization around $\bar{y}$ for $\bar{x}$ that is strictly differentiable at $\bar{y}$.

In parallel to Theorem 1E.3, it is possible also to state and prove a corresponding implicit function theorem with Lipschitz continuity. For that purpose, we need to introduce the concept of partial first-order approximations for functions of two variables.

Partial first-order estimators and approximations. For $f : \mathbb{R}^d \times \mathbb{R}^n \to \mathbb{R}^m$ and a point $(\bar{p}, \bar{x}) \in \operatorname{int} \operatorname{dom} f$, a function $h : \mathbb{R}^n \to \mathbb{R}^m$ is said to be an estimator of f with respect to x uniformly in p at $(\bar{p}, \bar{x})$ with constant μ if $h(\bar{x}) = f(\bar{x}, \bar{p})$ and

$$\widehat{\operatorname{clm}}_x(e; (\bar{p}, \bar{x})) \leq \mu < \infty \quad \text{for } e(p, x) = f(p, x) - h(x).$$

It is a strict estimator in this sense if the stronger condition holds that

$$\widehat{\operatorname{lip}}_x(e; (\bar{p}, \bar{x})) \leq \mu < \infty \quad \text{for } e(p, x) = f(p, x) - h(x).$$

In the case of $\mu = 0$, such an estimator is called a partial first-order approximation.

Theorem 1E.13 (implicit function theorem beyond differentiability). *Consider $f : \mathbb{R}^d \times \mathbb{R}^n \to \mathbb{R}^n$ and $(\bar{p}, \bar{x}) \in \operatorname{int} \operatorname{dom} f$ with $f(\bar{p}, \bar{x}) = 0$ and $\widehat{\operatorname{lip}}_p(f; (\bar{p}, \bar{x})) \leq \gamma < \infty$. Let h be a strict estimator of f with respect to x uniformly in p at $(\bar{p}, \bar{x})$ with constant μ. Suppose at the point $h(\bar{x}) = 0$ that h^{-1} has a Lipschitz continuous single-valued localization σ around 0 for $\bar{x}$ with $\operatorname{lip}(\sigma; 0) \leq \kappa$ for a constant κ such that $\kappa \mu < 1$. Then the solution mapping*

$$S : p \mapsto \{ x \in \mathbb{R}^n \,|\, f(p, x) = 0 \} \quad \text{for } p \in \mathbb{R}^d$$

has a Lipschitz continuous single-valued localization s around $\bar{p}$ for $\bar{x}$ with

$$\operatorname{lip}(s; \bar{p}) \leq \frac{\kappa \gamma}{1 - \kappa \mu}.$$

Moreover, when $\mu = 0$ the function $\eta(p) = \sigma(-f(p, \bar{x}))$ is a first-order approximation to s at $\bar{p}$: $s(p) = \eta(p) + o(|p - \bar{p}|)$.

This is a special instance of the combination of Theorem 2B.7 (for $\mu > 0$) and of Theorem 2B.8 (for $\mu = 0$) which we will prove in Chapter 2, so there is no purpose in giving a separate argument for it here. If $f(p, x)$ has the form $f(x) - p$, in which case $\widehat{\operatorname{lip}}_p(f; \bar{p}, \bar{x}) = 1$, taking $\gamma = 1$ we immediately obtain Theorem 1E.3.

Exercise 1E.14 (approximation criteria). *Consider $f : \mathbb{R}^d \times \mathbb{R}^n \to \mathbb{R}^m$ and $h : \mathbb{R}^d \times \mathbb{R}^n \to \mathbb{R}^m$ with $f(\bar{p}, \bar{x}) = h(\bar{p}, \bar{x})$, and the difference $e(p, x) = f(p, x) - h(p, x)$. Prove that*

(a) *If $\widehat{\operatorname{clm}}_x(e; (\bar{p}, \bar{x})) = 0$ and $\widehat{\operatorname{clm}}_p(e; (\bar{p}, \bar{x})) = 0$, then h is a first-order approximation to f at $(\bar{p}, \bar{x})$.*

(b) If $\widehat{\operatorname{lip}}_x(e;(\bar{p},\bar{x})) = 0$ and $\operatorname{clm}_p(e;(\bar{p},\bar{x})) = 0$, then h is a first-order approximation to f at $(\bar{p},\bar{x})$.

(c) If $\widehat{\operatorname{lip}}_x(e;(\bar{p},\bar{x})) = 0$ and $\widehat{\operatorname{lip}}_p(e;(\bar{p},\bar{x})) = 0$, then h is a strict first-order approximation to f at $(\bar{p},\bar{x})$.

Exercise 1E.15 (partial first-order approximation from differentiability). Let $f : \mathbb{R}^d \times \mathbb{R}^n \to \mathbb{R}^m$ be differentiable with respect to x in a neighborhood of $(\bar{p},\bar{x})$ and let f and $\nabla_x f$ be continuous in this neighborhood. Prove that the function $h(x) = f(\bar{p},\bar{x}) + \nabla_x f(\bar{p},\bar{x})(x - \bar{x})$ is a strict first-order approximation to f with respect to x uniformly in p at $(\bar{p},\bar{x})$. Based on this, derive the Dini classical implicit function theorem 1B.1 from 1E.13.

Exercise 1E.16 (the zero function as an approximation). Let $f : \mathbb{R}^d \times \mathbb{R}^n \to \mathbb{R}^m$ satisfy $\widehat{\operatorname{lip}}_x(f;(\bar{p},\bar{x})) < \infty$, and let $u : \mathbb{R}^d \times \mathbb{R}^m \to \mathbb{R}^k$ have a strict first-order approximation v with respect to y at $(\bar{p},\bar{y})$, where $\bar{y} := f(\bar{p},\bar{x})$. Show that the zero function is a strict first-order approximation with respect to x at $(\bar{p},\bar{x})$ to the function $(p,x) \mapsto u(p,f(p,x)) - v(f(p,x))$.

As another illustration of applicability of Theorem 1E.3 beyond first-order approximations, we sketch now a proof of the quadratic convergence of Newton's method for solving equations, a method we used in Proof I of the classical inverse function theorem 1A.1.

Consider the equation $g(x) = 0$ for a continuously differentiable function $g : \mathbb{R}^n \to \mathbb{R}^n$ with a solution $\bar{x}$ at which the Jacobian $\nabla g(\bar{x})$ is nonsingular. Newton's method consists in choosing a starting point x^0 possibly close to $\bar{x}$ and generating a sequence of points $x^1, x^2, \ldots$ according to the rule

$$(7) \qquad x^{k+1} = x^k - \nabla g(x^k)^{-1} g(x^k), \quad k = 0, 1, \ldots.$$

According to the classical inverse function theorem 1A.1, g^{-1} has a smooth single-valued localization around 0 for $\bar{x}$. Consider the function $f(x) = \nabla g(x^0)(x - \bar{x})$ for which $f(\bar{x}) = g(\bar{x}) = 0$ and $f(x) - g(x) = -g(x) + \nabla g(x^0)(x - \bar{x})$. An easy calculation shows that the Lipschitz modulus of $e = f - g$ at $\bar{x}$ can be made arbitrarily small by making x^0 close to $\bar{x}$. However, *this modulus must be nonzero* — but less than $|\nabla g(\bar{x})^{-1}|^{-1}$, the Lipschitz modulus of the single-valued localization of g^{-1} around 0 for $\bar{x}$, if one wants to choose x^0 as an arbitrary starting point from an open neighborhood of $\bar{x}$. Here Theorem 1E.3 comes into play with $h = g$ and $\bar{y} = 0$, saying that f^{-1} has a Lipschitz continuous single-valued localization s around 0 for $\bar{x}$ with Lipschitz constant, say, μ. (In the simple case considered this also follows directly from the fact that if $\nabla g(\bar{x})$ is nonsingular at $\bar{x}$, then $\nabla g(x)$ is likewise nonsingular for all x in a neighborhood of $\bar{x}$, see Fact 2 in Section 1A.) Hence, the Lipschitz constant μ and the neighborhood V of 0 where s is Lipschitz continuous can be determined before the choice of x^0, which is to be selected so that $\nabla g(x^0)(x^0 - \bar{x}) - g(x^0)$ is in V. Noting for the iteration (7) that $x^1 = s(\nabla g(x^0)(x^0 - \bar{x}) - g(x^0))$ and $\bar{x} = s(g(\bar{x}))$, and using the smoothness of g, we obtain

$$|x^1 - \bar{x}| \leq \mu|g(\bar{x}) - g(x^0) - \nabla g(x^0)(\bar{x} - x^0)| \leq c|x^0 - \bar{x}|^2$$

for a suitable constant c. This kind of argument works for any k, and in that way, through induction, we obtain quadratic convergence for Newton's iteration (7).

In Chapter 6 we will present, in a broader framework of generalized equations in possibly infinite-dimensional spaces, a detailed proof of this quadratic convergence of Newton's method and study its stability properties.

1F. Selections of Multi-valued Inverses

Consider a function f which acts between Euclidean spaces of possibly different dimensions, say $f : \mathbb{R}^n \to \mathbb{R}^m$. What can be said then about the inverse mapping f^{-1}? The case of $f(x) = Ax + b$ with $A \in \mathbb{R}^{m \times n}$ and $b \in \mathbb{R}^m$ gives an indication of the differences that must be expected: when $m < n$, the equation $Ax + b = y$ either has no solution x or a continuum of solutions, so that the existence of single-valued localizations is totally hopeless. Anyway, though, if A has full rank m, we know at least that $f^{-1}(y)$ will be nonempty for every y.

Do we really always have to assume that $m = n$ in order to get a single-valued localization of the inverse? Specifically, consider a function f acting from $\mathbb{R}^n$ to $\mathbb{R}^m$ with $m \leq n$ and a point $\bar{x}$ in the interior of the domain of f. Suppose that f is continuous in an open neighborhood U of $\bar{x}$ and the inverse f^{-1} has a single-valued localization around $f(\bar{x})$ for $\bar{x}$. It turns out that in this case we necessarily must have $m = n$. This is due to a basic result in topology, the following version of which, stated here without proof, will serve our purposes.

Theorem 1F.1 (Brouwer invariance of domain theorem). *Let $O \subset \mathbb{R}^n$ be open and for $m \leq n$ let $f : O \to \mathbb{R}^m$ be continuous and such that f^{-1} is single-valued on $f(O)$. Then $f(O)$ is open, f^{-1} is continuous on $f(O)$, and $m = n$.*

This topological result reveals that, for continuous functions f, the dimension of the domain space has to agree with the dimension of the range space, if there is to be any hope of an inverse function theorem claiming the existence of a single-valued localization of f^{-1}. Of course, in the theorems already viewed for differentiable functions f, the dimensions were forced to agree because of a rank condition on the Jacobian matrix, but we see now that this limitation has a deeper source than a matrix condition.

Brouwer's invariance of domain theorem helps us to obtain the following characterization of the existence of a Lipschitz continuous single-valued localization of the inverse:

Theorem 1F.2 (invertibility characterization). *For a function $f : \mathbb{R}^n \to \mathbb{R}^n$ that is continuous around $\bar{x}$, the inverse f^{-1} has a Lipschitz continuous single-valued localization around $f(\bar{x})$ for $\bar{x}$ if and only if, in some neighborhood U of $\bar{x}$, there is a* constant $c > 0$ such that

(1) $$c|x' - x| \leq |f(x') - f(x)| \quad \text{for all } x', x \in U.$$

Proof. Let (1) hold. There is no loss of generality in supposing that U is open and f is continuous on U. For any $y \in f(U) := \{f(x) \mid x \in U\}$, we have from (1) that if both $f(x) = y$ and $f(x') = y$ with x and x' in U, then $x = x'$; in other words, the mapping s which takes $y \in f(U)$ to $U \cap f^{-1}(y)$ is actually a function on $f(U)$. Moreover $|s(y') - s(y)| \leq (1/c)|y' - y|$ by (1), so that this function is Lipschitz continuous relative to $f(U)$.

But this is not yet enough to get us to the desired conclusion about f^{-1}. For that, we need to know that s, or some restriction of s, is a localization of f^{-1} around $f(\bar{x})$ for $\bar{x}$, with graph equal to $(V \times U) \cap \text{gph } f^{-1}$ for some neighborhood V of $f(\bar{x})$. Brouwer's invariance of domain theorem 1F.1 enters here: as applied to the restriction of f to the open set U, it tells us that $f(U)$ is open. We can therefore take $V = f(U)$ and be done.

Conversely, suppose that f^{-1} has a Lipschitz continuous single-valued localization around $f(\bar{x})$ for $\bar{x}$, its domain being a neighborhood V of $f(\bar{x})$. Let κ be a Lipschitz constant for s on V. Because f is continuous around $\bar{x}$, there is a neighborhood U of $\bar{x}$ such that $f(U) \subset V$. For x and x' in U, we have $s(f(x)) = x$ and $s(f(x')) = x'$, hence $|x' - x| \leq \kappa |f(x') - f(x)|$. Thus, (1) holds for any $c > 0$ small enough such that $c\kappa \leq 1$. □

We will now re-prove the classical inverse function theorem in a somewhat different formulation having an important extra feature, which is here derived from Brouwer's invariance of domain theorem 1F.1.

Theorem 1F.3 (inverse function theorem for local diffeomorphism). *Let $f : \mathbb{R}^n \to \mathbb{R}^n$ be continuously differentiable in a neighborhood of a point $\bar{x}$ and let the Jacobian $\nabla f(\bar{x})$ be nonsingular. Then for some open neighborhood U of $\bar{x}$ there exists an open neighborhood V of $\bar{y} := f(\bar{x})$ and a continuously differentiable function $s : V \to U$ which is one-to-one from V onto U and which satisfies $s(y) = f^{-1}(y) \cap U$ for all $y \in V$. Moreover, the Jacobian of s is given by*

$$\nabla s(y) = \nabla f(s(y))^{-1} \quad \text{for every } y \in V.$$

Proof. First, we utilize a simple observation from linear algebra: for a nonsingular $n \times n$ matrix A, one has $|Ax| \geq |x|/|A^{-1}|$ for every $x \in \mathbb{R}^n$. Thus, let $c > 0$ be such that $|\nabla f(\bar{x})u| \geq 2c|u|$ for every $u \in \mathbb{R}^n$ and choose $a > 0$ to have, on the basis of (b) in Fact 1 of Section 1A, that

$$|f(x') - f(x) - \nabla f(\bar{x})(x' - x)| \leq c|x' - x| \quad \text{for every } x', x \in \mathbb{B}_a(\bar{x}).$$

Using the triangle inequality, for any $x', x \in \mathbb{B}_a(\bar{x})$ we then have

$$|f(x') - f(x)| \geq |\nabla f(\bar{x})(x' - x)| - c|x' - x| \geq 2c|x' - x| - c|x' - x| \geq c|x' - x|.$$

1 Functions Defined Implicitly by Equations

We can therefore apply 1F.2, obtaining that there is an open neighborhood U of $\bar{x}$ relative to which f is continuous and f^{-1} is single-valued on $V := f(U)$. Brouwer's theorem 1F.1 then tells us that V is open. Then the mapping $s : V \to U$ whose graph is gph $s = $ gph $f^{-1} \cap (V \times U)$ is the claimed single-valued localization of f^{-1} and the rest is argued through Step 3 in Proof I of 1A.1. □

A continuously differentiable function f acting between some open sets U and V in $\mathbb{R}^n$ and having the property that the inverse mapping f^{-1} is continuously differentiable, is called a *diffeomorphism* (or $\mathscr{C}^1$ diffeomorphism) between U and V. Theorem 1F.3 simply says that when $\nabla f(\bar{x})$ is nonsingular, then f is a diffeomorphism[6] relative to open neighborhoods U of $\bar{x}$ and V of $f(\bar{x})$.

Exercise 1F.4 (implicit function version). *Let $f : \mathbb{R}^d \times \mathbb{R}^n \to \mathbb{R}^n$ be continuously differentiable in a neighborhood of $(\bar{p}, \bar{x})$ and such that $f(\bar{p}, \bar{x}) = 0$, and let $\nabla_x f(\bar{p}, \bar{x})$ be nonsingular. Then for some open neighborhood U of $\bar{x}$ there exists an open neighborhood Q of $\bar{p}$ and a continuously differentiable function $s : Q \to U$ such that $\{(p, s(p)) \mid p \in Q\} = \{(p, x) \mid f(p, x) = 0\} \cap (Q \times U)$; that is, s is a single-valued localization of the solution mapping $S(p) = \{x \mid f(p, x) = 0\}$ with associated open neighborhoods Q for $\bar{p}$ and U for $\bar{x}$. Moreover, the Jacobian of s is given by*

$$\nabla s(p) = -\nabla_x f(p, s(p))^{-1} \nabla_p f(p, s(p)) \quad \text{for every } p \in Q.$$

Guide. Apply 1F.3 in the same way as 1A.1 is used in the proof of 1B.1. □

Exercise 1F.5. *Derive Theorem 1D.9 from 1F.2.*

Guide. By 1D.7, strict differentiability is equivalent to the assumption in Fact 1 of Section 1A; then repeat the argument in the proof of 1F.3. □

Brouwer's invariance of domain theorem tells us that, for a function $f : \mathbb{R}^n \to \mathbb{R}^m$ with $m < n$, the inverse f^{-1} fails to have a localization which is single-valued. In this case, however, although multi-valued, f^{-1} may "contain" a function with the properties of the single-valued localization for the case $m = n$. Such functions are generally called *selections* and their formal definition is as follows.

Selections. *Given a set-valued mapping $F : \mathbb{R}^n \rightrightarrows \mathbb{R}^m$ and a set $D \subset $ dom F, a function $w : \mathbb{R}^n \to \mathbb{R}^m$ is said to be a selection of F on D if dom $w \supset D$ and $w(x) \in F(x)$ for all $x \in D$. If $(\bar{x}, \bar{y}) \in $ gph F, D is a neighborhood of $\bar{x}$ and w is a selection on D which satisfies $w(\bar{x}) = \bar{y}$, then w is said to be a local selection of F around $\bar{x}$ for $\bar{y}$.*

A selection of the inverse f^{-1} of a function $f : \mathbb{R}^n \to \mathbb{R}^m$ might provide a *left inverse* or a *right inverse* to f. A left inverse to f on D is a selection $l : \mathbb{R}^m \to \mathbb{R}^n$ of f^{-1} on $f(D)$ such that $l(f(x)) = x$ for all $x \in D$. Analogously, a right inverse to f on D is a selection $r : \mathbb{R}^m \to \mathbb{R}^n$ of f^{-1} on $f(D)$ such that $f(r(y)) = y$ for all $y \in f(D)$. Commonly known are the right and the left inverses of the linear mapping from $\mathbb{R}^n$ to $\mathbb{R}^m$ represented by a matrix $A \in \mathbb{R}^{m \times n}$ that is of full rank. When $m \leq n$, the

---
[6] Theorem 1F.3 can of course be proved directly, without resorting to Brouwer's theorem 1F.1.

right inverse of the mapping corresponds to $A^\mathsf{T}(AA^\mathsf{T})^{-1}$, while when $m \geq n$, the left inverse[7] corresponds to $(A^\mathsf{T}A)^{-1}A^\mathsf{T}$. For $m = n$ they coincide and equal the inverse. In general, of course, whenever a mapping f is one-to-one from a set C to $f(C)$, any left inverse to f on C is also a right inverse, and vice versa, and the restriction of such an inverse to $f(C)$ is uniquely determined.

The following result can be viewed as an extension of the classical inverse function theorem 1A.1 for selections.

Theorem 1F.6 (inverse selections when $m \leq n$). *Let $f : \mathbb{R}^n \to \mathbb{R}^m$, where $m \leq n$, be k times continuously differentiable in a neighborhood of $\bar{x}$ and suppose that its Jacobian $\nabla f(\bar{x})$ is full rank m. Then, for $\bar{y} = f(\bar{x})$, there exists a local selection s of f^{-1} around $\bar{y}$ for $\bar{x}$ which is k times continuously differentiable in a neighborhood V of $\bar{y}$ and whose Jacobian satisfies*

$$(2) \qquad \nabla s(\bar{y}) = A^\mathsf{T}(AA^\mathsf{T})^{-1}, \quad \text{where } A := \nabla f(\bar{x}).$$

Proof. There are various ways to prove this; here we apply the classical inverse function theorem. Since A has rank m, the $m \times m$ matrix AA^T is nonsingular. Then the function $\varphi : \mathbb{R}^m \to \mathbb{R}^m$ defined by $\varphi(u) = f(A^\mathsf{T}u)$ is k times continuously differentiable in a neighborhood of the point $\bar{u} := (A^\mathsf{T}A)^{-1}A\bar{x}$, its Jacobian $\nabla \varphi(\bar{u}) = AA^\mathsf{T}$ is nonsingular, and $\varphi(\bar{u}) = \bar{y}$. Then, from Theorem 1A.1 supplemented by Proposition 1B.5, it follows that φ^{-1} has a single-valued localization σ at $\bar{y}$ for $\bar{u}$ which is k times continuously differentiable near $\bar{y}$ with Jacobian $\nabla \sigma(\bar{y}) = (AA^\mathsf{T})^{-1}$. But then the function $s(y) = A^\mathsf{T}\sigma(y)$ satisfies $s(\bar{y}) = \bar{x}$ and $f(s(y)) = y$ for all y near $\bar{y}$ and is k times continuously differentiable near $\bar{y}$ with Jacobian satisfying (2). Thus, $s(y)$ is a solution of the equation $f(x) = y$ for y close to $\bar{y}$ and x close to $\bar{x}$, but perhaps *not the only solution* there, as it would be in the classical inverse function theorem. Therefore, s is a local selection of f^{-1} around $\bar{y}$ for $\bar{x}$ with the desired properties. □

When $m = n$ the Jacobian becomes nonsingular and the right inverse of A in (2) is just A^{-1}. The uniqueness of the localization can be obtained much as in Step 2 of Proof I of the classical theorem 1A.1.

Exercise 1F.7 (parameterization of solution sets). *Let $M = \{x \,|\, f(x) = 0\}$ for a function $f : \mathbb{R}^n \to \mathbb{R}^m$, where $n - m = d > 0$. Let $\bar{x} \in M$ be a point around which f is k times continuously differentiable, and suppose that the Jacobian $\nabla f(\bar{x})$ has full rank m. Then for some open neighborhood U of $\bar{x}$ there is an open neighborhood O of the origin in $\mathbb{R}^d$ and a k times continuously differentiable function $s : O \to U$ which is one-to-one from O onto $M \cap U$, such that the Jacobian $\nabla s(0)$ has full rank d and*

$$\nabla f(\bar{x})w = 0 \text{ if and only if there exists } q \in \mathbb{R}^d \text{ with } \nabla s(0)q = w.$$

---

[7] The left inverse and the right inverse are particular cases of the *Moore-Penrose pseudo-inverse* A^+ of a matrix A. For more on this, including the singular-value decomposition, see Golub and Van Loan [1996].

1 Functions Defined Implicitly by Equations   51

Guide. Choose an $d \times n$ matrix B such that the matrix

$$\begin{pmatrix} \nabla f(\bar{x}) \\ B \end{pmatrix}$$

is nonsingular. Consider the function

$$\bar{f} : (p,x) \mapsto \begin{pmatrix} f(x) \\ B(x-\bar{x}) - p \end{pmatrix} \quad \text{for } (p,x) \text{ near } (0,\bar{x}),$$

and apply 1F.6 (with a modification parallel to Proposition 1B.5) to the equation $\bar{f}(p,x) = (0,0)$, obtaining for the solution mapping of this equation a localization s with $\bar{f}(p,s(p)) = (0,0)$, i.e., $Bs(p) = p + B\bar{x}$ and $f(s(p)) = 0$. Show that this function s has the properties claimed.   □

Exercise 1F.8 (strictly differentiable selections). *Let $f : \mathbb{R}^n \to \mathbb{R}^m$, where $m \leq n$, be strictly differentiable at $\bar{x}$ with Jacobian $A := \nabla f(\bar{x})$ of full rank. Then there exists a local selection s of the inverse f^{-1} around $\bar{y} := f(\bar{x})$ for $\bar{x}$ which is strictly differentiable at $\bar{y}$ and with Jacobian $\nabla s(\bar{y})$ satisfying (2).*

Guide. Mimic the proof of 1F.6 taking into account 1D.9.   □

Exercise 1F.9 (implicit selections). *Consider a function $f : \mathbb{R}^d \times \mathbb{R}^n \to \mathbb{R}^m$, where $m \leq n$, along with the associated solution mapping*

$$S : p \mapsto \{ x \in \mathbb{R}^n \,|\, f(p,x) = 0 \} \quad \text{for } p \in \mathbb{R}^d.$$

Let $f(\bar{p},\bar{x}) = 0$, so that $\bar{x} \in S(\bar{p})$. Assume that f is strictly differentiable at $(\bar{p},\bar{x})$ and suppose further that the partial Jacobian $\nabla_x f(\bar{p},\bar{x})$ is of full rank m. Then the mapping S has a local selection s around $\bar{p}$ for $\bar{x}$ which is strictly differentiable at $\bar{p}$ with Jacobian

(3) $$\nabla s(\bar{p}) = A^\top (AA^\top)^{-1} \nabla_p f(\bar{p},\bar{x}), \quad \text{where } A = \nabla_x f(\bar{p},\bar{x}).$$

Guide. Use 1F.8 and the argument in the proof of 1B.1.   □

1G. Selections from Nonstrict Differentiability

Even in the case when a function f maps $\mathbb{R}^n$ to itself, the inverse f^{-1} may fail to have a single-valued localization at $\bar{y} = f(\bar{x})$ for $\bar{x}$ if f is not strictly differentiable but merely differentiable at $\bar{x}$ with Jacobian $\nabla f(\bar{x})$ nonsingular. As when $m < n$, we have to deal with just a local selection of f^{-1}.

Theorem 1G.1 (inverse selections from nonstrict differentiability). *Let $f : \mathbb{R}^n \to \mathbb{R}^n$ be continuous in a neighborhood of a point $\bar{x} \in \operatorname{int} \operatorname{dom} f$ and differentiable at $\bar{x}$ with $\nabla f(\bar{x})$ nonsingular. Then, for $\bar{y} = f(\bar{x})$, there exists a local selection of f^{-1} around $\bar{y}$ for $\bar{x}$ which is continuous at $\bar{y}$. Moreover, every local selection s of f^{-1} around $\bar{y}$ for $\bar{x}$ which is continuous at $\bar{y}$ has the property that*

(1) $\qquad s$ *is differentiable at $\bar{y}$ with Jacobian $\nabla s(\bar{y}) = \nabla f(\bar{x})^{-1}$.*

The verification of this claim relies on the following fixed point theorem, which we state here without proof.

Theorem 1G.2 (Brouwer fixed point theorem). *Let Q be a compact and convex set in $\mathbb{R}^n$, and let $\Phi : \mathbb{R}^n \to \mathbb{R}^n$ be a function which is continuous on Q and maps Q into itself. Then there exists a point $x \in Q$ such that $\Phi(x) = x$.*

Proof of Theorem 1G.1. Without loss of generality, we can suppose that $\bar{x} = 0$ and $f(\bar{x}) = 0$. Let $A := \nabla f(0)$ and choose a neighborhood U of $0 \in \mathbb{R}^n$. Take $c \geq |A^{-1}|$. Choose any $\alpha \in (0, c^{-1})$. From the assumed differentiability of f, there exists $a > 0$ such that $x \in a\mathbb{B}$ implies $|f(x) - Ax| \leq \alpha|x|$. By making a smaller if necessary, we can arrange that f is continuous in $a\mathbb{B}$ and $a\mathbb{B} \subset U$. Let $b = a(1 - c\alpha)/c$ and pick any $y \in b\mathbb{B}$. Consider the function

$$\Phi_y : x \mapsto x - A^{-1}(f(x) - y) \quad \text{for } x \in a\mathbb{B}.$$

This function is of course continuous on the compact and convex set $a\mathbb{B}$. Furthermore, for any $x \in a\mathbb{B}$ we have

$$|\Phi_y(x)| = |x - A^{-1}(f(x) - y)| = |A^{-1}(Ax - f(x) + y)| \leq |A^{-1}|(|Ax - f(x)| + |y|)$$
$$\leq c|Ax - f(x)| + c|y| \leq c\alpha|x| + cb \leq c\alpha a + ca(1 - c\alpha)/c = a,$$

so Φ_y maps $a\mathbb{B}$ into itself. Then, by Brouwer's fixed point theorem 1G.2, there exists a point $x \in a\mathbb{B}$ such that $\Phi_y(x) = x$. Note that, in contrast to the contraction mapping principle 1A.2, this point may be not unique in $a\mathbb{B}$. But $\Phi_y(x) = x$ if and only if $f(x) = y$. For each $y \in b\mathbb{B}$, $y \neq 0$, we pick one $x \in a\mathbb{B}$ such that $x = \Phi_y(x)$; then $x \in f^{-1}(y)$. For $y = 0$ we take $x = 0$, which is clearly in $f^{-1}(0)$. Denoting this x by $s(y)$, we deduce the existence of a local selection $s : b\mathbb{B} \to a\mathbb{B}$ of f^{-1} around 0 for 0, also having the property that for any neighborhood U of 0 there exists $b > 0$ such that $s(y) \in U$ for $y \in b\mathbb{B}$, that is, s is continuous at 0.

Let s be a local selection of f^{-1} around 0 for 0 that is continuous at 0. Choose c, α and a as in the beginning of the proof. Then there exists $b' > 0$ with the property that $s(y) \in f^{-1}(y) \cap a\mathbb{B}$ for every $y \in b'\mathbb{B}$. This can be written as

$$s(y) = A^{-1}(As(y) - f(s(y)) + y) \quad \text{for every } y \in b'\mathbb{B},$$

which gives

$$|s(y)| \leq |A^{-1}|(|As(y) - f(s(y))| + |y|) \leq c\alpha|s(y)| + c|y|,$$

that is,

(2) $$|s(y)| \le \frac{c}{1-c\alpha}|y| \quad \text{for all } y \in b'\mathbb{B}.$$

In particular, s is calm at 0. But we have even more. Choose any $\varepsilon > 0$. The differentiability of f with $\nabla f(0) = A$ furnishes the existence of $\tau \in (0,a)$ such that

(3) $$|f(x) - Ax| \le \frac{(1-c\alpha)\varepsilon}{c^2}|x| \quad \text{whenever } |x| \le \tau.$$

Let $\delta = \min\{b', \tau(1-c\alpha)/c\}$. Then $\delta \le b'$, so that on $\delta\mathbb{B}$ we have our local selection s of f^{-1} satisfying (2) and consequently

$$|s(y)| \le \frac{c}{1-c\alpha}\delta \le \frac{c}{1-c\alpha}\frac{(1-c\alpha)\tau}{c} = \tau \quad \text{when } |y| \le \delta.$$

Taking norms in the identity

$$s(y) - A^{-1}y = -A^{-1}\big(f(s(y)) - As(y)\big),$$

and using (2) and (3), we obtain for $|y| \le \delta$ that

$$|s(y) - A^{-1}y| \le |A^{-1}||f(s(y)) - As(y)| \le \frac{c(1-c\alpha)\varepsilon}{c^2}|s(y)| \le \frac{c(1-c\alpha)\varepsilon c}{c^2(1-c\alpha)}|y| = \varepsilon|y|.$$

Having demonstrated that for any $\varepsilon > 0$ there exists $\delta > 0$ such that $|s(y) - A^{-1}y| \le \varepsilon|y|$ when $|y| \le \delta$, we conclude that (1) holds, as claimed. $\square$

In order to gain more insight into what Theorem 1G.1 does or does not say, think about the case where the assumptions of the theorem hold and f^{-1} has a localization at $\bar{y}$ for $\bar{x}$ that avoids multi-valuedness. This localization must actually be single-valued around $\bar{y}$, coinciding in some neighborhood with the local selection s in the theorem. Then we have a result which appears to be fully analogous to the classical inverse function theorem, but its shortcoming is the need to guarantee that a localization of f^{-1} without multi-valuedness does exist. That, in effect, is what strict differentiability of f at $\bar{x}$, in contrast to just ordinary differentiability, is able to provide. An illustration of how inverse multi-valuedness can indeed come up when the differentiability is not strict has already been encountered in Example 1E.5 with $\alpha \in (0,1)$. Observe that in this example there are infinitely many (even uncountably many) local selections of the inverse f^{-1} and, as the theorem says, each is continuous and even differentiable at 0, but also each selection is discontinuous at infinitely many points near zero.

We can now partially extend Theorem 1G.1 to the case when $m \le n$.

Theorem 1G.3 (differentiable inverse selections). *Let $f : \mathbb{R}^n \to \mathbb{R}^m$ be continuous in a neighborhood of a point $\bar{x} \in \mathrm{int\,dom\,} f$ and differentiable at $\bar{x}$ with $A := \nabla f(\bar{x})$ of full rank m. Then, for $\bar{y} = f(\bar{x})$, there exists a local selection s of f^{-1} around $\bar{y}$ for $\bar{x}$ which satisfies (1).*

Comparing 1G.1 with 1G.3, we see that the equality $m = n$ gives us not only the existence of a local selection which is differentiable at $\bar{y}$ but also that every local selection which is continuous at $\bar{y}$, whose existence is assured also for $m < n$, is differentiable at $\bar{y}$ with the same Jacobian. Of course, if we assume in addition that f is strictly differentiable, we obtain strict differentiability of s at $\bar{y}$. To get this last result, however, we do not have to resort to Brouwer's fixed point theorem 1G.2.

Theorem 1G.1 is in fact a special case of a still broader result in which f does not need to be differentiable.

Theorem 1G.4 (inverse selections from first-order approximation). *Let $f : \mathbb{R}^n \to \mathbb{R}^m$ be continuous around $\bar{x}$ with $f(\bar{x}) = \bar{y}$, and let $h : \mathbb{R}^n \to \mathbb{R}^m$ be a first-order approximation of f at $\bar{x}$ which is continuous around $\bar{x}$. Suppose h^{-1} has a Lipschitz continuous local selection σ around $\bar{y}$ for $\bar{x}$. Then f^{-1} has a local selection s around $\bar{y}$ for $\bar{x}$ for which σ is a first-order approximation at $\bar{y}$: $s(y) = \sigma(y) + o(|y - \bar{y}|)$.*

Proof. We follow the proof of 1G.1 with some important modifications. Without loss of generality, take $\bar{x} = 0$, $\bar{y} = 0$. Let U be a neighborhood of the origin in $\mathbb{R}^n$. Let $\gamma > 0$ be such that σ is Lipschitz continuous on $\gamma \mathbb{B}$, and let $c > 0$ be a constant for this. Choose α such that $0 < \alpha < c^{-1}$ and $a > 0$ with $a\mathbb{B} \subset U$ and such that $\alpha a \leq \gamma/2$, f and h are continuous on $a\mathbb{B}$, and

(4) $\qquad |e(x)| \leq \alpha |x|$ for all $x \in a\mathbb{B}$, where $e(x) = f(x) - h(x)$.

Let

(5) $$0 < b \leq \min\left\{\frac{a(1-c\alpha)}{c}, \frac{\gamma}{2}\right\}.$$

For $x \in a\mathbb{B}$ and $y \in b\mathbb{B}$ we have

(6) $\qquad |y - e(x)| \leq \alpha a + b \leq \gamma.$

Fix $y \in b\mathbb{B}$ and consider the function

(7) $$\Phi_y : x \mapsto \sigma(y - e(x)) \quad \text{for } x \in a\mathbb{B}.$$

This function is of course continuous on $a\mathbb{B}$; moreover, from (6), the Lipschitz continuity of σ on $\gamma \mathbb{B}$ with constant c, the fact that $\sigma(0) = 0$, and the choice of b in (5), we obtain

$$|\Phi_y(x)| = |\sigma(y - e(x))| = |\sigma(y - e(x)) - \sigma(0)| \leq c(\alpha a + b) \leq a \quad \text{for all } x \in a\mathbb{B}.$$

Hence, by Brouwer's fixed point theorem 1G.2, there exists $x \in a\mathbb{B}$ with $x = \sigma(y - e(x))$. Then $h(x) = h(\sigma(y - e(x))) = y - e(x)$, that is, $f(x) = y$. For each $y \in b\mathbb{B}$, $y \neq 0$ we pick one such fixed point x of the function Φ_y in (7) in $a\mathbb{B}$ and call it $s(y)$; for $y = 0$ we set $s(0) = 0 \in f^{-1}(0)$. The function s is a local selection of f^{-1} around 0 for 0 which is, moreover, continuous at 0, since for an arbitrary neighborhood U of 0 we found $b > 0$ such that $s(y) \in U$ whenever $|y| \leq b$. Also, for any $y \in b\mathbb{B}$ we have

(8) $$s(y) = \sigma(y - e(s(y))).$$

From the continuity of s at 0 there exists $b' \in (0,b)$ such that $|s(y)| \le a$ for all $y \in b'\mathbb{B}$. For $y \in b'\mathbb{B}$, we see from (4), (8), the Lipschitz continuity of σ with constant c and the equality $\sigma(0) = 0$ that

$$|s(y)| = |\sigma(y - e(s(y))) - \sigma(0)| \le c\alpha|s(y)| + c|y|.$$

Hence, since $c\alpha < 1$,

(9) $$|s(y)| \le \frac{c}{1 - \alpha c}|y| \quad \text{when } |y| \le b'.$$

Now, let $\varepsilon > 0$. By the assumption that h is a first-order approximation of f at 0, there exists $\tau \in (0,a)$ such that

(10) $$|e(x)| \le \frac{(1 - \alpha c)\varepsilon}{c^2}|x| \quad \text{whenever } |x| \le \tau.$$

Finally, taking $b' > 0$ smaller if necessary and using (9) and (10), for any y with $|y| \le b'$ we obtain

$$|s(y) - \sigma(y)| = |\sigma(y - e(s(y))) - \sigma(y)|$$
$$\le c|e(s(y))| \le c\frac{(1 - \alpha c)\varepsilon}{c^2}|s(y)|$$
$$\le c\frac{(1 - \alpha c)\varepsilon}{c^2}\frac{c}{1 - \alpha c}|y| = \varepsilon|y|.$$

Since for any $\varepsilon > 0$ we found $b' > 0$ for which this holds when $|y| \le b'$, the proof is complete. □

Proof of Theorem 1G.3. Apply Theorem 1G.4 with $h(x) = f(\bar{x}) + A(x - \bar{x})$ and $\sigma(y) = A^T(AA^T)^{-1}y$. □

We state next as an exercise an implicit function counterpart of 1G.3.

Exercise 1G.5 (differentiability of a selection). Consider a function $f : \mathbb{R}^d \times \mathbb{R}^n \to \mathbb{R}^m$ with $m \le n$, along with the solution mapping

$$S : p \mapsto \{x \,|\, f(p,x) = 0\} \quad \text{for } p \in \mathbb{R}^d.$$

Let $f(\bar{p},\bar{x}) = 0$, so that $\bar{x} \in S(\bar{p})$, and suppose f is continuous around $(\bar{p},\bar{x})$ and differentiable at $(\bar{p},\bar{x})$. Assume further that $\nabla_x f(\bar{p},\bar{x})$ has full rank m. Prove that the mapping S has a local selection s around $\bar{p}$ for $\bar{x}$ which is differentiable at $\bar{p}$ with Jacobian

$$\nabla s(\bar{y}) = A^T(AA^T)^{-1}\nabla_p f(\bar{p},\bar{x}), \quad \text{where } A = \nabla_x f(\bar{p},\bar{x}).$$

The existence of a local selection of the inverse of a function f around $\bar{y} = f(\bar{x})$ for $\bar{x}$ implies in particular that $f^{-1}(y)$ is nonempty for all y in a neighborhood of $\bar{y} = f(\bar{x})$. This weaker property has even deeper significance and is defined next.

Openness. *A function $f : \mathbb{R}^n \to \mathbb{R}^m$ is said to be open at $\bar{x}$ if $\bar{x} \in \text{int dom } f$ and for every neighborhood U of $\bar{x}$ the set $f(U)$ is a neighborhood of $f(\bar{x})$.*

Thus, f is open at $\bar{x}$ if for every open neighborhood U of $\bar{x}$ there is an open neighborhood V of $\bar{y} = f(\bar{x})$ such that $f^{-1}(y) \cap U \neq \emptyset$ for every $y \in V$. In particular this corresponds to the localization of f^{-1} relative to V and U being nonempty-valued on V, but goes further than referring just to a single such localization at $\bar{y}$ for $\bar{x}$. It actually requires the existence of a nonempty-valued graphical localization for every neighborhood U of $\bar{x}$, no matter how small. From 1G.3 we obtain the following basic result about openness:

Corollary 1G.6 (Jacobian criterion for openness). *For a function $f : \mathbb{R}^n \to \mathbb{R}^m$, where $m \leq n$, suppose that f is continuous around $\bar{x}$ and differentiable at $\bar{x}$ with $\nabla f(\bar{x})$ being of full rank m. Then f is open at $\bar{x}$.*

There is much more to say about openness of functions and set-valued mappings, and we will explore this in detail in Chapters 3 and 5.

1 Functions Defined Implicitly by Equations

Commentary

Although functions given implicitly by equations had been considered earlier by Descartes, Newton, Leibnitz, Lagrange, Bernoulli, Euler, and others, Cauchy [1831] is credited by historians to be the first who stated and rigorously proved an implicit function theorem — for analytic functions, by using his calculus of residuals and limits. As we mentioned in the preamble to this chapter, Dini [1877/78] gave the form of the implicit function theorem for continuously differentiable functions which is now used in most calculus books; in his proof he relied on the mean value theorem. More about early history of the implicit function theorem can be found in historical notes of the paper of Hurwicz and Richter [2003] and in the book of Krantz and Parks [2002].

Proof I of the classical inverse function theorem, 1A.1, goes back to Goursat [1903][8]. Apparently not aware of Dini's theorem and inspired by Picard's successive approximation method for proving solution existence of differential equations, Goursat stated an implicit function theorem under assumptions weaker than in Dini's theorem, and supplied it with a new path-breaking proof. With updated notation, Goursat's proof employs the iterative scheme

(1) $$x^{k+1} = x^k - A^{-1}f(p,x^k), \text{ where } A = \nabla_x f(\bar{p},\bar{x}).$$

This scheme would correspond to Newton's method for solving $f(p,x) = 0$ with respect to x if A were replaced by A^k giving the partial derivative at (p,x^k) instead of $(\bar{p},\bar{x})$. But Goursat proved anyway that for each p near enough to $\bar{p}$ the sequence $\{x^k\}$ is convergent to a unique point $x(p)$ close to $\bar{x}$, and furthermore that the function $p \mapsto x(p)$ is continuous at $\bar{p}$. Behind the scene, as in Proof I of Theorem 1A.1, is the contraction mapping idea. An updated form of Goursat's implicit function theorem is given in Theorem 1B.6. In the functional analysis text by Kantorovich and Akilov [1964], Goursat's iteration is called a "modified Newton's method."

The rich potential in this proof was seen by Lamson [1920], who used the iterations in (1) to generalize Goursat's theorem to what are now known as Banach spaces[9]. Especially interesting for our point of view in the present book is the fact that Lamson was motivated by an optimization problem, namely the problem of Lagrange in the calculus of variations with equality constraints, for which he proved a Lagrange multiplier rule by way of his implicit function theorem.

Lamson's work was extended in a significant way by Hildebrand and Graves [1927], who also investigated differentiability properties of the implicit function. They first stated a contraction mapping result (their Theorem 1), in which the only difference with the statement of Theorem 1A.2 is the presence of a superfluous parameter. The contraction mapping principle, as formulated in 1A.3, was published

---

[8] Edouard Jean-Baptiste Goursat (1858–1936). Goursat paper from 1903 is available at http://www.numdam.org/.

[9] The name "Banach spaces" for normed linear spaces that are complete was coined by Fréchet, according to Hildebrand and Graves [1927]; we deal with Banach spaces in Chapter 5.

five years earlier in Banach [1922] (with some easily fixed typos), but the idea behind the contraction mapping was evidently known to Goursat, Picard and probably even earlier. Hildebrand and Graves [1927] cited in their paper Banach's work [1922], but only in the context of the definition of a Banach space. Further, based on their parameterized formulation of 1A.2, they established an implicit function theorem in the classical form of 1B.1 (their Theorem 4) for functions acting in linear metric spaces. More intriguing, however, for its surprising foresight, is their Theorem 3, called by the authors a "neighborhood theorem," where they do not assume differentiability; they say "only an approximate differential ... is required." In this, they are far ahead of their time. (We will see a similar picture with Graves' theorem later in Section 5D.) Because of the importance of this result of Hildebrand and Graves, we provide a statement of it here in finite dimensions with some adjustments in terminology and notation.

Theorem (Hildebrand–Graves). *Let $Q \subset \mathbb{R}^d$ and consider a function $f : Q \times \mathbb{R}^n \to \mathbb{R}^n$ along with a point $\bar{x} \in \mathbb{R}^n$. Suppose there are a positive constant a, a linear bounded mapping $A : \mathbb{R}^n \to \mathbb{R}^n$ which is invertible, and a positive constant M with $M|A^{-1}| < 1$ such that*

(a) for all $p \in Q$ and $x, x' \in \mathbb{B}_a(\bar{x})$, one has $|f(p,x) - f(p,x') - A(x - x')| \le M|x - x'|$;

(b) for every $p \in Q$, one has $|A^{-1}||f(p,\bar{x})| \le a(1 - M|A^{-1}|)$.

Then the solution mapping $S : p \mapsto \{x \mid f(p,x) = 0\}$ is single-valued on Q [when its values are restricted to a neighborhood of $\bar{x}$].

The phrase in brackets in the last sentence is our addition: Hildebrand and Graves apparently overlooked the fact, which is still overlooked by some writers, that the implicit function theorem is about a *graphical localization* of the solution mapping. If we assume in addition that f is Lipschitz continuous with respect to p uniformly in x, we will obtain, according to 1E.13, that the solution mapping has a Lipschitz continuous single-valued localization. When f is assumed strictly differentiable at $(\bar{p},\bar{x})$, by taking $A = D_x f(\bar{p},\bar{x})$ we come to 1D.13.

The classical implicit function theorem is present in many of the textbooks in calculus and analysis written in the last hundred years. The proofs on the introductory level are mainly variations of Dini's proof, depending on the material covered prior to the theorem's statement. Interestingly enough, in his text Goursat [1904] applies the mean value theorem, as in Dini's proof, and not the contraction mapping iteration he introduced in his paper of 1903. Similar proofs are given as early as the 30s in Courant [1988] and most recently in Fitzpatrick [2006]. In more advanced textbooks from the second half of the last century, such as in the popular texts of Apostol [1962], Schwartz [1967] and Dieudonné [1969], it has become standard to use the contraction mapping principle.

The material of Chapters 1C and 1D is mostly known, but the way it is presented is new. We were not able to identify a calculus text in which the inverse function theorem is given in the symmetric form 1C.3. First-order approximations of functions were introduced in Robinson [1991].

1 Functions Defined Implicitly by Equations 59

Theorem 1E.3 can be viewed as an extension of the Hildebrand–Graves theorem (see above) where the "approximate differential" is not required to be a linear mapping; we will get back to this result in Chapter 2 and also later in the book. The statement of 1E.8(a) is sometimes called the Banach lemma, see, e.g., Noble and Daniel [1977]. Kahan [1966] and many after him attribute Theorem 1E.9 to Gastinel, without giving a reference. This result can be also found in the literature as the "Eckart–Young theorem" with the citation of Eckart and Young [1936], which however is a related but different kind of result, concerning the distance from a matrix from another matrix with lower rank. That result in turn is much older still and is currently referred to as the Schmidt–Eckart–Young–Mirsky theorem on singular value decomposition. For history see Chipman [1997] and the book by Stewart and Sun [1990].

On the other hand, 1E.9 can be derived with the help of the Schmidt–Eckart–Young–Mirsky theorem, inasmuch as the latter implies that the distance from a nonsingular matrix to the set of rank-one matrices is equal to the smallest singular value, which is the reciprocal to the norm of the inverse. Hence the distance from a nonsingular matrix to the set of singular matrices is not greater than the reciprocal to the norm of the inverse. Combining this with 1E.7 gives us the radius equality. More about stability of perturbed inversions will be presented in Chapters 5 and 6.

Brouwer's invariance of domain theorem 1F.1 can be found, e.g., in Spanier [1966], while Brouwer's fixed point theorem 1G.2 is given in Dunford and Schwartz [1958]. Theorem 1F.2 is from Kummer [1991]. Theorem 1G.3 slightly extends a result in Halkin [1974]; for extensions in other directions, see Hurwicz and Richter [2003].

Chapter 2
Implicit Function Theorems for Variational Problems

Solutions mappings in the classical setting of the implicit function theorem concern problems in the form of parameterized equations. The concept can go far beyond that, however. In any situation where some kind of problem in x depends on a parameter p, there is the mapping S that assigns to each p the corresponding set of solutions x. The same questions then arise about the extent to which a localization of S around a pair $(\bar{p},\bar{x})$ in its graph yields a function s which might be continuous or differentiable, and so forth.

This chapter moves into that much wider territory in replacing equation-solving problems by more complicated problems termed "generalized equations." Such problems arise variationally in constrained optimization, models of equilibrium, and many other areas. An important feature, in contrast to ordinary equations, is that functions obtained implicitly from their solution mappings typically lack differentiability, but often exhibit Lipschitz continuity and sometimes combine that with the existence of one-sided directional derivatives.

The first task is to explain "generalized equations" and their special case, somewhat confusingly termed "variational inequality" problems, which arises from the variational geometry of sets expressing constraints. Problems of optimization and the Lagrange multiplier conditions characterizing their solutions provide key examples. Convexity of sets and functions enters as a valuable ingredient.

From that background, the chapter proceeds to Robinson's implicit function theorem for parameterized variational inequalities and several of its extensions. Subsequent sections introduce concepts of ample parameterization and semidifferentiability, building toward major results in 2E for variational inequalities over convex sets that are polyhedral. A follow-up in 2F looks at a type of "monotonicity" and its consequences, after which, in 2G, a number of applications in optimization are worked out.

2A. Generalized Equations and Variational Problems

By a *generalized equation* in $\mathbb{R}^n$ will be meant a condition on x of the form

(1) $\qquad f(x) + F(x) \ni 0, \text{ or equivalently } -f(x) \in F(x),$

for a function $f : \mathbb{R}^n \to \mathbb{R}^m$ and a (generally) set-valued mapping $F : \mathbb{R}^n \rightrightarrows \mathbb{R}^m$. The name refers to the fact that (1) reduces to an ordinary equation $f(x) = 0$ when F is the *zero mapping* (with $F(x)$ containing 0 and nothing else, for every x), which we indicate in notation by $F \equiv 0$. Any x satisfying (1) is a *solution* to (1).

Generalized equations take on importance in many situations, but an especially common and useful type arises from normality conditions with respect to convex sets.

Normal cones. *For a convex set $C \subset \mathbb{R}^n$ and a point $x \in C$, a vector v is said to be normal to C at x if $\langle v, x' - x \rangle \leq 0$ for all $x' \in C$. The set of all such vectors v is called the normal cone to C at x and is denoted by $N_C(x)$. For $x \notin C$, $N_C(x)$ is taken to be the empty set. The normal cone mapping is thus defined as*

$$N_C : x \mapsto \begin{cases} N_C(x) & \text{for } x \in C, \\ \emptyset & \text{otherwise.} \end{cases}$$

The term *cone* refers to a set of vectors which contains 0 and contains with any of its elements v all positive multiples of v. For each $x \in C$, the normal cone $N_C(x)$ is indeed a cone in this sense. Moreover it is closed and convex. The normal cone mapping $N_C : x \mapsto N_C(x)$ has $\text{dom } N_C = C$. When C is closed, $\text{gph } N_C$ is a closed subset of $\mathbb{R}^n \times \mathbb{R}^n$.

Variational inequalities. *For a function $f : \mathbb{R}^n \to \mathbb{R}^n$ and a closed convex set $C \subset \text{dom } f$, the generalized equation*

(2) $\qquad f(x) + N_C(x) \ni 0, \text{ or equivalently } -f(x) \in N_C(x),$

is called the variational inequality for f and C.

Note that, because $N_C(x) = \emptyset$ when $x \notin C$, a solution x to (2) must be a point of C. The name of this condition originated from the fact that, through the definition of the normal vectors to C, (2) is equivalent to having

(3) $\qquad x \in C \text{ and } \langle f(x), x' - x \rangle \geq 0 \text{ for all } x' \in C.$

Instead of contemplating a system of infinitely many linear inequalities, however, it is usually better to think in terms of the properties of the set-valued mapping N_C, which the formulation in (2) helps to emphasize.

When $x \in \text{int } C$, the only normal vector at x is 0, and the condition in (2) just becomes $f(x) = 0$. Indeed, the variational inequality (2) is totally the same as the equation $f(x) = 0$ in the case of $C = \mathbb{R}^n$, which makes $N_C \equiv 0$. In general, though, (2) imposes a relationship between $f(x)$ and the boundary behavior of C at x.

There is a simple connection between the normal cone mapping N_C and the projection mapping P_C in 1D: one has

(4) $$v \in N_C(x) \iff P_C(x+v) = x.$$

Interestingly, this *projection rule* for normals means for the mappings N_C and P_C that

(5) $$N_C = P_C^{-1} - I, \qquad P_C = (I + N_C)^{-1}.$$

A consequence of (5) is that the variational inequality (2) can actually be written as an equation, namely

(6) $$f(x) + N_C(x) \ni 0 \iff P_C(x - f(x)) - x = 0.$$

It should be kept in mind, though, that this doesn't translate the *solving* of variational inequalities into the classical framework of solving nonlinear equations. There, "linearizations" are essential, but P_C often fails to be differentiable, so linearizations generally aren't available for the equation in (6), regardless of the degree of differentiability of f. Other approaches can sometimes be brought in, however, depending on the nature of the set C. Anyway, the characterization in (6) has the advantage of leading quickly to a criterion for the existence of a solution to a variational inequality in a basic case.

Theorem 2A.1 (solutions to variational inequalities). *For a function $f : \mathbb{R}^n \to \mathbb{R}^n$ and a nonempty, closed convex set $C \subset \text{dom } f$ relative to which f is continuous, the set of solutions to the variational inequality (2) is always closed. It is sure to be nonempty when C is bounded.*

Proof. Let $M(x) = P_C(x - f(x))$. Because C is nonempty, closed and convex, the projection mapping P_C is, by 1D.5, a continuous function from $\mathbb{R}^n$ to C. Then M is a continuous function from C to C under our continuity assumption on f. According to (6), the set of solutions x to (2) is the same as the set of points $x \in C$ such that $M(x) = x$, which is closed. When C is bounded, we can apply Brouwer's fixed point theorem, 1G.2, to conclude the existence of at least one such point x. □

Other existence theorems which don't require C to be bounded can also be given, especially for situations in which f has a property called "monotonicity." This will be taken up in 2F.

The examples and properties to which the rest of this section is devoted will help to indicate the scope of the variational inequality concept. They will also lay the foundations for the generalizations of the implicit function theorem that we are aiming at.

Exercise 2A.2 (some normal cone formulas).
 (a) If M is a linear subspace of $\mathbb{R}^n$, then $N_M(x) = M^\perp$ for every $x \in M$, where $M^\perp$ is the orthogonal complement of M.
 (b) The unit Euclidean ball $\mathbb{B}$ has $N_\mathbb{B}(x) = \{0\}$ when $|x| < 1$, but $N_\mathbb{B}(x) = \{\lambda x \mid \lambda \geq 0\}$ when $|x| = 1$.
 (c) The nonnegative orthant $\mathbb{R}^n_+ = \{x = (x_1, \ldots, x_n) \mid x_j \geq 0 \text{ for } j = 1, \ldots, n\}$ has

$$(v_1, \ldots, v_n) \in N_{\mathbb{R}^n_+}(x_1, \ldots, x_n) \iff v \leq 0, \ v \perp x$$
$$\iff \begin{cases} v_j \leq 0 & \text{for } j \text{ with } x_j = 0, \\ v_j = 0 & \text{for } j \text{ with } x_j > 0. \end{cases}$$

Guide. The projection rule (4) provides an easy way of identifying the normal vectors in these examples. □

The formula in 2A.2(c) comes up far more frequently than might be anticipated. A variational inequality (2) in which $C = \mathbb{R}^n_+$ is called a *complementarity problem*; one has

$$-f(x) \in N_{\mathbb{R}^n_+}(x) \iff x \geq 0, \ f(x) \geq 0, \ x \perp f(x).$$

Here the common notation is adopted that a vector inequality like $x \geq 0$ is to be taken componentwise, and that $x \perp y$ means $\langle x, y \rangle = 0$. Many variational inequalities can be recast, after some manipulation, as complementarity problems, and the numerical methodology for solving such problems has therefore received especially much attention.

The orthogonality relation in 2A.2(a) extends to a "polarity" relation for cones which has a major presence in our subject.

Proposition 2A.3 (polar cone). *Let K be a closed, convex cone in $\mathbb{R}^n$ and let K^* be its polar, defined by*

$$(7) \qquad K^* = \{y \mid \langle x, y \rangle \leq 0 \text{ for all } x \in K\}.$$

Then K^* is likewise a closed, convex cone, and its polar $(K^*)^*$ is in turn K. Furthermore, the normal vectors to K and K^* are related by

$$(8) \qquad y \in N_K(x) \iff x \in N_{K^*}(y) \iff x \in K, \ y \in K^*, \ \langle x, y \rangle = 0.$$

Proof. First consider any $x \in K$ and $y \in N_K(x)$. From the definition of normality in (7) we have $\langle y, x' - x \rangle \leq 0$ for all $x' \in K$, so the maximum of $\langle y, x' \rangle$ over $x' \in K$ is attained at x. Because K contains all positive multiples of each of its vectors, this comes down to having $\langle y, x \rangle = 0$ and $\langle y, x' \rangle \leq 0$ for all $x' \in K$. Therefore $N_K(x) = \{y \in K^* \mid y \perp x\}$.

It's elementary that K^* is a cone which is closed and convex, with $(K^*)^* \supset K$. Consider any $z \notin K$. Let $x = P_K(z)$ and $y = z - x$. Then $y \neq 0$ and $P_K(x+y) = x$, hence $y \in N_K(x)$, so that $y \in K^*$ and $y \perp x$. We have $\langle y, z \rangle = \langle y, y \rangle > 0$, which confirms that $z \notin (K^*)^*$. Therefore $(K^*)^* = K$. The formula for normals to K must hold then equally for K^* through symmetry: $N_{K^*}(y) = \{x \in K \mid x \perp y\}$ for any $y \in K^*$. This establishes the normality relations that have been claimed. □

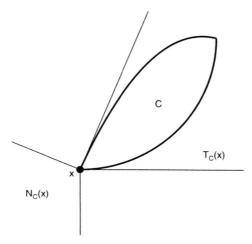

Fig. 2.1 Tangent and normal cone to a convex set.

Polarity has a basic role in relating the normal vectors to a convex set to its "tangent vectors."

Tangent cones. For a set $C \subset \mathbb{R}^n$ (not necessarily convex) and a point $x \in C$, a vector v is said to be *tangent* to C at x if

$$\frac{1}{\tau^k}(x^k - x) \to v \text{ for some } x^k \to x,\ x^k \in C,\ \tau^k \searrow 0.$$

The set of all such vectors v is called the *tangent cone* to C at x and is denoted by $T_C(x)$. For $x \notin C$, $T_C(x)$ is taken to be the empty set.

Although we will mainly be occupied with normal cones to convex sets at present, tangent cones to convex sets and even nonconvex sets will be put to serious use later in the book.

Exercise 2A.4. *The tangent cone $T_C(x)$ to a closed, convex set C at a point $x \in C$ is the closed, convex cone that is polar to the normal cone $N_C(x)$: one has*

(9) $$T_C(x) = N_C(x)^*, \qquad N_C(x) = T_C(x)^*.$$

Guide. The second of the equations (9) comes immediately from the definition of $N_C(x)$, and the first is then obtained from Proposition 2A.3. □

Variational inequalities are instrumental in capturing conditions for optimality in problems of minimization or maximization and even "equilibrium" conditions such as arise in games and models of conflict. To explain this motivation, it will be helpful to be able to appeal to the convexity of functions, at least in part.

Convex functions. A function $g : \mathbb{R}^n \to \mathbb{R}$ is said to be *convex relative to a convex set C* (or just *convex*, when $C = \mathbb{R}^n$) if

$$g((1-\tau)x + \tau x') \leq (1-\tau)g(x) + \tau g(x') \text{ for all } \tau \in (0,1) \text{ when } x, x' \in C.$$

It is *strictly convex* if this holds with strict inequality for $x \neq x'$. It is *strongly convex* with constant μ when $\mu > 0$ and, for every $x, x' \in C$,

$$g((1-\tau)x + \tau x') \leq (1-\tau)g(x) + \tau g(x') - \mu\tau(1-\tau)|x-x'|^2 \text{ for all } \tau \in (0,1).$$

A function g is *concave*, *strictly concave* or *strongly concave*, if $-g$ is convex, strictly convex or strongly convex, respectively. It is *affine relative to C* when the inequality is an equation, which corresponds to g being simultaneously convex and concave relative to C.

The following are the standard criteria for convexity or strict convexity of g which can be obtained from the definitions in terms of the *gradient vectors*

$$\nabla g(x) = \left[\frac{\partial g}{\partial x_j}(x_1,\ldots,x_n)\right]_{j=1}^n$$

and the *Hessian matrices*

$$\nabla^2 g(x) = \left[\frac{\partial^2 g}{\partial x_i \partial x_j}(x_1,\ldots,x_n)\right]_{i,j=1}^{n,n}.$$

Exercise 2A.5 (characterizations of convexity).

(a) A differentiable function $g : \mathbb{R}^n \to \mathbb{R}$ on an open convex set O is convex if and only if

$$g(x') \geq g(x) + \langle \nabla g(x), x' - x \rangle \text{ for all } x, x' \in O.$$

It is strictly convex if and only if this inequality is always strict when $x' \neq x$. It is strongly convex with constant μ, where $\mu > 0$, if and only if

$$g(x') \geq g(x) + \langle \nabla g(x), x' - x \rangle + \frac{\mu}{2}|x' - x|^2 \text{ for all } x, x' \in O.$$

(b) A twice differentiable function g on an open convex set O is convex if and only if $\nabla^2 g(x)$ is positive semidefinite for every $x \in O$. It is strictly convex if $\nabla^2 g(x)$ is positive definite for every $x \in O$. (This sufficient condition for strict convexity is not necessary, however, in general.) It is strongly convex with constant μ if and only if $\mu > 0$ and $\langle \nabla^2 g(x)w, w \rangle \geq \mu$ for all $x \in O$ and $w \in \mathbb{R}^n$ with $|w| = 1$.

Guide. Because the definition of convexity revolves only around points that are collinear, the convexity of g can be verified by showing that, for arbitrary $x \in O$ and

$w \in \mathbb{R}^n$, the function $\varphi(t) = g(x+tw)$ is convex on the interval $\{t \mid x+tw \in O\}$. The conditions for this on $\varphi'(t)$ and $\varphi''(t)$, known from basic calculus, can be applied by expressing these derivatives in terms of the gradients and Hessian of g. This approach can be used to verify all the claims. □

Optimization problems. In this chapter and later, we consider optimization problems which, for a given *objective function* $g : \mathbb{R}^n \to \mathbb{R}$ and a given *constraint set* $C \subset \mathbb{R}^n$, take the form

$$\text{minimize } g(x) \text{ over all } x \in C.$$

The greatest lower bound of the objective function g on C, namely $\inf_{x \in C} g(x)$, is the *optimal value* in the problem, which may or may not be attained, however, and could even be infinite. If it is attained at a point $\bar{x}$, then $\bar{x}$ is said to furnish a *global minimum*, or just a minimum, and to be a *globally optimal solution*; the set of such points is denoted as $\operatorname{argmin}_{x \in C} g(x)$. A point $x \in C$ is said to furnish a *local minimum* of g relative to C and to be a *locally optimal solution* when, at least, $g(x) \leq g(x')$ for every $x' \in C$ belonging to some neighborhood of x. A global or local *maximum* of g corresponds to a global or local minimum of $-g$.

In the context of variational inequalities, the gradient mapping $\nabla g : \mathbb{R}^n \to \mathbb{R}^n$ associated with a differentiable function $g : \mathbb{R}^n \to \mathbb{R}$ will be a focus of attention. Observe that

$$\nabla^2 g(x) = \nabla f(x) \text{ when } f(x) = \nabla g(x).$$

Theorem 2A.6 (basic variational inequality for minimization). *Let $g : \mathbb{R}^n \to \mathbb{R}$ be differentiable on an open convex set O, and let C be a closed convex subset of O. In minimizing g over C, the variational inequality*

(10) $$\nabla g(x) + N_C(x) \ni 0, \text{ or equivalently } -\nabla g(x) \in N_C(x),$$

is necessary for x to furnish a local minimum. It is both necessary and sufficient for a global minimum if g is convex.

Proof. Along with $x \in C$, consider any other point $x' \in C$ and the function $\varphi(t) = g(x+tw)$ with $w = x' - x$. From convexity we have $x + tw \in C$ for all $t \in [0,1]$. If a local minimum of g occurs at x relative to C, then φ must have a local minimum at 0 relative to $[0,1]$, and consequently $\varphi'(0) \geq 0$. But $\varphi'(0) = \langle \nabla g(x), w \rangle$. Hence $\langle \nabla g(x), x' - x \rangle \geq 0$. This being true for arbitrary $x' \in C$, we conclude through the characterization of (2) in (3) that $-\nabla g(x) \in N_C(x)$.

In the other direction, if g is convex and $-\nabla g(x) \in N_C(x)$ we have for every $x' \in C$ that $\langle \nabla g(x), x' - x \rangle \geq 0$, but also $g(x') - g(x) \geq \langle \nabla g(x), x' - x \rangle$ by the convexity criterion in 2A.5(a). Hence $g(x') - g(x) \geq 0$ for all $x' \in C$, and we have a global minimum at x. □

To illustrate the condition in Theorem 2A.6, we may use it to reconfirm the projection rule for normal vectors in (4), which can be stated equivalently as saying that $P_C(z) = x$ if and only if $z - x \in N_C(x)$. Consider any nonempty, closed, convex set $C \subset \mathbb{R}^n$ and any point $z \in \mathbb{R}^n$. Let $g(x) = \frac{1}{2}|x-z|^2$, which has $\nabla g(x) = x - z$ and $\nabla^2 g(x) \equiv I$, implying strict convexity. The projection $x = P_C(z)$ is the solution to the problem of minimizing g over C. The variational inequality (10) characterizes it by the relation $-(x - z) \in N_C(x)$, which is exactly what was targeted.

According to Theorem 2A.6, minimizing a differentiable convex function g over a closed, convex set C is equivalent to solving a type of variational inequality (2) in which f is the gradient mapping ∇g. When $C = \mathbb{R}^n$, so that we are dealing with unconstrained minimization, this is equivalent to solving $f(x) = 0$ for $f = \nabla g$. The notion of a variational inequality thus makes it possible to pass from unconstrained minimization to constrained minimization. Whether the problem is constrained or unconstrained, there is no guarantee that the minimum will be attained at a unique point (although nonuniqueness is impossible when g is strictly convex, at least), but still, local uniqueness dominates the picture conceptually. For that reason, it does make sense to be thinking of the task as one of "solving a generalized equation."

When g is not convex, solving the variational inequality (2) is no longer equivalent to minimization over C, but nevertheless it has a strong association with identifying a local minimum. Anyway, there's no need really to insist on a minimum. Just as the equation $\nabla g(x) = 0$ describes, in general, a "stationary point" of g (unconstrained), the variational inequality (10) can be viewed as describing a constrained version of a stationary point, which could be of interest in itself.

The minimization rule in Theorem 2A.6 can be employed to deduce a rule for determining normal vectors to intersections of convex sets, as in the second part of the following proposition.

Proposition 2A.7 (normals to products and intersections).

(a) *If $C = C_1 \times C_2$ for closed, convex sets $C_1 \subset \mathbb{R}^{n_1}$ and $C_2 \subset \mathbb{R}^{n_2}$, then for any $x = (x_1, x_2) \in C$ one has $N_C(x) = N_{C_1}(x_1) \times N_{C_2}(x_2)$.*

(b) *If $C = C_1 \cap C_2$ for closed, convex sets C_1 and C_2 in $\mathbb{R}^n$, then the formula*

$$N_C(x) = N_{C_1}(x) + N_{C_2}(x) = \{v_1 + v_2 \mid v_1 \in N_{C_1}(x), v_2 \in N_{C_2}(x)\}$$

holds for any $x \in C$ such that there is no $v \neq 0$ with $v \in N_{C_1}(x)$ and $-v \in N_{C_2}(x)$. This condition is fulfilled in particular for every $x \in C$ if $C_1 \cap \operatorname{int} C_2 \neq \emptyset$ or $C_2 \cap \operatorname{int} C_1 \neq \emptyset$.

Proof. To prove (a), we note that, by definition, a vector $v = (v_1, v_2)$ belongs to $N_C(x)$ if and only if, for every $x' = (x_1', x_2')$ in $C_1 \times C_2$ we have $0 \geq \langle v, x' - x \rangle = \langle v_1, x_1' - x_1 \rangle + \langle v_2, x_2' - x_2 \rangle$. That's the same as having $\langle v_1, x_1' - x_1 \rangle \leq 0$ for all $x_1' \in C_1$ and $\langle v_2, x_2' - x_2 \rangle \leq 0$ for all $x_2' \in C_2$, or in other words, $v_1 \in N_{C_1}(x_1)$ and $v_2 \in N_{C_2}(x_2)$.

In proving (b), it is elementary that if $v = v_1 + v_2$ with $v_1 \in N_{C_1}(x)$ and $v_2 \in N_{C_2}(x)$, then for every x' in $C_1 \cap C_2$ we have both $\langle v_1, x' - x \rangle \leq 0$ and $\langle v_2, x' - x \rangle \leq 0$, so that $\langle v, x' - x \rangle \leq 0$. Thus, $N_C(x) \supset N_{C_1}(x) + N_{C_2}(x)$.

The opposite inclusion takes more work to establish. Fix any $x \in C$ and $v \in N_C(x)$. As we know from (4), this corresponds to x being the projection of $x + v$ on C, which we can elaborate as follows: (x, x) is the unique solution to the problem

$$\text{minimize } |x_1 - (x+v)|^2 + |x_2 - (x+v)|^2 \text{ over all } (x_1, x_2) \in C_1 \times C_2 \text{ with } x_1 = x_2.$$

Consider for $k = 1, 2, \ldots$ the version of this minimization problem in which the constraint $x_1 = x_2$ is relaxed by a penalty expression dependent on k:

(11) $\quad$ minimize $|x_1 - (x+v)|^2 + |x_2 - (x+v)|^2 + k|x_1 - x_2|^2$
over all $(x_1, x_2) \in C_1 \times C_2$.

The expression being minimized here is nonnegative and, as seen from the case of $x_1 = x_2 = x$, has minimum no greater than $2|v|^2$. It suffices therefore in the minimization to consider only points x_1 and x_2 such that $|x_1 - (x+v)|^2 + |x_2 - (x+v)|^2 \leq 2|v|^2$ and $k|x_1 - x_2|^2 \leq 2|v|^2$. For each k, therefore, the minimum in (11) is attained by some (x_1^k, x_2^k), and these pairs form a bounded sequence such that $x_1^k - x_2^k \to 0$. Any accumulation point of this sequence must be of the form $(\bar{x}, \bar{x})$ and satisfy $|\bar{x} - (x+v)|^2 + |\bar{x} - (x+v)|^2 \leq 2|v|^2$, or in other words $|\bar{x} - (x+v)| \leq |v|$. But by the projection rule (4), x is the unique closest point of C to $x+v$, the distance being $|v|$, so this inequality implies $\bar{x} = x$. Therefore, $(x_1^k, x_2^k) \to (x, x)$.

We investigate next the necessary condition for optimality provided for problem (11). Invoking the formula in (a) for the normal cone to $C_1 \times C_2$ at (x_1^k, x_2^k), we see that it requires

$$-2[x_1^k - (x+v) + k(x_1^k - x_2^k)] \in N_{C_1}(x_1^k),$$
$$-2[x_2^k - (x+v) - k(x_1^k - x_2^k)] \in N_{C_2}(x_2^k),$$

or equivalently, for $w^k = k(x_2^k - x_1^k)$,

(12) $\quad v + (x - x_1^k) + w^k \in N_{C_1}(x_1^k)$ and $v + (x - x_2^k) - w^k \in N_{C_2}(x_2^k)$.

Two cases have to be analyzed now separately. In the first case, we suppose that the sequence of vectors w^k is bounded and therefore has an accumulation point w. Let $v_1^k = v + (x - x_1^k) + w^k$ and $v_2^k = v + (x - x_2^k) - w^k$, so that, through (4), we have $P_{C_1}(x_1^k + v_1^k) = x_1^k$ and $P_{C_2}(x_2^k + v_2^k) = x_2^k$. Since $x_1^k \to x$ and $x_2^k \to x$, the sequences of vectors v_1^k and v_2^k have accumulation points $v_1 = v - w$ and $v_2 = v + w$ which satisfy $v_1 + v_2 = 2v$. By the continuity of the projection mappings coming from 1D.5, we get $P_{C_1}(x + v_1) = x$ and $P_{C_2}(x + v_2) = x$. By (6), these relations mean $v_1 \in N_{C_1}(x)$ and $v_2 \in N_{C_2}(x)$ and hence $2v \in N_{C_1}(x) + N_{C_2}(x)$. Since the sum of cones is a cone, we get $v \in N_{C_1}(x) + N_{C_2}(x)$. Thus $N_C(x) \subset N_{C_1}(x) + N_{C_2}(x)$, and since we have already shown the opposite inclusion, we have equality.

In the second case, we suppose that the sequence of vectors w^k is unbounded. By passing to a subsequence if necessary, we can reduce this to having $0 < |w^k| \to \infty$ with $w^k/|w^k|$ converging to some $\bar{v} \neq 0$. Let

$$\bar{v}_1^k = [v + (x - x_1^k) + w^k]/|w^k| \text{ and } \bar{v}_2^k = [v + (x - x_2^k) - w^k]/|w^k|.$$

Then $\bar{v}_1^k \to \bar{v}$ and $\bar{v}_2^k \to -\bar{v}$. By (12) we have $\bar{v}_1^k \in N_{C_1}(x_1^k)$ and $\bar{v}_2^k \in N_{C_2}(x_2^k)$, or equivalently through (4), the projection relations $P_{C_1}(x_1^k + \bar{v}_1^k) = x_1^k$ and $P_{C_2}(x_2^k + \bar{v}_2^k) = x_1^k$. In the limit we get $P_{C_1}(x + \bar{v}) = x$ and $P_{C_2}(x - \bar{v}) = x$, so that $\bar{v} \in N_{C_1}(x)$ and $-\bar{v} \in N_{C_2}(x)$. This contradicts our assumption in (b), and we see thereby that the second case is impossible. $\square$

We turn now to minimization over sets C that might not be convex and are specified by systems of constraints which have to be handled with Lagrange multipliers. This will lead us to other valuable examples of variational inequalities, after some elaborations.

Theorem 2A.8 (Lagrange multiplier rule). *Let $X \subset \mathbb{R}^n$ and $D \subset \mathbb{R}^m$ be nonempty, closed, convex sets, and consider the problem*

(13) $$\text{minimize } g_0(x) \text{ over } C = \{x \in X \mid g(x) \in D\},$$

for $g(x) = (g_1(x), \ldots, g_m(x))$, where the functions $g_i : \mathbb{R}^n \to \mathbb{R}$, $i = 0, 1, \ldots, m$ are continuously differentiable. Let x be a point of C at which the following constraint qualification condition is fulfilled:

(14) $\quad$ *there is no $y \in N_D(g(x))$, $y \neq 0$, such that $-y\nabla g(x) \in N_X(x)$.*

If g_0 has a local minimum relative to C at x, then

(15) $\quad$ *there exists $y \in N_D(g(x))$ such that $-[\nabla g_0(x) + y\nabla g(x)] \in N_X(x)$.*

Proof. Assume that a local minimum occurs at x. Let X' and D' be compact, convex sets which coincide with X and D in neighborhoods of x and $g(x)$, respectively, and are small enough that $g_0(x') \geq g_0(x)$ for all $x' \in X'$ having $g(x') \in D'$. Consider the auxiliary problem

(16) $$\begin{array}{l}\text{minimize } g_0(x') + \tfrac{1}{2}|x'-x|^2 \\ \text{over all } (x', u') \in X' \times D' \text{ satisfying } g(x') - u' = 0.\end{array}$$

Obviously the unique solution to this is $(x', u') = (x, g(x))$. Next, for $k \to \infty$, consider the following sequence of problems, which replace the equation in (16) by a penalty expression:

(17) $$\text{minimize } g_0(x') + \frac{1}{2}|x'-x|^2 + \frac{k}{2}|g(x') - u'|^2 \text{ over all } (x', u') \in X' \times D'.$$

For each k let (x_k, u_k) give the minimum in this relaxed problem (the minimum being attained because the functions are continuous and the sets X' and D' are compact). The minimum value in (17) can't be greater than the minimum value in (16), as seen by taking (x', u') to be the unique solution $(x, g(x))$ to (16). It's apparent then that the only possible cluster point of the bounded sequence $\{(x_k, u_k)\}_{k=1}^{\infty}$ as $k \to \infty$ is $(x, g(x))$. Thus, $(x_k, u_k) \to (x, g(x))$.

Next we apply the optimality condition in Theorem 2A.6 to problem (17) at its solution (x_k, u_k). We have $N_{X' \times D'}(x_k, u_k) = N_{X'}(x_k) \times N_{D'}(u_k)$, and on the other hand $N_{X'}(x_k) = N_X(x_k)$ and $N_{D'}(u_k) = N_D(u_k)$ by the choice of X' and D', at least when k is sufficiently large. The variational inequality condition in Theorem 2A.6 comes down in this way to

(18) $$\begin{cases} -[\nabla g_0(x_k) + (x_k - x) + k(g(x_k) - u_k)\nabla g(x_k)] \in N_X(x_k), \\ k(g(x_k) - u_k) \in N_D(u_k). \end{cases}$$

By passing to subsequences if necessary, we can reduce the rest of the analysis to distinguishing between case (A), where the norms of the vectors $k(g(x_k) - u_k) \in \mathbb{R}^m$ stay bounded as $k \to \infty$, and case (B), where these norms go to ∞.

In case (A) we can arrange, by passing again to a subsequence if necessary, that the sequence of vectors $k(g(x_k) - u_k)$ converges to some y. Then y satisfies the desired relations in (18), inasmuch as $(x_k, u_k) \to (x, g(x))$ and the graphs of the mappings N_X and N_D are closed.

In case (B) we look at the vectors $y_k = k(g(x_k) - u_k)/\rho_k$ with $\rho_k = k|g(x_k) - u_k| \to \infty$, which have $|y_k| = 1$ and, from (18), satisfy

(19) $$-\rho_k^{-1}(\nabla g_0(x_k) + (x_k - x)) - y_k \nabla g(x_k) \in N_X(x_k), \qquad y_k \in N_D(u_k).$$

(Here we use the fact that any positive multiple of a vector in $N_X(x_k)$ or $N_D(u_k)$ is another such vector.) By passing to a subsequence, we can arrange that the sequence of vectors y_k converges to some y, necessarily with $|y| = 1$. In this limit, (19) turns into the relation in (14), which has been forbidden to hold for any $y \neq 0$. Hence case (B) is impossible under our assumptions, and we are left with the conclusion (15) obtained from case (A). □

In the first-order optimality condition (15), y is said to be a *Lagrange multiplier vector* associated with x. More can be said about this condition by connecting it with the *Lagrangian function* for problem (13), which is defined by

(20) $$L(x, y) = g_0(x) + \langle y, g(x) \rangle = g_0(x) + y_1 g_1(x) + \cdots + y_m g_m(x)$$

for $y = (y_1, \ldots, y_m)$.

Theorem 2A.9 (Lagrangian variational inequalities). *In the minimization problem (13), suppose that the set D is a cone, and let Y be the polar cone D^*,*

$$Y = \{ y \mid \langle u, y \rangle \leq 0 \text{ for all } u \in D \}.$$

Then, in terms of the Lagrangian L in (20), the condition on x and y in (15) can be written in the form

(21) $$-\nabla_x L(x, y) \in N_X(x), \qquad \nabla_y L(x, y) \in N_Y(y),$$

which furthermore can be identified with the variational inequality

(22) $$-f(x, y) \in N_{X \times Y}(x, y) \text{ for } f(x, y) = (\nabla_x L(x, y), -\nabla_y L(x, y)).$$

The existence of $y \in Y$ satisfying this variational inequality with x is thus necessary for the local optimality of x in problem (13) when the constraint qualification (14) is fulfilled. If $L(\cdot, y)$ is convex on X when $y \in Y$, the existence of a y satisfying this variational inequality with x is moreover sufficient for x to give a global minimum in problem (13), without any need for invoking (14).

Proof. We have $\nabla_x L(x,y) = \nabla g_0(x) + y\nabla g(x)$ and $\nabla_y L(x,y) = g(x)$. The N_X condition in (15) amounts therefore to the first condition in (21). The choice of $Y = D^*$ makes it possible through the polarity rule for normal vectors in (8) to express the N_D condition in (15) as $g(x) \in N_Y(y)$ and identify it with the second condition in (21), while deducing from it also that $\langle y, g(x) \rangle = 0$, hence $L(x,y) = g_0(x)$. The recasting of (21) as the variational inequality in (22) comes out of the product rule in 2A.7(a).

When the function $L(\cdot, y)$ is convex on X, the condition $-\nabla_x L(x,y) \in N_X(x)$ implies through Theorem 2A.6 that $L(x',y) \geq L(x,y)$ for all $x' \in X$, where it may be recalled that $L(x,y) = g_0(x)$ because $\langle y, g(x) \rangle = 0$. Thus, $L(x',y) \geq g_0(x)$ for all $x' \in X$. On the other hand, since $y \in Y$ and $Y = D^*$, we have $\langle y, g(x') \rangle \leq 0$ when $g(x') \in D$. Therefore $g_0(x') \geq L(x',y) \geq g_0(x)$ for all x' satisfying the constraints in (13). It follows that all such x' have $g_0(x') \geq g_0(x)$, so x furnishes the global minimum in problem (13). □

Application to nonlinear programming. Theorems 2A.8 and 2A.9 cover the case of a standard problem of *nonlinear programming*, where the task is to

$$(23) \qquad \text{minimize } g_0(x) \text{ over all } x \text{ satisfying } g_i(x) \begin{cases} \leq 0 & \text{for } i \in [1,s], \\ = 0 & \text{for } i \in [s+1,m]. \end{cases}$$

This problem[1] corresponds in (13) to taking $X = \mathbb{R}^n$ and having D be the closed, convex cone in $\mathbb{R}^m$ consisting of all $u = (u_1, \ldots, u_m)$ such that $u_i \leq 0$ for $i \in [1,s]$ but $u_i = 0$ for $i \in [s+1,m]$. The polar cone $Y = D^*$ is $Y = \mathbb{R}^s_+ \times \mathbb{R}^{m-s}$. The optimality condition in (18) can equally well be placed then in the Lagrangian framework in (21), corresponding to the variational inequality (22). The requirements it imposes on x and y come out as

$$(24) \qquad \begin{aligned} y \in \mathbb{R}^s_+ \times \mathbb{R}^{m-s}, \; g_i(x) &\begin{cases} \leq 0 & \text{for } i \in [1,s] \text{ with } y_i = 0, \\ = 0 & \text{for all other } i \in [1,m], \end{cases} \\ \nabla g_0(x) + y_1 \nabla g_1(x) + \cdots + y_m \nabla g_m(x) &= 0. \end{aligned}$$

These are the *Karush–Kuhn–Tucker conditions* for the nonlinear programming problem (23). According to Theorem 2A.8, the existence of y satisfying these conditions with x is necessary for the local optimality of x under the constraint qualification (14), which insists on the nonexistence of $y \neq 0$ satisfying the same conditions with the term $\nabla g_0(x)$ suppressed. The existence of y satisfying (24) is sufficient for the global optimality of x by Theorem 2A.9 as long as $L(x,y)$ is convex as a function of $x \in \mathbb{R}^n$ for each fixed $y \in \mathbb{R}^s_+ \times \mathbb{R}^{m-s}$, which is equivalent to having

$$g_0, g_1, \ldots, g_s \text{ convex, but } g_{s+1}, \ldots, g_m \text{ affine.}$$

---

[1] In (23) and later in the book $[1,s]$ denotes the set of integers $\{1, 2, \ldots, s\}$.

Then (23) is a problem of *convex programming*. The Karush–Kuhn–Tucker conditions correspond then to a saddle point property, as indicated next.

Exercise 2A.10 (variational inequality for a saddle point). *Let $X \subset \mathbb{R}^n$ and $Y \subset \mathbb{R}^m$ be any nonempty, closed, convex sets, and let L be a $\mathscr{C}^1$ function on $\mathbb{R}^n \times \mathbb{R}^m$ such that $L(\cdot, y)$ is a convex function on X for each $y \in Y$, and $L(x, \cdot)$ is a concave function on Y for each $x \in X$. The variational inequality (22) is equivalent to having (x, y) be a saddle point of L with respect to $X \times Y$ in the sense that*

$$x \in X, \ y \in Y, \text{ and } L(x', y) \geq L(x, y) \geq L(x, y') \text{ for all } x' \in X, \ y' \in Y.$$

Guide. Rely on the equivalence between (21) and (22), plus Theorem 2A.6. □

A saddle point as defined in Exercise 2A.10 represents an equilibrium in the two-person zero-sum game in which Player 1 chooses $x \in X$, Player 2 chooses $y \in Y$, and then Player 1 pays the amount $L(x, y)$ (possibly negative) to Player 2. Other kinds of equilibrium can likewise be captured by other variational inequalities.

For example, in an N-person game there are players $1, \ldots, N$, with Player k having a nonempty strategy set X_k. Each Player k chooses some $x_k \in X_k$, and is then obliged to pay—to an abstract entity (not necessarily another player)—an amount which depends not only on x_k but also on the choices of all the other players; this amount can conveniently be denoted by

$$L_k(x_k, x_{-k}), \text{ where } x_{-k} = (x_1, \ldots, x_{k-1}, x_{k+1}, \ldots, x_N).$$

(The game is *zero-sum* if $\sum_{k=1}^N L_k(x_k, x_{-k}) = 0$.) A choice of strategies $x_k \in X_k$ for $k = 1, \ldots, N$ is said to furnish a *Nash equilibrium* if

$$L_k(x'_k, x_{-k}) \geq L_k(x_k, x_{-k}) \text{ for all } x'_k \in X_k, \ k = 1, \ldots, N.$$

A saddle point as in Exercise 2A.10 corresponds to the case of this where $N = 2$, so x_{-1} and x_{-2} are just x_2 and x_1 respectively, and one has $L_2(x_2, x_1) = -L_1(x_1, x_2)$.

Exercise 2A.11 (variational inequality for a Nash equilibrium). *In an N-person game as described, suppose that X_k is a closed, convex subset of $\mathbb{R}^{n_k}$ and that $L(x_k, x_{-k})$ is differentiable with respect to x_k for every k. Then for $x = (x_1, \ldots, x_N)$ to furnish a Nash equilibrium, it must solve the variational inequality (2) for f and C in the case of*

$$C = X_1 \times \cdots \times X_N, \quad f(x) = f(x_1, \ldots, x_N) = \Big(\nabla_{x_1} L_1(x_1, x_{-1}), \ldots, \nabla_{x_N} L_N(x_N, x_{-N})\Big).$$

This necessary condition is sufficient for a Nash equilibrium if, in addition, the functions $L_k(\cdot, x_{-k})$ on $\mathbb{R}^{n_k}$ are convex.

Guide. Make use of the product rule for normals in 2A.7(a) and the optimality condition in Theorem 2A.6. □

Finally, we look at a kind of generalized equation (1) that is *not* a variational inequality (2), but nonetheless has importance in many situations:

(25) $$(g_1(x),\ldots,g_m(x)) \in D,$$

which is (1) for $f(x) = -(g_1(x),\ldots,g_m(x))$, $F(x) \equiv D$. Here D is a subset of $\mathbb{R}^m$; the format has been chosen to be that of the constraints in problem (13), or as a special case, problem (23).

Although (25) would reduce to an equation, pure and simple, if D consists of a single point, the applications envisioned for it lie mainly in situations where inequality constraints are involved, and there is little prospect or interest in a solution being locally unique. In the study of generalized equations with parameters, to be taken up next in 2B, our attention will at first be concentrated on issues parallel to those in Chapter 1. Only later, in Chapter 3, will generalized equations like (25) be brought in.

The example in (25) also brings a reminder about a feature of generalized equations which dropped out of sight in the discussion of the variational inequality case. In (2), necessarily f had to go from $\mathbb{R}^n$ to $\mathbb{R}^n$, whereas in (25), and in (1), f may go from $\mathbb{R}^n$ to a space $\mathbb{R}^m$ of different dimension.

2B. Implicit Function Theorems for Generalized Equations

With the concept of a generalized equation, and in particular that of a variational inequality problem at our disposal, we are ready to embark on a broad exploration of implicit function theorems beyond those in Chapter 1. The object of study is now a *parameterized* generalized equation

(1) $$f(p,x) + F(x) \ni 0$$

for a function $f : \mathbb{R}^d \times \mathbb{R}^n \to \mathbb{R}^m$ and a set-valued mapping $F : \mathbb{R}^n \rightrightarrows \mathbb{R}^m$. Specifically, we consider the properties of the *solution mapping* $S : \mathbb{R}^p \rightrightarrows \mathbb{R}^n$ defined by

(2) $$S : p \mapsto \{x \mid f(p,x) + F(x) \ni 0\} \quad \text{for } p \in \mathbb{R}^d.$$

The questions we will concentrate on answering, for now, are nevertheless the same as in Chapter 1. To what extent might S be single-valued and possess various properties of continuity or some type of differentiability?

In a landmark paper[2], S. M. Robinson studied the solution mapping S in the case of a parameterized variational inequality, where $m = n$ and F is a normal cone mapping $N_C : \mathbb{R}^n \rightrightarrows \mathbb{R}^n$:

(3) $$f(p,x) + N_C(x) \ni 0, \text{ with } C \subset \mathbb{R}^n \text{ convex, closed and nonempty.}$$

---

[2] Cf. Robinson [1980].

His results were, from the very beginning, stated in abstract spaces, and we will come to that in Chapter 5. Here, we confine the exposition to Euclidean spaces, but the presentation is tailored in such a way that, for readers who are familiar with some basic functional analysis, the expansion of the framework from Euclidean spaces to general Banach spaces is straightforward. The original formulation of Robinson's theorem, up to some rewording to fit this setting, is as follows.

Theorem 2B.1 (Robinson implicit function theorem). *For the solution mapping S to a parameterized variational inequality (3), consider a pair $(\bar{p},\bar{x})$ with $\bar{x} \in S(\bar{p})$. Assume that:*

(a) *$f(p,x)$ is differentiable with respect to x in a neighborhood of the point $(\bar{p},\bar{x})$, and both $f(p,x)$ and $\nabla_x f(p,x)$ depend continuously on (p,x) in this neighborhood;*

(b) *the inverse G^{-1} of the set-valued mapping $G : \mathbb{R}^n \rightrightarrows \mathbb{R}^n$ defined by*

$$(4) \qquad G(x) = f(\bar{p},\bar{x}) + \nabla_x f(\bar{p},\bar{x})(x - \bar{x}) + N_C(x), \quad \text{with } G(\bar{x}) \ni 0,$$

has a Lipschitz continuous single-valued localization σ around 0 for $\bar{x}$ with

$$\text{lip}(\sigma;0) \leq \kappa.$$

Then S has a single-valued localization s around $\bar{p}$ for $\bar{x}$ which is continuous at $\bar{p}$, and moreover for every $\varepsilon > 0$ there is a neighborhood Q of $\bar{p}$ such that

$$(5) \qquad |s(p') - s(p)| \leq (\kappa + \varepsilon)|f(p',s(p)) - f(p,s(p))| \text{ for all } p',p \in Q.$$

An extended version of this result will be stated shortly as Theorem 2B.5, so we can postpone the discussion of its proof until then. Instead, we can draw some immediate conclusions from the estimate (5) which rely on additional assumptions about partial calmness and Lipschitz continuity properties of $f(p,x)$ with respect to p and the modulus notation for such properties that was introduced in 1C and 1D.

Corollary 2B.2 (calmness of solutions). *In the setting of Theorem 2B.1, if f is calm with respect to p at $(\bar{p},\bar{x})$, having $\text{clm}_p(f;(\bar{p},\bar{x})) \leq \lambda$, then s is calm at $\bar{p}$ with $\text{clm}(s;\bar{p}) \leq \kappa\lambda$.*

Corollary 2B.3 (Lipschitz continuity of solutions). *In the setting of Theorem 2B.1, if f is Lipschitz continuous with respect to p uniformly in x around $(\bar{p},\bar{x})$, having $\widehat{\text{lip}}_p(f;(\bar{p},\bar{x})) \leq \lambda$, then s is Lipschitz continuous around $\bar{p}$ with $\text{lip}(s;\bar{p}) \leq \kappa\lambda$.*

Differentiability of the localization s around $\bar{p}$ can't be deduced from the estimate in (4), not to speak of continuous differentiability around $\bar{p}$, and in fact differentiability may fail. Elementary one-dimensional examples of variational inequalities exhibit solution mappings that are not differentiable, usually in connection with the "solution trajectory" hitting or leaving the boundary of the set C. For such mappings, weaker concepts of differentiability are available. We will touch upon this in 2D.

In the special case where the variational inequality treated by Robinson's theorem reduces to the equation $f(p,x) = 0$ (namely with $C = \mathbb{R}^n$, so $N_C \equiv 0$), the invertibility condition on the mapping G in assumption (b) of Robinson's theorem comes down to the nonsingularity of the Jacobian $\nabla_x f(\bar{p},\bar{x})$ in the Dini classical implicit function theorem 1B.1. But because of the absence of an assertion about the differentiability of s, Theorem 2B.1 falls short of yielding all the conclusions of that theorem. It could, though, be used as an intermediate step in a proof of Theorem 1B.1, which we leave to the reader as an exercise.

Exercise 2B.4. *Supply a proof of the classical implicit function theorem 1B.1 based on Robinson's theorem 2B.1.*

Guide. In the case $C = \mathbb{R}^n$, so $N_C \equiv 0$, use the Lipschitz continuity of the single-valued localization s following from Corollary 2B.3 to show that s is continuously differentiable around $\bar{p}$ when f is continuously differentiable near $(\bar{p},\bar{x})$. □

The invertibility property in assumption (b) of 2B.1 is what Robinson called "strong regularity" of the generalized equation (3). A related term, "strong metric regularity," will be employed in Chapter 3 for set-valued mappings in reference to the existence of Lipschitz continuous single-valued localizations of their inverses.

In the extended version of Theorem 2B.1 which we present next, the differentiability assumptions on f are replaced by assumptions about an estimator h for $f(\bar{p},\cdot)$, which could in particular be a first-order approximation in the x argument. This mode of generalization was initiated in 1E.3 and 1E.13 for equations, but now we use it for a generalized equation (1). In contrast to Theorem 2B.1, which was concerned with the case of a variational inequality (3), the mapping $F : \mathbb{R}^n \rightrightarrows \mathbb{R}^m$ need not be of form N_C and the dimensions n and m could in principle be different. Remarkably, no *direct* assumptions need be made about F, but certain properties of F will implicitly underlie the "invertibility" condition imposed jointly on F and the estimator h.

Theorem 2B.5 (Robinson theorem extended beyond differentiability). *For a generalized equation (1) and its solution mapping S in (2), let $\bar{p}$ and $\bar{x}$ be such that $\bar{x} \in S(\bar{p})$. Assume that:*

(a) $f(\cdot,\bar{x})$ is continuous at $\bar{p}$, and h is a strict estimator of f with respect to x uniformly in p at $(\bar{p},\bar{x})$ with constant μ;

(b) the inverse G^{-1} of the mapping $G = h + F$, for which $G(\bar{x}) \ni 0$, has a Lipschitz continuous single-valued localization σ around 0 for $\bar{x}$ with $\mathrm{lip}(\sigma;0) \leq \kappa$ for a constant κ such that $\kappa\mu < 1$.

Then S has a single-valued localization s around $\bar{p}$ for $\bar{x}$ which is continuous at $\bar{p}$, and moreover for every $\varepsilon > 0$ there is a neighborhood Q of $\bar{p}$ such that

$$(6) \quad |s(p') - s(p)| \leq \frac{\kappa + \varepsilon}{1 - \kappa\mu} |f(p',s(p)) - f(p,s(p))| \quad \text{for all } p',p \in Q.$$

Theorem 2B.1 follows at once from Theorem 2B.5 by taking F to be N_C and h to be the linearization of $f(\cdot,\bar{p})$ given by $h(x) = f(\bar{p},\bar{x}) + \nabla_x f(\bar{p},\bar{x})(x - \bar{x})$, and employing 1E.15. More generally, h could be a strict first-order approximation: the case when $\mu = 0$. That case, which has further implications, will be taken up later. However, Theorem 2B.5 is able to extract information from much weaker relationships between f and h than strict first-order approximation, and this information can still have important consequences for the behavior of solutions to a generalized equation, as seen in this pattern already for equations in 1E.

Our proof of Theorem 2B.5 will proceed through an intermediate stage in which we isolate a somewhat lengthy statement as a lemma. Proving the lemma requires an appeal to the contraction mapping principle, as formulated in Theorem 1A.2.

Lemma 2B.6. *Consider a function $\varphi : \mathbb{R}^d \times \mathbb{R}^n \to \mathbb{R}^m$ and a point $(\bar{p}, \bar{x}) \in \text{int dom } \varphi$ and let the scalars $v \geq 0$, $b \geq 0$, $a > 0$, and the set $Q \subset \mathbb{R}^d$ be such that $\bar{p} \in Q$ and*

(7) $\quad \begin{cases} |\varphi(p,x') - \varphi(p,x)| \leq v|x - x'| & \text{for all } x', x \in \mathbb{B}_a(\bar{x}) \text{ and } p \in Q, \\ |\varphi(p,\bar{x}) - \varphi(\bar{p},\bar{x})| \leq b & \text{for all } p \in Q. \end{cases}$

Consider also a set-valued mapping $M : \mathbb{R}^m \rightrightarrows \mathbb{R}^n$ with $(\bar{y}, \bar{x}) \in \text{gph } M$ where $\bar{y} := \varphi(\bar{p}, \bar{x})$, such that for each $y \in \mathbb{B}_{va+b}(\bar{y})$ the set $M(y) \cap \mathbb{B}_a(\bar{x})$ consists of exactly one point, denoted by $r(y)$, and suppose that the function

(8) $\quad\quad\quad r : y \mapsto M(y) \cap \mathbb{B}_a(\bar{x}) \ \text{ for } y \in \mathbb{B}_{va+b}(\bar{y})$

is Lipschitz continuous on $\mathbb{B}_{va+b}(\bar{y})$ with a Lipschitz constant λ. In addition, suppose that
 (a) $\lambda v < 1$;
 (b) $\lambda v a + \lambda b \leq a$.
Then for each $p \in Q$ the set $\{x \in \mathbb{B}_a(\bar{x}) \mid x \in M(\varphi(p,x))\}$ consists of exactly one point, and the associated function

(9) $\quad\quad\quad s : p \mapsto \{x \mid x = M(\varphi(p,x)) \cap \mathbb{B}_a(\bar{x})\} \ \text{ for } p \in Q$

satisfies

(10) $\quad |s(p') - s(p)| \leq \dfrac{\lambda}{1 - \lambda v}|\varphi(p', s(p)) - \varphi(p, s(p))| \ \text{ for all } p', p \in Q.$

Proof. Fix $p \in Q$ and consider the function $\Phi_p : \mathbb{R}^n \to \mathbb{R}^n$ defined by

$$\Phi_p : x \mapsto r(\varphi(p,x)) \ \text{ for } x \in \mathbb{B}_a(\bar{x}).$$

First, note that for $x \in \mathbb{B}_a(\bar{x})$ from (7) one has $|\bar{y} - \varphi(p,x)| \leq b + va$, thus, by (8), $\mathbb{B}_a(\bar{x}) \subset \text{dom } \Phi_p$. Next, if $x \in \mathbb{B}_a(\bar{x})$, we have from the identity $\bar{x} = r(\varphi(\bar{p},\bar{x}))$, the Lipschitz continuity of r, and conditions (7) and (b) that

$$|\Phi_p(\bar{x}) - \bar{x}| = |r(\varphi(p,\bar{x})) - r(\varphi(\bar{p},\bar{x}))| \leq \lambda|\varphi(p,\bar{x}) - \varphi(\bar{p},\bar{x})| \leq \lambda b \leq a(1 - \lambda v).$$

For any $x', x \in \mathbb{B}_a(\bar{x})$ we obtain

$$|\Phi_p(x') - \Phi_p(x)| = |r(\varphi(p,x')) - r(\varphi(p,x))| \leq \lambda |\varphi(p,x') - \varphi(p,x)| \leq \lambda v |x' - x|,$$

that is, Φ_p is Lipschitz continuous in $\mathbb{B}_a(\bar{x})$ with constant $\lambda v < 1$, from condition (a). We are in position then to apply the contraction mapping principle 1A.2 and to conclude from it that Φ_p has a unique fixed point in $\mathbb{B}_a(\bar{x})$.

Denoting that fixed point by $s(p)$, and doing this for every $p \in Q$, we get a function $s : Q \to \mathbb{B}_a(\bar{x})$. But having $x = \Phi_p(x)$ is equivalent to having $x = r(\varphi(p,x)) = M(\varphi(p,x)) \cap \mathbb{B}_a(\bar{x})$. Hence s is the function in (9). Moreover, since $s(p) = r(\varphi(p,s(p)))$, we have from the Lipschitz continuity of r and (7) that, for any $p', p \in Q$,

$$\begin{aligned}
|s(p') - s(p)| &= |r(\varphi(p',s(p'))) - r(\varphi(p,s(p)))| \\
&\leq |r(\varphi(p',s(p'))) - r(\varphi(p',s(p)))| + |r(\varphi(p',s(p))) - r(\varphi(p,s(p)))| \\
&\leq \lambda |\varphi(p',s(p')) - \varphi(p',s(p))| + \lambda |\varphi(p',s(p)) - \varphi(p,s(p))| \\
&\leq \lambda v |s(p') - s(p)| + \lambda |\varphi(p',s(p)) - \varphi(p,s(p))|.
\end{aligned}$$

Since $\lambda v < 1$, we see that s satisfies (10), as needed. □

It's worth noting that the version of the contraction mapping principle utilized in proving Lemma 2B.6 (namely, Theorem 1A.2 with $X = \mathbb{R}^n$ equipped with the metric induced by the Euclidean norm $|\cdot|$) can in turn be derived from Lemma 2B.6. For that, the data in the lemma need to be specified as follows: $d = m = n$, $v = \lambda$, a unchanged, $b = |\Phi(\bar{x}) - \bar{x}|$, $\bar{p} = 0$, $Q = \mathbb{B}_b(0)$, $\varphi(p,x) = \Phi(x) + p$, $\bar{y} = \Phi(\bar{x})$, $M(y) = y + \bar{x} - \Phi(\bar{x})$, and consequently the λ in 1A.2 is 1. All the conditions of Lemma 2B.6 hold for such data under the assumptions of the contraction mapping principle 1A.2. Hence for $p = \Phi(\bar{x}) - \bar{x} \in Q$ the set $\{x \in \mathbb{B}_a(\bar{x}) \,|\, x = M(\varphi(p,x)) = \Phi(x)\}$ consists of exactly one point; that is, Φ has a unique fixed point in $\mathbb{B}_a(\bar{x})$. Thus, Lemma 2B.6 is actually equivalent[3] to the form of the contraction mapping principle used in its proof.

Proof of Theorem 2B.5. For an arbitrary $\varepsilon > 0$, choose any $\lambda > \mathrm{lip}(\sigma;0)$ and $v > \mu$ such that $\lambda v < 1$ and

(11) $$\frac{\lambda}{1 - \lambda v} \leq \frac{\kappa + \varepsilon}{1 - \kappa \mu},$$

as is possible under the assumption that $\kappa \mu < 1$. Let a, b and c be positive numbers such that

$$|\sigma(y) - \sigma(y')| \leq \lambda |y - y'| \quad \text{for } y, y' \in \mathbb{B}_{va+b}(0),$$

$$|e(p,x') - e(p,x)| \leq v |x - x'| \quad \text{for } x, x' \in \mathbb{B}_a(\bar{x}) \text{ and } p \in \mathbb{B}_c(\bar{p}),$$

---

[3] Lemma 2B.6 can be stated in a complete metric space X and then it will be equivalent to the standard formulation of the contraction mapping principle in Theorem 1A.2. There is no point, of course, in giving a fairly complicated equivalent formulation of a classical result unless, as in our case, this formulation would bring some insights and dramatically simplify the proofs of later results.

2 Implicit Function Theorems for Variational Problems   79

where $e(p,x) = f(p,x) - h(x)$, and

(12) $$|f(p,\bar{x}) - f(\bar{p},\bar{x})| \leq b \text{ for } p \in I\!B_c(\bar{p}).$$

Take b smaller if necessary so that $b\lambda < a(1 - \lambda v)$, and accordingly adjust c to ensure having (12). Now apply Lemma 2B.6 with $r = \sigma$, $M = (h+F)^{-1}$, $\bar{y} = 0$ and $\varphi = -e$, keeping the rest of the notation the same. It's straightforward to check that the estimates in (7) and the conditions (a) and (b) hold for the function in (8). Then, through the conclusion of Lemma 2B.6 and the observation that

(13) $$x \in (h+F)^{-1}(-e(p,x)) \iff x \in S(p),$$

we obtain that the solution mapping S in (2) has a single-valued localization s around $\bar{p}$ for $\bar{x}$. Due to (11), the inequality in (6) holds for $Q = I\!B_c(\bar{p})$. That estimate implies the continuity of s at $\bar{p}$, in particular. □

From Theorem 2B.5 we obtain a generalization of Theorem 1E.13, the result in Chapter 1 about implicit functions without differentiability, in which the function f is replaced now by the sum $f + F$ for an *arbitrary* set-valued mapping F. The next statement, 2B.7, covers most of this generalization; the final part of 1E.13 (giving special consequences when $\mu = 0$) will be addressed in the follow-up statement, 2B.8.

Theorem 2B.7 (implicit function theorem for generalized equations). *Consider a function* $f : I\!R^d \times I\!R^n \to I\!R^n$ *and a mapping* $F : I\!R^n \rightrightarrows I\!R^n$ *with* $(\bar{p},\bar{x}) \in$ int dom f *and* $f(\bar{p},\bar{x}) + F(\bar{x}) \ni 0$, *and suppose that* $\widehat{\text{lip}}_p(f;(\bar{p},\bar{x})) \leq \gamma < \infty$. *Let* h *be a strict estimator of* f *with respect to* x *uniformly in* p *at* $(\bar{p},\bar{x})$ *with constant* μ. *Suppose that* $(h+F)^{-1}$ *has a Lipschitz continuous single-valued localization* σ *around* 0 *for* $\bar{x}$ *with* lip$(\sigma;0) \leq \kappa$ *for a constant* κ *such that* $\kappa\mu < 1$. *Then the solution mapping*

$$S : p \mapsto \{x \in I\!R^n \,|\, f(p,x) + F(x) \ni 0\} \quad \text{for } p \in I\!R^d$$

has a Lipschitz continuous single-valued localization s *around* $\bar{p}$ *for* $\bar{x}$ *with*

$$\text{lip}(s;\bar{p}) \leq \frac{\kappa\gamma}{1-\kappa\mu}.$$

For the case of 2B.7 with $\mu = 0$, in which case h is a partial first-order approximation of f with respect to x at $(\bar{p},\bar{x})$, much more can be said about the single-valued localization s. The details are presented in the next result, which extends the part of 1E.13 for this case, and with it, Corollaries 2B.2 and 2B.3. We see that, by adding some relatively mild assumptions about the function f (while still allowing F to be arbitrary!), we can develop a first-order approximation of the localized solution mapping s in Theorem 2B.5. This opens the way to obtain differentiability properties of s, for example.

Theorem 2B.8 (extended implicit function theorem with first-order approximations). *Specialize Theorem 2B.5 to the case where* $\mu = 0$ *in 2B.5(a), so that* h *is* a strict first-order approximation of f with respect to x uniformly in p at $(\bar{p},\bar{x})$. Then, with the localization σ in 2B.5(b) we have the following additions to the conclusions of Theorem 2B.5:

(a) If $\text{clm}_p(f;(\bar{p},\bar{x})) < \infty$ then the single-valued localization s of the solution mapping S in (2) is calm at $\bar{p}$ with

(14) $$\text{clm}(s;\bar{p}) \leq \text{lip}(\sigma;0) \cdot \text{clm}_p(f;(\bar{p},\bar{x})).$$

(b) If $\widehat{\text{lip}}_p(f;(\bar{p},\bar{x})) < \infty$, then the single-valued localization s of the solution mapping S in (2) is Lipschitz continuous near $\bar{p}$ with

(15) $$\text{lip}(s;\bar{p}) \leq \text{lip}(\sigma;0) \cdot \widehat{\text{lip}}_p(f;(\bar{p},\bar{x})).$$

(c) If, along with (a), f has a first-order approximation r with respect to p at $(\bar{p},\bar{x})$, then, for Q as in (6), the function $\eta : Q \to \mathbb{R}^n$ defined by

(16) $$\eta(p) = \sigma(-r(p) + f(\bar{p},\bar{x})) \text{ for } p \in Q$$

is a first-order approximation at $\bar{x}$ to the single-valued localization s.

(d) If, in addition to (b)(c), σ is affine, i.e., $\sigma(y) = \bar{x} + Ay$ for some $n \times m$ matrix A, and furthermore the first-order approximation r is strict with respect to p uniformly in x at $(\bar{p},\bar{x})$, then η is a strict first-order approximation of s at $\bar{p}$ in the form

(17) $$\eta(p) = \bar{x} + A(-r(p) + f(\bar{p},\bar{x})) \text{ for } p \in Q.$$

Proof. Let the constants a and c be as in the proof of Theorem 2B.5; then $Q = \mathbb{B}_c(\bar{p})$. Let $U = \mathbb{B}_a(\bar{x})$. For $p \in Q$, from (13) we have

(18) $$s(p) = \sigma(-e(p,s(p))) \text{ for } e(p,x) = f(p,x) - h(x)$$

along with $\bar{x} = s(\bar{p}) = \sigma(0)$. Let κ equal $\text{lip}(\sigma;0)$ and consider for any $\varepsilon > 0$ the estimate in (6) with $\mu = 0$. Let $p' \in Q$, $p' \neq \bar{p}$ and $p = \bar{p}$ in (6) and divide both sides of (6) by $|p' - \bar{p}|$. Taking the limsup as $p' \to \bar{p}$ and $\varepsilon \to 0$ gives us (14).

Under the assumptions of (b), in a similar way, by letting $p', p \in Q$, $p' \neq p$ in (6), and dividing both sides of (6) by $|p' - p|$ and passing to the limit, we obtain (15). Observe that (15) follows directly from 2B.7.

Consider now any $\lambda > \text{clm}(s;\bar{p})$ and $\varepsilon > 0$. Make the neighborhoods Q and U smaller if necessary so that for all $p \in Q$ and $x \in U$ we have $|s(p) - s(\bar{p})| \leq \lambda|p - \bar{p}|$ and

(19) $$|e(p,x) - e(p,\bar{x})| \leq \varepsilon|x - \bar{x}|, \qquad |f(p,\bar{x}) - r(p)| \leq \varepsilon|p - \bar{p}|,$$

and furthermore so that the points $-e(p,x)$ and $-r(p) + f(\bar{p},\bar{x})$ are contained in a neighborhood of 0 on which the function σ is Lipschitz continuous with Lipschitz constant $\kappa + \varepsilon = \text{lip}(\sigma;0) + \varepsilon$. Then, for $p \in Q$, we get by way of (18), along with the first inequality in (19) and the fact that $e(\bar{p},\bar{x}) = 0$, the estimate that

2 Implicit Function Theorems for Variational Problems

$$\begin{aligned}
|s(p) - \eta(p)| &= |s(p) - \sigma(-r(p) + f(\bar{p}, \bar{x}))| \\
&= |\sigma(-e(p, s(p))) - \sigma(-r(p) + f(\bar{p}, \bar{x}))| \\
&\leq (\kappa + \varepsilon)(|-e(p, s(p)) + e(p, \bar{x})| + |f(p, \bar{x}) - r(p)|) \\
&\leq (\kappa + \varepsilon)\varepsilon|s(p) - \bar{x}| + (\kappa + \varepsilon)\varepsilon|p - \bar{p}| \leq \varepsilon(\kappa + \varepsilon)(\lambda + 1)|p - \bar{p}|.
\end{aligned}$$

Since ε can be arbitrarily small and also $s(\bar{p}) = \bar{x} = \sigma(0) = \eta(\bar{p})$, the function η defined in (16) is a first-order approximation of s at $\bar{p}$.

Moving on to part (d) of the theorem, suppose that the assumptions in (b)(c) are satisfied and also $\sigma(y) = \bar{x} + Ay$. Again, choose any $\varepsilon > 0$ and further adjust the neighborhoods Q of $\bar{p}$ and U of $\bar{x}$ so that

$$(20) \quad \begin{array}{ll} |e(p, x) - e(p, x')| \leq \varepsilon|x - x'| & \text{for all } x, x' \in U \text{ and } p \in Q, \\ |f(p', x) - r(p') - f(p, x) + r(p)| \leq \varepsilon|p' - p| & \text{for all } x \in U \text{ and } p', p \in Q, \end{array}$$

and moreover $s(p) \in U$ for $p \in Q$. By part (b), the single-valued localization s is Lipschitz continuous near $\bar{p}$; let $\lambda > \text{lip}(s; \bar{p})$ and shrink Q even more if necessary so as to ensure that s is Lipschitz continuous with constant λ on Q. For $p, p' \in Q$, using (17), (18) and (20), we obtain

$$\begin{aligned}
|s(p) - s(p') - \eta(p) + \eta(p')| &= |s(p) - s(p') - A(-r(p) + r(p'))| \\
&= |A(-e(p, s(p)) + e(p', s(p')) + r(p) - r(p'))| \\
&\leq |A||-e(p, s(p)) + e(p, s(p'))| \\
&\quad + |A||f(p', s(p')) - r(p') - f(p, s(p')) + r(p)| \\
&\leq |A|(\varepsilon|s(p) - s(p')| + \varepsilon|p' - p|) \\
&\leq |A|\varepsilon(\lambda + 1)|p - p'|.
\end{aligned}$$

Since ε can be arbitrarily small, we see that the first-order approximation of s furnished by η is strict, and the proof is complete. $\square$

Note that the assumption in part (d), that the localization σ of $G^{-1} = (h + F)^{-1}$ around 0 for $\bar{x}$ is affine, can be interpreted as a sort of differentiability condition on G^{-1} at 0 with A giving the derivative mapping.

Corollary 2B.9 (utilization of strict differentiability). *Suppose in the generalized equation (1) with solution mapping S given by (2), that $\bar{x} \in S(\bar{p})$ and f is strictly differentiable at $(\bar{p}, \bar{x})$. Assume that the inverse G^{-1} of the mapping*

$$G(x) = f(\bar{p}, \bar{x}) + \nabla_x f(\bar{p}, \bar{x})(x - \bar{x}) + F(x), \quad \text{with } G(\bar{x}) \ni 0,$$

has a Lipschitz continuous single-valued localization σ around 0 for $\bar{x}$. Then not only do the conclusions of Theorem 2B.5 hold for a solution localization s, but also there is a first-order approximation η to s at $\bar{p}$ given by

$$\eta(p) = \sigma\big(-\nabla_p f(\bar{p}, \bar{x})(p - \bar{p})\big).$$

Moreover, if $F \equiv 0$, then the first-order approximation η is strict and given by

$$\eta(p) = \bar{x} - \nabla_x f(\bar{p},\bar{x})^{-1} \nabla_p f(\bar{p},\bar{x})(p - \bar{p}), \tag{21}$$

so that s is strictly differentiable at $\bar{p}$.

Proof. In this case Theorem 2B.8 is applicable with h taken to be the linearization of $f(\bar{p},\cdot)$ at $\bar{x}$ and r taken to be the linearization of $f(\cdot,\bar{x})$ at $\bar{p}$. When $F \equiv 0$, we get $\sigma(y) = \bar{x} + \nabla_x f(\bar{p},\bar{x})^{-1} y$, so that η as defined in (16) achieves the form in (21). Having a strict first-order approximation by an affine function means strict differentiability. □

The second part of Corollary 2B.9 shows how the implicit function theorem for equations as stated in Theorem 1D.13 is covered as a special case of Theorem 2B.7.

In the case of the generalized equation (1) where $f(p,x) = g(x) - p$ for a function $g: \mathbb{R}^n \to \mathbb{R}^m$ ($d = m$), so that

$$S(p) = \{x \mid p \in g(x) + F(x)\} = (g + F)^{-1}(p), \tag{22}$$

the inverse function version of Theorem 2B.8 has the following symmetric form.

Theorem 2B.10 (inverse version). *In the framework of the solution mapping (22), consider any pair $(\bar{p},\bar{x})$ with $\bar{x} \in S(\bar{p})$. Let h be any strict first-order approximation to g at $\bar{x}$. Then $(g+F)^{-1}$ has a Lipschitz continuous single-valued localization s around $\bar{p}$ for $\bar{x}$ if and only if $(h+F)^{-1}$ has such a localization σ around $\bar{p}$ for $\bar{x}$, in which case σ is a first-order approximation of s at $\bar{p}$ and*

$$\operatorname{lip}(s;\bar{p}) = \operatorname{lip}(\sigma;\bar{p}). \tag{23}$$

If, in addition, σ is affine, $\sigma(y) = \bar{x} + Ay$, then s is strictly differentiable at $\bar{p}$ with $Ds(\bar{p}) = A$.

Proof. For the "if" part, suppose that $(h+F)^{-1}$ has a localization σ as described. Then, from (15) with $f(p,x) = -p + g(x)$ we get $\operatorname{lip}(s;\bar{p}) \leq \operatorname{lip}(\sigma;0)$. The "only if" part is completely analogous because g and h play symmetric roles in the statement, and yields $\operatorname{lip}(\sigma;\bar{p}) \leq \operatorname{lip}(s';0)$ for some single-valued localization s' of S. The localizations s and s' have to agree graphically around $(0,\bar{x})$, so we pass to a smaller localization, again called s, and get the equality in (23). Through the observation that $r(p) = g(\bar{x}) - p + \bar{p}$, the rest follows from Theorem 2B.8(d). □

We also can modify the results presented so far in this section in the direction indicated in Section 1F, where we considered local selections instead of single-valued localizations. We state such a result here as an exercise.

Exercise 2B.11 (implicit selections). Let $S = \{x \in \mathbb{R}^n \mid f(p,x) + F(x) \ni 0\}$ for a function $f: \mathbb{R}^d \times \mathbb{R}^n \to \mathbb{R}^m$ and a mapping $F: \mathbb{R}^n \rightrightarrows \mathbb{R}^m$, along with a pair $(\bar{p},\bar{x})$

such that $\bar{x} \in S(\bar{p})$, and suppose that $\widehat{\mathrm{lip}}_p(f;(\bar{p},\bar{x})) \leq \gamma < \infty$. Let h be a strict first-order approximation of f with respect to x at $(\bar{p},\bar{x})$ for which $(h+F)^{-1}$ has a Lipschitz continuous local selection σ around 0 for $\bar{x}$ with $\mathrm{lip}(\sigma;0) \leq \kappa$. Then S has a Lipschitz continuous local selection s around $\bar{p}$ for $\bar{x}$ with

$$\mathrm{lip}(s;\bar{p}) \leq \kappa\gamma.$$

If in addition f has a first-order approximation r with respect to p at $(\bar{p},\bar{x})$, then there exists a neighborhood Q of $\bar{p}$ such that the function

$$\eta : p \mapsto \sigma(-r(p) + f(\bar{p},\bar{x})) \quad \text{for } p \in Q$$

is a first-order approximation of s at $\bar{p}$.

Guide. First verify the following statement, which is a simple modification of Lemma 2B.6. For a function $\varphi : \mathbb{R}^d \times \mathbb{R}^n \to \mathbb{R}^m$ and a point $(\bar{p},\bar{x}) \in \mathrm{int\,dom\,}\varphi$, let the nonnegative scalars v, b, the positive scalar a, and the set $Q \subset \mathbb{R}^d$ be such that $\bar{p} \in Q$ and the conditions (7) hold. Consider also a set-valued mapping $M : \mathbb{R}^m \rightrightarrows \mathbb{R}^n$ with $(\bar{y},\bar{x}) \in \mathrm{gph}\,M$, where $\bar{y} := \varphi(\bar{p},\bar{x})$, and assume that there exists a Lipschitz continuous function r on $\mathbb{B}_{va+b}(\bar{y})$ such that

$$r(y) \in M(y) \cap \mathbb{B}_a(\bar{x}) \text{ for } y \in \mathbb{B}_{va+b}(\bar{y}) \quad \text{and} \quad r(\bar{y}) = \bar{x}.$$

In addition, suppose now that the Lipschitz constant λ for the function r is such that the conditions (a) and (b) in the statement of Lemma 2B.6 are fulfilled. Then for each $p \in Q$ the set $\{x \in \mathbb{B}_a(\bar{x}) \mid x \in M(\varphi(p,x))\}$ contains a point $s(p)$ such that the function $p \mapsto s(p)$ satisfies $s(\bar{p}) = \bar{x}$ and

$$(24) \quad |s(p') - s(p)| \leq \frac{\lambda}{1-\lambda v} |\varphi(p',s(p)) - \varphi(p,s(p))| \quad \text{for all } p',p \in Q.$$

Thus, the mapping $N := p \mapsto \{x \mid x \in M(\varphi(p,x))\} \cap \mathbb{B}_a(\bar{x})$ has a local selection s around $\bar{p}$ for $\bar{x}$ which satisfies (24). The difference from Lemma 2B.6 is that r is now required only to be a local selection of the mapping M with specified properties, and then we obtain a local selection s of N at $\bar{p}$ for $\bar{x}$. For the rest use the proofs of Theorems 2B.5 and 2B.8. □

2C. Ample Parameterization and Parametric Robustness

The results in 2B, especially the broad generalization of Robinson's theorem in 2B.5 and its complement in 2B.8 dealing with solution approximations, provide a substantial extension of the classical theory of implicit functions. Equations have been replaced by generalized equations, with variational inequalities as a particular case, and technical assumptions about differentiability have been greatly relaxed. Much of the rest of this chapter will be concerned with working out the consequences in situations where additional structure is available. Here, however, we reflect on the ways that parameters enter the picture and the issue of whether there are "enough" parameters, which emerges as essential in drawing good conclusions about solution mappings.

The differences in parameterization between an inverse function theorem and an implicit function theorem are part of a larger pattern which deserves, at this stage, a closer look. Let's start by considering a generalized equation without parameters,

$$(1) \qquad g(x) + F(x) \ni 0,$$

for a function $g : \mathbb{R}^n \to \mathbb{R}^m$ and a set-valued mapping $F : \mathbb{R}^n \rightrightarrows \mathbb{R}^m$. We can think of a *parameterization* as the choice of a function

$$(2) \qquad f : \mathbb{R}^d \times \mathbb{R}^n \to \mathbb{R}^m \text{ having } f(\bar{p}, x) \equiv g(x) \text{ for a particular } \bar{p} \in \mathbb{R}^d.$$

The specification of such a parameterization leads to an associated solution mapping

$$(3) \qquad S : p \mapsto \{x \mid f(p,x) + F(x) \ni 0\},$$

which we proceed to study around $\bar{p}$ and a point $\bar{x} \in S(\bar{p})$ for the presence of a nice localization σ. Different parameterizations yield different solution mappings, which may possess different properties according to the assumptions placed on f.

That's the general framework, but the special kind of parameterization that corresponds to the "inverse function" case has a fundamental role which is worth trying to understand more fully. In that case, we simply have $f(p,x) = g(x) - p$ in (2), so that in (1) we are solving $g(x) + F(x) \ni p$ and the solution mapping is $S = (g + F)^{-1}$. Interestingly, this kind of parameterization comes up even in obtaining "implicit function" results through the way that approximations are utilized. Recall that in Theorem 2B.5, for a function h which is "close" to $f(\bar{p}, \cdot)$ near $\bar{x}$, the mapping $(h + F)^{-1}$ having $\bar{x} \in (h + F)^{-1}(0)$ is required to have a Lipschitz continuous single-valued localization around 0 for $\bar{x}$. Only then are we able to deduce that the solution mapping S in (3) has a localization of such type at $\bar{p}$ for $\bar{x}$. In other words, the desired conclusion about S is obtained from an assumption about a simpler solution mapping in the "inverse function" category.

When S itself already belongs to that category, because $f(p,x) = g(x) - p$ and $S = (g + F)^{-1}$, another feature of the situation emerges. Then, as seen in Theorem 2B.10, the assumption made about $(h + F)^{-1}$ is not merely sufficient for obtaining the desired localization of S but actually necessary. This distinction was already observed in the classical setting. In the "symmetric" version of the inverse function theorem in 1C.3, the invertibility of a linearized mapping is both necessary and sufficient for the conclusion, whereas such invertibility acts only as a sufficient condition in the implicit function theorem 1B.2 (even though that theorem and the basic version of the inverse function theorem 1A.1 are equivalent to each other).

2 Implicit Function Theorems for Variational Problems

An implicit function theorem for generalized equations that exhibits necessity as well as sufficiency in its assumptions can nonetheless be derived. For this, the parameterization must be "rich" enough.

Ample parameterization. *A parameterization of the generalized equation (1) as in (2) will be called ample at $\bar{x}$ if $f(p,x)$ is strictly differentiable with respect to p uniformly in x at $(\bar{p},\bar{x})$ and the partial Jacobian $\nabla_p f(\bar{p},\bar{x})$ is of full rank:*

$$(4) \qquad \operatorname{rank} \nabla_p f(\bar{p},\bar{x}) = m.$$

The reason why the rank condition in (4) can be interpreted as ensuring the richness of the parameterization is that it can always be achieved through supplementary parameters. Any parameterization function f having the specified strict differentiability can be extended to a parameterization function $\tilde{f}$ with

$$(5) \qquad \tilde{f}(q,x) = f(p,x) - y, \quad q = (p,y),\ \bar{q} = (\bar{p},0),$$

which does satisfy the ampleness condition, since trivially $\operatorname{rank} \nabla_q \tilde{f}(\bar{q},\bar{x}) = m$. The generalized equation being solved then has solution mapping

$$(6) \qquad \widetilde{S} : (p,y) \mapsto \{x \mid f(p,x) + F(x) \ni y\}.$$

Results about localizations of $\widetilde{S}$ can be specialized to results about S by taking $y = 0$.

In order to arrive at the key result about ample parameterization, asserting an equivalence about the existence of several kinds of localizations, we need a lemma about local selections which is related to the results presented in Section 1F.

Lemma 2C.1. *Let $f : \mathbb{R}^d \times \mathbb{R}^n \to \mathbb{R}^m$ with $(\bar{p},\bar{x}) \in \operatorname{int} \operatorname{dom} f$ afford an ample parameterization of the generalized equation (1) at $\bar{x}$. Suppose that f has a strict first-order approximation $h : \mathbb{R}^n \to \mathbb{R}^m$ with respect to x uniformly in p at $(\bar{p},\bar{x})$. Then the mapping*

$$(7) \qquad \Psi : (x,y) \mapsto \{p \mid e(p,x) + y = 0\} \quad \text{for } (x,y) \in \mathbb{R}^n \times \mathbb{R}^m,$$

where $e(p,x) = f(p,x) - h(x)$, has a local selection ψ around $(\bar{x},0)$ for $\bar{p}$ which satisfies

$$(8a) \qquad \widehat{\operatorname{lip}}_x(\psi;(\bar{x},0)) = 0$$

and

$$(8b) \qquad \widehat{\operatorname{lip}}_y(\psi;(\bar{x},0)) < \infty.$$

Proof. Let $A = \nabla_p f(\bar{p},\bar{x})$; then $AA^\top$ is invertible. Without loss of generality, suppose $\bar{x} = 0$, $\bar{p} = 0$, and $f(0,0) = 0$; then $h(0) = 0$. Let $c = |A^\top(AA^\top)^{-1}|$. Let $0 < \varepsilon < 1/(2c)$ and choose a positive a such that for all $x, x' \in a\mathbb{B}$ and $p, p' \in a\mathbb{B}$ we have

(9) $$|e(p,x') - e(p,x)| \leq \varepsilon |x - x'|$$

and

(10) $$|f(p,x) - f(p',x) - A(p-p')| \leq \varepsilon |p - p'|.$$

For $b = a(1 - 2c\varepsilon)/c$, fix $x \in a\mathbb{B}$ and $y \in b\mathbb{B}$, and consider the mapping

$$\Phi_{x,y} : p \mapsto -A^\mathsf{T}(AA^\mathsf{T})^{-1}(e(p,x) + y - Ap) \quad \text{for } p \in a\mathbb{B}.$$

Through (9) and (10), keeping in mind that $e(0,0) = 0$, we see that

$$|\Phi_{x,y}(0)| \leq c|e(0,x) + y| \leq c|e(0,x) - e(0,0)| + c|y| \leq c\varepsilon a + cb = a(1 - c\varepsilon),$$

and for any $p, p' \in a\mathbb{B}$

$$|\Phi_{x,y}(p) - \Phi_{x,y}(p')| \leq c|f(p,x) - f(p',x) - A(p-p')| \leq c\varepsilon |p - p'|.$$

The contraction mapping principle 1A.2 then applies, and we obtain from it the existence of a unique $p \in a\mathbb{B}$ such that

(11) $$p = -A^\mathsf{T}(AA^\mathsf{T})^{-1}(e(p,x) + y - Ap).$$

We denote by $\psi(x,y)$ the unique solution in $a\mathbb{B}$ of this equation for $x \in a\mathbb{B}$ and $y \in b\mathbb{B}$. Multiplying both sides of (11) by A and simplifying, we get $e(p,x) + y = 0$. This means that for each $(x,y) \in a\mathbb{B} \times b\mathbb{B}$ the equation $e(p,x) + y = 0$ has $\psi(x,y)$ as a solution. From (11), we know that

(12) $$\psi(x,y) = -A^\mathsf{T}(AA^\mathsf{T})^{-1}(f(\psi(x,y),x) - h(x) + y - A\psi(x,y)).$$

Let $x, x' \in a\mathbb{B}$ and $y, y' \in b\mathbb{B}$. Using (9) and (10) we have

$$\begin{aligned}|\psi(x,y) - \psi(x',y)| &\leq c|e(\psi(x,y),x) - e(\psi(x,y),x')| \\ &\quad + c|f(\psi(x,y),x') - f(\psi(x',y),x') - A(\psi(x,y) - \psi(x',y))| \\ &\leq c\varepsilon |x - x'| + c\varepsilon |\psi(x,y) - \psi(x',y)|.\end{aligned}$$

Hence

$$|\psi(x,y) - \psi(x',y)| \leq \frac{c\varepsilon}{1 - c\varepsilon}|x - x'|.$$

Since ε can be arbitrarily small, we conclude that (8a) holds. Analogously, from (12) and using again (9) and (10) we obtain

$$\begin{aligned}|\psi(x,y) &- \psi(x,y')| \\ &\leq c|f(\psi(x,y),x) + y - A\psi(x,y) - f(\psi(x,y'),x) - y' + A\psi(x,y')| \\ &\leq c|y - y'| + c\varepsilon |\psi(x,y) - \psi(x,y')|,\end{aligned}$$

and then

2 Implicit Function Theorems for Variational Problems

$$|\psi(x,y) - \psi(x,y')| \leq \frac{c}{1-c\varepsilon}|y-y'|$$

which gives us (8b). □

We are now ready to present the first main result of this section:

Theorem 2C.2 (equivalences from ample parameterization). *Let f parameterize the generalized equation (1) as in (2). Suppose the parameterization is ample at $(\bar{p},\bar{x})$, and let h be a strict first-order approximation of f with respect to x uniformly in p at $(\bar{p},\bar{x})$. Then the following properties are equivalent.*

(a) *S in (3) has a Lipschitz continuous single-valued localization around $\bar{p}$ for $\bar{x}$.*

(b) *$(h+F)^{-1}$ has a Lipschitz continuous single-valued localization around 0 for $\bar{x}$.*

(c) *$(g+F)^{-1}$ has a Lipschitz continuous single-valued localization around 0 for $\bar{x}$.*

(d) *$\widetilde{S}$ in (6) has a Lipschitz continuous single-valued localization around $(\bar{p},0)$ for $\bar{x}$.*

Proof. If the mapping $(h+F)^{-1}$ has a Lipschitz continuous single-valued localization around 0 for $\bar{x}$, then from Theorem 2B.5 together with Theorem 2B.10 we may conclude that the other three mappings likewise have such localizations at the respective reference points. In other words, (b) is sufficient for (a) and (d). Also, (b) is equivalent to (c) inasmuch as g and h are first-order approximations to each other (Theorem 2B.10). Since (d) implies (a), the issue is whether (b) is necessary for (a).

Assume that (a) holds with a Lipschitz localization s around $\bar{p}$ for $\bar{x}$ and choose $\lambda > \text{lip}(s;\bar{p})$. Let $\nu > 0$ be such that $\lambda \nu < 1$, and consider a Lipschitz continuous local selection ψ of the mapping Ψ in Lemma 2C.1. Then there exist positive a, b and c such that $\lambda \nu a + \lambda b < a$,

$$S(p) \cap \mathbb{B}_a(\bar{x}) = s(p) \quad \text{for } p \in \mathbb{B}_{\nu a+b}(\bar{p}),$$

$$|s(p) - s(p')| \leq \lambda |p - p'| \quad \text{for } p, p' \in \mathbb{B}_{\nu a+b}(\bar{p}),$$

$$h(x) - f(\psi(x,y),x) = y \quad \text{for } y \in \mathbb{B}_c(0), x \in \mathbb{B}_a(\bar{x}),$$

$$|\psi(x,y) - \psi(x',y)| \leq \nu|x-x'| \quad \text{for } x,x' \in \mathbb{B}_a(\bar{x}) \text{ and } y \in \mathbb{B}_c(0),$$

the last from (8a), and

$$|\psi(y,\bar{x}) - \psi(0,\bar{x})| \leq b \quad \text{for } y \in \mathbb{B}_c(0).$$

We now apply Lemma 2B.6 with $\varphi(p,x) = \psi(x,y)$ for $p = y$ and $M(p) = S(p)$, thereby obtaining that the mapping

$$\mathbb{B}_c(0) \ni y \mapsto \{x \in \mathbb{B}_a(\bar{x}) \,|\, x \in S(\psi(x,y))\}$$

is a function which is Lipschitz continuous on $\mathbb{B}_c(0)$. Noting that

$$(h+F)^{-1}(y) \cap \mathbb{B}_a(\bar{x}) = \{x \,|\, x = S(\psi(x,y)) \cap \mathbb{B}_a(\bar{x})\},$$

we conclude that $(h+F)^{-1}$ has a Lipschitz continuous single-valued localization around 0 for $\bar{x}$. Thus, (a) implies (b). □

The strict differentiability property with respect to p which is assumed in the definition of ample parameterization is satisfied of course when f is strictly differentiable with respect to (p,x) at $(\bar{p},\bar{x})$. Then, moreover, the linearization of $f(\bar{p},\cdot)$ at $\bar{x}$, which is the same as the linearization of g at $\bar{x}$, can be taken as the function h. This leads to a statement about an entire class of parameterizations.

Theorem 2C.3 (parametric robustness). *Consider the generalized equation (1) under the assumption that $\bar{x}$ is a point where g is strictly differentiable. Let $h(x) = g(\bar{x}) + \nabla g(\bar{x})(x-\bar{x})$. Then the following statements are equivalent.*

(a) $(h+F)^{-1}$ *has a Lipschitz continuous single-valued localization around* 0 *for* $\bar{x}$.

(b) *For every parameterization (2) in which f is strictly differentiable at $(\bar{x},\bar{p})$, the mapping S in (3) has a Lipschitz continuous single-valued localization around $\bar{p}$ for $\bar{x}$.*

Proof. The implication from (a) to (b) already follows from Theorem 2B.8. The focus is on the reverse implication. This is valid because, among the parameterizations covered by (b), there will be some that are ample. For instance, one could pass from a given one to an ample parameterization in the mode of (5). For the solution mapping for such a parameterization, we have the implication from (a) to (b) in Theorem 2C.2. That specializes to what we want. □

A solution $\bar{x}$ to the generalized equation (1) is said to be *parametrically robust* when the far-reaching property in (b) of Theorem 2C.3 holds. In that terminology, Theorem 2C.3 gives a criterion for parametric robustness.

2D. Semidifferentiable Functions

The notion of a first-order approximation of a function at a given point has already served us for various purposes as a substitute for differentiability, where the approximation is a linearization. We now bring in an intermediate concept in which linearity is replaced by positive homogeneity.

A function $\varphi : \mathbb{R}^n \to \mathbb{R}^m$ is *positively homogeneous* if $0 = \varphi(0)$ and $\varphi(\lambda w) = \lambda \varphi(w)$ for all $w \in \text{dom } \varphi$ and $\lambda > 0$. These conditions mean geometrically that the graph of φ is a cone in $\mathbb{R}^n \times \mathbb{R}^m$. A linear function is positively homogeneous in particular, of course. The graph of a linear function $\varphi : \mathbb{R}^n \to \mathbb{R}^m$ is actually a subspace of $\mathbb{R}^n \times \mathbb{R}^m$.

2 Implicit Function Theorems for Variational Problems

Semiderivatives. A function $f : \mathbb{R}^n \to \mathbb{R}^m$ is said to be *semidifferentiable*[4] at $\bar{x}$ if it has a first-order approximation at $\bar{x}$ of the form $h(x) = f(\bar{x}) + \varphi(x - \bar{x})$ with φ continuous and positively homogeneous; when the approximation is strict, f is *strictly semidifferentiable* at $\bar{x}$. Either way, the function φ, necessarily unique, is called the *semiderivative* of f at $\bar{x}$ and denoted by $Df(\bar{x})$, so that $h(x) = f(\bar{x}) + Df(\bar{x})(x - \bar{x})$.

In a first-order approximation we have by definition that $\mathrm{clm}\,(f - h)(\bar{x}) = 0$, which in the "strict" case is replaced by $\mathrm{lip}\,(f - h)(\bar{x}) = 0$. The uniqueness of the semiderivative, when it exists, comes from the fact that any two first-order approximations $f(\bar{x}) + \varphi(x - \bar{x})$ and $f(\bar{x}) + \psi(x - \bar{x})$ of f at $\bar{x}$ must have $\mathrm{clm}\,(\varphi - \psi)(0) = 0$, and under positive homogeneity that cannot hold without having $\varphi = \psi$. The uniqueness can also be gleaned through comparison with directional derivatives.

One-sided directional derivatives. For $f : \mathbb{R}^n \to \mathbb{R}^m$, a point $\bar{x} \in \mathrm{dom}\, f$ and a vector $w \in \mathbb{R}^n$, the limit

$$(1) \qquad f'(\bar{x}; w) = \lim_{t \searrow 0} \frac{f(\bar{x} + tw) - f(\bar{x})}{t},$$

when it exists, is the *(one-sided) directional derivative* of f at $\bar{x}$ for w; here $t \searrow 0$ means that $t \to 0$ with $t > 0$. If this directional derivative exists for every w, f is said to be *directionally differentiable* at $\bar{x}$.

Note that $f'(\bar{x}; w)$ is positively homogeneous in the w argument. This comes out of the limit definition itself. Directional differentiability is weaker than semidifferentiability in general, but equivalent to it in the presence of Lipschitz continuity, as we demonstrate next.

Proposition 2D.1. *If a function $f : \mathbb{R}^n \to \mathbb{R}^m$ is semidifferentiable at $\bar{x}$, then f is in particular directionally differentiable at $\bar{x}$ and has*

$$(2) \qquad Df(\bar{x})(w) = f'(\bar{x}; w) \text{ for all } w,$$

so that the first-order approximation in the definition of semidifferentiability has the form

$$h(x) = f(\bar{x}) + f'(\bar{x}; x - \bar{x}).$$

When $\mathrm{lip}\,(f; \bar{x}) < \infty$, directional differentiability at $\bar{x}$ in turn implies semidifferentiability at $\bar{x}$.

Proof. Having $\mathrm{clm}\,(f - h)(\bar{x}) = 0$ for $h(x) = f(\bar{x}) + \varphi(x - \bar{x})$ as in the definition of semidifferentiability entails having $[f(\bar{x} + tw) - h(\bar{x} + tw)]/t \to 0$ as $t \searrow 0$ with $h(\bar{x} + tw) = f(\bar{x}) + t\varphi(w)$. Thus, $\varphi(w)$ must be $f'(\bar{x}; w)$.

For the converse claim, consider $\lambda > \mathrm{lip}\,(f; \bar{x})$ and observe that for any u and v,

---

[4] Also called Bouligand differentiable or B-differentiable under the additional assumption of Lipschitz continuity, see the book Facchinei and Pang [2003].

(3) $\quad |f'(\bar{x};u) - f'(\bar{x};v)| = \lim_{\substack{t \to 0 \\ t > 0}} \frac{1}{t}|f(\bar{x}+tu) - f(\bar{x}+tv)| \leq \lambda |u-v|.$

Next, consider an arbitrary sequence $u_k \to 0$ and, without loss of generality, assume that $u_k/|u_k| \to \bar{u}$ with $|\bar{u}| = 1$. Letting $t_k = |u_k|$ and using the positive homogeneity of the directional derivative, we obtain

$$0 \leq \frac{1}{|u_k|}|f(\bar{x}+u_k) - f(\bar{x}) - f'(\bar{x};u_k)|$$

$$\leq \frac{1}{t_k}\bigg(|f(\bar{x}+u_k) - f(\bar{x}+t_k\bar{u})| + |f'(\bar{x};t_k\bar{u}) - f'(\bar{x};u_k)|$$

$$+ |f(\bar{x}+t_k\bar{u}) - f(\bar{x}) - f'(\bar{x};t_k\bar{u})|\bigg)$$

$$\leq 2\lambda \left|\frac{u_k}{t_k} - \bar{u}\right| + \left|\frac{1}{t_k}(f(\bar{x}+t_k\bar{u}) - f(\bar{x})) - f'(\bar{x};\bar{u})\right|,$$

where in the final inequality we invoke (3). Since u_k is arbitrarily chosen, we conclude by passing to the limit as $k \to \infty$ that for $h(x) = f(\bar{x}) + f'(\bar{x}; x - \bar{x})$ we do have clm$(f-h;\bar{x}) = 0$. □

When the semiderivative $Df(\bar{x}) : \mathbb{R}^n \to \mathbb{R}^m$ is linear, semidifferentiability turns into differentiability, and strict semidifferentiability turns into strict differentiability. The connections known between $Df(\bar{x})$ and the calmness modulus and Lipschitz modulus of f at $\bar{x}$ under differentiability can be extended to semidifferentiability by adopting the definition that

$$|\varphi| = \sup_{|x|\leq 1} |\varphi(x)| \text{ for a positively homogeneous function } \varphi.$$

We then have clm$(Df(\bar{x});0) = |Df(\bar{x})|$ and consequently clm$(f;\bar{x}) = |Df(\bar{x})|$, which in the case of strict semidifferentiability becomes lip$(f;\bar{x}) = |Df(\bar{x})|$. Thus in particular, semidifferentiability of f at $\bar{x}$ implies that clm$(f;\bar{x}) < \infty$, while strict semidifferentiability at $\bar{x}$ implies that lip$(f;\bar{x}) < \infty$.

Exercise 2D.2 (alternative characterization of semidifferentiability). *For a function $f : \mathbb{R}^n \to \mathbb{R}^m$ and a point $\bar{x} \in \text{dom } f$, semidifferentiability is equivalent to the existence for every $w \in \mathbb{R}^n$ of*

(4) $\quad \lim_{\substack{t \searrow 0 \\ w' \to w}} \frac{f(\bar{x}+tw') - f(\bar{x})}{t}.$

Guide. Directional differentiability of f at $\bar{x}$ corresponds to the difference quotient functions $\Delta_t(w) = [f(\bar{x}+tw) - f(\bar{x})]/t$ converging pointwise to something, namely $f'(\bar{x};\cdot)$, as $t \searrow 0$. Show that the existence of the limits in (4) means that these functions converge to $f'(\bar{x};\cdot)$ not just pointwise, but uniformly on bounded sets. Glean from that the equivalence with having a first-order approximation as in the definition of semidifferentiability. □

Examples.
1) The function $f(x) = e^{|x|}$ for $x \in \mathbb{R}$ is not differentiable at 0, but it is semidifferentiable there and its semiderivative is given by $Df(0) : w \mapsto |w|$. This is actually a case of strict semidifferentiability. Away from 0, f is of course continuously differentiable (hence strictly differentiable).

2) The function $f(x_1, x_2) = \min\{x_1, x_2\}$ on $\mathbb{R}^2$ is continuously differentiable at every point away from the line where $x_1 = x_2$. On that line, f is strictly semidifferentiable with

$$Df(x_1, x_2)(w_1, w_2) = \min\{w_1, w_2\}.$$

3) A function of the form $f(x) = \max\{f_1(x), f_2(x)\}$, with f_1 and f_2 continuously differentiable from $\mathbb{R}^n$ to $\mathbb{R}$, is strictly differentiable at all points x where $f_1(x) \neq f_2(x)$ and semidifferentiable where $f_1(x) = f_2(x)$, the semiderivative being given there by

$$Df(x)(w) = \max\{Df_1(x)(w), Df_2(x)(w)\}.$$

However, f might not be strictly semidifferentiable at such points; see Example 2D.5 below.

The semiderivative obeys standard calculus rules, such as semidifferentiation of a sum, product and ratio, and, most importantly, the chain rule. We pose the verification of these rules as exercises.

Exercise 2D.3. *Let f be semidifferentiable at $\bar{x}$ and let g be Lipschitz continuous and semidifferentiable at $\bar{y} := f(\bar{x})$. Then $g \circ f$ is semidifferentiable at $\bar{x}$ and*

$$D(g \circ f)(\bar{x}) = Dg(\bar{y}) \circ Df(\bar{x}).$$

Guide. Apply Proposition 1E.1 and observe that a composition of positively homogeneous functions is positively homogeneous. □

Exercise 2D.4. *Let f be strictly semidifferentiable at $\bar{x}$ and g be strictly differentiable at $f(\bar{x})$. Then $g \circ f$ is strictly semidifferentiable at $\bar{x}$.*

Guide. Apply 1E.2. □

Example 2D.5. The functions f and g in Exercise 2D.4 cannot exchange places: the composition of a strictly semidifferentiable function with a strictly differentiable function is not always strictly semidifferentiable. For a counterexample, consider the function $f : \mathbb{R}^2 \to \mathbb{R}$ given by

$$f(x_1, x_2) = \min\{x_1^3, x_2\} \quad \text{for } (x_1, x_2) \in \mathbb{R}^2.$$

According to 2D.3, the function f is semidifferentiable at $(0,0)$ with semiderivative $Df(0,0)(w_1, w_2) = \min\{0, w_2\}$. To see that f is not strictly semidifferentiable at $(0,0)$, however, observe for the function $g = f - Df(0,0)$ that

$$\frac{|g(x_1', x_2') - g(x_1, x_2)|}{|(x_1', x_2') - (x_1, x_2)|} = \frac{1}{1 + 2\varepsilon} \quad \text{for } (x_1, x_2) = (-\varepsilon, -\varepsilon^3/2) \text{ and } (x_1', x_2') = (-\varepsilon, \varepsilon^4).$$

As ε goes to 0 this ratio tends to 1, and therefore $\operatorname{lip}(f - Df(0,0); (0,0)) \geq 1$.

Our aim now is to forge out of Theorem 2B.8 a result featuring semiderivatives. For this purpose, we note that if $f(p, x)$ is (strictly) semidifferentiable at $(\bar{p}, \bar{x})$ jointly in its two arguments, it is also "partially (strictly) semidifferentiable" in these arguments separately. In denoting the semiderivative of $f(\bar{p}, \cdot)$ at $\bar{x}$ by $D_x f(\bar{p}, \bar{x})$ and the semiderivative of $f(\cdot, \bar{x})$ at $\bar{p}$ by $D_p f(\bar{p}, \bar{x})$, we have

$$D_x f(\bar{p}, \bar{x})(w) = Df(\bar{p}, \bar{x})(0, w), \qquad D_p f(\bar{p}, \bar{x})(q) = Df(\bar{p}, \bar{x})(q, 0).$$

In contrast to the situation for differentiability, however, $Df(\bar{p}, \bar{x})(q, w)$ isn't necessarily the sum of these two partial semiderivatives.

Theorem 2D.6 (implicit function theorem utilizing semiderivatives). *Let $\bar{x} \in S(\bar{p})$ for the solution mapping*

$$S: p \mapsto \{x \in \mathbb{R}^n \mid f(p, x) + F(x) \ni 0\}$$

associated with a choice of $F: \mathbb{R}^n \rightrightarrows \mathbb{R}^m$ and $f: \mathbb{R}^d \times \mathbb{R}^n \to \mathbb{R}^m$ such that f is strictly semidifferentiable at $(\bar{p}, \bar{x})$. Suppose that the inverse G^{-1} of the mapping

$$G(x) = f(\bar{p}, \bar{x}) + D_x f(\bar{p}, \bar{x})(x - \bar{x}) + F(x), \quad \text{with } G(\bar{x}) \ni 0,$$

has a Lipschitz continuous single-valued localization σ around 0 for $\bar{x}$ which is semidifferentiable at 0. Then S has a Lipschitz continuous single-valued localization s around $\bar{p}$ for $\bar{x}$ which is semidifferentiable at $\bar{p}$ with its semiderivative given by

$$Ds(\bar{p}) = D\sigma(0) \circ (-D_p f(\bar{p}, \bar{x})).$$

Proof. First, note that $s(\bar{p}) = \sigma(0) = \bar{x}$ and the function r in Theorem 2B.8 may be chosen as $r(p) = f(\bar{x}, \bar{p}) + D_p f(\bar{p}, \bar{x})(p - \bar{p})$. Then we have

$$|s(p) - s(\bar{p}) - (D\sigma(0) \circ (-D_p f(\bar{p}, \bar{x})))(p - \bar{p})| \leq |s(p) - \sigma(-r(p) + r(\bar{p}))|$$
$$+ |\sigma(-D_p f(\bar{p}, \bar{x})(p - \bar{p})) - \sigma(0) - D\sigma(0)(-D_p f(\bar{p}, \bar{x})(p - \bar{p}))|.$$

According to Theorem 2B.8 the function $p \mapsto \sigma(-r(p) + r(\bar{p}))$ is a first-order approximation to s at $\bar{p}$, hence the first term on the right side of this inequality is of order $o(|p - \bar{p}|)$ when p is close to $\bar{p}$. The same is valid for the second term, since σ is assumed to be semidifferentiable at 0. It remains to observe that the composition of positively homogeneous mappings is positively homogeneous. $\square$

An important class of semidifferentiable functions will be brought in next.

Piecewise smooth functions. *A function $f : \mathbb{R}^n \to \mathbb{R}^m$ is said to be piecewise smooth on an open set $O \subset \text{dom } f$ if it is continuous on O and for each $x \in O$ there is a finite collection $\{f_i\}_{i \in I}$ of smooth ($\mathscr{C}^1$) functions defined on a neighborhood of x such that, for some $\varepsilon > 0$, one has*

$$(5) \qquad f(y) \in \{f_i(y) \mid i \in I\} \quad \text{when } |y - x| < \varepsilon.$$

The collection $\{f_i\}_{i \in I(x)}$, where $I(x) = \{i \in I \mid f(x) = f_i(x)\}$, is said then to furnish a local representation *of f at x. A local representation in this sense is* minimal *if no proper subcollection of it forms a local representation of f at x.*

Note that a local representation of f at x characterizes f on a neighborhood of x, and minimality means that this would be lost if any of the functions f_i were dropped.

The piecewise smoothness terminology finds its justification in the following observation.

Proposition 2D.7 (decomposition of piecewise smooth functions). *Let f be piecewise smooth on an open set O with a minimal local representation $\{f_i\}_{i \in I}$ at a point $\bar{x} \in O$. Then for each $i \in I(\bar{x})$ there is an open set O_i such that $\bar{x} \in \text{cl } O_i$ and $f(x) = f_i(x)$ on O_i.*

Proof. Let $\varepsilon > 0$ be as in (5) with $x = \bar{x}$ and assume that $\mathbb{B}_\varepsilon(\bar{x}) \subset O$. For each $i \in I(\bar{x})$, let $U_i = \{x \in \text{int}\mathbb{B}_\varepsilon(\bar{x}) \mid f(x) = f_i(x)\}$ and $O_i = \text{int}\mathbb{B}_\varepsilon(\bar{x}) \setminus \cup_{j \neq i} U_j$. Because f and f_i are continuous, U_i is closed relative to $\text{int}\mathbb{B}_\varepsilon(\bar{x})$ and therefore O_i is open. Furthermore $\bar{x} \in \text{cl } O_i$, for if not, the set $\cup_{j \neq i} U_j$ would cover a neighborhood of $\bar{x}$, and then f_i would be superfluous in the local representation, thus contradicting minimality. □

It's not hard to see from this fact that a piecewise smooth function on an open set O must be continuous on O and even locally Lipschitz continuous, since each of the $\mathscr{C}^1$ functions f_i in a local representation is locally Lipschitz continuous, in particular. Semidifferentiability in this situation takes only a little more effort to confirm.

Proposition 2D.8 (semidifferentiability of piecewise smooth functions). *If a function $f : \mathbb{R}^n \to \mathbb{R}^m$ is piecewise smooth on an open set $O \subset \text{dom } f$, then f is semidifferentiable on O. Furthermore, the semiderivative function $Df(\bar{x})$ at any point $\bar{x} \in O$ is itself piecewise smooth, in fact with local representation composed by the linear functions $\{Df_i(\bar{x})\}_{i \in I(\bar{x})}$ when f has local representation $\{f_i\}_{i \in I}$ around $\bar{x}$.*

Proof. We apply the criterion in 2D.2. Consider, for any $\bar{x} \in O$ and $w \in \mathbb{R}^n$, sequences $t_k \searrow 0$ and $w_k \to w$. Invoke a local representation as in (5). Note that because $f(\bar{x} + t_k w_k) \to f(\bar{x})$, we must eventually have $I(\bar{x} + t_k w_k) \subset I(\bar{x})$, so that the difference quotient $\Delta_k = [f(\bar{x} + t_k w_k) - f(\bar{x})]/t_k$ coincides with the difference quotient $\Delta_k^i = [f_i(\bar{x} + t_k w_k) - f_i(\bar{x})]/t_k$ for some $i \in I(\bar{x})$. Since $I(\bar{x})$ is finite, every subsequence of $\{\Delta_k\}_{k=1}^\infty$ must have a subsubsequence coinciding with a subsequence of $\{\Delta_k^i\}_{k=1}^\infty$ for some $i \in I(\bar{x})$ and therefore converging to $\nabla f_i(\bar{x}) \cdot w$ for that i. A sequence with the property that every subsequence contains a convergent subsubsequence necessarily converges as a whole. Thus the sequence $\{\Delta_k\}_{k=1}^{\infty}$ converges with its limit equaling $\nabla f_i(\bar{x}) \cdot w$ for at least one $i \in I(\bar{x})$. This confirms semidifferentiability and establishes that the semiderivative is a selection from $\{\nabla f_i(\bar{x}) \cdot w \mid i \in I(\bar{x})\}$. □

The functions in the examples given after 2D.2 are not only semidifferentiable but also piecewise smooth. Of course, a semidifferentiable function does not have to be piecewise smooth, e.g., when it is a selection of infinitely many, but not finitely many, smooth functions.

A more elaborate example of a piecewise smooth function is the projection mapping P_C on a nonempty, convex and closed set $C \subset \mathbb{R}^n$ specified by finitely many inequalities.

Exercise 2D.9 (piecewise smoothness of special projection mappings). *For a convex set C of the form*

$$C = \{x \in \mathbb{R}^n \mid g_i(x) \leq 0,\ i = 1, \ldots, m\}$$

for convex functions g_i of class $\mathscr{C}^2$ on $\mathbb{R}^n$, let $\bar{x}$ be a point of C at which the gradients $\nabla g_i(\bar{x})$ associated with the active constraints, i.e., the ones with $g_i(\bar{x}) = 0$, are linearly independent. Then there is an open neighborhood O of $\bar{x}$ such that the projection mapping P_C is piecewise smooth on O.

Guide. Since in a sufficiently small neighborhood of $\bar{x}$ the inactive constraints remain inactive, one can assume without loss of generality that $g_i(\bar{x}) = 0$ for all $i = 1, \ldots, m$. Recall that because C is nonempty, closed and convex, P_C is a Lipschitz continuous function from $\mathbb{R}^n$ onto C (see 1D.5). For each u around $\bar{x}$ the projection $P_C(u)$ is the unique solution to the problem of minimizing $\frac{1}{2}|x-u|^2$ in x subject to $g_i(x) \leq 0$ for $i = 1, \ldots, m$. The associated Lagrangian variational inequality (Theorem 2A.9) tells us that when u belongs to a small enough neighborhood of $\bar{x}$, the point x solves the problem if and only if x is feasible and there is a subset J of the index set $\{1, 2, \ldots, m\}$ and Lagrange multipliers $y_i \geq 0$, $i \in J$, such that

(6) $$\begin{cases} x + \sum_{i \in J} y_i \nabla g_i(x)^\mathsf{T} = u, \\ g_i(x) = 0, \quad i \in J. \end{cases}$$

The linear independence of the gradients of the active constraint gradients yields that the Lagrange multiplier vector y is unique, hence it is zero for $u = x = \bar{x}$. For each fixed subset J of the index set $\{1, 2, \ldots, m\}$ the Jacobian of the function on the left of (6) at $(\bar{x}, 0)$ is

$$Q = \begin{pmatrix} I_n + \sum_{i \in J} y_i \nabla^2 g_i(\bar{x}) & \nabla g_J(\bar{x})^\mathsf{T} \\ \nabla g_J(\bar{x}) & 0 \end{pmatrix},$$

where

2 Implicit Function Theorems for Variational Problems

$$\nabla g_J(\bar{x}) = \left[\frac{\partial g_i}{\partial x_j}(\bar{x})\right]_{i\in J, j\in\{1,\ldots,n\}} \quad \text{and } I_n \text{ is the } n\times n \text{ identity matrix.}$$

Since $\nabla g_J(\bar{x})$ has full rank, the matrix Q is nonsingular and then we can apply the classical inverse function theorem (Theorem 1A.1) to the equation (6), obtaining that its solution mapping $u \mapsto (x_J(u), y_J(u))$ has a smooth single-valued localization around $u = \bar{x}$ for $(x,y) = (\bar{x}, 0)$. There are finitely many subsets J of $\{1,\ldots,m\}$, and for each u close to $\bar{x}$ we have $P_C(u) = x_J(u)$ for some J. Thus, the projection mapping P_C is a selection of finitely many smooth functions. □

Exercise 2D.10. *For a set of the form $C = \{x \in \mathbb{R}^n \mid Ax = b \in \mathbb{R}^m\}$, if the $m \times n$ matrix A has linearly independent rows, then the projection mapping is given by*

$$P_C(x) = (I - A^\mathsf{T}(AA^\mathsf{T})^{-1}A)x + A^\mathsf{T}(AA^\mathsf{T})^{-1}b.$$

Guide. The optimality condition (6) in this case leads to the system of equations

$$\begin{pmatrix} x \\ b \end{pmatrix} = \begin{pmatrix} I & A^\mathsf{T} \\ A & 0 \end{pmatrix} \begin{pmatrix} P_C(x) \\ \lambda \end{pmatrix}.$$

Use the identity

$$\begin{pmatrix} I & A^\mathsf{T} \\ A & 0 \end{pmatrix} = \begin{pmatrix} I - A^\mathsf{T}(AA^\mathsf{T})^{-1}A & A^\mathsf{T}(AA^\mathsf{T})^{-1} \\ (AA^\mathsf{T})^{-1}A & -(AA^\mathsf{T})^{-1} \end{pmatrix}^{-1}$$

to reach the desired conclusion. □

2E. Variational Inequalities with Polyhedral Convexity

In this section we apply the theory presented in the preceding sections of this chapter to the parameterized variational inequality

(1) $$f(p,x) + N_C(x) \ni 0$$

where $f : \mathbb{R}^d \times \mathbb{R}^n \to \mathbb{R}^n$ and C is a nonempty, closed and convex subset of $\mathbb{R}^n$. The corresponding solution mapping $S : \mathbb{R}^p \rightrightarrows \mathbb{R}^n$, with

(2) $$S(p) = \{x \mid f(p,x) + N_C(x) \ni 0\},$$

has already been the direct subject of Theorem 2B.1, the implicit function theorem of Robinson. From there we moved on to broader results about solution mappings to generalized equations, but now wish to summarize what those results mean back in the variational inequality setting, and furthermore to explore special features which emerge under additional assumptions on the set C.

Theorem 2E.1 (solution mappings for parameterized variational inequalities). *For a variational inequality (1) and its solution mapping (2), let $\bar{p}$ and $\bar{x}$ be such that $\bar{x} \in S(\bar{p})$. Assume that*
  (a) *f is strictly differentiable at $(\bar{p},\bar{x})$;*
  (b) *the inverse G^{-1} of the mapping*

(3) $\qquad G(x) = f(\bar{p},\bar{x}) + \nabla_x f(\bar{p},\bar{x})(x - \bar{x}) + N_C(x), \quad \text{with } G(\bar{x}) \ni 0,$

has a Lipschitz continuous single-valued localization σ around 0 for $\bar{x}$.

  Then S has a Lipschitz continuous single-valued localization s around $\bar{p}$ for $\bar{x}$ with

$$\text{lip}(s;\bar{p}) \leq \text{lip}(\sigma;0) \cdot |\nabla_p f(\bar{p},\bar{x})|,$$

and this localization s has a first-order approximation η at $\bar{p}$ given by

(4) $\qquad \eta(p) = \sigma(-\nabla_p f(\bar{p},\bar{x})(p - \bar{p})).$

Moreover, under the ample parameterization condition

$$\text{rank } \nabla_p f(\bar{p},\bar{x}) = n,$$

the existence of a Lipschitz continuous single-valued localization s of S around $\bar{p}$ for $\bar{p}$ not only follows from but also necessitates the existence of a localization σ of G^{-1} having the properties described.

Proof. This comes from the application to S of the combination of Theorem 2B.5 and its specialization in Corollary 2B.9, together with the ample parameterization result in Theorem 2C.2. □

If the localization σ that is assumed to exist in Theorem 2E.1 is actually linear, the stronger conclusion is obtained that s is differentiable at $\bar{p}$. But that's a circumstance which can hardly be guaranteed without supposing, for instance, that C is an affine set (given by a system of linear equations). In some situations, however, s could be at least piecewise smooth, as the projection mapping in 2D.9.

Our special goal here is trying to understand better the circumstances in which the existence of a single-valued localization σ of G^{-1} around 0 for $\bar{x}$ of the kind assumed in (b) of Theorem 2E.1 is assured. It's clear from the formula for G in (3) that everything hinges on how a normal cone mapping $N_C : \mathbb{R}^n \rightrightarrows \mathbb{R}^n$ may relate to an affine function $x \mapsto a + Ax$. The key lies in the local geometry of the graph of N_C. We will be able to make important progress in analyzing this geometry by restricting our attention to the following class of sets C.

Polyhedral convex sets. *A set C in $\mathbb{R}^n$ is said to be polyhedral convex when it can be expressed as the intersection of finitely many closed half-spaces and/or hyperplanes.*

In other words, C is a polyhedral convex set when it can be described by a finite set of constraints $f_i(x) \leq 0$ or $f_i(x) = 0$ on affine functions $f_i : \mathbb{R}^n \to \mathbb{R}$. Since an equation $f_i(x) = 0$ is equivalent to the pair of inequalities $f_i(x) \leq 0$ and $-f_i(x) \leq 0$, a polyhedral convex set C is characterized by having a (nonunique) representation of the form

$$(5) \qquad C = \big\{ x \,\big|\, \langle b_i, x \rangle \leq \alpha_i \text{ for } i = 1, \ldots, m \big\}.$$

Any such set must obviously be closed. The empty set $\emptyset$ and the whole space $\mathbb{R}^n$ are regarded as polyhedral convex sets, in particular.

Polyhedral convex *cones* are characterized by having a representation (5) in which $\alpha_i = 0$ for all i. A basic fact about polyhedral convex cones is that they can equally well be represented in another way, which we recall next.

Theorem 2E.2 (Minkowski–Weyl). *A set $K \subset \mathbb{R}^n$ is a polyhedral convex cone if and only if there is a collection of vectors $b_1, \ldots, b_m$ such that*

$$(6) \qquad K = \big\{ y_1 b_1 + \cdots + y_m b_m \,\big|\, y_i \geq 0 \text{ for } i = 1, \ldots, m \big\}.$$

It is easy to see that the cone K^* that is polar to a cone K having a representation of the kind in (6) consists of the vectors x satisfying $\langle b_i, x \rangle \leq 0$ for $i = 1, \ldots, m$. The polar of a polyhedral convex cone having such an inequality representation must therefore have the representation in (6), inasmuch as $(K^*)^* = K$ for any closed, convex cone K. This fact leads to a special description of the tangent and normal cones to a polyhedral convex set.

Theorem 2E.3 (variational geometry of polyhedral convex sets). *Let C be a polyhedral convex set represented as in (5). Let $x \in C$ and $I(x) = \big\{ i \,\big|\, \langle b_i, x \rangle = \alpha_i \big\}$, this being the set of indices of the constraints in (5) that are active at x. Then the tangent and normal cones to C at x are polyhedral convex, with the tangent cone having the representation*

$$(7) \qquad T_C(x) = \big\{ w \,\big|\, \langle b_i, w \rangle \leq 0 \text{ for } i \in I(x) \big\}$$

and the normal cone having the representation

$$(8) \qquad N_C(x) = \Big\{ v \,\Big|\, v = \sum_{i=1}^m y_i b_i \text{ with } y_i \geq 0 \text{ for } i \in I(x),\, y_i = 0 \text{ for } i \notin I(x) \Big\}.$$

Furthermore, the tangent cone has the properties that

$$(9) \qquad W \cap [C - x] = W \cap T_C(x) \text{ for some neighborhood } W \text{ of } 0$$

and

$$(10) \qquad T_C(x) \supset T_C(\bar{x}) \text{ for all } x \text{ in some neighborhood } U \text{ of } \bar{x}.$$

Proof. The formula (7) for $T_C(x)$ follows from (5) just by applying the definition of the tangent cone in 2A. Then from (7) and the preceding facts about polyhedral cones and polarity, utilizing also the relation in 2A(8), we obtain (8). The equality (9) is deduced simply by comparing (5) and (7). To obtain (10), observe that $I(x) \subset I(\bar{x})$ for x close to $\bar{x}$ and then the inclusion follows from (7). □

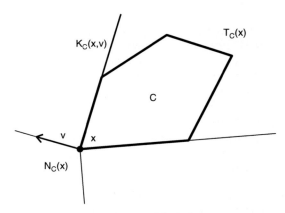

Fig. 2.2 Tangent, normal and critical cones to a polyhedral set.

The normal cone mapping N_C associated with a polyhedral convex set C has a special property which will be central to our analysis. It revolves around the following notion.

Critical cone. *For a convex set C, any $x \in C$ and any $v \in N_C(x)$, the critical cone to C at x for v is*
$$K_C(x,v) = \{w \in T_C(x) \mid w \perp v\}.$$

If C is polyhedral, then $K_C(x,v)$ is polyhedral as well, as seen immediately from the representation in (7).

Lemma 2E.4 (reduction lemma). *Let C be a polyhedral convex set in $\mathbb{R}^n$, and let*
$$\bar{x} \in C, \quad \bar{v} \in N_C(\bar{x}), \quad K = K_C(\bar{x}, \bar{v}).$$

The graphical geometry of the normal cone mapping N_C around $(\bar{x}, \bar{v})$ reduces then to the graphical geometry of the normal cone mapping N_K around $(0,0)$, in the sense that
$$O \cap [\text{gph } N_C - (\bar{x}, \bar{v})] = O \cap \text{gph } N_K \quad \text{for some neighborhood } O \text{ of } (0,0).$$

In other words, one has

(11) $\quad \bar{v} + u \in N_C(\bar{x} + w) \iff u \in N_K(w) \quad \text{for } (w, u) \text{ sufficiently near to } (0,0).$

2 Implicit Function Theorems for Variational Problems

Proof. Since we are only involved with local properties of C around one of its points $\bar{x}$, and $C - \bar{x}$ agrees with the cone $T_C(\bar{x})$ around 0 by Theorem 2E.3, we can assume without loss of generality that $\bar{x} = 0$ and C is a cone, and $T_C(\bar{x}) = C$. Then, in terms of the polar cone C^* (which likewise is polyhedral on the basis of Theorem 2E.2), we have the characterization from 2A.3 that

(12) $\quad v \in N_C(w) \iff w \in N_{C^*}(v) \iff w \in C, \ v \in C^*, \ \langle v, w \rangle = 0.$

In particular for our focus on the geometry of gph N_C around $(0, \bar{v})$, we have from (12) that

(13) $\quad N_{C^*}(\bar{v}) = \{ w \in C \,|\, \langle \bar{v}, w \rangle = 0 \} = K.$

We know on the other hand from 2E.3 that $U \cap [C^* - \bar{v}] = U \cap T_{C^*}(\bar{v})$ for some neighborhood U of 0, where moreover $T_{C^*}(\bar{v})$ is polar to $N_{C^*}(\bar{v})$, hence equal to K^* by (13). Thus, there is a neighborhood O of $(0,0)$ such that

(14) $\quad$ for $(w, u) \in O: \quad \bar{v} + u \in N_C(w) \iff w \in C, \ u \in K^*, \ \langle \bar{v} + u, w \rangle = 0.$

This may be compared with the fact that

(15) $\quad u \in N_K(w) \iff w \in K, \ u \in K^*, \ \langle u, w \rangle = 0.$

Our goal (in the context of $\bar{x} = 0$) is to show that (14) reduces to (11), at least when the neighborhood O in (14) is chosen still smaller, if necessary. Because of (15), this comes down to demonstrating that $\langle \bar{v}, w \rangle = 0$ in the circumstances of (14).

We can take C to be represented by

(16) $\quad C = \{ w \,|\, \langle b_i, w \rangle \leq 0 \text{ for } i = 1, \ldots, m \},$

in which case, as observed after 2E.2, the polar C^* is represented by

(17) $\quad C^* = \{ y_1 b_1 + \cdots + y_m b_m \,|\, y_i \geq 0 \text{ for } i = 1, \ldots, m \}.$

The relations in (12) can be coordinated with these representations as follows. For each index set $I \subset \{1, \ldots, m\}$, consider the polyhedral convex cones

$$W_I = \{ w \in C \,|\, \langle b_i, w \rangle = 0 \text{ for } i \in I \}, \quad V_I = \{ \textstyle\sum_{i \in I} y_i b_i \text{ with } y_i \geq 0 \},$$

with $W_\emptyset = C$ and $V_\emptyset = \{0\}$. Then $v \in N_C(w)$ if and only if, for some I, one has $w \in W_I$ and $v \in V_I$. In other words, gph N_C is the union of the finitely many polyhedral convex cones $G_I = W_I \times V_I$ in $\mathbb{R}^n \times \mathbb{R}^n$.

Among these cones G_I, we will only be concerned with the ones containing $(0, \bar{v})$. Let $\mathscr{I}$ be the collection of index sets $I \subset \{1, \ldots, m\}$ having that property. According to (9) in 2E.3, there exists for each $I \in \mathscr{I}$ a neighborhood O_I of $(0, 0)$ such that $O_I \cap [G_I - (0, \bar{v})] = O_I \cap T_{G_I}(0, \bar{v})$. Furthermore, $T_{G_I}(0, \bar{v}) = W_I \times T_{V_I}(\bar{v})$. This has the crucial consequence that when $\bar{v} + u \in N_C(w)$ with (w, u) near enough to $(0, 0)$, we also have $\bar{v} + \tau u \in N_C(w)$ for all $\tau \in [0,1]$. Since having $\bar{v} + \tau u \in N_C(w)$ entails having $\langle \bar{v} + \tau u, w \rangle = 0$ through (12), this implies that $\langle \bar{v}, w \rangle = -\tau \langle u, w \rangle$ for all $\tau \in [0,1]$. Hence $\langle \bar{v}, w \rangle = 0$, as required. We merely have to shrink the neighborhood O in (14) to lie within every O_I for $I \in \mathscr{I}$. □

Example 2E.5. The nonnegative orthant $\mathbb{R}^n_+$ is a polyhedral convex cone in $\mathbb{R}^n$, since it consists of the vectors $x = (x_1, \ldots, x_n)$ satisfying the linear inequalities $x_j \geq 0$, $j = 1, \ldots, n$. For $v = (v_1, \ldots, v_n)$, one has

$$v \in N_{\mathbb{R}^n_+}(x) \iff x_j \geq 0,\ v_j \leq 0,\ x_j v_j = 0 \text{ for } j = 1, \ldots, n.$$

Thus, whenever $v \in N_{\mathbb{R}^n_+}(x)$ one has in terms of the index sets

$$J_1 = \{j \,|\, x_j > 0,\ v_j = 0\},$$
$$J_2 = \{j \,|\, x_j = 0,\ v_j = 0\},$$
$$J_3 = \{j \,|\, x_j = 0,\ v_j < 0\}$$

that the vectors $w = (w_1, \ldots, w_n)$ belonging to the critical cone to $\mathbb{R}^n_+$ at x for v are characterized by

$$w \in K_{\mathbb{R}^n_+}(x,v) \iff \begin{cases} w_j \text{ free} & \text{for } j \in J_1, \\ w_j \geq 0 & \text{for } j \in J_2, \\ w_j = 0 & \text{for } j \in J_3. \end{cases}$$

In the developments ahead, we will make use of not only critical cones but also certain subspaces.

Critical subspaces. The smallest linear subspace that includes the critical cone $K_C(x,v)$ will be denoted by $K_C^+(x,v)$, whereas the smallest linear subspace that is included in $K_C(x,v)$ will be denoted by $K_C^-(x,v)$, the formulas being

(18) $\quad K_C^+(x,v) = K_C(x,v) - K_C(x,v) = \{w - w' \,|\, w, w' \in K_C(x,v)\},$
$\quad\quad\ \ K_C^-(x,v) = K_C(x,v) \cap [-K_C(x,v)] = \{w \in K_C(x,v) \,|\, -w \in K_C(x,v)\}.$

The formulas follow from the fact that $K_C(x,v)$ is already a convex cone. Obviously, $K_C(x,v)$ is itself a subspace if and only if $K_C^+(x,v) = K_C^-(x,v)$.

Theorem 2E.6 (affine-polyhedral variational inequalities). *For an affine function $x \mapsto a + Ax$ from $\mathbb{R}^n$ into $\mathbb{R}^n$ and a polyhedral convex set $C \subset \mathbb{R}^n$, consider the variational inequality*

$$a + Ax + N_C(x) \ni 0.$$

Let $\bar{x}$ be a solution and let $\bar{v} = -a - A\bar{x}$, so that $\bar{v} \in N_C(\bar{x})$, and let $K = K_C(\bar{x}, \bar{v})$ be the associated critical cone. Then for the mappings

(19) $\quad\quad G(x) = a + Ax + N_C(x) \text{ with } G(\bar{x}) \ni 0,$
$\quad\quad\quad\ \ G_0(w) = Aw + N_K(w) \text{ with } G_0(0) \ni 0,$

2 Implicit Function Theorems for Variational Problems

the following properties are equivalent:

(a) G^{-1} *has a Lipschitz continuous single-valued localization* σ *around* 0 *for* $\bar{x}$,
(b) G_0^{-1} *is a single-valued mapping with all of* $\mathbb{R}^n$ *as its domain*,

in which case G_0^{-1} *is necessarily Lipschitz continuous globally and the function* $\sigma(v) = \bar{x} + G_0^{-1}(v)$ *furnishes the localization in* (a). *Moreover, in terms of critical subspaces* $K^+ = K_C^+(\bar{x}, \bar{v})$ *and* $K^- = K_C^-(\bar{x}, \bar{v})$, *the following condition is sufficient for* (a) *and* (b) *to hold:*

(20) $\qquad w \in K^+, \quad Aw \perp K^-, \quad \langle w, Aw \rangle \leq 0 \quad \Longrightarrow \quad w = 0.$

Proof. According to reduction lemma 2E.4, we have, for (w, u) in some neighborhood of $(0,0)$, that $\bar{v} + u \in N_C(\bar{x} + w)$ if and only if $u \in N_K(w)$. In the change of notation from u to $v = u + Aw$, this means that, for (w, v) in a neighborhood of $(0,0)$, we have $v \in G(\bar{x} + w)$ if and only if $v \in G_0(w)$. Thus, the existence of a Lipschitz continuous single-valued localization σ of G^{-1} around 0 for $\bar{x}$ as in (a) corresponds to the existence of a Lipschitz continuous single-valued localization σ_0 of G_0^{-1} around 0 for 0; the relationship is given by $\sigma(v) = \bar{x} + \sigma_0(v)$. But whenever $v \in G_0(w)$ we have $\lambda v \in G_0(\lambda w)$ for all $\lambda > 0$, i.e., the graph of G_0 is a cone. Therefore, when σ_0 exists it can be scaled arbitrarily large and must correspond to G_0^{-1} being a single-valued mapping with all of $\mathbb{R}^n$ as its domain.

We claim next that when G_0^{-1} is single-valued everywhere it is necessarily Lipschitz continuous. This comes out of the argument pursued in the proof of 2E.4 in analyzing the graph of N_C, which applies equally well to N_K, inasmuch as K is a polyhedral convex cone. Specifically, the graph of N_K is the union of finitely many polyhedral convex cones in $\mathbb{R}^n \times \mathbb{R}^n$. The same also holds then for the graphs of G_0 and G_0^{-1}. It remains only to observe that if a single-valued mapping has its graph composed of the union of finitely many polyhedral convex sets it has to be Lipschitz continuous (prove or see 3D.6).

This leaves us with verifying that the condition in (20) is sufficient for G_0^{-1} to be single-valued with all of $\mathbb{R}^n$ as its domain. We note in preparation for this that

(21) $\qquad (K^+)^\perp = K^* \cap (-K^*) = (K^*)^-, \qquad (K^-)^\perp = K^* - K^* = (K^*)^+.$

We first argue that if $w_1 \in G_0^{-1}(v)$ and $w_2 \in G_0^{-1}(v)$, then $v - Aw_1 \in N_K(w_1)$ and $v - Aw_2 \in N_K(w_2)$. This entails having

$$w_1 \in K, \quad v - Aw_1 \in K^*, \quad \langle w_1, v - Aw_1 \rangle = 0,$$
$$w_2 \in K, \quad v - Aw_2 \in K^*, \quad \langle w_2, v - Aw_2 \rangle = 0,$$

with $\langle w_1, v - Aw_2 \rangle \leq 0$ and $\langle w_2, v - Aw_1 \rangle \leq 0$. Then $w_1 - w_2 \in K - K = K^+$ and $-A(w_1 - w_2) \in K^* - K^* = (K^-)^\perp$, with $\langle w_1 - w_2, A(w_1 - w_2) \rangle = \langle w_1 - w_2, [v - Aw_2] - [v - Aw_2] \rangle \leq 0$. Under our condition (20), these relations require $w_1 - w_2 = 0$. Thus, (20) guarantees that $G_0^{-1}(v)$ can never contain more than a single w.

Working toward showing that (20) guarantees also that $\text{dom}\, G_0^{-1} = \mathbb{R}^n$, we next consider the case where $\text{dom}\, G_0^{-1}$ omits some point $\tilde{v}$ and analyze what that would imply. Again we utilize the fact that the graph of G_0^{-1} is the union of finitely many polyhedral convex cones in $\mathbb{R}^n \times \mathbb{R}^n$. Under the mapping $(v,w) \to v$, each of them projects onto a cone in $\mathbb{R}^n$; the union of these cones is dom G_0^{-1}. Since the image of a polyhedral convex cone under a linear transformation is another polyhedral convex cone, in consequence of 2E.2 (since the image of a cone generated by finitely many vectors is another such cone), and polyhedral convex cones are closed sets in particular, this ensures that dom G_0^{-1} is closed. Then there is certain to exist a point $v_0 \in$ dom G_0^{-1} that is closest to $\tilde{v}$; for all $\tau > 0$ sufficiently small, we have $v_0 + \tau(\tilde{v} - v_0) \notin$ dom G_0^{-1}. For each of the polyhedral convex cones D in the finite union making up dom G_0^{-1}, if $v_0 \in D$, then v_0 must be the projection $P_D(\tilde{v})$, so that $\tilde{v} - v_0$ must belong to $N_D(v_0)$ (cf. relation (4) in 2A). It follows that, for some neighborhood U of v_0, we have

$$\text{(22)} \qquad \langle \tilde{v} - v_0, u - v_0 \rangle \leq 0 \text{ for all } u \in U \cap \text{dom } G_0^{-1}.$$

Consider any $w_0 \in G_0^{-1}(v_0)$; this means $v_0 - Aw_0 \in N_K(w_0)$. Let K_0 be the critical cone to K at w_0 for $v_0 - Aw_0$:

$$\text{(23)} \qquad K_0 = \{ w' \in T_K(w_0) \, | \, w' \perp (v_0 - Aw_0) \}.$$

In the line of argument already pursued, the geometry of the graph of N_K around $(w_0, v_0 - Aw_0)$ can be identified with that of the graph of N_{K_0} around $(0,0)$. Equivalently, the geometry of the graph of $G_0^{-1} = (A + N_K)^{-1}$ around (v_0, w_0) can be identified with that of $(A + N_{K_0})^{-1}$ around $(0,0)$; for (v',w') near enough to $(0,0)$, we have $w_0 + w' \in G_0^{-1}(v_0 + v')$ if and only if $w' \in (A + N_{K_0})^{-1}(v')$. Because of (22) holding for the neighborhood U of v_0, this implies that $\langle \tilde{v} - v_0, v' \rangle \leq 0$ for all $v' \in$ dom $(A + N_{K_0})^{-1}$ close to 0. Thus,

$$\text{(24)} \qquad \langle \tilde{v} - v_0, Aw' + u' \rangle \leq 0 \text{ for all } w' \in K_0 \text{ and } u' \in N_{K_0}(w').$$

The case of $w' = 0$ has $N_{K_0}(w') = K_0^*$, so (24) implies in particular that $\langle \tilde{v} - v_0, u' \rangle \leq 0$ for all $u' \in K_0^*$, so that $\tilde{v} - v_0 \in (K_0^*)^* = K_0$. On the other hand, since $u' = 0$ is always one of the elements of $N_{K_0}(w')$, we must have from (24) that $\langle \tilde{v} - v_0, Aw' \rangle \leq 0$ for all $w' \in K_0$. Here $\langle \tilde{v} - v_0, Aw' \rangle = \langle A^\top(\tilde{v} - v_0), w' \rangle$ for all $w' \in K_0$, so this means $A^\top(\tilde{v} - v_0) \in K_0^*$. In summary, (24) requires, among other things, having

$$\text{(25)} \qquad \begin{array}{c} \tilde{v} - v_0 \in K_0 \text{ and } A^\top(\tilde{v} - v_0) \in K_0^*, \\ \text{hence in particular } \langle A^\top(\tilde{v} - v_0), \tilde{v} - v_0 \rangle \leq 0. \end{array}$$

We observe now from the formula for K_0 in (23) that $K_0 \subset T_K(w_0)$, where furthermore $T_K(w_0)$ is the cone generated by the vectors $w - w_0$ with $w \in K$ and hence lies in $K - K$. Therefore $K_0 \subset K^+$. On the other hand, because $T_K(w_0)$ and $N_K(w_0)$ are polar to each other by 2E.3, we have from (23) that K_0 is polar to the cone comprised by all differences $v - \tau(v_0 - Aw_0)$ with $v \in N_K(w_0)$ and $\tau \geq 0$, which is again polyhedral. That cone of differences must then in fact be K_0^*. Since we have taken v_0 and w_0 to satisfy $v_0 - Aw_0 \in N_K(w_0)$, and also $N_K(w_0) \subset K^*$, it follows that

$K_0^* \subset K^* - K^* = (K^-)^\perp$. Thus, (25) implies that $\tilde{v} - v_0 \in K^+$, $A^T(\tilde{v} - v_0) \in (K^-)^\perp$, with $\langle A^T(\tilde{v} - v_0), \tilde{v} - v_0 \rangle \leq 0$. In consequence, (25) would be impossible if we knew that

(26) $\qquad w \in K^+, \quad A^T w \perp K^-, \quad \langle A^T w, w \rangle \leq 0 \implies w = 0.$

Our endgame will be to demonstrate that (26) is actually equivalent to condition (20).

Of course, $\langle A^T w, w \rangle$ is the same as $\langle w, Aw \rangle$. For additional comparison between (20) and (26), we can simplify matters by expressing $\mathbb{R}^n$ as $W_1 \times W_2 \times W_3$ for the linear subspaces $W_1 = K^-$, $W_3 = (K^+)^\perp$, and W_2 the orthogonal complement of K^- within K^+. Any vector $w \in \mathbb{R}^n$ corresponds then to a triple (w_1, w_2, w_3) in this product, and there are linear transformations $A_{ij} : W_j \to W_i$ such that

$$Aw \longleftrightarrow (A_{11}w_1 + A_{12}w_2 + A_{13}w_3, A_{21}w_1 + A_{22}w_2 + A_{23}w_3, A_{31}w_1 + A_{32}w_2 + A_{33}w_3).$$

In this schematic, (20) has the form

(27) $\quad A_{11}w_1 + A_{12}w_2 = 0, \quad \langle w_2, A_{21}w_1 + A_{22}w_2 \rangle \leq 0 \implies w_1 = 0, \; w_2 = 0,$

whereas (26) has the form

(28) $\quad A_{11}^T w_1 + A_{21}^T w_2 = 0, \quad \langle w_2, A_{12}^T w_1 + A_{22}^T w_2 \rangle \leq 0 \implies w_1 = 0, \; w_2 = 0.$

In particular, through the choice of $w_2 = 0$, (27) insists that the only w_1 with $A_{11}w_1 = 0$ is $w_1 = 0$. Thus, A_{11} must be nonsingular. Then the initial equation in (27) can be solved for w_1, yielding $w_1 = -A_{11}^{-1}A_{12}w_2$, and this expression can be substituted into the inequality, thereby reducing the condition to

$$\langle w_2, (A_{22} - A_{21}A_{11}^{-1}A_{12})w_2 \rangle \leq 0 \implies w_2 = 0.$$

In the same manner, (28) comes out as the nonsingularity of A_{11}^T and the property that

$$\langle w_2, (A_{22}^T - A_{12}^T(A_{11}^T)^{-1}A_{21}^T)w_2 \rangle \leq 0 \implies w_2 = 0.$$

Since the nonsingularity of A_{11}^T is equivalent to that of A_{11}, and

$$A_{22}^T - A_{12}^T(A_{11}^T)^{-1}A_{21}^T = (A_{22} - A_{21}A_{11}^{-1}A_{12})^T,$$

the equivalence of (27) and (28) is now evident. $\qquad\square$

Example 2E.7.

(a) When the critical cone K in Theorem 2E.6 is a subspace, the condition in (20) reduces to the nonsingularity of the linear transformation $K \ni w \mapsto P_K(Aw)$, where P_K is the projection onto K.

(b) When the critical cone K in Theorem 2E.6 is pointed, in the sense that $K \cap (-K) = \{0\}$, the condition in (20) reduces to the requirement that $\langle w, Aw \rangle > 0$ for all nonzero $w \in K^+$.

(c) Condition (20) always holds when A is the identity matrix.

Theorem 2E.6 tells us that, for a polyhedral convex set C, the assumption in (b) of Theorem 2E.1 is equivalent to the critical cone $K = K_C(\bar{x}, -f(\bar{p},\bar{x}))$ being such that the inverse G_0^{-1} of the mapping $G_0 : w \mapsto \nabla_x f(\bar{x},\bar{p})w + N_K(w)$ is single-valued from all of $\mathbb{R}^n$ into itself and hence Lipschitz continuous globally. Furthermore, Theorem 2E.6 provides a sufficient condition for this to hold. In putting these facts together with observations about the special nature of G_0, we obtain a powerful fact which distinguishes variational inequalities with polyhedral convexity from other variational inequalities.

Theorem 2E.8 (localization criterion under polyhedral convexity). *For a variational inequality (1) and its solution mapping (2) under the assumption that C is polyhedral convex and f is strictly differentiable at $(\bar{p},\bar{x})$, with $\bar{x} \in S(\bar{p})$, let*

$$A = \nabla_x f(\bar{p},\bar{x}) \quad \text{and} \quad K = K_C(\bar{x},\bar{v}) \text{ for } \bar{v} = -f(\bar{p},\bar{x}).$$

Suppose that for each $u \in \mathbb{R}^n$ there is a unique solution $w = \bar{s}(u)$ to the auxiliary variational inequality $Aw - u + N_K(w) \ni 0$, this being equivalent to saying that

(29) $\qquad \bar{s} = (A + N_K)^{-1}$ *is everywhere single-valued,*

in which case the mapping $\bar{s}$ is Lipschitz continuous globally. (A sufficient condition for this assumption to hold is the property in (20) with respect to the critical subspaces $K^+ = K_C^+(\bar{x},\bar{v})$ and $K^- = K_C^-(\bar{x},\bar{v})$.)

Then S has a Lipschitz continuous single-valued localization s around $\bar{p}$ for $\bar{x}$ which is semidifferentiable with

(30) $\qquad \text{lip}(s;\bar{p}) \leq \text{lip}(\bar{s};0) \cdot |\nabla_p f(\bar{p},\bar{x})|, \qquad Ds(\bar{p})(q) = \bar{s}(-\nabla_p f(\bar{p},\bar{x})q).$

Moreover, under the ample parameterization condition, rank $\nabla_p f(\bar{p},\bar{x}) = n$, condition (29) is not only sufficient but also necessary for a Lipschitz continuous single-valued localization of S around $\bar{p}$ for $\bar{x}$.

Proof. We merely have to combine the observation made before this theorem's statement with the statement of Theorem 2E.1. According to formula (4) in that theorem for the first-order approximation η of s at $\bar{p}$, we have $\eta(\bar{p}+q) - \bar{x} = \bar{s}(-\nabla_p f(\bar{p},\bar{x})q)$. Because K is a cone, the mapping N_K is positively homogeneous, and the same is true then for $A + N_K$ and its inverse, which is $\bar{s}$. Thus, the function $q \mapsto \bar{s}(-\nabla_p f(\bar{p},\bar{x})q)$ gives a first-order approximation to $s(\bar{p}+q) - s(\bar{p})$ at $q = 0$ that is positively homogeneous. We conclude that s is semidifferentiable at $\bar{p}$ with this function furnishing its semiderivative, as indicated in (30). □

As a special case of Example 2E.7(a), if $C = \mathbb{R}^n$ the result in Theorem 2E.8 reduces once more to a version of the classical implicit function theorem. Further insights into solution mappings associated with variational inequalities will be gained in Chapter 4.

Exercise 2E.9. *Prove that the projection mapping P_C associated with a polyhedral convex set C is Lipschitz continuous and semidifferentiable everywhere, with its semiderivative being given by*

$$DP_C(x)(u) = P_K(u) \text{ for } K = K_C(P_C(x), x - P_C(x)).$$

Guide. Use the relation between the projection mapping and the normal cone mapping given in formula 2A(4). □

Additional facts about critical cones, which will be useful later, can be developed from the special geometric structure of polyhedral convex sets.

Proposition 2E.10 (local behavior of critical cones and subspaces). *Let $C \subset \mathbb{R}^n$ be a polyhedral convex set, and let $\bar{v} \in N_C(\bar{x})$. Then the following properties hold:*
   (a) $K_C(x,v) \subset K_C^+(\bar{x}, \bar{v})$ *for all* $(x,v) \in$ gph N_C *in some neighborhood of* $(\bar{x}, \bar{v})$.
   (b) $K_C(x,v) = K_C^+(\bar{x}, \bar{v})$ *for some* $(x,v) \in$ gph N_C *in each neighborhood of* $(\bar{x}, \bar{v})$.

Proof. By appealing to 2E.3 as in the proof of 2E.4, we can reduce to the case where $\bar{x} = 0$ and C is a cone. Theorem 2E.2 then provides a representation in terms of a collection of nonzero vectors $b_1, \ldots, b_m$, in which C consists of all linear combinations $y_1 b_1 + \cdots + y_m b_m$ with coefficients $y_i \geq 0$, and the polar cone C^* consists of all v such that $\langle b_i, v \rangle \leq 0$ for all i. We know from 2A.3 that, at any $x \in C$, the normal cone $N_C(x)$ is formed by the vectors $v \in C^*$ such that $\langle x, v \rangle = 0$, so that $N_C(x)$ is the cone that is polar to the one comprised of all vectors $w - \lambda x$ with $w \in C$ and $\lambda \geq 0$. Since the latter cone is again polyhedral (in view of Theorem 2E.2), hence closed, it must in turn be the cone polar to $N_C(x)$ and therefore equal to $T_C(x)$. Thus,

$$T_C(x) = \{y_1 b_1 + \cdots + y_m b_m - \lambda x \mid y_i \geq 0, \lambda \geq 0\} \text{ for any } x \in C.$$

On the other hand, in the notation

(31) $\quad I(v) = \{i \mid \langle b_i, v \rangle = 0\},$
$\quad\quad F(v) = \{y_1 b_1 + \cdots + y_m b_m \mid y_i \geq 0 \text{ for } i \in I(v), y_i = 0 \text{ for } i \notin I(v)\},$

we see that for $v \in C^*$ we have $F(v) = \{x \in C \mid v \in N_C(x)\}$, i.e.,

$$v \in N_C(x) \iff x \in F(v).$$

Then too, for such x and v, the critical cone $K_C(x,v) = \{w \in T_C(x) \mid \langle w, v \rangle = 0\}$ we have

(32) $\quad\quad\quad K_C(x,v) = \{w - \lambda x \mid w \in F(v), \lambda \geq 0\},$

and actually $K_C(\bar{x}, \bar{v}) = F(\bar{v})$ (inasmuch as $\bar{x} = 0$). In view of the fact, evident from (31), that

$$I(v) \subset I(\bar{v}) \text{ and } F(v) \subset F(\bar{v}) \text{ for all } v \text{ near enough to } \bar{v},$$

we have, for v in some neighboorhood of $\bar{v}$, that

$$x \in F(\bar{v}) \text{ and } K_C(x,v) \subset \{w - \lambda x \mid w \in F(\bar{v}), \lambda \geq 0\} \text{ when } v \in N_C(x).$$

In that case $K_C(x,v) \subset F(\bar{v}) - F(\bar{v}) = K_C(\bar{x},\bar{v}) - K_C(\bar{x},\bar{v}) = K_C^+(\bar{x},\bar{v})$, so (a) is valid.

To confirm (b), it will be enough now to demonstrate that, arbitrarily close to $\bar{x} = 0$, we can find a vector $\tilde{x}$ for which $K_C(\tilde{x},\bar{v}) = F(\bar{v}) - F(\bar{v})$. Here $F(\bar{v})$ consists by definition of all nonnegative linear combinations of the vectors b_i with $i \in I(\bar{v})$, whereas $F(\bar{v}) - F(\bar{v})$ is the subspace consisting of all linear combinations. For arbitrary $\varepsilon > 0$, let $\tilde{x} = \tilde{y}_1 b_1 + \cdots + \tilde{y}_m b_m$ with $\tilde{y}_i = \varepsilon$ for $i \in I(\bar{v})$ but $\tilde{y}_i = 0$ for $i \notin I(\bar{v})$. Then $K_C(\tilde{x},\bar{v})$, equaling $\{w - \lambda \tilde{x} \mid w \in F(\bar{v}), \lambda \geq 0\}$ by (32), consists of all linear combinations of the vectors b_i for $i \in I(\bar{v})$ in which the coefficients have the form $y_i - \lambda \varepsilon$ with $y_i \geq 0$ and $\lambda \geq 0$. Can any given choice of coefficients y_i' for $i \in I(\bar{v})$ be obtained in this manner? Yes, by taking λ high enough that $y_i' + \lambda \varepsilon \geq 0$ for all $i \in I(\bar{v})$ and then setting $y_i = y_i' + \lambda \varepsilon$. This completes the argument. □

2F. Variational Inequalities with Monotonicity

Our attention shifts now from special properties of the set C in a variational inequality to special properties of the function f and their effect on solutions.

Monotone functions. A function $f : \mathbb{R}^n \to \mathbb{R}^n$ is said to be *monotone* on a set $C \subset \text{dom } f$ if C is convex and

(1) $$\langle f(x') - f(x), x' - x \rangle \geq 0 \text{ for all } x, x' \in C.$$

It is *strongly monotone* on C if there exists $\mu > 0$ such that

(2) $$\langle f(x') - f(x), x' - x \rangle \geq \mu |x' - x|^2 \text{ for all } x, x' \in C.$$

Specifically, then f is strongly monotone on C with constant μ.

The name "monotonicity" comes from the following characterization of the defining property.

Exercise 2F.1 (monotonicity along line segments). *Monotonicity of f on a convex set $C \subset \text{dom } f$ means that, for every $\hat{x} \in C$ and $w \in \mathbb{R}^n$ with $|w| = 1$, the function $\varphi(\tau) = \langle f(\hat{x} + \tau w), w \rangle$ is nondecreasing over the (interval of) τ values such that $\hat{x} + \tau w \in C$. Strong monotonicity with constant $\mu > 0$ corresponds to the condition that $\varphi(\tau') - \varphi(\tau) \geq \mu |\tau' - \tau|$ when $\tau' > \tau$.*

An affine function $f(x) = a + Ax$ with $a \in \mathbb{R}^n$ and $A \in \mathbb{R}^{n \times n}$ is monotone on $\mathbb{R}^n$ if and only if $\langle w, Aw \rangle \geq 0$ for all w, i.e., A is positive semidefinite. It is strongly monotone if and only if $\langle w, Aw\rangle > 0$ for all $w \neq 0$, i.e., A is positive definite. These terms make no requirement of symmetry on A. It may be recalled that any square matrix A can be written as a sum $A_s + A_a$ in which A_s is symmetric ($A_s^* = A_s$) and A_a is antisymmetric ($A_a^* = -A_a$), namely with $A_s = \frac{1}{2}[A + A^*]$ and $A_a = \frac{1}{2}[A - A^*]$; then $\langle w, Aw\rangle = \langle w, A_s w\rangle$. The monotonicity of $f(x) = a + Ax$ thus depends only on the symmetric part A_s of A; the antisymmetric part A_a can be anything.

For differentiable functions f that aren't affine, monotonicity has a similar characterization with respect to the Jacobian matrices $\nabla f(x)$.

Exercise 2F.2 (monotonicity from derivatives). *For a function $f: \mathbb{R}^n \to \mathbb{R}^n$ that is continuously differentiable on an open convex set $O \subset$ dom f, verify the following facts.*

(a) *A necessary and sufficient condition for f to be monotone on O is the positive semidefiniteness of $\nabla f(x)$ for all $x \in O$.*

(b) *If $\nabla f(x)$ is positive definite at every point x of a closed, bounded, convex set $C \subset O$, then f is strongly monotone on C.*

(c) *If C is a convex subset of O such that $\langle \nabla f(x)w, w\rangle \geq 0$ for every $x \in C$ and $w \in C - C$, then f is monotone on C.*

(d) *If C is a convex subset of O such that $\langle \nabla f(x)w, w\rangle \geq \mu |w|^2$ for every $x \in C$ and $w \in C - C$, where $\mu > 0$, then f is strongly monotone on C with constant μ.*

Guide. Derive this from the characterizations in 2F.1 by investigating the derivatives of the function $\varphi(\tau)$ introduced there. In proving (c), argue by way of the mean value theorem that $\langle f(x') - f(x), x' - x\rangle$ equals $\langle \nabla f(\tilde{x})(x' - x), x' - x\rangle$ for some point $\tilde{x}$ on the line segment joining x with x'. ☐

Exercise 2F.3 (gradient connections).

(a) *Let g be continuously differentiable from an open set $O \subset \mathbb{R}^n$ to $\mathbb{R}$, and let C be a convex subset of O. Show that the function $f(x) = \nabla g(x)$ is monotone on C if and only if g is convex on C. Show further that f is strongly monotone on C with constant μ if and only if g is strongly convex on C with constant μ.*

(b) *Let h be continuously differentiable from a product $O_1 \times O_2$ of open sets $O_1 \subset \mathbb{R}^{n_1}$ and $O_2 \subset \mathbb{R}^{n_2}$ to $\mathbb{R}$, and let $C_1 \subset O_1$ and $C_2 \subset O_2$ be convex. Show that the function*

$$f(x_1, x_2) = (\nabla_{x_1} h(x_1, x_2), -\nabla_{x_2} h(x_1, x_2))$$

is monotone on $C_1 \times C_2$ if and only if $h(x_1, x_2)$ is convex with respect to $x_1 \in C_1$ for fixed $x_2 \in C_2$, and on the other hand concave with respect to $x_2 \in C_2$ for fixed $x_1 \in C_1$.

Guide. Derive (a) from the characterization of the convexity and strong convexity of g in 2A.5. Proceed similarly in (b), applying also the corresponding characterization of concavity. ☐

We are ready now to develop some special results for variational inequalities

(3) $$f(x) + N_C(x) \ni 0,$$

in which attention is devoted to the case when f is monotone. We work with the basic perturbation scheme in which $f(x)$ is replaced by $f(x) - p$ for a parameter vector $p \in \mathbb{R}^n$. The solution mapping is then

(4) $$S(p) = \{x \,|\, p - f(x) \in N_C(x)\} = (f + N_C)^{-1}(p),$$

with the solution set to (3) then being $S(0)$.

Theorem 2F.4 (solution convexity for monotone variational inequalities). *For a function $f : \mathbb{R}^n \to \mathbb{R}^n$ and a nonempty closed convex set $C \subset \text{dom } f$ relative to which f is monotone and continuous, the solution mapping S in (4) is closed and convex valued. In particular, therefore, the set of solutions (if any) to the variational inequality (3) is not only closed but also convex.*

Proof. It suffices to deal with $S(0)$, since $S(p) = (f_p + N_C)^{-1}(0)$ for $f_p(x) = f(x) - p$ (which is monotone and continuous like f).

The closedness of $S(0)$ already follows from Theorem 2A.1. To see the convexity, consider any two points x_0 and x_1 in $S(0)$. We have $-f(x_0) \in N_C(x_0)$ and $-f(x_1) \in N_C(x_1)$; this is equivalent to

(5) $$\langle f(x_0), x - x_0 \rangle \geq 0 \text{ and } \langle f(x_1), x - x_1 \rangle \geq 0 \text{ for all } x \in C.$$

Let $\bar{x} = (1 - \lambda)x_0 + \lambda x_1$ for any $\lambda \in (0, 1)$. Then $\bar{x} \in C$ by convexity. Consider an arbitrary point $\tilde{x} \in C$. The goal is to show that $\langle f(\bar{x}), \tilde{x} - \bar{x} \rangle \geq 0$, which will confirm that $-f(\bar{x}) \in N_C(\bar{x})$, i.e., that $\bar{x} \in S(0)$.

Taking $t \in (0, 1)$ as a parameter, let $x(t) = \bar{x} + t(\tilde{x} - \bar{x})$ and note that the convexity of C ensures $x(t) \in C$. From the monotonicity of f and the first inequality in (5) we have

$$0 \leq \langle f(x(t)) - f(x_0), x(t) - x_0 \rangle + \langle f(x_0), x(t) - x_0 \rangle = \langle f(x(t)), x(t) - x_0 \rangle.$$

In parallel from the second inequality in (5), we have $0 \leq \langle f(x(t)), x(t) - x_1 \rangle$. Therefore

$$0 \leq (1 - \lambda)\langle f(x(t)), x(t) - x_0 \rangle + \lambda \langle f(x(t)), x(t) - x_1 \rangle$$
$$= \langle f(x(t)), x(t) - (1 - \lambda)x_0 - \lambda x_1 \rangle,$$

where the final expression equals $\langle f(x(t)), x(t) - \bar{x} \rangle = t\langle f(x(t)), \tilde{x} - \bar{x} \rangle$, since $x(t) - \bar{x} = t[\tilde{x} - \bar{x}]$. Thus, $0 \leq \langle f(x(t)), \tilde{x} - \bar{x} \rangle$. Because $x(t) \to \bar{x}$ as $t \to 0$, and f is continuous, we conclude that $\langle f(\bar{x}), \tilde{x} - \bar{x} \rangle \geq 0$, as required. □

In order to add nonemptiness of the solution set to the conclusions of Theorem 2F.4 we need an existence theorem for the variational inequality (3). There is already such a result in 2A.1, but only for bounded sets C. The following result goes beyond that boundedness restriction, without yet imposing any monotonicity assumption on f. When combined with monotonicity, it will have particularly powerful consequences.

Theorem 2F.5 (solution existence for variational inequalities without boundedness). *Consider a function $f : {I\!\!R}^n \to {I\!\!R}^n$ and a nonempty closed convex set $C \subset \mathrm{dom}\, f$ relative to which f is continuous (but not necessarily monotone). Suppose there exist $\hat{x} \in C$ and $\rho > 0$ such that*

(6) $\qquad$ *there is no $x \in C$ with $|x - \hat{x}| \geq \rho$ and $\langle f(x), x - \hat{x}\rangle \leq 0$.*

Then the variational inequality (3) has a solution, and every solution x of (3) satisfies $|x - \hat{x}| < \rho$.

Proof. Any solution x to (3) would have $\langle f(x), x - \hat{x}\rangle \leq 0$ in particular, and then necessarily $|x - \hat{x}| < \rho$ under (6). Hence it will suffice to show that (6) guarantees the existence of at least one solution x to (3) with $|x - \hat{x}| < \rho$.

Let $C_\rho = \{x \in C \,|\, |x - \hat{x}| \leq \rho\}$ and consider the modified variational inequality (3) in which C is replaced by C_ρ. According to Theorem 2A.1, this modified variational inequality has a solution $\bar{x}$. We have $\bar{x} \in C_\rho$ and $-f(\bar{x}) \in N_{C_\rho}(\bar{x})$. From 2A.7(b) we know that $N_{C_\rho}(\bar{x}) = N_C(\bar{x}) + N_B(\bar{x})$ for the ball $B = {I\!\!B}_\rho(\hat{x}) = \{x \,|\, |x - \hat{x}| \leq \rho\}$. Thus,

(7) $\qquad -f(\bar{x}) - w \in N_C(\bar{x})$ for some $w \in N_B(\bar{x})$.

By demonstrating that this implies $w = 0$, we will be able to see that $\bar{x}$ actually satisfies (3).

The normal cone formula for the unit ball in 2A.2(b) extends in an elementary way to the ball B and indicates that w can only be nonzero if $|\bar{x} - \hat{x}| = \rho$ and $w = \lambda[\bar{x} - \hat{x}]$ for some $\lambda > 0$. The normality relation in (7), requiring $0 \geq \langle -f(\bar{x}) - w, x - \bar{x}\rangle$ for all $x \in C$, can be invoked then in the case of $x = \hat{x}$ to obtain $0 \geq \langle -f(\bar{x}) - \lambda[\bar{x} - \hat{x}], \hat{x} - \bar{x}\rangle$, which simplifies to $\langle f(\bar{x}), \bar{x} - \hat{x}\rangle \leq -\lambda \rho^2$. But this is impossible under (6). $\square$

The assumption in (6) is fulfilled trivially when C is bounded, and in that way Theorem 2A.1 is seen to be covered by Theorem 2F.5.

Corollary 2F.6 (uniform local existence). *Consider a function $f : {I\!\!R}^n \to {I\!\!R}^n$ and a nonempty closed convex set $C \subset \mathrm{dom}\, f$ relative to which f is continuous (but not necessarily monotone). Suppose there exist $\hat{x} \in C$, $\rho > 0$ and $\eta > 0$ such that*

(8) $\qquad$ *there is no $x \in C$ with $|x - \hat{x}| \geq \rho$ and $\langle f(x), x - \hat{x}\rangle / |x - \hat{x}| \leq \eta$.*

Then the solution mapping S in (4) has the property that

$$\emptyset \neq S(v) \subset \{x \in C \,|\, |x - \hat{x}| < \rho\} \quad \text{when } |v| \leq \eta.$$

Proof. The stronger assumption here ensures that assumption (6) of Theorem 2F.5 is fulfilled by the function $f_v(x) = f(x) - v$ for every v with $|v| \leq \eta$. Since $S(v) = (f_v + N_C)^{-1}(0)$, this leads to the desired conclusion. □

We can proceed now to take advantage of monotonicity of f on C through the property in 2F.1 and the observation that

$$\langle f(x), x - \hat{x}\rangle / |x - \hat{x}| = \langle f(\hat{x} + \tau w), w\rangle \quad \text{when } x = \hat{x} + \tau w \text{ with } \tau > 0, |w| = 1.$$

Then, for any vector w such that $\hat{x} + \tau w \in C$ for all $\tau \in (0, \infty)$, the expression $\langle f(\hat{x} + \tau w), w\rangle$ is nondecreasing as a function of $\tau \in (0, \infty)$ and thus has a limit (possibly ∞) as $\tau \to \infty$.

Theorem 2F.7 (solution existence for monotone variational inequalities). *Consider a function $f : \mathbb{R}^n \to \mathbb{R}^n$ and a nonempty closed convex set $C \subset \text{dom } f$ relative to which f is continuous and monotone. Let $\hat{x} \in C$ and let W consist of the vectors w with $|w| = 1$ such that $\hat{x} + \tau w \in C$ for all $\tau \in (0, \infty)$, if any.*

(a) If $\lim_{\tau \to \infty} \langle f(\hat{x} + \tau w), w\rangle > 0$ for every $w \in W$, then the solution mapping S in (4) is nonempty-valued on a neighborhood of 0.

(b) If $\lim_{\tau \to \infty} \langle f(\hat{x} + \tau w), w\rangle = \infty$ for every $w \in W$, then the solution mapping S in (4) is nonempty-valued on all of $\mathbb{R}^n$.

Proof. To establish (a), we aim at showing that the limit criterion it proposes is enough to guarantee the condition (8) in Corollary 2F.6. Suppose the latter didn't hold. Then there would be a sequence of points $x_k \in C$ and a sequence of scalars $\eta_k > 0$ such that

$$\langle f(x_k), x_k - \hat{x}\rangle / |x_k - \hat{x}| \leq \eta_k \quad \text{with } |x_k - \hat{x}| \to \infty, \eta_k \to 0.$$

Equivalently, in terms of $\tau_k = |x_k - \hat{x}|$ and $w_k = \tau_k^{-1}(x_k - \hat{x})$ we have $\langle f(\hat{x} + \tau_k w_k), w_k\rangle \leq \eta_k$ with $|w_k| = 1$, $\hat{x} + \tau_k w_k \in C$ and $\tau_k \to \infty$. Without loss of generality we can suppose that $w_k \to w$ for a vector w again having $|w| = 1$. Then for any $\tau > 0$ and k high enough that $\tau_k \geq \tau$, we have from the convexity of C that $\hat{x} + \tau w_k \in C$ and from the monotonicity of f that $\langle f(\hat{x} + \tau w_k), w_k\rangle \leq \eta_k$. On taking the limit as $k \to \infty$ and utilizing the closedness of C and the continuity of f, we get $\hat{x} + \tau w \in C$ and $\langle f(\hat{x} + \tau w), w\rangle \leq 0$. This being true for any $\tau > 0$, we see that $w \in W$ and the limit condition in (a) is violated. The validity of the claim in (a) is thereby confirmed.

The condition in (b) not only implies the condition in (a) but also, by a slight extension of the argument, guarantees that the criterion in Corollary 2F.6 holds for every $\varepsilon > 0$. □

Exercise 2F.8 (Jacobian criterion for existence and uniqueness). *Let $f : \mathbb{R}^n \to \mathbb{R}^n$ and $C \subset \mathbb{R}^n$ be such that f is continuously differentiable on C and monotone relative to C. Fix $\hat{x} \in C$ and let W consist of the vectors w with $|w| = 1$ such that $\hat{x} + \tau w \in C$ for all $\tau \in (0, \infty)$. Suppose there exists $\mu > 0$ such that $\langle \nabla f(x) w, w\rangle \geq \mu$ for every*

$w \in W$ and $x \in C$, if any, when $x \in C$. Then the solution mapping S in (4) is single-valued on all of $\mathbb{R}^n$.

Guide. Argue through the mean value theorem as applied to $\varphi(\tau) = \langle f(\hat{x} + \tau w), w \rangle$ that $\varphi(\tau) = \tau \langle \nabla f(\hat{x} + \theta w) w, w \rangle + \langle f(\hat{x}), w \rangle$ for some $\theta \in (0, \tau)$. Work toward applying the criterion in Theorem 2F.7(b). □

In the perspective of 2F.2(b), the result in 2F.8 seems to come close to invoking strong monotonicity of f in the case where f is continuously differentiable. However, it only involves special vectors w, not every nonzero $w \in \mathbb{R}^n$. For instance, in the affine case where $f(x) = Ax + b$ and $C = \mathbb{R}^n_+$, the criterion obtained from 2F.8 by choosing $\hat{x} = 0$ is simply that $\langle Aw, w \rangle > 0$ for every $w \in \mathbb{R}^n_+$ with $|w| = 1$, whereas strong monotonicity of f would require this for w in $\mathbb{R}^n$, not just $\mathbb{R}^n_+$. In fact, full strong monotonicity has bigger implications than those in 2F.8.

Theorem 2F.9 (variational inequalities with strong monotonicity). *For a function $f : \mathbb{R}^n \to \mathbb{R}^n$ and a nonempty closed convex set $C \subset \operatorname{dom} f$, suppose that f is continuous relative to C and strongly monotone on C with constant $\mu > 0$ in the sense of (2). Then the solution mapping S in (4) is single-valued on all of $\mathbb{R}^n$ and moreover Lipschitz continuous with constant μ^{-1}.*

Proof. The strong monotonicity condition in (2) implies for an arbitrary choice of $\hat{x} \in C$ and $w \in \mathbb{R}^n$ with $|w| = 1$ that $\langle f(\hat{x} + \tau w) - f(\hat{x}), \tau w \rangle \geq \mu \tau^2$ when $\hat{x} + \tau w \in C$. Then $\langle f(\hat{x} + \tau w), w \rangle \geq \langle f(\hat{x}), w \rangle + \mu \tau$, from which it's clear that the limit criterion in Theorem 2F.7(b) is satisfied, so that S is single-valued on all of $\mathbb{R}^n$.

Consider now any two vectors v_0 and v_1 in $\mathbb{R}^n$ and the corresponding solutions $x_0 = S(v_0)$ and $x_1 = S(v_1)$. We have $v_0 - f(x_0) \in N_C(x_0)$ and $v_1 - f(x_1) \in N_C(x_1)$, hence in particular $\langle v_0 - f(x_0), x_1 - x_0 \rangle \leq 0$ and $\langle v_1 - f(x_1), x_0 - x_1 \rangle \leq 0$. The second of these inequalities can also be written as $0 \leq \langle v_1 - f(x_1), x_1 - x_0 \rangle$, and from this we see that $\langle v_0 - f(x_0), x_1 - x_0 \rangle \leq \langle v_1 - f(x_1), x_1 - x_0 \rangle$, which is equivalent to

$$\langle f(x_1) - f(x_0), x_1 - x_0 \rangle \leq \langle v_1 - v_0, x_1 - x_0 \rangle.$$

Since $\langle f(x_1) - f(x_0), x_1 - x_0 \rangle \geq \mu |x_1 - x_0|^2$ by our assumption of strong monotonicity, while $\langle v_1 - v_0, x_1 - x_0 \rangle \leq |v_1 - v_0||x_1 - x_0|$, it follows that $|x_1 - x_0| \leq \mu^{-1} |v_1 - v_0|$. This verifies the claimed Lipschitz continuity with constant μ^{-1}. □

We extend our investigations now to the more broadly parameterized variational inequalities of the form

(9) $$f(p, x) + N_C(x) \ni 0$$

and their solution mappings

(10) $$S(p) = \{ x \mid f(p, x) + N_C(x) \ni 0 \},$$

with the aim of drawing on the achievements in Section 2E in the presence of monotonicity properties of f with respect to x.

Theorem 2F.10 (strong monotonicity and strict differentiability). *For a variational inequality (9) and its solution mapping (10) in the case of a nonempty, closed, convex set C, let $\bar{x} \in S(\bar{p})$ and assume that f is strictly differentiable at $(\bar{p},\bar{x})$. Suppose for some $\mu > 0$ that*

(11) $$\langle \nabla_x f(\bar{p},\bar{x})w, w \rangle \geq \mu |w|^2 \text{ for all } w \in C - C.$$

Then S has a Lipschitz continuous single-valued localization s around $\bar{p}$ for $\bar{x}$ with

(12) $$\text{lip}(s;\bar{p}) \leq \mu^{-1} |\nabla_p f(\bar{p},\bar{x})|.$$

Proof. We apply Theorem 2E.1, observing that its assumption (b) is satisfied on the basis of Theorem 2F.9 and the criterion in 2F.2(d) for strong monotonicity with constant μ. Theorem 2F.9 tells us moreover that the Lipschitz constant for the localization σ in Theorem 2E.1 is no more than μ^{-1}, and we then obtain (12) from Theorem 2E.1. □

Theorem 2F.10 can be compared to the result in Theorem 2E.8. That result requires C to be polyhedral but allows (11) to be replaced by a weaker condition in terms of the critical cone $K = K_C(\bar{x},\bar{v})$ for $\bar{v} = -f(\bar{p},\bar{x})$. Specifically, instead of asking the inequality in (11) to hold for all $w \in C - C$, one only asks it to hold for all $w \in K - K$ such that $\nabla_x f(\bar{p},\bar{x})w \perp K \cap (-K)$. The polyhedral convexity leads in this case to the further conclusion that the localization is semidifferentiable.

2G. Consequences for Optimization

Several types of variational inequalities are closely connected with problems of optimization. These include the basic condition for minimization in Theorem 2A.6 and the Lagrange condition in Theorem 2A.9, in particular. In this section we investigate what the general results obtained for variational inequalities provide in such cases.

Recall from Theorem 2A.6 that in minimizing a continuously differentiable function g over a nonempty, closed, convex set $C \subset \mathbb{R}^n$, the variational inequality

(1) $$\nabla g(x) + N_C(x) \ni 0$$

stands as a necessary condition for x to furnish a local minimum. When g is convex relative to C, it is sufficient for x to furnish a global minimum, but in the absence of convexity, an x satisfying (1) might not even correspond to a local minimum. However, there is an important case beyond convexity, which we will draw on later, in which an x satisfying (1) can be identified through additional criteria as yielding a local minimum.

In elucidating this case, we will appeal to the fact noted in 2A.4 that the normal cone $N_C(x)$ and the tangent cone $T_C(x)$ are polar to each other, so that (1) can be

2 Implicit Function Theorems for Variational Problems

written equivalently in the form

(2) $$\langle \nabla g(x), w \rangle \geq 0 \text{ for all } w \in T_C(x).$$

This gives a way to think about the first-order condition for a local minimum of g in which the vectors w in (2) making the inequality hold as an equation can be anticipated to have a special role. In fact, those vectors w comprise the critical cone $K_C(x, -\nabla g(x))$ to C at x with respect to the vector $-\nabla g(x)$ in $N_C(x)$, as defined in Section 2E:

(3) $$K_C(x, -\nabla g(x)) = \{ w \in T_C(x) \,|\, \langle \nabla g(x), w \rangle = 0 \}.$$

When C is polyhedral, at least, this critical cone is able to serve in the expression of second-order necessary and sufficient conditions for the minimization of g over C.

Theorem 2G.1 (second-order optimality on a polyhedral convex set). *Let C be a polyhedral convex set in $\mathbb{R}^n$ and let $g : \mathbb{R}^n \to \mathbb{R}$ be twice continuously differentiable on C. Let $\bar{x} \in C$ and $\bar{v} = -\nabla g(\bar{x})$.*

(a) (necessary condition) If g has a local minimum with respect to C at $\bar{x}$, then $\bar{x}$ satisfies the variational inequality (1) and has $\langle w, \nabla^2 g(\bar{x}) w \rangle \geq 0$ for all $w \in K_C(\bar{x}, \bar{v})$.

(b) (sufficient condition) If $\bar{x}$ satisfies the variational inequality (1) and has $\langle w, \nabla^2 g(\bar{x}) w \rangle > 0$ for all nonzero $w \in K_C(\bar{x}, \bar{v})$, then g has a local minimum relative to C at $\bar{x}$, indeed a strong local minimum in the sense of there being an $\varepsilon > 0$ such that

(4) $$g(x) \geq g(\bar{x}) + \frac{\varepsilon}{2} |x - \bar{x}|^2 \text{ for all } x \in C \text{ near } \bar{x}.$$

Proof. The necessity emerges through the observation that for any $x \in C$ the function $\varphi(t) = g(\bar{x} + tw)$ for $w = x - \bar{x}$ has $\varphi'(0) = \langle \nabla g(\bar{x}), w \rangle$ and $\varphi''(0) = \langle w, \nabla^2 g(\bar{x}) w \rangle$. From one-dimensional calculus, it is known that if φ has a local minimum at 0 relative to $[0, 1]$, then $\varphi'(0) \geq 0$ and, in the case of $\varphi'(0) = 0$, also has $\varphi''(0) \geq 0$. Having $\varphi'(0) = 0$ corresponds to having $w \in K_C(\bar{x}, -\nabla g(\bar{x}))$.

Conversely, if $\varphi'(0) \geq 0$ and in the case of $\varphi'(0) = 0$ also has $\varphi''(0) > 0$, then φ has a local minimum relative to $[0, 1]$ at 0. That one-dimensional sufficient condition is inadequate for concluding (b), however, because (b) requires a neighborhood of $\bar{x}$ relative to C, not just a separate neighborhood relative to each line segment proceeding from x into C.

To get (b), we have to make use of the properties of the second-order Taylor expansion of g at $\bar{x}$ which are associated with g being twice differentiable there: the error expression

$$e(w) = g(\bar{x} + w) - g(\bar{x}) - \langle \nabla g(\bar{x}), w \rangle - \tfrac{1}{2} \langle w, \nabla^2 g(\bar{x}) w \rangle$$

is of type $o(|w|^2)$. It will help to translate this into the notation where $w = tz$ with $t \geq 0$ and $|z| = 1$. Let $Z = \{ z \,|\, |z| = 1 \}$. To say that $e(w)$ is of type $o(|w|^2)$ is to say that the functions

$$f_t(z) = e(tz)/t^2 \text{ for } t > 0$$

converge to 0 on Z uniformly as $t \to 0$.

We furthermore need to rely on the tangent cone property at the end of 2E.3, which is available because C is polyhedral: there is a neighborhood W of the origin in $\mathbb{R}^n$ such that, as long as $w \in W$, we have $\bar{x} + w \in C$ if and only if $w \in T_C(\bar{x})$. Through this, it will be enough to show, on the basis of the assumption in (b), that the inequality in (4) holds for $w \in T_C(\bar{x})$ when $|w|$ is sufficiently small. Equivalently, it will be enough to produce an $\varepsilon > 0$ for which

(5) $\quad t^{-2}[g(\bar{x} + tz) - g(\bar{x})] \geq \varepsilon$ for all $z \in Z \cap T_C(\bar{x})$ when t is sufficiently small.

The assumption in (b) entails (in terms of $w = tz$) the existence of $\varepsilon > 0$ such that $\langle z, \nabla^2 g(\bar{x}) z \rangle > \varepsilon$ when $z \in Z \cap K_C(\bar{x}, \bar{v})$. This inequality also holds then for all z in some open set containing $Z \cap K_C(\bar{x}, \bar{v})$. Let Z_0 be the intersection of the complement of that open set with $Z \cap T_C(\bar{x})$. Since (1) is equivalent to (2), we have for $z \in Z_0$ that $\langle \nabla g(\bar{x}), z \rangle > 0$. Because Z_0 is compact, we actually have an $\eta > 0$ such that $\langle \nabla g(\bar{x}), z \rangle > \eta$ for all $z \in Z_0$. We see then, in writing

$$t^{-2}[g(\bar{x} + tz) - g(\bar{x})] = f_t(z) + t^{-1}\langle \nabla g(\bar{x}), z \rangle + \tfrac{1}{2}\langle z, \nabla^2 g(\bar{x}) z \rangle$$

and referring to the uniform convergence of the functions f_t to 0 on Z as $t \to 0$, that for t sufficiently small the left side is at least $t^{-1}\eta$ when $z \in Z_0$ and at least ε when $z \in Z \cap T_C(\bar{x})$ but $z \notin Z_0$. By taking t small enough that $t^{-1}\eta > \varepsilon$, we get (5) as desired. $\square$

When $\bar{x}$ belongs to the interior of C, as for instance when $C = \mathbb{R}^n$ (an extreme case of a polyhedral convex set), the first-order condition in (1) is simply $\nabla g(\bar{x}) = 0$. The second-order conditions in 2G.1 then specify positive semidefiniteness of $\nabla^2 g(\bar{x})$ for necessity and positive definiteness for sufficiency. Second-order conditions for a minimum can also be developed for convex sets that aren't polyhedral, but not in such a simple form. When the boundary of C is "curved," the tangent cone property in 2E.3 fails, and the critical cone $K_C(\bar{x}, \bar{v})$ for $\bar{v} = -\nabla g(\bar{x})$ no longer captures the local geometry adequately. Second-order optimality with respect to nonpolyhedral and even nonconvex sets specified by constraints as in nonlinear programming will be addressed later in this section (Theorem 2G.6).

Stationary points. *An x satisfying (1), or equivalently (2), will be called a stationary point of g with respect to minimizing over C, regardless of whether or not it furnishes a local or global minimum.*

Stationary points attract attention for their own sake, due to the role they have in the design and analysis of minimization algorithms, for example. Our immediate plan is to study how stationary points, as "quasi-solutions" in problems of minimization over a convex set, behave under perturbations. Along with that, we will clarify circumstances in which a stationary point giving a local minimum continues to give a local minimum when perturbed by not too much.

2 Implicit Function Theorems for Variational Problems

Moving in that direction, we look now at parameterized problems of the form

(6) $$\text{minimize } g(p,x) \text{ over all } x \in C,$$

where $g : \mathbb{R}^d \times \mathbb{R}^n \to \mathbb{R}$ is twice continuously differentiable with respect to x (not necessarily convex), and C is a nonempty, closed, convex subset of $\mathbb{R}^n$. In the pattern already seen, the variational inequality

(7) $$\nabla_x g(p,x) + N_C(x) \ni 0$$

provides for each p a first-order condition which x must satisfy if it furnishes a local minimum, but only describes, in general, the stationary points x for the minimization in (6). If C is polyhedral the question of a local minimum can be addressed through the second-order conditions provided by Theorem 2G.1 relative to the critical cone

(8) $$K_C(x, -\nabla_x g(p,x)) = \{ w \in T_C(x) \,|\, w \perp \nabla_x g(p,x) \}.$$

The basic object of interest to us for now, however, is the *stationary point mapping* $S : \mathbb{R}^d \rightrightarrows \mathbb{R}^n$ defined by

(9) $$S(p) = \{ x \,|\, \nabla_x g(p,x) + N_C(x) \ni 0 \}.$$

With respect to a choice of $\bar{p}$ and $\bar{x}$ such that $\bar{x} \in S(\bar{p})$, it will be useful to consider alongside of (6) an auxiliary problem with parameter $v \in \mathbb{R}^n$ in which $g(\bar{p}, \cdot)$ is essentially replaced by its second-order expansion at $\bar{x}$:

(10) $$\begin{array}{l} \text{minimize } \bar{g}(w) - \langle v, w \rangle \text{ over all } w \in W, \\ \text{where } \begin{cases} \bar{g}(w) = g(\bar{p},\bar{x}) + \langle \nabla_x g(\bar{p},\bar{x}), w \rangle + \tfrac{1}{2} \langle w, \nabla^2_{xx} g(\bar{p},\bar{x}) w \rangle, \\ W = \{ w \,|\, \bar{x} + w \in C \} = C - \bar{x}. \end{cases} \end{array}$$

The subtraction of $\langle v, w \rangle$ "tilts" $\bar{g}$, and is referred to therefore as a *tilt perturbation*. When $v = 0$, $\bar{g}$ itself is minimized.

For this auxiliary problem the basic first-order condition comes out to be the parameterized variational inequality

(11) $$\nabla_x g(\bar{p},\bar{x}) + \nabla^2_{xx} g(\bar{p},\bar{x}) w - v + N_W(w) \ni 0, \text{ where } N_W(w) = N_C(\bar{x}+w).$$

The stationary point mapping for the problem in (10) is accordingly the mapping $\bar{S} : \mathbb{R}^n \rightrightarrows \mathbb{R}^n$ defined by

(12) $$\bar{S}(v) = \{ w \,|\, \nabla_x g(\bar{p},\bar{x}) + \nabla^2_{xx} g(\bar{p},\bar{x}) w + N_W(w) \ni v \}.$$

The points $w \in \bar{S}(v)$ are sure to furnish a minimum in (10) if, for instance, the matrix $\nabla^2_{xx} g(\bar{p},\bar{x})$ is positive semidefinite, since that corresponds to the convexity of the "tilted" function being minimized. For polyhedral C, Theorem 2G.1 could be brought in for further analysis of a local minimum in (10). Note that $0 \in \bar{S}(0)$.

Theorem 2G.2 (parameterized minimization over a convex set). *Suppose in the preceding notation, with $\bar{x} \in S(\bar{p})$, that*
 (a) $\nabla_x g$ *is strictly differentiable at* $(\bar{p}, \bar{x})$, *and*
 (b) $\bar{S}$ *has a Lipschitz continuous single-valued localization $\bar{s}$ around 0 for 0.*
Then S has a Lipschitz continuous single-valued localization s around $\bar{p}$ for $\bar{x}$ with

$$\mathrm{lip}\,(s; \bar{p}) \leq \mathrm{lip}\,(\bar{s}; 0) \cdot |\nabla^2_{xp} g(\bar{p}, \bar{x})|,$$

and s has a first-order approximation η at $\bar{p}$ given by

$$(13) \qquad \eta(p) = \bar{x} + \bar{s}\Big(-\nabla^2_{xp} g(\bar{p}, \bar{x})(p - \bar{p})\Big).$$

On the other hand, (b) is necessary for S to have a Lipschitz continuous single-valued localization around $\bar{p}$ for $\bar{x}$ when the $n \times d$ matrix $\nabla^2_{xp} g(\bar{p}, \bar{x})$ has rank n.

Under the additional assumption that C is polyhedral, condition (b) is equivalent to the condition that, for the critical cone $K = K_C(\bar{x}, -\nabla_x g(\bar{p}, \bar{x}))$, the mapping

$$v \mapsto \bar{S}_0(v) = \big\{ w \,\big|\, \nabla^2_{xx} g(\bar{p}, \bar{x}) w + N_K(w) \ni v \big\} \quad \text{is everywhere single-valued.}$$

Moreover, a sufficient condition for this can be expressed in terms of the critical subspaces $K_C^+(\bar{x}, \bar{v}) = K_C(\bar{x}, \bar{v}) - K_C(\bar{x}, \bar{v})$ and $K_C^-(\bar{x}, \bar{v}) = K_C(\bar{x}, \bar{v}) \cap [-K_C(\bar{x}, \bar{v})]$ for $\bar{v} = -\nabla_x g(\bar{p}, \bar{x})$, namely

$$(14) \qquad \langle w, \nabla^2_{xx} g(\bar{p}, \bar{x}) w \rangle > 0 \quad \begin{cases} \text{for every nonzero } w \in K_C^+(\bar{x}, \bar{v}) \\ \text{with } \nabla^2_{xx} g(\bar{p}, \bar{x}) w \perp K_C^-(\bar{x}, \bar{v}). \end{cases}$$

Furthermore, in this case the localization s is semidifferentiable at $\bar{p}$ with semiderivative given by

$$Ds(\bar{p})(q) = \bar{s}(-\nabla^2_{xp} g(\bar{p}, \bar{x}) q).$$

Proof. We apply Theorem 2E.1 with $f(p,x) = \nabla_x g(p,x)$. The mapping G in that result coincides with $\nabla \bar{g} + N_C$, so that G^{-1} is $\bar{S}$. Assumptions (a) and (b) cover the corresponding assumptions in 2E.1, with $\sigma(v) = \bar{x} + \bar{s}(v)$, and then (13) follows from 2E(4). In the polyhedral case we also have Theorems 2E.6 and 2E.8 at our disposal, and this gives the rest. □

Theorem 2G.3 (stability of a local minimum on a polyhedral convex set). *Suppose in the setting of the parameterized minimization problem in (6) and its stationary point mapping S in (9) that C is polyhedral and $\nabla_x g(p,x)$ is strictly differentiable with respect to (p,x) at $(\bar{p}, \bar{x})$, where $\bar{x} \in S(\bar{p})$. With respect to the critical subspace $K_C^+(\bar{x}, \bar{v})$ for $\bar{v} = -\nabla_x g(\bar{p}, \bar{x})$, assume that*

$$(15) \qquad \langle w, \nabla^2_{xx} g(\bar{p}, \bar{x}) w \rangle > 0 \text{ for every nonzero } w \in K_C^+(\bar{x}, \bar{v}).$$

Then S has a localization s not only with the properties laid out in Theorem 2G.2, but also with the property that, for every p in some neighborhood of $\bar{p}$, the point

$x = s(p)$ furnishes a strong local minimum in (6). Moreover, (15) is necessary for the existence of a localization s with all these properties, when the $n \times d$ matrix $\nabla^2_{xp} g(\bar{p}, \bar{x})$ has rank n.

Proof. Obviously (15) implies (14), which ensures according to Theorem 2G.2 that S has a Lipschitz continuous single-valued localization s around $\bar{p}$ for $\bar{x}$. Applying 2E.10(a), we see then that

$$K_C(x, -\nabla_x g(p, x)) \subset K_C^+(\bar{x}, -\nabla_x g(\bar{p}, \bar{x})) = K_C^+(\bar{x}, \bar{v})$$

when $x = s(p)$ and p is near enough to $\bar{p}$. Since the matrix $\nabla^2_{xx} g(p, x)$ converges to $\nabla^2_{xx} g(\bar{p}, \bar{x})$ as (p, x) tends to $(\bar{p}, \bar{x})$, it follows that $\langle w, \nabla^2_{xx} g(p, x) w \rangle > 0$ for all nonzero $w \in K_C(x, -\nabla_x g(p, x))$ when $x = s(p)$ and p is close enough to $\bar{p}$. Since having $x = s(p)$ corresponds to having the first-order condition in (7), we conclude that from Theorem 2G.1 that x furnishes a strong local minimum in this case.

Arguing now toward the necessity of (15) under the rank condition on $\nabla^2_{xp} g(\bar{p}, \bar{x})$, we suppose S has a Lipschitz continuous single-valued localization s around $\bar{p}$ for $\bar{x}$ such that $x = s(p)$ gives a local minimum when p is close enough to $\bar{p}$. For any $x \in C$ near $\bar{x}$ and $v \in N_C(x)$, the rank condition gives us a p such that $v = -\nabla_x g(p, x)$; this follows e.g. from 1F.6. Then $x = s(p)$ and, because we have a local minimum, it follows that $\langle w, \nabla^2_{xx} g(p, x) w \rangle \geq 0$ for every nonzero $w \in K_C(x, v)$. We know from 2E.10(b) that $K_C(x, v) = K_C^+(\bar{x}, \bar{v})$ for choices of x and v arbitrarily close to $(\bar{x}, \bar{v})$, where $\bar{v} = -\nabla_x g(\bar{p}, \bar{x})$. Through the continuous dependence of $\nabla^2_{xx} g(p, x)$ on (p, x), we therefore have

(16) $\langle w, Aw \rangle \geq 0$ for all $w \in K_C^+(\bar{x}, \bar{v})$, where $A = \nabla^2_{xx} g(\bar{p}, \bar{x})$ is symmetric.

For this reason, we can only have $\langle w, Aw \rangle = 0$ if $Aw \perp K_C^+(\bar{x}, \bar{v})$, i.e., $\langle w', Aw \rangle = 0$ for all $w' \in K_C^+(\bar{x}, \bar{v})$.

On the other hand, because the rank condition corresponds to the ample parameterization property, we know from Theorem 2E.8 that the existence of the single-valued localization s requires for A and the critical cone $K = K_C(\bar{x}, \bar{v})$ that the mapping $(A + N_K)^{-1}$ be single-valued. This would be impossible if there were a nonzero w such that $Aw \perp K_C^+(\bar{x}, \bar{v})$, because we would have $\langle w', Aw \rangle = 0$ for all $w' \in K$ in particular (since $K \subset K_C^+(\bar{x}, \bar{v})$), implying that $-Aw \in N_K(w)$. Then $(A + N_K)^{-1}(0)$ would contain w along with 0, contrary to single-valuedness. Thus, the inequality in (16) must be strict when $w \neq 0$. □

Next we provide a complementary, global result for the special case of a tilted strongly convex function.

Proposition 2G.4 (tilted minimization of strongly convex functions). *Let $g : \mathbb{R}^n \to \mathbb{R}$ be continuously differentiable on an open set O, and let $C \subset O$ be a nonempty, closed, convex set on which g is strongly convex with constant $\mu > 0$. Then for each $v \in \mathbb{R}^n$ the problem*

(17) $\text{minimize } g(x) - \langle v, x \rangle \text{ over } x \in C$

has a unique solution $S(v)$, and the solution mapping S is Lipschitz continuous on $\mathbb{R}^n$ (globally) with Lipschitz constant μ^{-1}.

Proof. Let g_v denote the function being minimized in (17). Like g, this function is continuously differentiable and strongly convex on C with constant μ; we have $\nabla g_v(x) = \nabla g(x) - v$. According to Theorem 2A.6, the condition $\nabla g_v(x) + N_C(x) \ni 0$, or equivalently $x \in (\nabla g + N_C)^{-1}(v)$, is both necessary and sufficient for x to furnish the minimum in (17). The strong convexity of g makes the mapping $f = \nabla g$ strongly monotone on C with constant μ; see 2F.3(a). The conclusion follows now by applying Theorem 2F.9 to this mapping f. □

When the function g in Proposition 2G.4 is twice continuously differentiable, the strong monotonicity can be identified through 2A.5(b) with the inequality $\langle \nabla^2 g(x)w, w \rangle \geq \mu |w|^2$ holding for all $x \in C$ and $w \in C - C$.

Exercise 2G.5. In the setting of Theorem 2G.2, condition (b) is fulfilled in particular if there exists $\mu > 0$ such that

$$(18) \qquad \langle \nabla_{xx}^2 g(\bar{p},\bar{x})w, w \rangle \geq \mu |w|^2 \quad \text{for all } w \in C - C,$$

and then $\mathrm{lip}(s;\bar{p}) \leq \mu^{-1}$. If C is polyhedral, the additional conclusion holds that, for all p in some neighborhood of $\bar{p}$, there is a strong local minimum in problem (6) at the point $x = s(p)$.

Guide. Apply Proposition 2G.4 to the function $\bar{g}$ in the auxiliary minimization problem (10). Get from this that $\bar{s}$ coincides with $\bar{S}$, which is single-valued and Lipschitz continuous on $\mathbb{R}^n$ with Lipschitz constant μ^{-1}. In the polyhedral case, also apply Theorem 2G.3, arguing that (18) entails (15). □

Observe that because $C - C$ is a convex set containing 0, the condition in (18) holds for all $w \in C - C$ if it holds for all $w \in C - C$ with $|w|$ sufficiently small.

We turn now to minimization over sets which need not be convex but are specified by a system of constraints. A first-order necessary condition for a minimum in that case was developed in a very general manner in Theorem 2A.8. Here, we restrict ourselves to the most commonly treated problem of nonlinear programming, where the format is to

$$(19) \qquad \text{minimize } g_0(x) \text{ over all } x \text{ satisfying } g_i(x) \begin{cases} \leq 0 & \text{for } i \in [1,s], \\ = 0 & \text{for } i \in [s+1,m]. \end{cases}$$

In order to bring second-order conditions for optimality into the picture, we assume that the functions $g_0, g_1, \ldots, g_m$ are *twice* continuously differentiable on $\mathbb{R}^n$.

The basic first-order condition in this case has been worked out in detail in Section 2A as a consequence of Theorem 2A.8. It concerns the existence, relative to x, of a multiplier vector $y = (y_1, \ldots, y_m)$ fulfilling the Karush–Kuhn–Tucker conditions:

(20) $\quad y \in I\!\!R_+^s \times I\!\!R^{m-s}, \; g_i(x) \begin{cases} \leq 0 & \text{for } i \in [1,s] \text{ with } y_j = 0, \\ = 0 & \text{for all other } i \in [1,m], \end{cases}$
$\nabla g_0(x) + y_1 \nabla g_1(x) + \cdots + y_m \nabla g_m(x) = 0.$

This existence is necessary for a local minimum at x as long as x satisfies the constraint qualification requiring that the same conditions, but with the term $\nabla g_0(x)$ suppressed, can't be satisfied with $y \neq 0$. It is sufficient for a global minimum at x if $g_0, g_1, \ldots, g_s$ are convex and $g_{s+1}, \ldots, g_m$ are affine. However, we now wish to take a second-order approach to local sufficiency, rather than rely on convexity for global sufficiency.

The key for us will be the fact, coming from Theorem 2A.9, that (20) can be identified in terms of the Lagrangian function

(21) $\quad L(x,y) = g_0(x) + y_1 g_1(x) + \cdots + y_m g_m(x)$

with a certain variational inequality for a continuously differentiable function $f : I\!\!R^n \times I\!\!R^m \to I\!\!R^n \times I\!\!R^m$ and polyhedral convex cone $E \subset I\!\!R^n \times I\!\!R^m$, namely

(22) $\quad f(x,y) + N_E(x,y) \ni (0,0), \text{ where } \begin{cases} f(x,y) = (\nabla_x L(x,y), -\nabla_y L(x,y)), \\ E = I\!\!R^n \times [I\!\!R_+^s \times I\!\!R^{m-s}]. \end{cases}$

Because our principal goal is to illustrate the application of the results in the preceding sections, rather than push consequences for optimization theory to the limit, we will only deal with this variational inequality under an assumption of linear independence for the gradients of the active constraints. A constraint in (20) is *inactive* at x if it is an inequality constraint with $g_i(x) < 0$; otherwise it is *active* at x.

Theorem 2G.6 (second-order optimality in nonlinear programming). *Let $\bar{x}$ be a point satisfying the constraints in (19). Let $I(\bar{x})$ be the set of indices i of the active constraints at $\bar{x}$, and suppose that the gradients $\nabla g_i(\bar{x})$ for $i \in I(\bar{x})$ are linearly independent. Let K consist of the vectors $w \in I\!\!R^n$ satisfying*

(23) $\quad \langle \nabla g_i(\bar{x}), w \rangle \begin{cases} \leq 0 & \text{for } i \in I(\bar{x}) \text{ with } i \leq s, \\ = 0 & \text{for all other } i \in I(\bar{x}) \text{ and also for } i = 0. \end{cases}$

(a) (necessary condition) *If $\bar{x}$ furnishes a local minimum in problem (19), then a multiplier vector $\bar{y}$ exists such that $(\bar{x}, \bar{y})$ not only satisfies the variational inequality (22) but also has*

(24) $\quad \langle w, \nabla_{xx}^2 L(\bar{x}, \bar{y}) w \rangle \geq 0 \text{ for all } w \in K.$

(b) (sufficient condition) *If a multiplier vector $\bar{y}$ exists such that $(\bar{x}, \bar{y})$ satisfies the conditions in (20), or equivalently (22), and if (24) holds with strict inequality when $w \neq 0$, then $\bar{x}$ furnishes a local minimum in (19). Indeed, it furnishes a strong local minimum in the sense of there being an $\varepsilon > 0$ such that*

(25) $\quad g_0(x) \geq g_0(\bar{x}) + \frac{\varepsilon}{2} |x - \bar{x}|^2 \text{ for all } x \text{ near } \bar{x} \text{ satisfying the constraints.}$

Proof. The linear independence of the gradients of the active constraints guarantees, among other things, that $\bar{x}$ satisfies the constraint qualification under which (22) is necessary for local optimality.

In the case of a local minimum, as in (a), we do therefore have the variational inequality (22) fulfilled by $\bar{x}$ and some vector $\bar{y}$; and of course (22) holds by assumption in (b). From this point on, therefore, we can concentrate on just the second-order parts of (a) and (b) in the framework of having $\bar{x}$ and $\bar{y}$ satisfying (20). In particular then, we have

$$(26) \qquad -\nabla g_0(\bar{x}) = \bar{y}_1 \nabla g_1(\bar{x}) + \cdots + \bar{y}_m \nabla g_m(\bar{x}),$$

where the multiplier vector $\bar{y}$ is moreover uniquely determined by the linear independence of the gradients of the active constraints and the stipulation in (20) that inactive constraints get coefficient 0.

Actually, the inactive constraints play no role around $\bar{x}$, so we can just as well assume, for simplicity of exposition in our local analysis, that *every* constraint is active at $\bar{x}$: we have $g_i(\bar{x}) = 0$ for $i = 1,\ldots,m$. Then, on the level of first-order conditions, we just have the combination of (20), which corresponds to $\nabla_x L(\bar{x},\bar{y}) = 0$, and the requirement that $\bar{y}_i \geq 0$ for $i = 1,\ldots,s$. In this simplified context, let

$$(27) \quad T = \text{ set of all } w \in \mathbb{R}^n \text{ satisfying } \langle \nabla g_i(\bar{x}), w \rangle \begin{cases} \leq 0 & \text{for } i = 1,\ldots,s, \\ = 0 & \text{for } i = s+1,\ldots,m, \end{cases}$$

so that the cone K described by (23) can be expressed in the form

$$(28) \qquad K = \{ w \in T \mid \langle \nabla g_0(\bar{x}), w \rangle = 0 \}.$$

The rest of our argument will rely heavily on the classical inverse function theorem, 1A.1. Our assumption that the vectors $\nabla g_i(\bar{x})$ for $i = 1,\ldots,m$ are linearly independent in $\mathbb{R}^n$ entails of course that $m \leq n$. These vectors can be supplemented, if necessary, by vectors a_k for $k = 1,\ldots,n-m$ so as to form a basis for $\mathbb{R}^n$. Then, by setting $g_{m+k}(x) = \langle a_k, x - \bar{x} \rangle$, we get functions g_i for $i = m+1,\ldots,n$ such that for

$$g : \mathbb{R}^n \to \mathbb{R}^n \text{ with } g(x) = (g_1(x),\ldots,g_m(x), g_{m+1}(x),\ldots,g_n(x))$$

we have $g(\bar{x}) = 0$ and $\nabla g(\bar{x})$ nonsingular. We can view this as providing, at least locally around $\bar{x}$, a change of coordinates $g(x) = u = (u_1,\ldots,u_n)$, $x = s(u)$ (for a localization s of g^{-1} around 0 for $\bar{x}$) in which $\bar{x}$ corresponds to 0 and the constraints in (19) correspond to linear constraints

$$u_i \leq 0 \text{ for } i = 1,\ldots,s, \qquad u_i = 0 \text{ for } i = s+1,\ldots,m$$

(with no condition on u_i for $i = m+1,\ldots,n$), which specify a *polyhedral convex* set D in $\mathbb{R}^n$. Problem (19) is thereby transformed in a local sense into minimizing over this set D the twice continuously differentiable function $f(u) = g_0(s(u))$, and we are concerned with whether or not there is a local minimum at $\bar{u} = 0$. The necessary

2 Implicit Function Theorems for Variational Problems

and sufficient conditions in Theorem 2G.1 are applicable to this and entail having $-f(0)$ belong to $N_D(0)$. It will be useful to let $\tilde{y}$ stand for $(\bar{y}, 0, \ldots, 0) \in \mathbb{R}^n$.

The inverse function theorem reveals that the Jacobian $\nabla s(0)$ is $\nabla g(\bar{x})^{-1}$. We have $\nabla f(0) = \nabla g_0(\bar{x}) \nabla s(0)$ by the chain rule, and on the other hand $-\nabla g_0(\bar{x}) = \tilde{y} \nabla g(\bar{x})$ by (26), and therefore $\nabla f(0) = -\tilde{y}$. The vectors w belonging to the set T in (27) correspond one-to-one with the vectors $z \in D$ through $\nabla g(\bar{x})w = z$, and under this, through (28), the vectors $w \in K$ correspond to the vectors $z \in D$ such that $\langle z, \tilde{y} \rangle = 0$, i.e., the vectors in the critical cone $K_D(0, \tilde{y}) = K_D(0, -\nabla f(0))$.

The second-order conditions in Theorem 2G.1, in the context of the transformed version of problem (19), thus revolve around the nonnegativity or positivity of $\langle z, \nabla^2 f(0)z \rangle$ for vectors $z \in K_D(0, \tilde{y})$. It will be useful that this is the same as the nonnegativity or positivity or $\langle z, \nabla^2 h(0)z \rangle$ for the function

$$h(u) = f(u) + \langle \tilde{y}, u \rangle = f(u) + \langle \bar{y}, Pu \rangle = L(s(u), \bar{y}),$$

where P is the projection from $(u_1, \ldots, u_m, u_{m+1}, \ldots, u_n)$ to $(u_1, \ldots, u_m)$. Furthermore,

$$\langle z, \nabla^2 h(0)z \rangle = \varphi''(0) \text{ for the function } \varphi(t) = h(tz) = L(s(tz), \bar{y}).$$

Fix any nonzero $z \in K_D(0, \tilde{y})$ and the corresponding $w \in K$, given by $w = \nabla s(0)z = \nabla g(\bar{x})^{-1} z$. Our task is to demonstrate that actually

$$\varphi''(0) = \langle w, \nabla^2_{xx} L(\bar{x}, \bar{y}) w \rangle. \tag{29}$$

Let $x(t) = s(tz)$, so that $x(0) = \bar{x}$ and $x'(0) = w$. We have

$$\varphi(t) = L(x(t), \bar{y}),$$
$$\varphi'(t) = \langle \nabla_x L(x(t), \bar{y}), x'(t) \rangle,$$
$$\varphi''(t) = \langle w, \nabla^2_{xx} L(x(t), \bar{y}), x'(t) \rangle + \langle \nabla_x L(x(t), \bar{y}), x''(t) \rangle,$$

hence $\varphi''(0) = \langle w, \nabla^2_{xx} L(\bar{x}, \bar{y}), w \rangle + \langle \nabla_x L(\bar{x}, \bar{y}), x''(0) \rangle$. But $\nabla_x L(\bar{x}, \bar{y}) = 0$ from the first-order conditions. Thus, (29) holds, as claimed.

The final assertion of part (b) automatically carries over from the corresponding assertion of part (b) of Theorem 2G.1 under the local change of coordinates that we utilized. □

Exercise 2G.7. *In the context of Theorem 2G.6, let $\bar{y}$ be a multiplier associated with $\bar{x}$ through the first-order condition (22). Let $I_0(\bar{x}, \bar{y})$ be the set of indices $i \in I(\bar{x})$ such that $i \leq s$ and $\bar{y}_i = 0$. Then an equivalent description of the cone K in the second-order conditions is that*

$$w \in K \iff \langle \nabla g_i(\bar{x}), w \rangle \begin{cases} \leq 0 & \text{for } i \in I_0(\bar{x}, \bar{y}), \\ = 0 & \text{for } i \in I(\bar{x}) \setminus I_0(\bar{x}, \bar{y}). \end{cases}$$

Guide. Utilize the fact that $-\nabla g_0(\bar{x}) = \bar{y}_1 \nabla g_1(\bar{x}) + \cdots + \bar{y}_m \nabla g_m(\bar{x})$ with $\bar{y}_i \geq 0$ for $i = 1, \ldots, s$. □

The alternative description in 2G.7 lends insights in some situations, but it makes K appear to depend on $\bar{y}$, whereas in reality it doesn't.

Next we take up the study of a parameterized version of the nonlinear programming problem in the form

(30)   minimize $g_0(p,x)$ over all x satisfying $g_i(p,x) \begin{cases} \leq 0 & \text{for } i \in [1,s], \\ = 0 & \text{for } i \in [s+1,m], \end{cases}$

where the functions $g_0, g_1, \ldots, g_m$ are twice continuously differentiable from $\mathbb{R}^d \times \mathbb{R}^n$ to $\mathbb{R}$. The Lagrangian function is now

(31) $$L(p,x,y) = g_0(p,x) + y_1 g_1(p,x) + \cdots + y_m g_m(p,x)$$

and the variational inequality capturing the associated first-order conditions is
(32)
$$f(p,x,y) + N_E(x,y) \ni (0,0), \text{ where } \begin{cases} f(p,x,y) = (\nabla_x L(p,x,y), -\nabla_y L(p,x,y)), \\ E = \mathbb{R}^n \times [\mathbb{R}^s_+ \times \mathbb{R}^{m-s}]. \end{cases}$$

The pairs (x,y) satisfying this variational inequality are the Karush–Kuhn–Tucker pairs for the problem specified by p in (30). The x components of such pairs might or might not give a local minimum according to the circumstances in Theorem 2G.6 (or whether certain convexity assumptions are fulfilled), and indeed we are not imposing a linear independence condition on the constraint gradients in (30) of the kind on which Theorem 2G.6 was based. But these x's serve anyway as stationary points and we wish to learn more about their behavior under perturbations by studying the *Karush–Kuhn–Tucker mapping* $S : \mathbb{R}^d \to \mathbb{R}^n \times \mathbb{R}^m$ defined by

(33) $$S(p) = \{(x,y) \mid f(p,x,y) + N_E(x,y) \ni (0,0)\}.$$

Once more, an auxiliary problem will be important with respect to a choice of $\bar{p}$ and a pair $(\bar{x},\bar{y}) \in S(\bar{p})$. To formulate it, we let

$$\bar{g}_0(w) = L(\bar{p},\bar{x},\bar{y}) + \langle \nabla_x L(\bar{p},\bar{x},\bar{y}), w \rangle + \tfrac{1}{2} \langle w, \nabla^2_{xx} L(\bar{p},\bar{x},\bar{y}) w \rangle,$$
$$\bar{g}_i(w) = g_i(\bar{p},\bar{x}) + \langle \nabla_x g_i(\bar{p},\bar{x}), w \rangle \text{ for } i = 1,\ldots,m,$$

and introduce the notation

(34) $$\begin{aligned} I &= \{i \in [1,m] \mid g_i(\bar{p},\bar{x}) = 0\} \supset \{s+1,\ldots,m\}, \\ I_0 &= \{i \in [1,s] \mid g_i(\bar{p},\bar{x}) = 0 \text{ and } \bar{y}_i = 0\} \subset I, \\ I_1 &= \{i \in [1,s] \mid g_i(\bar{p},\bar{x}) < 0\}. \end{aligned}$$

The auxiliary problem, depending on a tilt parameter vector v but also now an additional parameter vector $u = (u_1,\ldots,u_m)$, is to

2 Implicit Function Theorems for Variational Problems   123

(35)
$$\text{minimize } \bar{g}_0(w) - \langle v, w \rangle \text{ over all } w \text{ satisfying}$$
$$\bar{g}_i(w) + u_i \begin{cases} = 0 & \text{for } i \in I \setminus I_0, \\ \leq 0 & \text{for } i \in I_0, \\ \text{free} & \text{for } i \in I_1, \end{cases}$$

where "free" means unrestricted. (The functions $\bar{g}_i$ for $i \in I_1$ play no role in this problem, but it will be more convenient to carry them through in this scheme than to drop them.)

In comparison with the auxiliary problem introduced earlier in (10) with respect to minimization over a set C, it's apparent that a second-order expansion of L rather than g_0 has entered, but merely first-order expansions of the constraint functions $g_1, \ldots, g_m$. In fact, only the quadratic part of the Lagrangian expansion matters, inasmuch as $\nabla_x L(\bar{p}, \bar{x}, \bar{y}) = 0$ by the first-order conditions. The Lagrangian for the problem in (35) depends on the parameter pair (v, u) and involves a multiplier vector $z = (z_1, \ldots, z_m)$:

$$\bar{g}_0(w) - \langle v, w \rangle + z_1[\bar{g}_1(w) + u_1] + \cdots + z_m[\bar{g}_m(w) + u_m] =: \bar{L}(w, z) - \langle v, w \rangle + \langle z, u \rangle.$$

The corresponding first-order conditions are given by the variational inequality

(36)
$$\bar{f}(w, z) - (v, u) + N_{\bar{E}}(w, z) \ni (0, 0), \quad \text{where}$$
$$\bar{f}(w, z) = (\nabla_w \bar{L}(w, z), -\nabla_z \bar{L}(w, z)), \quad \bar{E} = \mathbb{R}^n \times W, \text{ with}$$
$$z = (z_1, \ldots, z_m) \in W \iff z_i \begin{cases} \geq 0 & \text{for } i \in I_0, \\ = 0 & \text{for } i \in I_1, \end{cases}$$

which translate into the requirements that

(37)
$$\nabla^2_{xx} L(\bar{p}, \bar{x}, \bar{y}) w + z_1 \nabla_x g_1(\bar{p}, \bar{x}) + \cdots + z_m \nabla_x g_m(\bar{p}, \bar{x}) - v = 0,$$
$$\text{with } z_i \begin{cases} \geq 0 & \text{for } i \in I_0 \text{ having } \bar{g}_i(w) + u_i = 0, \\ = 0 & \text{for } i \in I_0 \text{ having } \bar{g}_i(w) + u_i < 0 \text{ and for } i \in I_1. \end{cases}$$

We need to pay heed to the auxiliary solution mapping $\bar{S} : \mathbb{R}^n \times \mathbb{R}^n \rightrightarrows \mathbb{R}^n \times \mathbb{R}^m$ defined by

(38) $\quad \bar{S}(v, u) = \{(w, z) \mid \bar{f}(w, z) + N_{\bar{E}}(w, z) \ni (v, u)\} = \{(w, z) \mid \text{satisfying (37)}\},$

which has
$$(0, 0) \in \bar{S}(0, 0).$$

The following subspaces will enter our analysis of the properties of the mapping $\bar{S}$:

(39)
$$M^+ = \{ w \in \mathbb{R}^n \mid w \perp \nabla_x g_i(\bar{p}, \bar{x}) \text{ for all } i \in I \setminus I_0 \},$$
$$M^- = \{ w \in \mathbb{R}^n \mid w \perp \nabla_x g_i(\bar{p}, \bar{x}) \text{ for all } i \in I \}.$$

Theorem 2G.8 (implicit function theorem for stationary points). *Let $(\bar{x},\bar{y}) \in S(\bar{p})$ for the mapping S in (33), constructed from functions g_i that are twice continuously differentiable. Assume for the auxiliary mapping $\bar{S}$ in (38) that*

(40) $\qquad \begin{cases} \bar{S} \text{ has a Lipschitz continuous single-valued} \\ \text{localization } \bar{s} \text{ around } (0,0) \text{ for } (0,0). \end{cases}$

Then S has a Lipschitz continuous single-valued localization s around $\bar{p}$ for $(\bar{x},\bar{y})$, and this localization s is semidifferentiable at $\bar{p}$ with semiderivative given by

(41) $\qquad Ds(\bar{p})(q) = \bar{s}(-Bq), \text{ where } B = \nabla_p f(\bar{p},\bar{x},\bar{y}) = \begin{bmatrix} \nabla_{xp}^2 L(\bar{p},\bar{x},\bar{y}) \\ -\nabla_p g_1(\bar{p},\bar{x}) \\ \vdots \\ -\nabla_p g_m(\bar{p},\bar{x}) \end{bmatrix}.$

Moreover the condition in (40) is necessary for the existence of a Lipschitz continuous single-valued localization of S around $\bar{p}$ for $(\bar{x},\bar{y})$ when the $(n+m) \times d$ matrix B has rank $n+m$. In particular, $\bar{S}$ is sure to satisfy (40) when the following conditions are both fulfilled:

(a) *the gradients $\nabla_x g_i(\bar{p},\bar{x})$ for $i \in I$ are linearly independent,*
(b) *$\langle w, \nabla_{xx}^2 L(\bar{p},\bar{x},\bar{y})w \rangle > 0$ for every nonzero $w \in M^+$ with $\nabla_{xx}^2 L(\bar{p},\bar{x},\bar{y})w \perp M^-$, with M^+ and M^- as in (39).*

On the other hand, (40) always entails at least (a).

Proof. This is obtained by applying 2E.1 with the additions in 2E.6 and 2E.8 to the variational inequality (32). Since $\nabla_y L(p,x,y) = g(p,x)$ for $g(p,x) = (g_1(p,x),\ldots,g_m(p,x))$, the Jacobian in question is

(42) $\qquad \nabla_{(x,y)} f(\bar{p},\bar{x},\bar{y}) = \begin{bmatrix} \nabla_{xx}^2 L(\bar{p},\bar{x},\bar{y}) & \nabla_x g(\bar{p},\bar{x})^T \\ -\nabla_x g(\bar{p},\bar{x}) & 0 \end{bmatrix}.$

In terms of polyhedral convex cone $Y = \mathbb{R}_+^s \times \mathbb{R}^{m-s}$, the critical cone to the polyhedral convex cone set E is

(43) $\qquad K_E(\bar{x},\bar{y},-f(\bar{p},\bar{x},\bar{y})) = \mathbb{R}^n \times W$

for the polyhedral cone W in (36). By taking A to be the matrix in (42) and K to be the cone in (43), the auxiliary mapping $\bar{S}$ can be identified with $(A+N_K)^{-1}$ in the framework of Theorem 2E.6. (The w and v in that result have here turned into pairs (w,z) and (v,u).)

This leads to all the conclusions except for establishing that (40) implies (a) and working out the details of the sufficient condition provided by Theorem 2E.6. To verify that (40) implies (a), consider any $\varepsilon > 0$ and let

$$v^\varepsilon = \sum_{i=1}^m z_i^\varepsilon \nabla_x g_i(\bar{p},\bar{x}), \qquad z_i^\varepsilon = \begin{cases} \varepsilon & \text{for } i \in I_0, \\ 0 & \text{otherwise.} \end{cases}$$

Then, as seen from the conditions in (37), we have $(0,z^\varepsilon) \in \bar{S}(v^\varepsilon,0)$. If (a) didn't hold, we would also have $\sum_{i=1}^m \zeta_i \nabla_x g(\bar{p},\bar{x}) = 0$ for some coefficient vector $\zeta \neq 0$ with $\zeta_i = 0$ when $i \in I_1$. Then for every $\delta > 0$ small enough that $\varepsilon + \delta \zeta_i \geq 0$ for all $i \in I_0$, we would also have $(0,z^\varepsilon + \delta \zeta) \in \bar{S}(v^\varepsilon,0)$. Since ε and δ can be chosen arbitrarily small, this would contradict the single-valuedness in (40). Thus, (a) is necessary for (40).

To come next to an understanding of what the sufficient condition in 2E.6 means here, we observe in terms of $Y = I\!R_+^s \times I\!R^{m-s}$ that

$$K_E^+(\bar{x},\bar{y},-f(\bar{p},\bar{x},\bar{y})) = I\!R^n \times K_Y^+(\bar{y},g(\bar{p},\bar{x})),$$
$$K_E^-(\bar{x},\bar{y},-f(\bar{p},\bar{x},\bar{y})) = I\!R^n \times K_Y^-(\bar{y},g(\bar{p},\bar{x})),$$

where

$$z \in K_Y^+(\bar{y},g(\bar{p},\bar{x})) \iff z_i = 0 \text{ for } i \in I_1,$$
$$z \in K_Y^-(\bar{y},g(\bar{p},\bar{x})) \iff z_i = 0 \text{ for } i \in I_0 \cup I_1.$$

In the shorthand notation

$$H = \nabla_{xx}^2 L(\bar{p},\bar{x},\bar{y}), \quad K^+ = K_E^+(\bar{x},\bar{y},-f(\bar{p},\bar{x},\bar{y})), \quad K^- = K_E^-(\bar{x},\bar{y},-f(\bar{p},\bar{x},\bar{y})),$$

our concern is to have $\langle (w,z), A(w,z) \rangle > 0$ for every $(w,z) \in K^+$ with $A(w,z) \perp K^-$ and $(w,z) \neq (0,0)$. It's clear from (41) that

$$\langle (w,z), A(w,z) \rangle = \langle w, Hw \rangle,$$
$$(w,z) \in K^+ \iff z_i = 0 \text{ for } i \in I_1,$$
$$A(w,z) \perp K^- \iff Hw + \nabla_x g(\bar{p},\bar{x})^\mathsf{T} z = 0 \text{ and } \nabla_x g(\bar{p},\bar{x}) w \perp K_Y^-.$$

Having $\nabla_x g(\bar{p},\bar{x}) w \perp K_Y^-$ corresponds to having $w \perp \nabla_x g_i(\bar{p},\bar{x})$ for all $i \in I \setminus I_0$, which means $w \in M^+$. On the other hand, having $Hw + \nabla_x g(\bar{p},\bar{x})^\mathsf{T} z = 0$ corresponds to having $Hw = -(z_1 \nabla_x g_1(\bar{p},\bar{x}) + \cdots + z_m \nabla_x g_m(\bar{p},\bar{x}))$. The sufficient condition in 2E.6 boils down, therefore, to the requirement that

$$\langle w, Hw \rangle > 0 \text{ when } (w,z) \neq (0,0), \ w \in M^+, \ Hw = -\sum_{i \in I} z_i \nabla_x g_i(\bar{p},\bar{x}).$$

In particular this requirement has to cover cases where $w = 0$ but $z \neq 0$. That's obviously equivalent to the linear independence in (a). Beyond that, we need only observe that expressing Hw in the manner indicated corresponds simply to having $Hw \perp M^-$. Thus, the sufficient condition in Theorem 2E.6 turns out to be the combination of (a) with (b). □

Our final topic concerns the conditions under which the mapping S in (33) describes perturbations not only of stationarity, but also of local minimuma.

Theorem 2G.9 (implicit function theorem for local minima). *Suppose in the setting of the parameterized nonlinear programming problem (30) for twice continuously differentiable functions g_i and its Karush–Kuhn–Tucker mapping S in (33) that the following conditions hold in the notation coming from (34):*
  (a) *the gradients $\nabla_x g_i(\bar{p},\bar{x})$ for $i \in I$ are linearly independent, and*
  (b) *$\langle w, \nabla^2_{xx} L(\bar{p},\bar{x},\bar{y})w \rangle > 0$ for every $w \neq 0$ in the subspace M^+ in (39).*

Then not only does S have a localization s with the properties laid out in Theorem 2G.8, but also, for every p in some neighborhood of $\bar{p}$, the x component of $s(p)$ furnishes a strong local minimum in (30). Moreover, (a) and (b) are necessary for this additional conclusion when $n+m$ is the rank of the $(n+m) \times d$ matrix B in (41).

Proof. Sufficiency. Condition (b) here is a stronger assumption than (b) of Theorem 2G.8, so we can be sure that (a) and (b) guarantee the existence of a localization s possessing the properties in that result. Moreover (b) implies satisfaction of the sufficient condition for a local minimum at $\bar{x}$ in Theorem 2G.6, inasmuch as the cone K in that theorem is obviously contained in the set of w such that $\langle \nabla g_i(\bar{p},\bar{x}), w \rangle = 0$ for all $i \in I \setminus I_0$. We need to demonstrate, however, that this local minimum property persists in passing from $\bar{p}$ to nearby p.

To proceed with that, denote the two components of $s(p)$ by $x(p)$ and $y(p)$, and let $I(p)$, $I_0(p)$ and $I_1(p)$ be the index sets which correspond to $x(p)$ as I, I_0 and I_1 do to $\bar{x}$, so that $I(p)$ consists of the indices $i \in \{1,\ldots,m\}$ with $g_i(p,x(p)) = 0$, and $I(p) \setminus I_0(p)$ consists of the indices $i \in I(p)$ having $y_i(p) > 0$ for inequality constraints, but $I_1(p)$ consists of the indices of the inequality constraints having $g_i(p,x(p)) < 0$. Consider the following conditions, which reduce to (a) and (b) when $p = \bar{p}$:

  (a(p)) the gradients $\nabla_x g_i(p,x(p))$ for $i \in I(p)$ are linearly independent,
  (b(p)) $\langle w, \nabla^2_{xx} L(p,x(p),y(p))w \rangle > 0$ for every $w \neq 0$ such that $w \perp \nabla_x g_i(p,x(p))$
for all $i \in I(p) \setminus I_0(p)$.

Since $x(p)$ and $y(p)$ tend toward $x(\bar{p}) = \bar{x}$ and $y(\bar{p}) = \bar{y}$ as $p \to \bar{p}$, the fact that $y_i(p) = 0$ for $i \in I_1(p)$ and the continuity of the g_i's ensure that

$$I(p) \subset I \text{ and } I(p) \setminus I_0(p) \supset I \setminus I_0 \text{ for } p \text{ near enough to } \bar{p}.$$

Through this and the fact that $\nabla_x g_i(p,x(p))$ tends toward $\nabla_x g_i(\bar{p},\bar{x})$ as p goes to $\bar{p}$, we see that the linear independence in (a) entails the linear independence in (a(p)) for p near enough to $\bar{p}$. Indeed, not only (a(p)) but also (b(p)) must hold, in fact in the stronger form that there exist $\varepsilon > 0$ and a neighborhood Q of $\bar{p}$ for which

$$\langle w, \nabla^2_{xx} L(p,x(p),y(p))w \rangle > \varepsilon$$

when $|w| = 1$ and $w \perp \nabla_x g_i(p,x(p))$ for all $i \in I(p) \setminus I_0(p)$. Indeed, otherwise there would be sequences of vectors $p_k \to \bar{p}$ and $w_k \to w$ violating this condition for $\varepsilon_k \to 0$, and this would lead to a contradiction of (b) in view of the continuous dependence of the matrix $\nabla^2_{xx} L(p,x(p),y(p))$ on p.

Of course, with both (a(p)) and (b(p)) holding when p is in some neighborhood of $\bar{p}$, we can conclude through Theorem 2G.6, as we did for $\bar{x}$, that $x(p)$ furnishes a strong local minimum for problem (30) for such p, since the cone

$$K(p) = \text{ set of } w \text{ satisfying } \begin{cases} \langle \nabla_x g_i(p,x(p)), w \rangle \leq 0 & \text{for } i \in I(p) \text{ with } i \leq s, \\ \langle \nabla_x g_i(p,x(p)), w \rangle = 0 & \text{for } i = s+1, \ldots, m \text{ and } i = 0 \end{cases}$$

lies in the subspace formed by the vectors w with $\langle \nabla_x g_i(p,x(p)), w \rangle = 0$ for all $i \in I(p) \setminus I_0(p)$.

Necessity. Suppose that S has a Lipschitz continuous single-valued localization s around $\bar{p}$ for $(\bar{x},\bar{y})$. We already know from Theorem 2G.8 that, under the rank condition in question, the auxiliary mapping $\bar{S}$ in (38) must have such a localization around $(0,0)$ for $(0,0)$, and that this requires the linear independence in (a). Under the further assumption now that $x(p)$ gives a local minimum in problem (30) when p is near enough to $\bar{p}$, we wish to deduce that (b) must hold as well. Having a local minimum at $x(p)$ implies that the second-order necessary condition for optimality in Theorem 2G.6 is satisfied with respect to the multiplier vector $y(p)$:

(44) $\qquad \langle w, \nabla^2_{xx} L(p,x(p),y(p))w \rangle \geq 0$ for all $w \in K(p)$ when p is near to $\bar{p}$.

We will now find a value of the parameter p close to $\bar{p}$ such that $(x(p),y(p)) = (\bar{x},\bar{y})$ and $K(p) = M^+$. If $I_0 = \emptyset$ there is nothing to prove. Let $I_0 \neq \emptyset$. The rank condition on the Jacobian $B = \nabla_p f(\bar{p},\bar{x},\bar{y})$ provides through Theorem 1F.6 (for $k=1$) the existence of $p(v,u)$, depending continuously on some (v,u) in a neighborhood of $(0, -g(\bar{p},\bar{x}))$, such that $f(p(v,u),\bar{x},\bar{y}) = (v,u)$, i.e., $\nabla_x L(p(v,u),\bar{x},\bar{y}) = v$ and $-g(p(v,u),\bar{x}) = u$. For an arbitrarily small $\varepsilon > 0$, let the vector u^ε have $u_i^\varepsilon = -\varepsilon$ for $i \in I_0$ but $u_i^\varepsilon = 0$ for all other i. Let $p^\varepsilon = p(0,u^\varepsilon)$. Then $\nabla_x L(p^\varepsilon,\bar{x},\bar{y}) = 0$ with $g_i(p^\varepsilon,\bar{x}) = 0$ for $i \in I \setminus I_0$ but $g_i(p^\varepsilon,\bar{x}) < 0$ for $i \in I_0$ as well as for $i \in I_1$. Thus, $I(p^\varepsilon) = I \setminus I_0$, $I_0(p^\varepsilon) = \emptyset$, $I_1(p^\varepsilon) = I_0 \cup I_1$, and $(\bar{x},\bar{y}) \in S(p^\varepsilon)$ and, moreover, $(\bar{x},\bar{y})$ furnishes a local minimum in (30) for $p = p^\varepsilon$, moreover with $K(p^\varepsilon)$ coming out to be the subspace

$$M^+(p^\varepsilon) = \{ w \mid w \perp \nabla_x g_i(p^\varepsilon,\bar{x}) \text{ for all } i \in I \setminus I_0 \}.$$

In consequence of (44) we therefore have

$$\langle w, \nabla^2_{xx} L(p^\varepsilon,\bar{x},\bar{y})w \rangle \geq 0 \text{ for all } w \in M^+(p^\varepsilon),$$

whereas we are asking in (b) for this to hold with strict inequality for $w \neq 0$ in the case of $M^+ = M^+(\bar{p})$.

We know that $p^\varepsilon \to \bar{p}$ as $\varepsilon \to 0$. Owing to (a) and the continuity of the functions g_i and their derivatives, the gradients $\nabla_x g_i(p^\varepsilon,\bar{x})$ for $i \in I$ must be linearly independent when ε is sufficiently small. It follows from this that any w in M^+ can be approximated as $\varepsilon \to 0$ by vectors w^ε belonging to the subspaces $M^+(p^\varepsilon)$. In the limit therefore, we have at least that

(45) $\qquad \langle w, \nabla^2_{xx} L(\bar{p},\bar{x},\bar{y})w \rangle \geq 0$ for all $w \in M^+$.

How are we to conclude strict inequality when $w \neq 0$? It's important that the matrix $H = \nabla_{xx}^2 L(\bar{p},\bar{x},\bar{y})$ is symmetric. In line with the positive semidefiniteness in (45), any $\bar{w} \in M^+$ with $\langle \bar{w}, H\bar{w} \rangle = 0$ must have $H\bar{w} \perp M^+$. But then in particular, the auxiliary solution mapping $\bar{S}$ in (38) would have $(t\bar{w},0) \in \bar{S}(0,0)$ for all $t \geq 0$, in contradiction to the fact, coming from Theorem 2G.8, that $\bar{S}(0,0)$ contains only $(0,0)$ in the current circumstances. □

Commentary

The basic facts about convexity, polyhedral sets, and tangent and normal cones given in Section 2A are taken mainly from Rockafellar [1970]. Robinson's implicit function theorem was stated and proved in Robinson [1980], where the author was clearly motivated by the problem of how the solutions of the standard nonlinear programming problem depend on parameters, and he pursued this goal in the same paper.

At that time it was already known from the work of Fiacco and McCormick [1968] that under the linear independence of the constraint gradients and the standard second-order sufficient condition, together with *strict complementarity slackness* at the reference point (which means that there are no inequality constraints satisfied as equalities that are associated with zero Lagrange multipliers), the solution mapping for the standard nonlinear programming problem has a *smooth* single-valued localization around the reference point. The proof of this result was based on the classical implicit function theorem, inasmuch as under strict complementarity slackness the Karush–Kuhn–Tucker system turns into a system of equations locally. Robinson looked at the case when the strict complementarity slackness is violated, which may happen, as already noted in 2B, when the "stationary point trajectory" hits the constraints. Based on his implicit function theorem, which actually reached far beyond his immediate goal, Robinson proved, still in his paper from 1980, that under a stronger form of the second-order sufficient condition, together with linear independence of the constraint gradients, the solution mapping of the standard nonlinear programming problem has a *Lipschitz continuous* single-valued localization around the reference point; see Theorem 2G.9 for an updated statement.

This result was a stepping stone to the subsequent extensive development of stability analysis in optimization, whose maturity came with the publication of the books Bank, Guddat, Klatte, Kummer and Tammer [1983], Levitin [1992], Bonnans and Shapiro [2000], Klatte and Kummer [2002] and Facchinei and Pang [2003].

Robinson's breakthrough in the stability analysis of nonlinear programming was in fact much needed for the emerging numerical analysis of variational problems more generally. In his paper from 1980, Robinson noted the thesis of his Ph.D. student Josephy [1979], who proved that strong regularity yields local quadratic convergence of Newton's method for solving variational inequalities, a method whose version for constrained optimization problems is well known as the sequential quadratic programming (SQP) method in nonlinear programming.

Quite a few years after Robinson's theorem was published, it was realized that the result could be used as a tool in the analysis of a variety of variational problems, and beyond. Alt [1990] applied it to optimal control, while in Dontchev and Hager [1993], and further in Dontchev [1995b], the statement of Robinson's theorem was observed actually to hold for generalized equations of the form 2B(1) for an *arbitrary* mapping F, not just a normal cone mapping. Variational inequalities thus serve as an example, not a limitation. Important applications, e.g. to convergence analysis of algorithms and discrete approximations to infinite-dimensional variational problems, came later. In the explosion of works in this area in the 80's and 90's Robinson's contribution, if not forgotten, was sometimes taken for granted. More about this will come out in Chapter 6.

The presentation of the material in Section 2B mainly follows Dontchev and Rockafellar [2009a], while that in Section 2C comes from Dontchev and Rockafellar [2001]. In Section 2D, we used some facts from the books of Facchinei and Pang [2003] (in particular, 2D.5) and Scholtes [1994]. Theorem 2E.6 is a particular case of a result in Dontchev and Rockafellar [1996].

Sections 2F and 2G give an introduction to the theory of monotone mappings which for its application to optimization problems goes back to Rockafellar [1976a] and [1976b]. Much more about this kind of monotonicity and its earlier history can be found in Chapter 12 of the book of Rockafellar and Wets [1998]. The stability analysis in 2G uses material from Dontchev and Rockafellar [1996,1998], but some versions of these results could be extracted from earlier works.

Chapter 3
Regularity Properties of Set-valued Solution Mappings

In the concept of a solution mapping for a problem dependent on parameters, whether formulated with equations or something broader like variational inequalities, we have always had to face the possibility that solutions might not exist, or might not be unique when they do exist. This goes all the way back to the setting of the classical implicit function theorem. In letting $S(p)$ denote the set of all x satisfying $f(p,x) = 0$, where f is a given function from $\mathbb{R}^d \times \mathbb{R}^n$ to $\mathbb{R}^m$, we cannot expect to be defining a *function* S from $\mathbb{R}^d$ to $\mathbb{R}^n$, even when $m = n$. In general, we only get a set-valued mapping S. However, this mapping S could have a single-valued localization s with properties of continuity or differentiability. The study of such localizations, as "subfunctions" within a set-valued mapping, has been our focus so far, but now we open up to a wider view.

There are plenty of reasons, already in the classical context, to be interested in localizations of solution mappings without insisting on single-valuedness. For instance, in the case of $S(p) = \{x \mid f(p,x) = 0\}$ with f going from $\mathbb{R}^d \times \mathbb{R}^n$ to $\mathbb{R}^m$ and $m < n$, it can be anticipated for a choice of $\bar{p}$ and $\bar{x}$ with $\bar{x} \in S(\bar{p})$, under assumptions on $\nabla_x f(\bar{p},\bar{x})$, that a graphical localization S_0 of S exists around $(\bar{p},\bar{x})$ such that $S_0(p)$ is an $(n-m)$-dimensional manifold which varies with p. What generalizations of the usual notions of continuity and differentiability might help in understanding, and perhaps quantifying, this dependence on p?

Such challenges in dealing with the dependence of a set on the parameters which enter its definition carry over to solution mappings to variational inequalities, and also to problems of a broader character centered on constraint systems. Just as the vector equation $f(p,x) = 0$ for

$$f : \mathbb{R}^d \times \mathbb{R}^n \to \mathbb{R}^m \text{ with } f(p,x) = (f_1(p,x), \ldots, f_m(p,x))$$

can be viewed as standing for a system of scalar equations

$$f_i(p,x) = 0 \text{ for } i = 1, \ldots, m,$$

we can contemplate vector representations of mixed systems of inequalities and equations like

(1) $\quad f_i(p,x) \begin{cases} \leq 0 & \text{for } i = 1,\ldots,s, \\ = 0 & \text{for } i = s+1,\ldots,m, \end{cases}$

which are important in optimization. Such a system takes the form

$$f(p,x) - K \ni 0 \quad \text{for } K = {I\!R}^s_- \times \{0\}^{m-s}.$$

In fact this is an instance of a parameterized generalized equation:

(2) $\quad f(p,x) + F(x) \ni 0$ with F a constant mapping, $F(x) \equiv -K$.

In studying the behavior of the corresponding solution mapping $S : {I\!R}^d \rightrightarrows {I\!R}^n$ given by

(3) $\quad S(p) = \{x \,|\, x \text{ satisfies } (2)\}$ (covering (1) as a special case),

we are therefore still, very naturally, in the realm of the "extended implicit function theory" we have been working to build up.

Here, F is not a normal cone mapping N_C, so we are not dealing with a variational inequality. The results in Chapter 2 for solution mappings to parameterized generalized equations would anyway, in principle, be applicable, but in this framework they miss the mark. The trouble is that those results focus on the prospects of finding single-valued localizations of a solution mapping, especially ones that exhibit Lipschitz continuity. For a solution mapping S as in (3), coming from a generalized equation as in (2), single-valued localizations are unlikely to exist at all (apart from the pure equation case with $m = n$) and aren't even a topic of serious concern. Rather, we are confronted with a "varying set" $S(p)$ which cannot be reduced locally to a "varying point." That could be the case even if, in (2), $F(x)$ is not a constant set but a sort of continuously moving or deforming set. What we are looking for is not so much a generalized implicit function theorem, but an *implicit mapping theorem*, the distinction being that "mappings" truly embrace set-valuedness.

To understand the behavior of such a solution mapping S, whether qualitatively or quantitatively, we have to turn to other concepts, beyond those in Chapter 2. Our immediate task, in Sections 3A, 3B, 3C and 3D, is to introduce the notions of Painlevé–Kuratowski convergence and Pompeiu–Hausdorff convergence for sequences of sets, and to utilize them in developing properties of continuity and Lipschitz continuity for set-valued mappings. In tandem with this, we gain important insights into the solution mappings (3) associated with constraint systems as in (1) and (2), especially for cases where f is affine. We also obtain by-products concerning the behavior of various mappings associated with problems of optimization.

In Section 3E, however, we open a broader investigation in which the Aubin property, serving as a sort of localized counterpart to Lipschitz continuity for set-valued mappings, is tied to the concept of metric regularity, which directly relates to estimates of distances to solutions. The natural context for this is the study of how properties of a set-valued mapping correspond to properties of its set-valued inverse, or in other words, the paradigm of the inverse function theorem. We are able nevertheless to return in Section 3F to the paradigm of the implicit function theorem, based on a stability property of metric regularity fully developed later in Chapter 5. Powerful results, applicable to fully set-valued solution mappings (2)(3) even when F is *not* just a constant mapping, are thereby obtained. Sections 3G and 3I then take these ideas back to situations where single-valuedness is available in a localization of a solution mapping, at least at the reference point, showing how previous results such as those in 2B can thereby be amplified. Section 3H reveals that a set-valued version of calmness does not similarly submit to the implicit function theorem paradigm.

3A. Set Convergence

Various continuity properties of set-valued mappings will be essential for the developments in this chapter. To lay the foundation for them we must first introduce two basic concepts: convergence of sets and distance between sets.

The set of all natural numbers $k = 1, 2, \ldots$, will be denoted by $\mathbb{N}$. The collection of all subsets N of $\mathbb{N}$ such that $\mathbb{N} \setminus N$ is finite will be denoted by $\mathcal{N}$, whereas the collection of all infinite $N \subset \mathbb{N}$ will be denoted by $\mathcal{N}^{\sharp}$. This scheme is designed for convenience in handling subsequences of a given sequence. For instance, if we have a sequence $\{x^k\}_{k=1}^{\infty}$ of points in $\mathbb{R}^n$, the notation $\{x^k\}_{k \in N}$ for either $N \in \mathcal{N}$ or $N \in \mathcal{N}^{\sharp}$ designates a subsequence. In the first case it is a subsequence which coincides with the full sequence beyond some k_0, whereas in the second case it is a general subsequence. Limits as $k \to \infty$ with $k \in N$ will be indicated by $\lim_{k \in N}$, or in terms of arrows by $\xrightarrow{N}$, and so forth.

For a sequence $\{r^k\}_{k=1}^{\infty}$ in $\mathbb{R}$, the limit as $k \to \infty$ may or may not exist—even though we always include ∞ and $-\infty$ as possible limit values in the obvious sense. However, the upper limit, or "limsup," and the lower limit, or "liminf," do always exist, as defined by

$$\limsup_{k \to \infty} r^k = \lim_{k \to \infty} \sup_{m \geq k} r^m,$$
$$\liminf_{k \to \infty} r^k = \lim_{k \to \infty} \inf_{m \geq k} r^m.$$

An alternative description of these values is that $\limsup_{k \to \infty} r^k$ is the highest r for which there exists $N \in \mathcal{N}^{\sharp}$ such that $r^k \xrightarrow{N} r$, whereas $\liminf_{k \to \infty} r^k$ is the lowest such r. The limit itself exists if and only if these upper and lower limits coincide. For simplicity, we often just write $\limsup_k$, $\liminf_k$ and $\lim_k$, with the understanding that this refers to $k \to \infty$.

In working with sequences of sets, a similar pattern is encountered in which "outer" and "inner" limits always exist and give a "limit" when they agree.

Outer and inner limits. *Consider a sequence $\{C^k\}_{k=1}^{\infty}$ of subsets of $\mathbb{R}^n$.*

(a) *The* outer limit *of this sequence, denoted by $\limsup_k C^k$, is the set of all $x \in \mathbb{R}^n$ for which*

there exist $N \in \mathcal{N}^{\sharp}$ and $x^k \in C^k$ for $k \in N$ such that $x^k \xrightarrow{N} x$.

(b) *The* inner limit *of this sequence, denoted by $\liminf_k C^k$, is the set of all $x \in \mathbb{R}^n$ for which*

there exist $N \in \mathcal{N}$ and $x^k \in C^k$ for $k \in N$ such that $x^k \xrightarrow{N} x$.

(c) *When the inner and outer limits are the same set C, this set is defined to be the* limit *of the sequence $\{C^k\}_{k=1}^{\infty}$:*

$$C = \lim_k C^k = \limsup_k C^k = \liminf_k C^k.$$

3 Regularity Properties of Set-valued Solution Mappings

In this case C^k is said to *converge to C in the sense of Painlevé–Kuratowski convergence*.

Note that although the outer and inner limit sets always exist by this definition, they might be empty. When $C^k \neq \emptyset$ for all k, these sets can be described equivalently in terms of the sequences $\{x^k\}_{k=1}^{\infty}$ that can be formed by selecting an $x_k \in C_k$ for each k: the set of all cluster points of such sequences is $\limsup_k C^k$, while the set of all limits of such sequences is $\liminf_k C^k$. Obviously $\liminf_k C^k \subset \limsup_k C^k$. When each C^k is a singleton, $\liminf_k C^k$ can at most be another singleton, but $\limsup_k C^k$ might have multiple elements.

Examples.
1) The sequence of doubletons $C^k = \{0, \frac{1}{k}\}$ in $\mathbb{R}$ has $\lim_k C^k = \{0\}$. Indeed, every sequence of elements $x_k \in C_k$ converges to 0.
2) The sequence of balls $\mathbb{B}(x^k, \rho^k)$ converges to $\mathbb{B}(x, \rho)$ when $x^k \to x$ and $\rho^k \to \rho$.
3) A sequence of sets C^k which alternates between two different closed sets D_1 and D_2, that is, $C^k = D_1$ when k is odd and $C^k = D_2$ when k is even, has $D_1 \cap D_2$ as its inner limit and $D_1 \cup D_2$ as its outer limit. Such a sequence is not convergent if $D_1 \neq D_2$.

Outer and inner limits can also be described with the help of neighborhoods:

(1a) $\quad \limsup_{k \to \infty} C^k = \left\{ x \,\Big|\, \forall \text{ neighborhood } V \text{ of } x, \exists N \in \mathcal{N}^\sharp, \forall k \in N : C^k \cap V \neq \emptyset \right\},$

(1b) $\quad \liminf_{k \to \infty} C^k = \left\{ x \,\Big|\, \forall \text{ neighborhood } V \text{ of } x, \exists N \in \mathcal{N}, \forall k \in N : C^k \cap V \neq \emptyset \right\}.$

Without loss of generality the neighborhoods in (1a,b) can be taken to be closed balls; then we obtain the following more transparent definitions:

(2a) $\quad \limsup_{k \to \infty} C^k = \{ x \,|\, \forall \varepsilon > 0, \exists N \in \mathcal{N}^\sharp : x \in C^k + \varepsilon \mathbb{B} \ (k \in N) \},$

(2b) $\quad \liminf_{k \to \infty} C^k = \{ x \,|\, \forall \varepsilon > 0, \exists N \in \mathcal{N} : x \in C^k + \varepsilon \mathbb{B} \ (k \in N) \}.$

Both the outer and inner limits of a sequence $\{C^k\}_{k \in N}$ are *closed sets*. Indeed, if $x \notin \limsup_k C^k$, then, from (2a), there exists $\varepsilon > 0$ such that for every $N \in \mathcal{N}^\sharp$ we have $x \notin C^k + \varepsilon \mathbb{B}$, that is, $\mathbb{B}(x, \varepsilon) \cap C^k = \emptyset$, for some $k \in N$. But then a neighborhood of x can meet C^k for finitely many k only. Hence no points in this neighborhood can be cluster points of sequences $\{x^k\}$ with $x^k \in C^k$ for infinitely many k. This implies that the complement of $\limsup_k C^k$ is an open set and therefore that $\limsup_k C^k$ is closed. An analogous argument works for $\liminf_k C^k$ (this could also be derived by the following Proposition 3A.1).

Recall from Section 1D that the distance from a point $x \in \mathbb{R}^n$ to a subset C of $\mathbb{R}^n$ is

$$d_C(x) = d(x,C) = \inf_{y \in C} |x-y|.$$

As long as C is closed, having $d(x,C) = 0$ is equivalent to having $x \in C$.

Proposition 3A.1 (distance function characterizations of limits). *Outer and inner limits of sequences of sets are described alternatively by the following formulas:*

(3a) $$\limsup_{k \to \infty} C^k = \Big\{ x \,\Big|\, \liminf_{k \to \infty} d(x, C^k) = 0 \Big\},$$

(3b) $$\liminf_{k \to \infty} C^k = \Big\{ x \,\Big|\, \lim_{k \to \infty} d(x, C^k) = 0 \Big\}.$$

Proof. If $x \in \limsup_k C^k$ then, by (2a), for any $\varepsilon > 0$ there exists $N \in \mathcal{N}^\sharp$ such that $d(x, C^k) \leq \varepsilon$ for all $k \in N$. But then, by the definition of the lower limit for a sequence of real numbers, as recalled in the beginning of this section, we have $\liminf_{k \to \infty} d(x, C^k) = 0$. The left side of (3a) is therefore contained in the right side. Conversely, if x is in the set on the right side of (3a), then there exists $N \in \mathcal{N}^\sharp$ and $x^k \in C^k$ for all $k \in N$ such that $x^k \xrightarrow{N} x$; then, by definition, x must belong to the left side of (3a).

If x is not in the set on the right side of (3b), then there exist $\varepsilon > 0$ and $N \in \mathcal{N}^\sharp$ such that $d(x, C^k) > \varepsilon$ for all $k \in N$. Then $x \notin C^k + \varepsilon \mathbb{B}$ for all $k \in N$ and hence by (2b) x is not in $\liminf_k C^k$. In a similar way, from (2b) we obtain that $x \notin \liminf_k C^k$ only if $\limsup_k d(x, C^k) > 0$. This gives us (3b). □

Observe that the distance to a set does not distinguish whether this set is closed or not. Therefore, in the context of convergence, there is no difference whether the sets in a sequence are closed or not. (But limits of all types are closed sets.)

More examples.
  1) The limit of the sequence of intervals $[k, \infty)$ as $k \to \infty$ is the empty set, whereas the limit of the sequence of intervals $[1/k, \infty)$ is $[0, \infty)$.
  2) More generally for monotone sequences of subsets $C^k \subset \mathbb{R}^n$, if $C^k \supset C^{k+1}$ for all $k \in \mathbb{N}$, then $\lim_k C^k = \bigcap_k \operatorname{cl} C^k$, whereas if $C^k \subset C^{k+1}$ for all k, then $\lim_k C^k = \operatorname{cl} \bigcup_k C^k$.
  3) The constant sequence $C^k = D$, where D is the set of vectors in $\mathbb{R}^n$ whose coordinates are rational numbers, converges not to D, which isn't closed, but to the closure of D, which is $\mathbb{R}^n$. More generally, if $C^k = C$ for all k, then $\lim_k C^k = \operatorname{cl} C$.

Theorem 3A.2 (characterization of Painlevé–Kuratowski convergence). *For a sequence C^k of sets in $\mathbb{R}^n$ and a closed set $C \subset \mathbb{R}^n$ one has:*
  (a) $C \subset \liminf_k C^k$ if and only if for every open set $O \subset \mathbb{R}^n$ with $C \cap O \neq \emptyset$ there exists $N \in \mathcal{N}$ such that $C^k \cap O \neq \emptyset$ for all $k \in N$;
  (b) $C \supset \limsup_k C^k$ if and only if for every compact set $B \subset \mathbb{R}^n$ with $C \cap B = \emptyset$ there exists $N \in \mathcal{N}$ such that $C^k \cap B = \emptyset$ for all $k \in N$;

3 Regularity Properties of Set-valued Solution Mappings

(c) $C \subset \liminf_k C^k$ if and only if for every $\rho > 0$ and $\varepsilon > 0$ there is an index set $N \in \mathcal{N}$ such that $C \cap \rho \mathbb{B} \subset C^k + \varepsilon \mathbb{B}$ for all $k \in N$;

(d) $C \supset \limsup_k C^k$ if and only if for every $\rho > 0$ and $\varepsilon > 0$ there is an index set $N \in \mathcal{N}$ such that $C^k \cap \rho \mathbb{B} \subset C + \varepsilon \mathbb{B}$ for all $k \in N$;

(e) $C \subset \liminf_k C^k$ if and only if $\limsup_k d(x, C^k) \leq d(x, C)$ for every $x \in \mathbb{R}^n$;

(f) $C \supset \limsup_k C^k$ if and only if $d(x, C) \leq \liminf_k d(x, C^k)$ for every $x \in \mathbb{R}^n$.

Thus, from (c)(d) $C = \lim_k C^k$ if and only if for every $\rho > 0$ and $\varepsilon > 0$ there is an index set $N \in \mathcal{N}$ such that

$$C^k \cap \rho \mathbb{B} \subset C + \varepsilon \mathbb{B} \text{ and } C \cap \rho \mathbb{B} \subset C^k + \varepsilon \mathbb{B} \text{ for all } k \in N.$$

Also, from (e)(f), $C = \lim_k C^k$ if and only if $\lim_k d(x, C^k) = d(x, C)$ for every $x \in \mathbb{R}^n$.

Proof. (a): Necessity comes directly from (1b). To show sufficiency, assume that there exists $x \in C \setminus \liminf_k C^k$. But then, by (1b), there exists an open neighborhood V of x such that for every $N \in \mathcal{N}$ there exists $k \in N$ with $V \cap C^k = \emptyset$ and also $V \cap C \neq \emptyset$. This is the negation of the condition on the right.

(b): Let $C \supset \limsup_k C^k$ and let there exist a compact set B with $C \cap B = \emptyset$, such that for any $N \in \mathcal{N}$ one has $C^k \cap B \neq \emptyset$ for some $k \in N$. But then there exist $N \in \mathcal{N}^\sharp$ and a convergent sequence $x^k \in C^k$ for $k \in N$ whose limit is not in C, a contradiction. Conversely, if there exists $x \in \limsup_k C^k$ which is not in C then, from (2a), a ball $\mathbb{B}_\varepsilon(x)$ with sufficiently small radius ε does not meet C yet meets C^k for infinitely many k; this contradicts the condition on the right.

Sufficiency in (c): Consider any point $x \in C$, and any $\rho > |x|$. For an arbitrary $\varepsilon > 0$, there exists, by assumption, an index set $N \in \mathcal{N}$ such that $C \cap \rho \mathbb{B} \subset C^k + \varepsilon \mathbb{B}$ for all $k \in N$. Then $x \in C^k + \varepsilon \mathbb{B}$ for all $k \in N$. By (2b), this yields $x \in \liminf_k C^k$. Hence, $C \subset \liminf_k C^k$.

Necessity of (c): It will be demonstrated that if the condition fails, there must be a point $\bar{x} \in C$ lying outside of $\liminf_k C^k$. To say that the condition fails is to say that there exist $\rho > 0$ and $\varepsilon > 0$, such that, for each $N \in \mathcal{N}$, the inclusion $C \cap \rho \mathbb{B} \subset C^k + \varepsilon \mathbb{B}$ is false for at least one $k \in N$. Then there is an index set $N_0 \in \mathcal{N}^\sharp$ such that this inclusion is false for all $k \in N_0$; there are points $x^k \in [C \cap \rho \mathbb{B}] \setminus [C^k + \varepsilon \mathbb{B}]$ for all $k \in N_0$. Such points form a bounded sequence in the closed set C with the property that $d(x^k, C^k) \geq \varepsilon$. A subsequence $\{x^k\}_{k \in N_1}$, for an index set $N_1 \in \mathcal{N}^\sharp$ within N_0, converges in that case to a point $\bar{x} \in C$. Since $d(x^k, C^k) \leq d(\bar{x}, C^k) + |\bar{x} - x^k|$, we must have

$$d(\bar{x}, C^k) \geq \varepsilon/2 \text{ for all } k \in N_1 \text{ large enough.}$$

It is impossible then for $\bar{x}$ to belong to $\liminf_k C^k$, because that requires $d(\bar{x}, C^k)$ to converge to 0, cf. (3b).

Sufficiency in (d): Let $\bar{x} \in \limsup_k C^k$; then for some $N_0 \in \mathcal{N}^\sharp$ there are points $x^k \in C^k$ such that $x^k \xrightarrow{N_0} \bar{x}$. Fix any $\rho > |\bar{x}|$, so that $x^k \in \rho \mathbb{B}$ for $k \in N_0$ large enough. By assumption, there exists for any $\varepsilon > 0$ an index set $N \in \mathcal{N}$ such that $C^k \cap \rho \mathbb{B} \subset C + \varepsilon \mathbb{B}$ when $k \in N$. Then for large enough $k \in N_0 \cap N$ we have $x^k \in C + \varepsilon \mathbb{B}$, hence $d(x^k, C) \leq \varepsilon$. Because $d(\bar{x}, C) \leq d(x^k, C) + |x^k - \bar{x}|$ and $x^k \xrightarrow{N_0} \bar{x}$, it follows from the arbitrary choice of ε that $d(\bar{x}, C) = 0$, which means $\bar{x} \in C$ (since C is closed).

Necessity in (d): Suppose to the contrary that one can find $\rho > 0$, $\varepsilon > 0$ and $N \in \mathcal{N}^\sharp$ such that, for all $k \in N$, there exists $x^k \in [C^k \cap \rho \mathbb{B}] \setminus [C + \varepsilon \mathbb{B}]$. The sequence $\{x^k\}_{k \in N}$ is then bounded, so it has a cluster point $\bar{x}$ which, by definition, belongs to $\limsup_k C^k$. On the other hand, since each x^k lies outside of $C + \varepsilon \mathbb{B}$, we have $d(x^k, C) \geq \varepsilon$ and, in the limit, $d(\bar{x}, C) \geq \varepsilon$. Hence $\bar{x} \notin C$, and therefore $\limsup_k C^k$ is not a subset of C.

(e): Sufficiency follows from (3b) by taking $x \in C$. To prove necessity, choose $x \in \mathbb{R}^n$ and let $y \in C$ be a projection of x on C: $|x - y| = d(x, C)$. By the definition of $\liminf$ there exist $N \in \mathcal{N}$ and $y^k \in C^k$, $k \in N$ such that $y^k \xrightarrow{N} y$. For such y^k we have $d(x, C^k) \leq |y^k - x|$, $k \in N$ and passing to the limit with $k \to \infty$ we get the condition on the right.

(f): Sufficiency follows from (3a) by taking $x \in \limsup_k C^k$. Choose $x \in \mathbb{R}^n$. If $x \in C$ there is nothing to prove. If not, note that for any nonnegative α the condition $d(x, C) > \alpha$ is equivalent to $C \cap \mathbb{B}_\alpha(x) = \emptyset$. But then from (b) there exists $N \in \mathcal{N}$ with $C^k \cap \mathbb{B}_\alpha(x) = \emptyset$ for $k \in N$, which is the same as $d(x, C^k) > \alpha$ for $k \in N$. This implies the condition on the right. $\square$

Observe that in parts (c)(d) of 3A.2 we can replace the phrase "for every ρ" by "there is some $\rho_0 \geq 0$ such that for every $\rho \geq \rho_0$".

Set convergence can also be characterized in terms of concepts of distance between sets.

Excess and Pompeiu–Hausdorff distance. *For sets C and D in $\mathbb{R}^n$, the excess of C beyond D is defined by*

$$e(C, D) = \sup_{x \in C} d(x, D),$$

where the convention is used that

$$e(\emptyset, D) = \begin{cases} 0 & \text{when } D \neq \emptyset, \\ \infty & \text{otherwise.} \end{cases}$$

The Pompeiu–Hausdorff distance between C and D is the quantity

$$h(C, D) = \max\{e(C, D), e(D, C)\}.$$

Equivalently, these quantities can be expressed by

$$e(C, D) = \inf\{\tau \geq 0 \,|\, C \subset D + \tau \mathbb{B}\}$$

and

$$h(C, D) = \inf\{\tau \geq 0 \,|\, C \subset D + \tau \mathbb{B}, D \subset C + \tau \mathbb{B}\}.$$

The excess and the Pompeiu–Hausdorff distance are illustrated in Fig. 3.1. They are unaffected by whether C and D are closed or not, but in the case of closed sets the infima in the alternative formulas are attained. Note that both $e(C, D)$ and $h(C, D)$ can sometimes be ∞ when unbounded sets are involved. For that reason in particular, the Pompeiu–Hausdorff distance does not furnish a metric on the space of nonempty closed subsets of $\mathbb{R}^n$, although it does on the space of nonempty closed subsets of a

3 Regularity Properties of Set-valued Solution Mappings

bounded set $X \subset \mathbb{R}^n$. Also note that $e(C, \emptyset) = \infty$ for any set C, including the empty set.

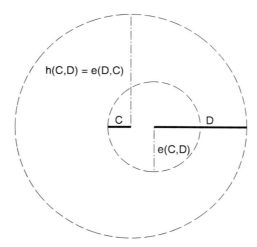

Fig. 3.1 Illustration of the excess and Pompeiu–Hausdorff distance.

Proposition 3A.3 (characterization of Pompeiu–Hausdorff distance). *For any nonempty sets C and D in $\mathbb{R}^n$, one has*

(4) $$h(C,D) = \sup_{x \in \mathbb{R}^n} |d(x,C) - d(x,D)|.$$

Proof. Since the distance to a set doesn't distinguish whether the set is closed or not, we may assume that C and D are nonempty closed sets.

According to 1D.4, for any $x \in \mathbb{R}^n$ we can pick $u \in C$ such that $d(x,u) = d(x,C)$. For any $v \in D$, the triangle inequality tells us that $d(x,v) \leq d(x,u) + d(u,v)$. Taking the infimum on both sides with respect to $v \in D$, we see that $d(x,D) \leq d(x,u) + d(u,D)$, where $d(u,D) \leq e(C,D)$. Therefore, $d(x,D) - d(x,C) \leq e(C,D)$, and by symmetry in exchanging the roles of C and D, also $d(x,C) - d(x,D) \leq e(D,C)$, so that
$$|d(x,C) - d(x,D)| \leq \max\{e(C,D), e(D,C)\} = h(C,D).$$
Hence "$\geq$" holds in (4).

On the other hand, since $d(x,C) = 0$ when $x \in C$, we have
$$e(C,D) = \sup_{x \in C} d(x,D) = \sup_{x \in C} |d(x,D) - d(x,C)| \leq \sup_{x \in \mathbb{R}^n} |d(x,D) - d(x,C)|$$
and likewise $e(D,C) \leq \sup_{x \in \mathbb{R}^n} |d(x,C) - d(x,D)|$, so that
$$\max\{e(C,D), e(D,C)\} \leq \sup_{x \in \mathbb{R}^n} |d(x,C) - d(x,D)|.$$

This confirms that "$\leq$" also holds in (4). □

Pompeiu–Hausdorff convergence. *A sequence of sets $\{C^k\}_{k=1}^{\infty}$ is said to converge with respect to Pompeiu–Hausdorff distance to a set C when C is closed and $h(C^k, C) \to 0$ as $k \to \infty$.*

From the definition of the Pompeiu-Hausdorff distance it follows that when a sequence C^k converges to C, then the set C must be nonempty and only finitely many C^k can be empty. Note that this is *not* the case when Painlevé–Kuratowski convergence is considered (see the first example after 3A.1). The following theorem exhibits the main relationship between these two types of convergence.

Theorem 3A.4 (Pompeiu–Hausdorff versus Painlevé–Kuratowski). *If a sequence of closed sets $\{C^k\}_{k=1}^{\infty}$ converges to C with respect to Pompeiu–Hausdorff distance then it also converges to C in the Painlevé–Kuratowski sense. The opposite implication holds if there is a bounded set X which contains C and every C^k.*

Proof. By definition, C^k converges to C with respect to Pompeiu–Hausdorff distance if and only if, for any $\varepsilon > 0$, there exists $N \in \mathcal{N}$ with

(5) $\qquad C^k \subset C + \varepsilon \mathbb{B} \text{ and } C \subset C^k + \varepsilon \mathbb{B} \quad \text{for all } k \in N.$

Since (5) implies (2a,b), the Painlevé–Kuratowski convergence of C^k to C then follows from the convergence with respect to Pompeiu–Hausdorff distance.

Suppose now that C^k converges to C in the Painlevé–Kuratowski sense, with C and every C^k included in a bounded set X. Then there exists $\rho_0 > 0$ such that $C^k = C^k \cap \rho \mathbb{B}$ and $C = C \cap \rho \mathbb{B}$ for every $\rho \geq \rho_0$. We obtain that for every $\rho > \rho_0$ and $k \in \mathbb{N}$,

$$C^k = C^k \cap \rho \mathbb{B} \subset C + \varepsilon \mathbb{B} \text{ and } C = C \cap \rho \mathbb{B} \subset C^k + \varepsilon \mathbb{B}.$$

But then, for every $\rho > 0$ we have

$$C^k \cap \rho \mathbb{B} \subset C^k \subset C + \varepsilon \mathbb{B} \text{ and } C \cap \rho \mathbb{B} \subset C \subset C^k + \varepsilon \mathbb{B} \text{ for } k \in \mathbb{N}$$

and hence (5) holds and we have convergence of C^k to C with respect to Pompeiu–Hausdorff distance. □

Exercise 3A.5 (convergence equivalence under boundedness). *For a sequence of sets C^k in $\mathbb{R}^n$ and a nonempty closed set C, the following are equivalent:*

(a) C^k converges to C in the Pompeiu–Hausdorff sense and C is bounded;

(b) C^k converges to C in the Painlevé–Kuratowski sense and there is a bounded set X along with an index set $N \in \mathcal{N}$ such that $C^k \subset X$ for all $k \in N$.

Theorem 3A.6 (conditions for Pompeiu–Hausdorff convergence). *A sequence C^k of sets in $\mathbb{R}^n$ is convergent with respect to Pompeiu–Hausdorff distance to a closed set $C \subset \mathbb{R}^n$ if both of the following conditions hold:*

3 Regularity Properties of Set-valued Solution Mappings

(a) for every open set $O \subset \mathbb{R}^n$ with $C \cap O \neq \emptyset$ there exists $N \in \mathcal{N}$ such that $C^k \cap O \neq \emptyset$ for all $k \in N$;

(b) for every open set $O \subset \mathbb{R}^n$ with $C \subset O$ there exists $N \in \mathcal{N}$ such that $C^k \subset O$ for all $k \in N$.

Moreover, condition (a) is always necessary for Pompeiu–Hausdorff convergence, while (b) is necessary when the set C is bounded.

Proof. Let (a)(b) hold and let the first inclusion in (5) be violated, that is, there exist $x \in C$, a scalar $\varepsilon > 0$ and a sequence $N \in \mathcal{N}^\sharp$ such that $x \notin C^k + \varepsilon \mathbb{B}$ for $k \in N$. Then an open neighborhood of x does not meet C^k for infinitely many k; this contradicts condition (a). Furthermore, (b) implies that for any $\varepsilon > 0$ there exists $N \in \mathcal{N}$ such that $C^k \subset C + \varepsilon \mathbb{B}$ for all $k \in N$, which is the second inclusion in (5). Then Pompeiu–Hausdorff convergence follows from (5).

According to 3A.2(a), condition (a) is equivalent to $C \subset \liminf_k C^k$, and hence it is necessary for Painlevé–Kuratowski convergence, and then also for Pompeiu–Hausdorff convergence. To show necessity of (b), let $C \subset O$ for some open set $O \subset \mathbb{R}^n$. For $k \in \mathbb{N}$ let there exist points $x^k \in C$ and y^k in the complement of O such that $|x^k - y^k| \to 0$ as $k \to \infty$. Since C is compact, there exists $N \in \mathcal{N}^\sharp$ and $x \in C$ such that $x^k \xrightarrow{N} x$, hence $y^k \xrightarrow{N} x$ as well. But then x must be also in the complement of O, which is impossible. The contradiction so obtained shows there is an $\varepsilon > 0$ such that $C + \varepsilon \mathbb{B} \subset O$; then, from (5), for some $N \subset \mathcal{N}$ we have $C^k \subset O$ for $k \in N$. □

Examples 3A.7 (unboundedness issues). As an illustration of the troubles that may occur when we deal with unbounded sets, consider first the sequence of bounded sets $C^k \subset \mathbb{R}^2$ in which C^k is the segment having one end at the origin and the other at the point $(\cos \frac{1}{k}, \sin \frac{1}{k})$; that is,

$$C^k = \left\{ x \in \mathbb{R}^2 \,\middle|\, x_1 = t \cos \frac{1}{k},\ x_2 = t \sin \frac{1}{k},\ 0 \leq t \leq 1 \right\}.$$

Both the Painlevé–Kuratowski and Pompeiu–Hausdorff limits exist and are equal to the segment having one end at the origin and the other at the point $(1,0)$. Also, both conditions (a) and (b) in 3A.6 are satisfied.

Let us now modify this example by taking as C^k, instead of a segment, the whole unbounded ray with its end at the origin. That is,

$$C^k = \left\{ x \in \mathbb{R}^2 \,\middle|\, x_1 = t \cos \frac{1}{k},\ x_2 = t \sin \frac{1}{k},\ t \geq 0 \right\}.$$

The Painlevé–Kuratowski limit is the ray $\{x \in \mathbb{R}^2 \,|\, x_1 \geq 0, x_2 = 0\}$, whereas the Pompeiu–Hausdorff limit fails to exist. In this case condition (a) in 3A.6 holds, whereas (b) is violated.

As another example demonstrating issues with unboundedness, consider the sequence of sets

$$C^k = \left\{ x \in \mathbb{R}^2 \,\middle|\, x_1 > 0,\ x_2 \geq \frac{1}{x_1} - \frac{1}{k} \right\},$$

which is obviously convergent with respect to Pompeiu–Hausdorff distance to the set $C = \{x \in \mathbb{R}^2 \,|\, x_1 > 0, x_2 \geq 1/x_1\}$ (choose $k > 1/\varepsilon$ in (5)). On the other hand, condition (b) in 3A.6 fails, since the open set $O = \{x \in \mathbb{R}^2 \,|\, x_1 > 0, x_2 > 0\}$ contains C but does not contain C^k for any k.

3B. Continuity of Set-valued Mappings

Continuity properties of a set-valued mapping $S : \mathbb{R}^m \rightrightarrows \mathbb{R}^n$ can be defined on the basis of Painlevé–Kuratowski set convergence. Alternatively they can be defined on the basis of Pompeiu–Hausdorff set convergence, which is the same in a context of boundedness but otherwise is more stringent and only suited to special situations, as explained at the end of Section 3A. Following the pattern of inner and outer limits used in introducing Painlevé–Kuratowski convergence, we let

$$\limsup_{y \to \bar{y}} S(y) = \bigcup_{y^k \to \bar{y}} \limsup_{k \to \infty} S(y^k)$$
$$= \left\{ x \,\middle|\, \exists y^k \to \bar{y},\, \exists x^k \to x \text{ with } x^k \in S(y^k) \right\}$$

and

$$\liminf_{y \to \bar{y}} S(y) = \bigcap_{y^k \to \bar{y}} \liminf_{k \to \infty} S(y^k)$$
$$= \left\{ x \,\middle|\, \forall y^k \to \bar{y},\, \exists N \in \mathcal{N},\, x^k \xrightarrow{N} x \text{ with } x^k \in S(y^k) \right\}.$$

In other words, the limsup is the set of all possible limits of sequences $x^k \in S(y^k)$ when $y^k \to \bar{y}$, while the liminf is the set of points x for which there exists a sequence $x^k \in S(y^k)$ when $y^k \to \bar{y}$ such that $x^k \to x$.

Semicontinuity and continuity. *A set-valued mapping $S : \mathbb{R}^m \rightrightarrows \mathbb{R}^n$ is outer semi-continuous (osc) at $\bar{y}$ when*

$$\limsup_{y \to \bar{y}} S(y) \subset S(\bar{y})$$

and inner semicontinuous (isc) at $\bar{y}$ when

$$\liminf_{y \to \bar{y}} S(y) \supset S(\bar{y}).$$

It is called Painlevé–Kuratowski continuous at $\bar{y}$ when it is both osc and isc at $\bar{y}$, as expressed by

$$\lim_{y \to \bar{y}} S(y) = S(\bar{y}).$$

On the other hand, S is called Pompeiu–Hausdorff continuous at $\bar{y}$ when

$S(\bar{y})$ is closed and $\lim_{y \to \bar{y}} h(S(y), S(\bar{y})) = 0$.

These terms are invoked *relative to a subset D in* $\mathbb{R}^m$ when the properties hold for limits taken with $y \to \bar{y}$ in D (but not necessarily for limits $y \to \bar{y}$ without this restriction). Continuity is taken to refer to Painlevé–Kuratowski continuity, unless otherwise specified.

For single-valued mappings both definitions of continuity reduce to the usual definition of continuity of a function. Note that when S is isc at $\bar{y}$ relative to D then there must exist a neighborhood V of $\bar{y}$ such that $D \cap V \subset \mathrm{dom}\, S$. When $D = \mathbb{R}^m$, this means $\bar{y} \in \mathrm{int}(\mathrm{dom}\, S)$.

Exercise 3B.1 (limit relations as equations).
(a) *Show that S is osc at $\bar{y}$ if and only if actually* $\limsup_{y \to \bar{y}} S(y) = S(\bar{y})$.
(b) *Show that, when $S(\bar{y})$ is closed, S is isc at $\bar{y}$ if and only if* $\liminf_{y \to \bar{y}} S(y) = S(\bar{y})$.

Although the closedness of $S(\bar{y})$ is automatic from this when S is continuous at $\bar{y}$ in the Painlevé–Kuratowski sense, it needs to be assumed directly for Pompeiu–Hausdorff continuity because the distance concept utilized for that concept is unable to distinguish whether sets are closed or not.

Recall that a set M is closed relative to a set D when any sequence $y^k \in M \cap D$ has its cluster points in M. A set M is open relative to D if the complement of M is closed relative to D. Also, recall that a function $f : \mathbb{R}^n \to \mathbb{R}$ is lower semicontinuous on a closed set $D \subset \mathbb{R}^n$ when the lower level set $\{x \in D \,|\, f(x) \leq \alpha\}$ is closed for every $\alpha \in \mathbb{R}$. We defined this property at the beginning of Chapter 1 for the case of $D = \mathbb{R}^n$ and for functions with values in $\mathbb{R}$, and now are merely echoing that for $D \subset \mathbb{R}^n$.

Theorem 3B.2 (characterization of semicontinuity). *For* $S : \mathbb{R}^m \rightrightarrows \mathbb{R}^n$, *a set* $D \subset \mathbb{R}^m$ *and* $\bar{y} \in \mathrm{dom}\, S$ *we have:*
(a) *S is osc at $\bar{y}$ relative to D if and only if for every $x \notin S(\bar{y})$ there are neighborhoods U of x and V of $\bar{y}$ such that $D \cap V \cap S^{-1}(U) = \emptyset$;*
(b) *S is isc at $\bar{y}$ relative to D if and only if for every $x \in S(\bar{y})$ and every neighborhood U of x there exists a neighborhood V of $\bar{y}$ such that $D \cap V \subset S^{-1}(U)$;*
(c) *S is osc at every $y \in \mathrm{dom}\, S$ if and only if $\mathrm{gph}\, S$ is closed;*
(d) *S is osc relative to a set $D \subset \mathbb{R}^m$ if and only if $S^{-1}(B)$ is closed relative to D for every compact set $B \subset \mathbb{R}^n$;*
(e) *S is isc relative to a set $D \subset \mathbb{R}^m$ if and only if $S^{-1}(O)$ is open relative to D for every open set $O \subset \mathbb{R}^n$;*
(f) *S is osc at $\bar{y}$ relative to a set $D \subset \mathbb{R}^m$ if and only if the distance function $y \mapsto d(x, S(y))$ is lower semicontinuous at $\bar{y}$ relative to D for every $x \in \mathbb{R}^n$;*
(g) *S is isc at $\bar{y}$ relative to a set $D \subset \mathbb{R}^m$ if and only if the distance function $y \mapsto d(x, S(y))$ is upper semicontinuous at $\bar{y}$ relative to D for every $x \in \mathbb{R}^n$.*

Thus, S is continuous relative to D at $\bar{y}$ if and only if the distance function $y \mapsto d(x, S(y))$ is continuous at $\bar{y}$ relative to D for every $x \in \mathbb{R}^n$.

Proof. Necessity in (a): Suppose that there exists $x \notin S(\bar{y})$ such that for any neighborhood U of x and neighborhood V of $\bar{y}$ we have $S(y) \cap U \neq \emptyset$ for some $y \in V \cap D$. But then there exists a sequence $y^k \to \bar{y}$, $y^k \in D$ and $x^k \in S(y^k)$ such that $x^k \to x$. This implies that $x \in \limsup_k S(y^k)$, hence $x \in S(\bar{y})$ since S is osc, a contradiction.

Sufficiency in (a): Let $x \notin S(\bar{y})$. Then there exists $\rho > 0$ such that $S(\bar{y}) \cap B_\rho(x) = \emptyset$; the condition in the second half of (a) then gives a neighborhood V of $\bar{y}$ such that for every $N \in \mathcal{N}$, every sequence $y^k \xrightarrow{N} \bar{y}$ with $y^k \in D \cap V$ has $S(y^k) \cap B_\rho(x) = \emptyset$. But in that case $d(x, S(y^k)) > \rho/2$ for all large k, which implies, by Proposition 3A.1 and the definition of limsup, that $x \notin \limsup_{y \to \bar{y}} S(y)$. This means that S is osc at $\bar{y}$.

Necessity in (b): Suppose that there exists $x \in S(\bar{y})$ such that for some neighborhood U of x and any neighborhood V of $\bar{y}$ we have $S(y) \cap U = \emptyset$ for some $y \in V \cap D$. Then there is a sequence y^k convergent to $\bar{y}$ in D such that for every sequence $x^k \to x$ one has $x^k \notin S(y^k)$. This means that $x \notin \liminf_{y \to \bar{y}} S(y)$. But then S is not isc at $\bar{y}$.

Sufficiency in (b): If S is not isc at $\bar{y}$ relative to D, then, according to 3A.2(a), there exist an infinite sequence $y^k \to \bar{y}$ in D, a point $x \in S(\bar{y})$ and an open neighborhood U of x such that $S(y^k) \cap U = \emptyset$ for infinitely many k. But then there exists a neighborhood V of $\bar{y}$ such that $D \cap V$ is not in $S^{-1}(U)$ which is the opposite of (b).

(c): S has closed graph if and only if for any $(y, x) \notin \text{gph } S$ there exist open neighborhoods V of y and U of x such that $V \cap S^{-1}(U) = \emptyset$. From (a), this comes down to S being osc at every $y \in \text{dom } S$.

(d): Every sequence in a compact set B has a convergent subsequence, and on the other hand, a set consisting of a convergent sequence and its limit is a compact set. Therefore the condition in the second part of (d) is equivalent to the condition that if $x^k \to \bar{x}$, $y^k \in S^{-1}(x^k)$ and $y^k \to \bar{y}$ with $y^k \in D$, one has $\bar{y} \in S^{-1}(\bar{x})$. But this is precisely the condition for S to be osc relative to D.

(e): Failure of the condition in (e) means the existence of an open set O and a sequence $y^k \to \bar{y}$ in D such that $\bar{y} \in S^{-1}(O)$ but $y^k \notin S^{-1}(O)$; that is, $S(\bar{y}) \cap O \neq \emptyset$ yet $S(y^k) \cap O = \emptyset$ for all k. This last property says that $\liminf_k S(y^k) \not\supset S(\bar{y})$, by 3A.2(a). Hence the condition in (e) fails precisely when S is not isc.

The equivalences in (f) and (g) follow from 3A.2(e) and 3A.2(f). $\square$

Theorem 3B.3 (characterization of Pompeiu–Hausdorff continuity). *A set-valued mapping $S : \mathbb{R}^m \rightrightarrows \mathbb{R}^n$ is Pompeiu–Hausdorff continuous at $\bar{y}$ if $S(\bar{y})$ is closed and both of the following conditions hold:*

(a) *for every open set $O \subset \mathbb{R}^n$ with $S(\bar{y}) \cap O \neq \emptyset$ there exists a neighborhood V of $\bar{y}$ such that $S(y) \cap O \neq \emptyset$ for all $y \in V$;*

(b) *for every open set $O \subset \mathbb{R}^n$ with $S(\bar{y}) \subset O$ there exists a neighborhood V of $\bar{y}$ such that $S(y) \subset O$ for all $y \in V$.*

Moreover, if S is Pompeiu–Hausdorff continuous at $\bar{y}$, then it is continuous at $\bar{y}$. On the other hand, when $S(\bar{y})$ is nonempty and bounded, Pompeiu–Hausdorff continuity of S at $\bar{y}$ reduces to continuity together with the existence of a neighborhood V of $\bar{y}$ such that $S(V)$ is bounded; in this case conditions (a) and (b) are not only sufficient but also necessary for continuity of S at $\bar{y}$.

3 Regularity Properties of Set-valued Solution Mappings

Observe that we can define inner semicontinuity of a mapping S at $\bar{y}$ in the Pompeiu–Hausdorff sense by $\lim_{y\to\bar{y}} e(S(\bar{y}), S(y)) = 0$, but this is simply equivalent to inner semicontinuity in the Painlevé–Kuratowski sense (compare 3B.2(b) and 3B.3(b)). In contrast, if we define outer semicontinuity in the Pompeiu–Hausdorff sense by $\lim_{y\to\bar{y}} e(S(y), S(\bar{y})) = 0$, we get a generally much more restrictive concept than outer semicontinuity in the Painlevé–Kuratowski sense.

We present next two applications of these concepts to mappings that play central roles in optimization.

Example 3B.4 (solution mapping for a system of inequalities). Consider a mapping defined implicitly by a parameterized system of inequalities, that is,

$$S : p \mapsto \{x \mid f_i(p,x) \leq 0, i = 1, \ldots, m\} \text{ for } p \in \mathbb{R}^d.$$

Assume that each f_i is a continuous real-valued function on $\mathbb{R}^d \times \mathbb{R}^n$. Then S is osc at any point of its domain. If moreover each f_i is convex in x for each p and $\bar{p}$ is such that there exists $\bar{x}$ with $f_i(\bar{p}, \bar{x}) < 0$ for each $i = 1, \ldots, m$, then S is continuous at $\bar{p}$.

Detail. The graph of S is the intersection of the sets $\{(p,x) \mid f_i(p,x) \leq 0\}$, which are closed by the continuity of f_i. Then gph S is closed, and the osc property comes from Theorem 3B.2(c). The isc part will follow from a much more general result (Robinson-Ursescu theorem) which we present in Chapter 5.

Applications in optimization. Consider the following general problem of minimization, involving a parameter p which ranges over a set $P \subset \mathbb{R}^d$, a function $f_0 : \mathbb{R}^d \times \mathbb{R}^n \to \mathbb{R}$, and a mapping $S_{\text{feas}} : P \rightrightarrows \mathbb{R}^n$:

minimize $f_0(p,x)$ over all $x \in \mathbb{R}^n$ satisfying $x \in S_{\text{feas}}(p)$.

Here f_0 is the *objective function* and S_{feas} is the *feasible set mapping* (with $S_{\text{feas}}(p)$ taken to be the empty set when $p \notin P$). In particular, S_{feas} could be specified by constraints in the manner of Example 3B.4, but we now allow it to be more general.

Our attention is focused now on two other mappings in this situation: the *optimal value* mapping acting from $\mathbb{R}^d$ to $\mathbb{R}$ and defined by

$$S_{\text{val}} : p \mapsto \inf_x \{f_0(p,x) \mid x \in S_{\text{feas}}(p)\} \text{ when the inf is finite,}$$

and the *optimal set* mapping acting from P to $\mathbb{R}^n$ and defined by

$$S_{\text{opt}} : p \mapsto \{x \in S_{\text{feas}}(p) \mid f_0(p,x) = S_{\text{val}}(p)\}.$$

Theorem 3B.5 (basic continuity properties of solution mappings in optimization). *In the preceding notation, let $\bar{p} \in P$ be fixed with the feasible set $S_{\text{feas}}(\bar{p})$ nonempty and bounded, and suppose that:*

(a) *the mapping S_{feas} is Pompeiu–Hausdorff continuous at $\bar{p}$ relative to P, or equivalently, S_{feas} is continuous at $\bar{p}$ relative to P with $S_{\text{feas}}(Q \cap P)$ bounded for some neighborhood Q of $\bar{p}$,*

(b) *the function f_0 is continuous relative to $P \times \mathbb{R}^n$ at $(\bar{p}, \bar{x})$ for every $\bar{x} \in S_{\text{feas}}(\bar{p})$.*

Then the optimal value mapping S_{val} is continuous at $\bar{p}$ relative to P, whereas the optimal set mapping S_{opt} is osc at $\bar{p}$ relative to P.

Proof. The equivalence in assumption (a) comes from the final statement in Theorem 3B.3. In particular (a) implies $S_{\text{feas}}(\bar{p})$ is closed, hence from boundedness actually compact. Then too, since $f_0(\bar{p}, \cdot)$ is continuous on $S_{\text{feas}}(\bar{p})$ by (b), the set $S_{\text{opt}}(\bar{p})$ is nonempty.

Let $\bar{x} \in S_{\text{opt}}(\bar{p})$. From (a) we get for any sequence $p^k \to \bar{p}$ in P the existence of a sequence of points x^k with $x^k \in S_{\text{feas}}(p^k)$ such that $x^k \to \bar{x}$ as $k \to \infty$. But then, for any $\varepsilon > 0$ there exists $N \in \mathcal{N}$ such that

$$S_{\text{val}}(p^k) \leq f_0(p^k, x^k) \leq f_0(\bar{p}, \bar{x}) + \varepsilon = S_{\text{val}}(\bar{p}) + \varepsilon \quad \text{for } k \in N.$$

This gives us

(1) $$\limsup_{p \to \bar{p}} S_{\text{val}}(p) \leq S_{\text{val}}(\bar{p}).$$

On the other hand, let us assume that

(2) $$\liminf_{p \to \bar{p}} S_{\text{val}}(p) < S_{\text{val}}(\bar{p}).$$

Then there exist $\varepsilon > 0$ and sequences $p_k \to \bar{p}$ in P and $x^k \in S_{\text{feas}}(p^k)$, $k \in \mathbb{N}$, such that

(3) $$f_0(p^k, x^k) < S_{\text{val}}(\bar{p}) - \varepsilon \quad \text{for all } k.$$

From (a) we see that $d(x^k, S_{\text{feas}}(\bar{p})) \to 0$ as $k \to \infty$. This provides the existence of a sequence of points $\bar{x}^k \in S_{\text{feas}}(\bar{p})$ such that $|x^k - \bar{x}^k| \to 0$ as $k \to \infty$. Because $S_{\text{feas}}(\bar{p})$ is compact, there must be some $\bar{x} \in S_{\text{feas}}(\bar{p})$ along with an index set $N \in \mathcal{N}^{\sharp}$ such that $\bar{x}^k \xrightarrow{N} \bar{x}$, in which case $x^k \xrightarrow{N} \bar{x}$ as well. Then, from the continuity of f_0 at $(\bar{p}, \bar{x})$, we have $f_0(\bar{p}, \bar{x}) \leq f_0(p^k, x^k) + \varepsilon$ for $k \in N$ and sufficiently large, which, together with (3), implies for such k that

$$S_{\text{val}}(\bar{p}) \leq f_0(\bar{p}, \bar{x}) \leq f_0(p^k, x^k) + \varepsilon < S_{\text{val}}(\bar{p}).$$

The contradiction obtained proves that (2) is false. Thus,

$$S_{\text{val}}(\bar{p}) \leq \liminf_{p \to \bar{p}} S_{\text{val}}(p),$$

which, combined with (1), gives us the continuity of the optimal value mapping S_{val} at $\bar{p}$ relative to P.

3 Regularity Properties of Set-valued Solution Mappings

To show that S_{opt} is osc at $\bar{p}$ relative to P, we use the equivalent condition in 3B.2(a). Suppose there exists $x \notin S_{\text{opt}}(\bar{p})$ such that for any neighborhoods U of x and Q of $\bar{p}$ there exist $u \in U$ and $p \in Q \cap P$ such that $u \in S_{\text{opt}}(p)$. This is the same as saying that there are sequences $p^k \to \bar{p}$ in P and $u^k \to x$ as $k \to \infty$ such that $u^k \in S_{\text{opt}}(p^k)$. Note that since $u^k \in S_{\text{feas}}(p^k)$ we have that $x \in S_{\text{feas}}(\bar{p})$. But then, as we already proved,

$$f_0(p^k, u^k) = S_{\text{val}}(p^k) \to S_{\text{val}}(\bar{p}) \text{ as } k \to \infty.$$

By continuity of f_0, the left side tends to $f_0(\bar{p}, x)$ as $k \to \infty$, which means that $x \in S_{\text{opt}}(\bar{p})$, a contradiction. □

Example 3B.6 (minimization over a fixed set). Let X be a nonempty, compact subset of $\mathbb{R}^n$ and let f_0 be a continuous function from $P \times X$ to $\mathbb{R}$, where P is a nonempty subset of $\mathbb{R}^d$. For each $p \in P$, let

$$S_{\text{val}}(p) = \min_{x \in X} f_0(p, x), \qquad S_{\text{opt}}(p) = \operatorname*{argmin}_{x \in X} f_0(p, x).$$

Then the function $S_{\text{val}} : P \to \mathbb{R}$ is continuous relative to P, and the mapping $S_{\text{opt}} : P \rightrightarrows \mathbb{R}^n$ is osc relative to P.

Detail. This exploits the case of Theorem 3B.5 where S_{feas} is the constant mapping $p \mapsto X$. □

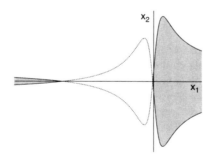

Fig. 3.2 The feasible set in Example 3B.7 for $p = 0.1$.

Example 3B.7 (Painlevé–Kuratowski continuity of the feasible set does not imply continuity of the optimal value). Consider the minimization of $f_0(x_1, x_2) = e^{x_1} + x_2^2$ on the set

$$S_{\text{feas}}(p) = \left\{ (x_1, x_2) \in \mathbb{R}^2 \;\middle|\; -\frac{x_1}{(1+x_1^2)} - p \leq x_2 \leq \frac{x_1}{(1+x_1^2)} + p \right\}.$$

For parameter value $p = 0$, the optimal value $S_{\text{val}}(0) = 1$ and occurs at $S_{\text{opt}}(0) = (0,0)$, but for $p > 0$ the asymptotics of the function $x_1/(1+x_1^2)$ open up a "phantom" portion of the feasible set along the negative x_1-axis, and the optimal value is 0, see Fig. 3.2. The feasible set nonetheless does depend continuously on p in $[0,\infty)$ in the Painlevé–Kuratowski sense.

3C. Lipschitz Continuity of Set-valued Mappings

A quantitative notation of continuity for set-valued mappings can be formulated with the help of the Pompeiu–Hausdorff distance between sets in the same way that Lipschitz continuity is defined for functions. It has important uses, although it suffers from shortcomings when the sets may be unbounded. Here we invoke the terminology that a set-valued mapping S is *closed-valued* on a set D when $S(y)$ is a closed set for each $y \in D$.

Lipschitz continuity of set-valued mappings. *A mapping $S : \mathbb{R}^m \rightrightarrows \mathbb{R}^n$ is said to be Lipschitz continuous relative to a (nonempty) set D in $\mathbb{R}^m$ if $D \subset \text{dom } S$, S is closed-valued on D, and there exists $\kappa \geq 0$ (Lipschitz constant) such that*

(1) $$h(S(y'), S(y)) \leq \kappa |y' - y| \quad \text{for all } y', y \in D,$$

or equivalently, there exists $\kappa \geq 0$ such that

(2) $$S(y') \subset S(y) + \kappa |y' - y| \mathbb{B} \quad \text{for all } y', y \in D.$$

When S is single-valued on D, we obtain from this definition the previous notion in Section 1D of Lipschitz continuity of a function.

One could contemplate defining Lipschitz continuity of a set-valued mapping S without requiring S to be closed-valued, relying in that case simply on (1) or (2). That might be workable, although κ in (2) could then be slightly larger than the κ in (1), but a fundamental objection arises. A mapping that is continuous necessarily does have closed values, so we would be in the position of having a concept of Lipschitz continuity which did not entail continuity. That is a paradox we prefer to avoid. The issue is absent for single-valued mappings, since they are trivially closed-valued.

Lipschitz continuity of a set-valued mapping can be characterized by a property which relates the distances to its value and the value of the inverse mapping:

Proposition 3C.1 (distance characterization of Lipschitz continuity). *Consider a closed-valued mapping $S : \mathbb{R}^m \rightrightarrows \mathbb{R}^n$ and a nonempty subset D of dom S. Then S is Lipschitz continuous relative to D with constant κ if and only if*

(3) $$d(x, S(y)) \leq \kappa d(y, S^{-1}(x) \cap D) \quad \text{for all } x \in \mathbb{R}^n \text{ and } y \in D.$$

Proof. Let S be Lipschitz continuous relative to D with a constant κ and let $x \in \mathbb{R}^n$ and $y \in D$. If $S^{-1}(x) \cap D = \emptyset$, the inequality (3) holds automatically. Let $S^{-1}(x) \cap D \neq \emptyset$ and choose $\varepsilon > 0$. Then there exists $y' \in S^{-1}(x) \cap D$ with $|y' - y| \leq d(y, S^{-1}(x) \cap D) + \varepsilon$. By (1),

$$d(x, S(y)) \leq h(S(y'), S(y)) \leq \kappa |y' - y| \leq \kappa d(y, S^{-1}(x) \cap D) + \kappa \varepsilon.$$

Since the left side of this inequality does not depend on ε, passing to zero with ε we conclude that (3) holds with the κ of (1).

Conversely, let (3) hold, let $y, y' \in D \subset \text{dom } S$ and let $x \in S(y)$. Then

$$d(x, S(y')) \leq \kappa d(y', S^{-1}(x) \cap D) \leq \kappa |y - y'|,$$

since $y \in S^{-1}(x) \cap D$. Taking the supremum with respect to $x \in S(y)$, we obtain $e(S(y), S(y')) \leq \kappa |y - y'|$ and, by symmetry, we get (1). □

For the inverse mapping $F = S^{-1}$ the property described in (3) can be written as

$$d(x, F^{-1}(y)) \leq \kappa d(y, F(x) \cap D) \quad \text{for all } x \in \mathbb{R}^n, \, y \in D,$$

and when gph F is closed this can be interpreted in the following manner. Whenever we pick a $y \in D$ and an $x \in \text{dom } F$, the distance from x to the set of solutions u of the inclusion $y \in F(u)$ is proportional to $d(y, F(x) \cap D)$, which measures the extent to which x itself fails to solve this inclusion. In Section 3E we will introduce a local version of this property which plays a major role in variational analysis and is known as "metric regularity."

The difficulty with the concept of Lipschitz continuity for set-valued mappings S with values $S(y)$ that may be unbounded comes from the fact that usually $h(C_1, C_2) = \infty$ when C_1 or C_2 is unbounded, the only exceptions being cases where both C_1 and C_2 are unbounded and "the unboundedness points in the same direction." For instance, when C_1 and C_2 are lines in $\mathbb{R}^2$, one has $h(C_1, C_2) < \infty$ only when these lines are parallel.

In the remainder of this section we consider a particular class of set-valued mappings, with significant applications in variational analysis, which are automatically Lipschitz continuous even when their values are unbounded sets.

Polyhedral convex mappings. *A mapping $S : \mathbb{R}^m \rightrightarrows \mathbb{R}^n$ is said to be polyhedral convex if its graph is a polyhedral convex set.*

Here it should be recalled from Section 2E that a set is polyhedral convex if it can be expressed as the intersection of a finite collection of closed half-spaces and/or hyperplanes.

Example 3C.2 (polyhedral convex mappings from linear constraint systems). A solution mapping S of the form in Example 3B.4 is polyhedral convex when the f_i there are all affine; furthermore, this continues to be true when some or all of the constraints are equations instead of inequalities.

In the notational context of elements $x \in S(y)$ for a mapping $S : \mathbb{R}^m \rightrightarrows \mathbb{R}^n$, polyhedral convexity of S is equivalent to the existence of a positive integer r, matrices $D \in \mathbb{R}^{r \times n}$, $E \in \mathbb{R}^{r \times m}$, and a vector $q \in \mathbb{R}^r$ such that

(4) $\qquad S(y) = \{x \in \mathbb{R}^n \,|\, Dx + Ey \leq q\}$ for all $y \in \mathbb{R}^m$.

Note for instance that any mapping S whose graph is a linear subspace is a polyhedral convex mapping.

Theorem 3C.3 (Lipschitz continuity of polyhedral convex mappings). *Any polyhedral convex mapping $S : \mathbb{R}^m \rightrightarrows \mathbb{R}^n$ is Lipschitz continuous relative to its domain.*

We will prove this theorem by using a fundamental result due to A. J. Hoffman regarding approximate solutions of systems of linear inequalities. For a vector $a = (a_1, a_2, \ldots, a_n) \in \mathbb{R}^n$, we use the vector notation that

$$a_+ = (\max\{0, a_1\}, \ldots, \max\{0, a_n\}).$$

Also, recall that the *convex hull* of a set $C \subset \mathbb{R}^n$, which will be denoted by co C, is the smallest convex set that includes C. (It can be identified as the intersection of all convex sets that include C, but also can be described as consisting of all linear combinations $\lambda_0 x_0 + \lambda_1 x_1 + \cdots + \lambda_n x_n$ with $x_i \in C$, $\lambda_i \geq 0$, and $\lambda_0 + \lambda_1 + \cdots + \lambda_n = 1$; this is Carathéodory's theorem.) The *closed convex hull* of C is the closure of the convex hull of C and denoted cl co C; it is the smallest closed convex set that contains C.

Lemma 3C.4 (Hoffman lemma). *For the set-valued mapping*

$$S : y \mapsto \{x \in \mathbb{R}^n \,|\, Ax \leq y\} \quad \text{for } y \in \mathbb{R}^m,$$

where A is a nonzero $m \times n$ matrix, there exists a constant L such that

(5) $\qquad d(x, S(y)) \leq L |(Ax - y)_+|$ *for every $y \in \operatorname{dom} S$ and every $x \in \mathbb{R}^n$.*

Proof. For any $y \in \operatorname{dom} S$ the set $S(y)$ is nonempty, convex and closed, hence any point $x \notin S(y)$ has a unique (Euclidean) projection $u = P_{S(y)}(x)$ on $S(y)$ (Proposition 1D.5):

(6) $\qquad u \in S(y), \ |u - x| = d(x, S(y)).$

As noted in Section 2A, the projection mapping satisfies

$$P_{S(y)} = (I + N_{S(y)})^{-1},$$

where $N_{S(y)}$ is the normal cone mapping to the convex set $S(y)$. In these terms, the problem of projecting x on $S(y)$ is equivalent to that of finding the unique $u \neq x$ such that

$$x \in u + N_{S(y)}(u).$$

3 Regularity Properties of Set-valued Solution Mappings

The formula in 2E(8) gives us a representation of the normal cone to a polyhedral convex set specified by affine inequalities, which here comes out as

$$N_{S(y)}(u) = \left\{ v \,\middle|\, v = \sum_{i=1}^{m} \lambda_i a_i \text{ with } \lambda_i \geq 0, \, \lambda_i(\langle a_i, u \rangle - y_i) = 0, \, i = 1, \ldots, m \right\},$$

where the a_i's are the rows of the matrix A regarded as vectors in $\mathbb{R}^n$. Thus, the projection u of x on $S(y)$, as described by (6), can be obtained by finding a pair (u, λ) such that

$$(7) \qquad \begin{cases} x - u - \sum_{i=1}^{m} \lambda_i a_i = 0, \\ \lambda_i \geq 0, \, \lambda_i(\langle a_i, u \rangle - y_i) = 0, \, i = 1, \ldots, m. \end{cases}$$

While the projection u exists and is unique, this variational inequality might not have a unique solution (u, λ) because the λ component might not be unique. But since $u \neq x$ (through our assumption that $x \notin S(y)$), we can conclude from the first relation in (7) that for any solution (u, λ) the vector $\lambda = (\lambda_1, \ldots, \lambda_m)$ is not the zero vector. Consider the family $\mathscr{J}$ of subsets J of $\{1, \ldots, m\}$ for which there are real numbers $\lambda_1, \ldots, \lambda_m$ with $\lambda_i > 0$ for $i \in J$ and $\lambda_i = 0$ for $i \notin J$ and such that (u, λ) satisfies (7). Of course, if $\langle a_i, u \rangle - y_i < 0$ for some i, then $\lambda_i = 0$ according to the second relation (complementarity) in (7), and then this i cannot be an element of any J. That is,

$$(8) \qquad J \in \mathscr{J} \text{ and } i \in J \implies \langle a_i, u \rangle = y_i \text{ and } \lambda_i > 0.$$

Since the set of vectors λ such that (u, λ) solves (7) does not contain the zero vector, we have $\mathscr{J} \neq \emptyset$.

We will now prove that there is a nonempty index set $\bar{J} \in \mathscr{J}$ for which there are no numbers $\beta_i, \, i \in \bar{J}$ satisfying

$$(9) \qquad \beta_i \geq 0, \, i \in \bar{J}, \quad \sum_{j \in \bar{J}} \beta_i > 0 \text{ and } \sum_{i \in \bar{J}} \beta_i a_i = 0.$$

On the contrary, suppose that for every $J \in \mathscr{J}$ this is not the case, that is, (9) holds with $\bar{J} = J$ for some $\beta_i, \, i \in J$. Let J' be a set in $\mathscr{J}$ with a minimal number of elements (J' might be not unique). Note that the number of elements in any J' is greater than 1. Indeed, if there were just one element i' in J', then we would have $\beta_{i'} a_{i'} = 0$ and $\beta_{i'} > 0$, hence $a_{i'} = 0$, and then, since (7) holds for (u, λ) such that $\lambda_i = \beta_i, \, i = i', \, \lambda_i = 0, \, i \neq i'$, from the first equality in (7) we would get $x = u$ which contradicts the assumption that $x \notin S(y)$. Since $J' \in \mathscr{J}$, there are $\lambda'_i > 0, \, i \in J'$ such that

$$(10) \qquad x - u = \sum_{i \in J'} \lambda'_i a_i.$$

By assumption, there are also real numbers $\beta'_i \geq 0, \, i \in J'$, such that

$$\text{(11)} \qquad \sum_{i \in J'} \beta'_i > 0 \quad \text{and} \quad \sum_{i \in J'} \beta'_i a_i = 0.$$

Multiplying both sides of the equality in (11) by a positive scalar t and adding to (10), we obtain

$$x - u = \sum_{i \in J'} (\lambda'_i - t\beta'_i) a_i.$$

Let

$$t_0 = \min_i \left\{ \frac{\lambda'_i}{\beta'_i} \;\middle|\; i \in J' \text{ with } \beta'_i > 0 \right\}.$$

Then for any $k \in J'$ for which this minimum is attained, we have

$$\lambda'_i - t_0 \beta'_i \geq 0 \quad \text{for every } i \in J' \setminus k \quad \text{and} \quad x - u = \sum_{i \in J' \setminus k} (\lambda'_i - t\beta'_i) a_i.$$

Thus, the vector $\lambda \in \mathbb{R}^m$ with components $\lambda'_i - t_0 \beta'_i$ when $i \in J'$ and $\lambda'_i = 0$ when $i \notin J'$ is such that (u, λ) satisfies (7). Hence, we found a nonempty index set $J'' \in \mathscr{J}$ having fewer elements than J', which contradicts the choice of J'. The contradiction obtained proves that there is a nonempty index set $\bar{J} \in \mathscr{J}$ for which there are no numbers β_i, $i \in J$, satisfying (8). In particular, the zero vector in $\mathbb{R}^n$ is not in the convex hull $\operatorname{co}\{a_j, j \in \bar{J}\}$.

Let $\bar{\lambda}_i > 0$, $i \in \bar{J}$, be the corresponding vector of multipliers such that, if we set $\bar{\lambda}_i = 0$ for $i \notin \bar{J}$, we have that $(u, \bar{\lambda})$ is a solution of (7). Since $\sum_{j \in \bar{J}} \bar{\lambda}_i a_i \neq 0$, because otherwise (9) would hold for $\beta_i = \bar{\lambda}_i$, we have

$$\gamma := \sum_{i \in \bar{J}} \bar{\lambda}_i > 0.$$

Because (7) holds with $(u, \bar{\lambda})$, using (7) and (8) we have

$$d(0, \operatorname{co}\{a_j, j \in \bar{J}\}) |x - u| \leq \left| \sum_{i \in \bar{J}} \frac{\bar{\lambda}_i}{\gamma} a_i \right| |x - u| = \frac{1}{\gamma} |x - u| |x - u|$$

$$= \left\langle \frac{1}{\gamma}(x - u), x - u \right\rangle = \left\langle \frac{1}{\gamma} \left(\sum_{i \in \bar{J}} \bar{\lambda}_i a_i \right), x - u \right\rangle$$

$$= \sum_{i \in \bar{J}} \frac{\bar{\lambda}_i}{\gamma} (\langle a_i, x \rangle - \langle a_i, u \rangle) = \sum_{i \in \bar{J}} \frac{\bar{\lambda}_i}{\gamma} (\langle a_i, x \rangle - y_i)$$

$$\leq \max_{i \in \bar{J}} \{(\langle a_i, x \rangle - y_i)_+\}.$$

Hence, for some constant c independent of x and y we have

$$d(x, S(y)) = |x - u| \leq c \max_{1 \leq i \leq m} \{(\langle a_i, x \rangle - y_i)_+\}.$$

This inequality remains valid (perhaps with a different constant c) after passing from the max vector norm to the equivalent Euclidean norm. This proves (5). □

3 Regularity Properties of Set-valued Solution Mappings

Proof of Theorem 3C.3. Let $y, y' \in \text{dom } S$ and let $x \in S(y)$. Since S is polyhedral, from the representation (4) we have $Dx + Ey - q \leq 0$ and then

(12) $$Dx + Ey' - q = Dx + Ey - q - Ey + Ey' \leq -Ey + Ey'.$$

Then from Lemma 3C.4 above we obtain the existence of a constant L such that

$$d(x, S(y')) \leq L |(Dx + Ey' - q)_+|,$$

and hence, by (12),

$$d(x, S(y')) \leq L |(E(y' - y))_+| \leq L |E(y - y')|.$$

Since x is arbitrarily chosen in $S(y)$, this leads to

$$e(S(y), S(y')) \leq \kappa |y - y'|$$

with $\kappa = L|E|$. The same must hold with the roles of y and y' reversed, and in consequence S is Lipschitz continuous on dom S. □

Applications to solution mappings in linear programming. Consider the following problem of linear programming in which y acts as a parameter:

(13) minimize $\langle c, x \rangle$ over all $x \in \mathbb{R}^n$ satisfying $Ax \leq y$.

Here c is a fixed vector in $\mathbb{R}^n$, A is a fixed matrix in $\mathbb{R}^{m \times n}$. Define the solution mappings associated with (13) as in Section 3B, that is, the feasible set mapping

(14) $$S_{\text{feas}} : y \mapsto \{x \,|\, Ax \leq y\},$$

the optimal value mapping

(15) $$S_{\text{val}} : y \mapsto \inf_x \{ \langle c, x \rangle \,|\, Ax \leq y\} \text{ when the inf is finite,}$$

and the optimal set mapping by

(16) $$S_{\text{opt}} : y \mapsto \{x \in S_{\text{feas}}(y) \,|\, \langle c, x \rangle = S_{\text{val}}(y)\}.$$

It is known from the theory of linear programming that $S_{\text{opt}}(y) \neq \emptyset$ when the infimum in (15) is finite (and only then).

Exercise 3C.5 (Lipschitz continuity of mappings in linear programming). *Establish that the mappings in (14), (15) and (16) are Lipschitz continuous relative to their domains, the domain in the case of (15) and (16) being the set D consisting of all y for which the infimum in (15) is finite.*

Guide. Derive the Lipschitz continuity of S_{feas} from Theorem 3C.3, out of the connection with Example 3C.2. Let κ be a Lipschitz constant for S_{feas}.

Next, for the case of S_{val}, consider any $y, y' \in D$ and any $x \in S_{\text{opt}}(y)$, which exists because S_{opt} is nonempty when $y \in D$. In particular we have $x \in S_{\text{feas}}(y)$. From the Lipschitz continuity of S_{feas}, there exists $x' \in S_{\text{feas}}(y')$ such that $|x - x'| \leq \kappa |y - y'|$. Use this along with the fact that $S_{\text{val}}(y') \leq \langle c, x' \rangle$ but $S_{\text{val}}(y) = \langle c, x \rangle$ to get a bound on $S_{\text{val}}(y') - S_{\text{val}}(y)$ which confirms the Lipschitz continuity claimed for S_{val}.

For the case of S_{opt}, consider the set-valued mapping

$$G : (y, t) \mapsto \{ x \in \mathbb{R}^n \mid Ax \leq y, \langle c, x \rangle \leq t \} \quad \text{for } (y, t) \in \mathbb{R}^m \times \mathbb{R}.$$

Confirm that this mapping is polyhedral convex and apply Theorem 3C.3 to it. Observe that $S_{\text{opt}}(y) = G(y, S_{\text{val}}(y))$ for $y \in D$ and invoke the Lipschitz continuity of S_{val}. □

3D. Outer Lipschitz Continuity

In this section we define a "one-point" property of set-valued mappings by fixing one of the points y and y' in the definition of Lipschitz continuity at its reference value $\bar{y}$. Then these points no longer play symmetric roles, so we use the excess instead of the Pompeiu–Hausdorff distance.

Outer Lipschitz continuity. *A mapping $S : \mathbb{R}^m \rightrightarrows \mathbb{R}^n$ is said to be outer Lipschitz continuous at $\bar{y}$ relative to a set D if $\bar{y} \in D \subset \text{dom } S$, $S(\bar{y})$ is a closed set, and there is a constant $\kappa \geq 0$ along with a neighborhood V of $\bar{y}$ such that*

(1) $\quad\quad\quad\quad e(S(y), S(\bar{y})) \leq \kappa |y - \bar{y}| \quad \text{for all } y \in V \cap D,$

or equivalently

(2) $\quad\quad\quad\quad S(y) \subset S(\bar{y}) + \kappa |y - \bar{y}| \mathbb{B} \quad \text{for all } y \in V \cap D.$

If S is outer Lipschitz continuous at every point $y \in D$ relative to D with the same κ, then S is said to be outer Lipschitz continuous relative to D.

It is clear that any mapping which is Lipschitz continuous relative to a set D with constant κ is also outer Lipschitz continuous relative to D with constant κ, but the converse may not be true. Also, outer Lipschitz continuity at a point $\bar{y}$ implies outer semicontinuity at $\bar{y}$. For single-valued mappings, outer Lipschitz continuity becomes the property of calmness which we considered in Section 1C. The examples in Section 1D show how very different this property is from the Lipschitz continuity.

The condition in the definition that the mapping is closed-valued at $\bar{y}$ could be dropped; but then the constant κ in (2) might be slightly larger than the one in

3 Regularity Properties of Set-valued Solution Mappings

(1), and furthermore outer Lipschitz continuity might not entail outer semicontinuity (where closed-valuedness is essential). Therefore, we hold back from such an extension.

We present next a result which historically was the main motivation for introducing the property of outer Lipschitz continuity and which complements Theorem 3C.3. It uses the following concept.

Polyhedral mappings. *A set-valued mapping* $S : \mathbb{R}^n \rightrightarrows \mathbb{R}^m$ *will be called polyhedral if* gph S *is the union of finitely many sets that are polyhedral convex in* $\mathbb{R}^n \times \mathbb{R}^m$.

Clearly, a polyhedral mapping has closed graph, since polyhedral convex sets are closed, and hence is osc and in particular closed-valued everywhere. Any polyhedral *convex* mapping as defined in 3C is obviously a polyhedral mapping, but the graph then is comprised of only one "piece," whereas now we are allowing a multiplicity of such polyhedral convex "pieces," which furthermore could overlap.

Theorem 3D.1 (outer Lipschitz continuity of polyhedral mappings). *Any polyhedral mapping* $S : \mathbb{R}^m \rightrightarrows \mathbb{R}^n$ *is outer Lipschitz continuous relative to its domain.*

Proof. Let gph $S = \bigcup_{i=1}^k G_i$ where the G_i's are polyhedral convex sets in $\mathbb{R}^m \times \mathbb{R}^n$. For each i define the mapping

$$S_i : y \mapsto \{x \mid (y,x) \in G_i\} \quad \text{for } y \in \mathbb{R}^m.$$

Then each S_i is Lipschitz continuous on its domain, according to Theorem 3C.3. Let $\bar{y} \in \mathrm{dom}\, S$ and let

$$\mathscr{I} = \{i \mid \text{there exists } x \in \mathbb{R}^n \text{ with } (\bar{y},x) \in G_i\}.$$

Then $\bar{y} \in \mathrm{dom}\, S_i$ for each $i \in \mathscr{I}$, and moreover,

$$(3) \qquad S(\bar{y}) = \bigcup_{i \in \mathscr{I}} S_i(\bar{y}).$$

For any $i \notin \mathscr{I}$, since the sets $\{\bar{y}\} \times \mathbb{R}^n$ and G_i are disjoint and polyhedral convex, there is a neighborhood V_i of $\bar{y}$ such that $(V_i \times \mathbb{R}^n) \cap G_i = \emptyset$. Let $V = \bigcap_{i \notin \mathscr{I}} V_i$. Then of course V is a neighborhood of $\bar{y}$ and we have

$$(4) \qquad (V \times \mathbb{R}^n) \cap \mathrm{gph}\, S \subset \bigcup_{i=1}^k G_i \setminus \bigcup_{i \notin \mathscr{I}} G_i \subset \bigcup_{i \in \mathscr{I}} G_i.$$

Let $y \in V$. If $S(y) = \emptyset$, then the relation (1) holds trivially. Let x be any point in $S(y)$. Then from (4),

$$(y,x) \in (V \times \mathbb{R}^n) \cap \mathrm{gph}\, S \subset \bigcup_{j \in \mathscr{I}} G_i,$$

hence for some $i \in \mathscr{I}$ we have $(y,x) \in G_i$, that is, $x \in S_i(y)$. Since each S_i is Lipschitz continuous and $\bar{y} \in \mathrm{dom}\, S_i$, with constant κ_i, say, we obtain by using (3) that

$$d(x,S(\bar{y})) \leq \max_i d(x,S_i(\bar{y})) \leq \max_i e(S_i(y),S_i(\bar{y})) \leq \max_i \kappa_i|y-\bar{y}|.$$

Since x is an arbitrary point in $S(y)$, we conclude that S is outer Lipschitz continuous at $\bar{y}$ with constant $\kappa := \max_i \kappa_i$. □

Exercise 3D.2 (polyhedrality of solution mappings to linear variational inequalities). *Given an $n \times n$ matrix A and a polyhedral convex set C in $\mathbb{R}^n$, show that the solution mapping of the linear variational inequality*

$$y \mapsto S(y) = \{x \,|\, y \in Ax + N_C(x)\} \quad \text{for } y \in \mathbb{R}^n$$

is polyhedral, and therefore it is outer Lipschitz continuous relative to its domain.

Guide. Any polyhedral convex set C is representable (in a non-unique manner) by a system of affine inequalities:

$$C = \{x \,|\, \langle a_i, x \rangle \leq \alpha_i \text{ for } i = 1, 2, \ldots, m\}.$$

We know from Section 2E that the normal cone to C at the point $x \in C$ is the set

$$N_C(x) = \left\{ u \,\Big|\, u = \sum_{i=1}^m y_i a_i, \; y_i \geq 0 \text{ for } i \in I(x), \; y_i = 0 \text{ for } i \notin I(x) \right\},$$

where $I(x) = \{i \,|\, \langle a_i, x\rangle = \alpha_i\}$ is the active index set for $x \in C$. The graph of the normal cone mapping N_C is not convex, unless C is a translate of a subspace, but it is the union, with respect to all possible subsets J of $\{1,\ldots,m\}$, of the polyhedral convex sets

$$\left\{ (x,u) \,\Big|\, u = \sum_{i=1}^m y_i a_i, \; \langle a_i,x\rangle = \alpha_i, \; y_i \geq 0 \text{ if } i \in J, \; \langle a_i,x\rangle < \alpha_i, \; y_i = 0 \text{ if } i \notin J \right\}.$$

It remains to observe that the graph of the sum $A + N_C$ is also the union of polyhedral convex sets. □

Outer Lipschitz continuity becomes automatically Lipschitz continuity when the mapping is inner semicontinuous, a property we introduced in the preceding section.

Theorem 3D.3 (isc criterion for Lipschitz continuity). *Consider a set-valued mapping $S : \mathbb{R}^m \rightrightarrows \mathbb{R}^n$ and a convex set $D \subset \mathrm{dom}\, S$ such that $S(y)$ is closed for every $y \in D$. Then S is Lipschitz continuous relative to D with constant κ if and only if S is both inner semicontinuous (isc) relative to D and outer Lipschitz continuous relative to D with constant κ.*

Proof. Let S be inner semicontinuous and outer Lipschitz continuous with constant κ, both relative to D. Choose $y,y' \in D$ and let $y_t = (1-t)y + ty'$. The assumed outer Lipschitz continuity together with the closedness of the values of S implies that for each $t \in [0,1]$ there exists a positive r_t such that

$$S(u) \subset S(y_t) + \kappa|u - y_t|\mathbb{B} \quad \text{for all } u \in D \cap \mathbb{B}_{r_t}(y_t).$$

3 Regularity Properties of Set-valued Solution Mappings

Let

(5) $\quad \tau = \sup\{ t \in [0,1] \,|\, S(y_s) \subset S(y) + \kappa|y_s - y|\mathbb{B} \text{ for each } s \in [0,t] \}.$

We will show that the supremum in (5) is attained at $\tau = 1$.

First, note that $\tau > 0$ because $r_0 > 0$. Since $S(y)$ is closed, the set $S(y) + \kappa|y_\tau - y|\mathbb{B}$ is closed too, thus its complement, denoted O, is open. Suppose that $y_\tau \in S^{-1}(O)$; then, applying Theorem 3B.2(e) to the isc mapping S, we obtain that there exists $\sigma \in [0, \tau)$ such that $y_\sigma \in S^{-1}(O)$ as well. But this is impossible since from $\sigma < \tau$ we have

$$S(y_\sigma) \subset S(y) + \kappa|y_\sigma - y|\mathbb{B} \subset S(y) + \kappa|y_\tau - y|\mathbb{B}.$$

Hence, $y_\tau \notin S^{-1}(O)$, that is, $S(y_\tau) \cap O = \emptyset$ and therefore $S(y_\tau)$ is a subset of $S(y) + \kappa|y_\tau - y|\mathbb{B}$. This implies that the supremum in (5) is attained.

Let us next prove that $\tau = 1$. If $\tau < 1$ there must exist $\eta \in (\tau, 1)$ with $|y_\eta - y_\tau| < r_\tau$ such that

(6) $\qquad\qquad S(y_\eta) \not\subset S(y) + \kappa|y_\eta - y|\mathbb{B}.$

But then, from the definition of r_τ,

$$S(y_\eta) \subset S(y_\tau) + \kappa|y_\eta - y_\tau|\mathbb{B} \subset S(y) + \kappa(|y_\eta - y_\tau| + |y_\tau - y|)\mathbb{B} = S(y) + \kappa|y_\eta - y|\mathbb{B},$$

where the final equality holds because y_τ is a point in the segment $[y, y_\eta]$. This contradicts (6), hence $\tau = 1$. Putting $\tau = 1$ into (5) results in $S(y') \subset S(y) + \kappa|y' - y|$. By the symmetry of y and y', we obtain that S is Lipschitz continuous relative to D.

Conversely, if S is Lipschitz continuous relative to D, then S is of course outer Lipschitz continuous. Let now $y \in D$ and let O be an open set such that $y \in S^{-1}(O)$. Then there is $x \in S(y)$ and $\varepsilon > 0$ such that $x \in S(y) \cap O$ and $x + \varepsilon\mathbb{B} \subset O$. Let $0 < \rho < \varepsilon/\kappa$ and pick a point $y' \in D \cap \mathbb{B}_\rho(y)$. Then

$$x \in S(y) \subset S(y') + \kappa|y - y'|\mathbb{B} \subset S(y') + \varepsilon\mathbb{B}.$$

Hence there exists $x' \in S(y')$ with $|x' - x| \leq \varepsilon$ and thus $x' \in S(y') \cap O$, that is $y' \in S^{-1}(O)$. This means that $S^{-1}(O)$ is open relative to D, and from Theorem 3B.2(e) we conclude that S is isc relative to D. □

We obtain from Theorems 3D.1 and 3D.3 some further insights.

Corollary 3D.4 (Lipschitz continuity of polyhedral mappings). *Let $S : \mathbb{R}^m \rightrightarrows \mathbb{R}^n$ be polyhedral and let $D \subset \text{dom } S$ be convex. Then S is isc relative to D if and only if S is actually Lipschitz continuous relative to D. Thus, for a polyhedral mapping, continuity relative to its domain implies Lipschitz continuity.*

Proof. This is immediate from 3D.3 in the light of 3D.1 and the fact that polyhedral mappings are osc and in particular closed-valued everywhere. □

Corollary 3D.5 (single-valued polyhedral mappings). *Let $S : \mathbb{R}^m \rightrightarrows \mathbb{R}^n$ be polyhedral and let $D \subset \text{dom } S$ be convex. If S is not multi-valued on D, then S must be a Lipschitz continuous function on D.*

Proof. It is sufficient to show that S is isc relative to D. Let $y \in D$ and O be an open set such that $y \in S^{-1}(O)$; then $x := S(y) \in O$. Since S is outer Lipschitz at y, there exists a neighborhood U of y such that if $y' \in U \cap D$ then $S(y') \in x + \kappa |y' - y| \mathbb{B}$. Taking U smaller if necessary so that $x' := S(y') \in x + \kappa |y' - y| \mathbb{B} \subset O$ for $y' \in U$, we obtain that for any $y' \in U \cap D$ one has $y' \in S^{-1}(x') \subset S^{-1}(O)$. But then $S^{-1}(O)$ must be open relative to D and, from Theorem 3B.2(e), S is isc relative to D. □

In the proof of 2E.6 we used the fact that if a function $f : \mathbb{R}^n \to \mathbb{R}^m$ with dom $f = \mathbb{R}^n$ has its graph composed by finitely many polyhedral convex sets, then it must be Lipschitz continuous. Now this is a particular case of the preceding result.

Corollary 3D.6. *If the solution mapping S of the linear variational inequality in Exercise 3D.2 is single-valued everywhere in $\mathbb{R}^n$, then it must be Lipschitz continuous globally.*

Exercise 3D.7 (distance characterization of outer Lipschitz continuity). *Prove that a mapping $S : \mathbb{R}^m \rightrightarrows \mathbb{R}^n$ is outer Lipschitz continuous at $\bar{y}$ relative to a set D with constant κ and neighborhood V if and only if $S(\bar{y})$ is closed and*

$$d(x, S(\bar{y})) \leq \kappa d(\bar{y}, S^{-1}(x) \cap D \cap V) \quad \text{for all } x \in \mathbb{R}^n.$$

Guide. Mimic the proof of 3C.1. □

In parallel with outer Lipschitz continuity we can introduce *inner Lipschitz continuity* of a set-valued mapping $S : \mathbb{R}^m \rightrightarrows \mathbb{R}^n$ relative to a set $D \subset \mathbb{R}^m$ at $\bar{y}$ when $\bar{y} \in D \subset \text{dom } S$, $S(\bar{y})$ is a closed set and there exist a constant $\kappa \geq 0$ and a neighborhood V of $\bar{y}$ such that

$$S(\bar{y}) \subset S(y) + \kappa |y - \bar{y}| \mathbb{B} \quad \text{for all } y \in V \cap D.$$

Inner Lipschitz continuity might be of interest on its own, but no significant application of this property in variational analysis has come to light, as yet. Even very simple polyhedral (nonconvex) mappings don't have this property (e.g., consider the mapping from $\mathbb{R}$ to $\mathbb{R}$, which graph is the union of the axes, and choose the origin as reference point) as opposed to the outer Lipschitz continuity which holds for every polyhedral mapping. In addition, a local version of this property does not obey the general implicit function theorem paradigm, as we will show in Section 3H. We therefore drop inner Lipschitz continuity from further consideration.

3E. Aubin Property, Metric Regularity and Linear Openness

A way to localize the concept of Lipschitz continuity of a set-valued mapping is to focus on a neighborhood of a reference point of the graph of the mapping and to use the Pompeiu–Hausdorff distance to a truncation of the mapping with such a neighborhood. More instrumental turns out to be to take the excess and truncate just one part of it. This leads us to the following definition.

Aubin property. *A mapping $S : \mathbb{R}^m \rightrightarrows \mathbb{R}^n$ is said to have the Aubin property at $\bar{y} \in \mathbb{R}^m$ for $\bar{x} \in \mathbb{R}^n$ if $\bar{x} \in S(\bar{y})$ and there is a constant $\kappa \geq 0$ together with neighborhoods U of $\bar{x}$ and V of $\bar{y}$ such that*

(1) $$e(S(y') \cap U, S(y)) \leq \kappa |y' - y| \quad \text{for all } y', y \in V,$$

or equivalently, there exist κ, U and V, as described, such that

(2) $$S(y') \cap U \subset S(y) + \kappa |y' - y| \mathbb{B} \quad \text{for all } y', y \in V.$$

The infimum of κ over all such combinations of κ, U and V is called the Lipschitz modulus of S at $\bar{y}$ for $\bar{x}$ and denoted by $\mathrm{lip}(S; \bar{y}|\bar{x})$. The absence of this property is signaled by $\mathrm{lip}(S; \bar{y}|\bar{x}) = \infty$.

It is not claimed that (1) and (2) are themselves equivalent, although that is true when $S(y)$ is closed for every $y \in V$. Nonetheless, the infimum furnishing $\mathrm{lip}(S; \bar{y}|\bar{x})$ is the same whichever formulation is adopted. When S is single-valued on a neighborhood of $\bar{y}$, then the Lipschitz modulus $\mathrm{lip}(S; \bar{y}|S(\bar{y}))$ equals the usual Lipschitz modulus $\mathrm{lip}(S; \bar{y})$ for functions.

In contrast to Lipschitz continuity, the Aubin property is tied to a particular point in the graph of the mapping. As an example, consider the set-valued mapping $S : \mathbb{R} \rightrightarrows \mathbb{R}$ defined as

$$S(y) = \begin{cases} \{0, 1 + \sqrt{y}\} & \text{for } y \geq 0, \\ 0 & \text{for } y < 0. \end{cases}$$

At 0, the value $S(0)$ consists of two points, 0 and 1. This mapping has the Aubin property at 0 for 0 but not at 0 for 1. Also, S is not Lipschitz continuous relative to any interval containing 0.

Observe that when a set-valued mapping S has the Aubin property at $\bar{y}$ for $\bar{x}$, then, for every point $(y,x) \in \mathrm{gph}\, S$ which is sufficiently close to $(\bar{y},\bar{x})$, it has the Aubin property at y for x as well. It is also important to note that the Aubin property of S at $\bar{y}$ for $\bar{x}$ implicitly requires $\bar{y}$ be an element of $\mathrm{int\, dom}\, S$; this is exhibited in the following proposition.

Proposition 3E.1 (local nonemptiness). *If $S : \mathbb{R}^m \rightrightarrows \mathbb{R}^n$ has the Aubin property at $\bar{y}$ for $\bar{x}$, then for every neighborhood U of $\bar{x}$ there exists a neighborhood V of $\bar{y}$ such that $S(y) \cap U \neq \emptyset$ for all $y \in V$.*

Proof. The inclusion (2) for $y' = \bar{y}$ yields

$$\bar{x} \in S(y) + \kappa|y - \bar{y}|B \quad \text{for every } y \in V,$$

which is the same as

$$(\bar{x} + \kappa|y - \bar{y}|B) \cap S(y) \neq \emptyset \quad \text{for every } y \in V.$$

That is, $S(y)$ intersects every neighborhood of $\bar{x}$ when y is sufficiently close to $\bar{y}$. □

The property displayed in Proposition 3E.1 is a local version of the inner semicontinuity. Further, if S is Lipschitz continuous relative to an open set D, then S has the Aubin property at any $y \in D \cap \text{int dom } S$ for any $x \in S(y)$. In particular, the inverse A^{-1} of a linear mapping A is Aubin continuous at any point provided that $\text{rge } A = \text{dom } A^{-1}$ has nonempty interior, that is, A is surjective. The converse is also true, since the inverse A^{-1} of a surjective linear mapping A is Lipschitz continuous on the whole space, by Theorem 3C.3, and hence A^{-1} has the Aubin property at any point.

Proposition 3E.2 (single-valued localization from Aubin property). *A set-valued mapping $S : \mathbb{R}^m \rightrightarrows \mathbb{R}^n$ has a Lipschitz continuous single-valued localization around $\bar{y}$ for $\bar{x}$ with constant κ if and only if it has a localization at $\bar{y}$ for $\bar{x}$ that is not multi-valued and has the Aubin property at $\bar{y}$ for $\bar{x}$ with constant κ.*

Proof. Let s be a localization of the second type, which in general is weaker. From Proposition 3E.1 we have $\bar{y} \in \text{int dom } S$, so s is a single-valued localization of S around $\bar{y}$ for $\bar{x}$. Let a and b be positive constants such that $y \mapsto s(y) := S(y) \cap B_a(\bar{x})$ is a function defined on $B_b(\bar{y})$ and let $b' > 0$ satisfy $b' < \min\{b, a/8\kappa\}$. Then for $y, y' \in B_{b'}(\bar{y})$ we have

$$d(s(y), S(y')) = d(S(y) \cap B_a(\bar{x}), S(y'))$$
$$\leq \kappa|y - y'| \leq \kappa|y - \bar{y}| + \kappa|y' - \bar{y}| \leq 2\kappa b' < a/4.$$

Hence, there exists $x' \in S(y')$ such that $|x' - s(y)| \leq d(s(y), S(y')) + a/4 \leq a/2$. Since $|s(y) - \bar{x}| = d(\bar{x}, S(y))$ and

$$|x' - \bar{x}| \leq |x' - s(y)| + |s(y) - \bar{x}| \leq a/2 + d(\bar{x}, S(y)) \leq a/2 + \kappa|y - \bar{y}| < a$$

we obtain $d(s(y), S(y') \cap B_a(\bar{x})) = d(s(y), S(y'))$ and therefore

$$\kappa|y - y'| \geq d(S(y) \cap B_a(\bar{x}), S(y'))$$
$$= d(s(y), S(y')) = d(s(y), S(y') \cap B_a(\bar{x})) = |s(y) - s(y')|.$$

Thus, s is a Lipschitz continuous single-valued localization of S around $\bar{y}$ for $\bar{x}$ with constant κ. □

Proposition 3E.2 is actually a special case of the following, more general result in which convexity enters, inasmuch as singletons are convex sets in particular.

3 Regularity Properties of Set-valued Solution Mappings 161

Theorem 3E.3 (Lipschitz continuity under truncation). *A set-valued mapping* $S : \mathbb{R}^m \rightrightarrows \mathbb{R}^n$ *whose values are convex sets has the Aubin property at $\bar{y}$ for $\bar{x}$ if and only if it has a Lipschitz continuous graphical localization (not necessarily single-valued) around $\bar{y}$ for $\bar{x}$, or in other words, there are neighborhoods U of $\bar{x}$ and V of $\bar{y}$ such that the truncated mapping $y \mapsto S(y) \cap U$ is Lipschitz continuous on V.*

Proof. The "if" part holds even without the convexity assumption. Indeed, if $y \mapsto S(y) \cap U$ is Lipschitz continuous on V we have

$$S(y') \cap U \subset S(y) \cap U + \kappa |y' - y| \mathbb{B} \subset S(y) + \kappa |y' - y| \mathbb{B} \quad \text{for all } y', y \in V,$$

that is, S has the desired Aubin property. For the "only if" part, suppose now that S has the Aubin property at $\bar{y}$ for $\bar{x}$ with constant κ, and let $a > 0$ and $b > 0$ be such that

(3) $\qquad S(y') \cap \mathbb{B}_a(\bar{x}) \subset S(y) + \kappa |y' - y| \mathbb{B} \quad \text{for all } y', y \in \mathbb{B}_b(\bar{y}).$

Adjust a and b so that, by 3E.1,

(4) $\qquad S(y) \cap \mathbb{B}_{a/2}(\bar{x}) \neq \emptyset \text{ for all } y \in \mathbb{B}_b(\bar{x}) \text{ and } b < \dfrac{a}{4\kappa}.$

Pick $y, y' \in \mathbb{B}_b(\bar{y})$ and let $x' \in S(y') \cap \mathbb{B}_a(\bar{x})$. Then from (3) there exists $x \in S(y)$ such that

(5) $\qquad |x - x'| \leq \kappa |y - y'|.$

If $x \in \mathbb{B}_a(\bar{x})$, there is nothing more to prove, so assume that $r := |x - \bar{x}| > a$. By (4), we can choose a point $\tilde{x} \in S(y) \cap \mathbb{B}_{a/2}(\bar{x})$. Since $S(y)$ is convex, there exists a point $z \in S(y)$ on the segment $[x, \tilde{x}]$ such that $|z - \bar{x}| = a$ and then $z \in S(y) \cap \mathbb{B}_a(\bar{x})$. We will now show that

(6) $\qquad |z - x'| \leq 5\kappa |y - y'|,$

which yields that the mapping $y \mapsto S(y) \cap \mathbb{B}_a(\bar{x})$ is Lipschitz continuous on $\mathbb{B}_b(\bar{y})$ with constant 5κ.

By construction, there exists $t \in (0, 1)$ such that $z = (1-t)x + t\tilde{x}$. Then

$$a = |z - \bar{x}| = |(1-t)(x - \bar{x}) + t(\tilde{x} - \bar{x})| \leq (1-t)r + t|\tilde{x} - \bar{x}|$$

and in consequence $t(r - |\tilde{x} - \bar{x}|) \leq r - a$. Since $\tilde{x} \in \mathbb{B}_{a/2}(\bar{x})$, we get

$$t \leq \frac{r-a}{r-a/2}.$$

Using the triangle inequality $|\tilde{x} - x| \leq |x - \bar{x}| + |\tilde{x} - \bar{x}| \leq r + a/2$, we obtain

(7) $\qquad |z - x| = t|\tilde{x} - x| \leq \dfrac{r-a}{r-a/2}(r + a/2).$

Also, in view of (4) and (5), we have that

(8) $\quad r = |x - \bar{x}| \leq |x' - x| + |x' - \bar{x}| \leq \kappa |y - y'| + a \leq \kappa 2b + a \leq \dfrac{3a}{2}.$

From (7), (8) and the inequality $r > a$ we obtain

(9) $\quad |z - x| \leq (r - a)\dfrac{r + a/2}{r - a/2} \leq (r - a)\dfrac{3a/2 + a/2}{a - a/2} = 4(r - a).$

Note that $d := r - a$ is exactly the distance from x to the ball $\mathbb{B}_a(\bar{x})$, hence $d \leq |x - x'|$ because $x' \in \mathbb{B}_a(\bar{x})$. Combining this with (9) and taking into account (6), we arrive at

$$|z - x'| \leq |z - x| + |x - x'| \leq 4d + |x' - x| \leq 5|x - x'| \leq 5\kappa |y - y'|.$$

But this is (6), and we are done. $\qquad\square$

The Aubin property could alternatively be defined with one variable "free," as shown in the next proposition.

Proposition 3E.4 (alternative description of Aubin continuity). *A mapping $S : \mathbb{R}^m \rightrightarrows \mathbb{R}^n$ has the Aubin property at $\bar{y}$ for $\bar{x}$ with constant κ if and only if there exist neighborhoods U of $\bar{x}$ and V of $\bar{y}$ such that*

(10) $\quad e(S(y') \cap U, S(y)) \leq \kappa |y' - y| \quad \text{for all } y' \in \mathbb{R}^m \text{ and } y \in V.$

Proof. Clearly, (10) implies (1). Assume (1) with corresponding U and V and choose positive a and b such that $\mathbb{B}_a(\bar{x}) \subset U$ and $\mathbb{B}_b(\bar{y}) \subset V$. Let $0 < a' < a$ and $0 < b' < b$ be such that

(11) $\quad 2\kappa b' + a' \leq \kappa b.$

For any $y \in \mathbb{B}_{b'}(\bar{y})$ we have from (1) that

$$d(\bar{x}, S(y)) \leq \kappa |y - \bar{y}| \leq \kappa b',$$

hence

(12) $\quad e(\mathbb{B}_{a'}(\bar{x}), S(y)) \leq \kappa b' + a'.$

Take any $y' \in \mathbb{R}^m$. If $y' \in \mathbb{B}_b(\bar{y})$ the inequality in (10) comes from (1) and there is nothing more to prove. Assume $|y' - \bar{y}| > b$. Then $|y - y'| > b - b'$ and from (11), $\kappa b' + a' \leq \kappa(b - b') \leq \kappa |y - y'|$. Using this in (12) we obtain

$$e(\mathbb{B}_{a'}(\bar{x}), S(y)) \leq \kappa |y' - y|$$

and since $S(y') \cap \mathbb{B}_{a'}(\bar{x})$ is obviously a subset of $\mathbb{B}_{a'}(\bar{x})$, we come again to (10). $\qquad\square$

The Aubin property of a mapping is characterized by Lipschitz continuity of the distance function associated with it.

3 Regularity Properties of Set-valued Solution Mappings

Theorem 3E.5 (distance function characterization of Aubin property). *For a mapping $S : \mathbb{R}^m \rightrightarrows \mathbb{R}^n$ with $(\bar{y},\bar{x}) \in$ gph S, let $s(y,x) = d(x,S(y))$. Then S has the Aubin property at $\bar{y}$ for $\bar{x}$ if and only if the function s is Lipschitz continuous with respect to y uniformly in x around $(\bar{y},\bar{x})$, in which case one has*

$$\text{lip}(S;\bar{y}|\bar{x}) = \widehat{\text{lip}}_y(s;(\bar{y},\bar{x})). \tag{13}$$

Proof. Let $\kappa > \text{lip}(S;\bar{y}|\bar{x})$. Then, from 3E.1, there exist positive constants a and b such that

$$\emptyset \neq S(y) \cap \mathbb{B}_a(\bar{x}) \subset S(y') + \kappa|y-y'|\mathbb{B} \quad \text{for all } y,y' \in \mathbb{B}_b(\bar{y}). \tag{14}$$

Without loss of generality, let $a/(4\kappa) \leq b$. Let $y \in \mathbb{B}_{a/(4\kappa)}(\bar{y})$ and $x \in \mathbb{B}_{a/4}(\bar{x})$ and let $\tilde{x}$ be a projection of x on cl $S(y)$. Using 1D.4(b) and (14) with $y = \bar{y}$ we have

$$|x-\tilde{x}| = d(x,S(y)) \leq |x-\bar{x}| + d(\bar{x},S(y))$$
$$\leq |x-\bar{x}| + e(S(\bar{y}) \cap \mathbb{B}_a(\bar{x}), S(y)) \leq \frac{a}{4} + \kappa|y-\bar{y}| \leq \frac{a}{4} + \kappa \frac{a}{4\kappa} = a/2.$$

Hence

$$|\bar{x}-\tilde{x}| \leq |\bar{x}-x| + |x-\tilde{x}| \leq \frac{a}{4} + \frac{a}{2} = \frac{3a}{4} < a.$$

This gives us that

$$|x-\tilde{x}| = d(x,S(y)) = d(x,S(y) \cap \mathbb{B}_a(\bar{x})). \tag{15}$$

Now, let $y' \in \mathbb{B}_{a/(4\kappa)}(\bar{y})$. The inclusion in (14) yields

$$d\left(x,S(y') + \kappa|y-y'|\mathbb{B}\right) \leq d(x,S(y) \cap \mathbb{B}_a(\bar{x})). \tag{16}$$

Using the fact that for any set C and for any $r \geq 0$ one has

$$d(x,C) - r \leq d(x,C + r\mathbb{B}),$$

from (15) and (16) we obtain

$$d(x,S(y')) - \kappa|y-y'| \leq d(x,S(y) \cap \mathbb{B}_a(\bar{x})) = d(x,S(y)).$$

By the symmetry of y and y', we conclude that $\widehat{\text{lip}}_y(s;(\bar{y},\bar{x})) \leq \kappa$. Since κ can be arbitrarily close to $\text{lip}(S;\bar{y}|\bar{x})$, it follows that

$$\widehat{\text{lip}}_y(s;(\bar{y},\bar{x})) \leq \text{lip}(S;\bar{y}|\bar{x}). \tag{17}$$

Conversely, let $\kappa > \widehat{\text{lip}}_y(s;(\bar{y},\bar{x}))$. Then there exist neighborhoods U and V of $\bar{x}$ and $\bar{y}$, respectively, such that $s(\cdot,x)$ is Lipschitz continuous relative to V with a constant κ for any given $x \in U$. Let $y,y' \in V$. Since $V \subset \text{dom } s(\cdot,x)$ for any $x \in U$ we have that $S(y') \cap U \neq \emptyset$. Pick any $x \in S(y') \cap U$; then $s(y',x) = 0$ and, by the

assumed Lipschitz continuity of $s(\cdot,x)$, we get
$$d(x,S(y)) = s(y,x) \leq s(y',x) + \kappa|y-y'| = \kappa|y-y'|.$$
Taking supremum with respect to $x \in S(y') \cap U$ on the left, we obtain that S has the Aubin property at $\bar{y}$ for $\bar{x}$ with constant κ. Since κ can be arbitrarily close to $\widehat{\text{lip}}_y(s;(\bar{y},\bar{x}))$, we get
$$\widehat{\text{lip}}_y(s;(\bar{y},\bar{x})) \geq \text{lip}\,(S;\bar{y}|\bar{x}).$$
This, combined with (17), gives us (13). □

The Aubin property of a mapping is closely tied with a property of its inverse, called *metric regularity*. The concept of metric regularity goes back to the classical Banach open mapping principle. We will devote most of Chapter 5 to studying the metric regularity of set-valued mappings acting in infinite-dimensional spaces.

Metric regularity. *A mapping $F : \mathbb{R}^n \rightrightarrows \mathbb{R}^m$ is said to be metrically regular at $\bar{x}$ for $\bar{y}$ when $\bar{y} \in F(\bar{x})$ and there is a constant $\kappa \geq 0$ together with neighborhoods U of $\bar{x}$ and V of $\bar{y}$ such that*

(18) $$d(x,F^{-1}(y)) \leq \kappa d(y,F(x)) \quad \text{for all } (x,y) \in U \times V.$$

The infimum of κ over all such combinations of κ, U and V is called the regularity modulus for F at $\bar{x}$ for $\bar{y}$ and denoted by $\text{reg}\,(F;\bar{x}|\bar{y})$. *The absence of metric regularity is signaled by* $\text{reg}\,(F;\bar{x}|\bar{y}) = \infty$.

Metric regularity is a valuable concept in its own right, especially for numerical purposes. For a general set-valued mapping F and a vector y, it gives an estimate for how far a point x is from being a solution to the generalized equation $F(x) \ni y$ in terms of the "residual" $d(y,F(x))$.

To be specific, let $\bar{x}$ be a solution of the inclusion $\bar{y} \in F(x)$, let F be metrically regular at $\bar{x}$ for $\bar{y}$, and let x_a and y_a be approximations to $\bar{x}$ and $\bar{y}$, respectively. Then from (18), the distance from x_a to the set of solutions of the inclusion $y_a \in F(x)$ is bounded by the constant κ times the residual $d(y_a,F(x_a))$. In applications, the residual is typically easy to compute or estimate, whereas finding a solution might be considerably more difficult. Metric regularity says that there exists a solution to the inclusion $y_a \in F(x)$ at distance from x_a proportional to the residual. In particular, if we know the rate of convergence of the residual to zero, then we will obtain the rate of convergence of approximate solutions to an exact one.

Proposition 3C.1 for a mapping S, when applied to $F = S^{-1}$, $F^{-1} = S$, ties the Lipschitz continuity of F^{-1} relative to a set D to a condition resembling (18), but with $F(x)$ replaced by $F(x) \cap D$ on the right, and with $U \times V$ replaced by $\mathbb{R}^n \times D$. We demonstrate now that metric regularity of F in the sense of (18) corresponds to the Aubin property of F^{-1} for the points in question.

3 Regularity Properties of Set-valued Solution Mappings

Theorem 3E.6 (equivalence of metric regularity and the inverse Aubin property). *A set-valued mapping $F : \mathbb{R}^n \rightrightarrows \mathbb{R}^m$ is metrically regular at $\bar{x}$ for $\bar{y}$ with a constant κ if and only if its inverse $F^{-1} : \mathbb{R}^m \rightrightarrows \mathbb{R}^n$ has the Aubin property at $\bar{y}$ for $\bar{x}$ with constant κ, i.e. there exist neighborhoods U of $\bar{x}$ and V of $\bar{y}$ such that*

(19) $$e(F^{-1}(y') \cap U, F^{-1}(y)) \leq \kappa |y' - y| \quad \text{for all } y', y \in V.$$

Thus,

(20) $$\operatorname{lip}(F^{-1}; \bar{y}|\bar{x}) = \operatorname{reg}(F; \bar{x}|\bar{y}).$$

Proof. Let $\kappa > \operatorname{reg}(F; \bar{x}|\bar{y})$; then there are positive constants a and b such that (18) holds with $U = \mathbb{B}_a(\bar{x})$, $V = \mathbb{B}_b(\bar{y})$ and with this κ. Without loss of generality, assume $b < a/\kappa$. Choose $y, y' \in \mathbb{B}_b(\bar{y})$. If $F^{-1}(y) \cap \mathbb{B}_a(\bar{x}) = \emptyset$, then $d(\bar{x}, F^{-1}(y)) \geq a$. But then the inequality (18) with $x = \bar{x}$ yields

$$a \leq d(\bar{x}, F^{-1}(y)) \leq \kappa d(y, F(\bar{x})) \leq \kappa |y - \bar{y}| \leq \kappa b < a,$$

a contradiction. Hence there exists $x \in F^{-1}(y) \cap \mathbb{B}_a(\bar{x})$, and for any such x we have from (18) that

(21) $$d(x, F^{-1}(y')) \leq \kappa d(y', F(x)) \leq \kappa |y - y'|.$$

Taking the supremum with respect to $x \in F^{-1}(y) \cap \mathbb{B}_a(\bar{x})$ we obtain (19) with $U = \mathbb{B}_a(\bar{x})$ and $V = \mathbb{B}_b(\bar{y})$, and therefore

(22) $$\operatorname{reg}(F; \bar{x}|\bar{y}) \geq \operatorname{lip}(F^{-1}; \bar{y}|\bar{x}).$$

Conversely, suppose there are neighborhoods U of $\bar{x}$ and V of $\bar{y}$ along with a constant κ such that (19) holds. Take U and V smaller if necessary so that, according to Proposition 3E.4, we have

(23) $$e(F^{-1}(y') \cap U, F^{-1}(y)) \leq \kappa |y' - y| \quad \text{for all } y' \in \mathbb{R}^m \text{ and } y \in V.$$

Let $x \in U$ and $y \in V$. If $F(x) \neq \emptyset$, then for any $y' \in F(x)$ we have $x \in F^{-1}(y') \cap U$. From (23), we obtain

$$d(x, F^{-1}(y)) \leq e(F^{-1}(y') \cap U, F^{-1}(y)) \leq \kappa |y - y'|.$$

This holds for any $y' \in F(x)$, hence, by taking the infimum with respect to $y' \in F(x)$ in the last expression we get

$$d(x, F^{-1}(y)) \leq \kappa d(y, F(x)).$$

(If $F(x) = \emptyset$, then because of the convention $d(y, \emptyset) = \infty$, this inequality holds automatically.) Hence, F is metrically regular at $\bar{x}$ for $\bar{y}$ with a constant κ. Then we have $\kappa \geq \operatorname{reg}(F; \bar{x}|\bar{y})$ and hence $\operatorname{reg}(F; \bar{x}|\bar{y}) \leq \operatorname{lip}(F^{-1}; \bar{y}|\bar{x})$. This inequality together with (22) results in (20). □

Observe that metric regularity of F at $\bar{x}$ for $\bar{y}$ does *not* require that $\bar{x} \in$ int dom F. Indeed, if $\bar{x}$ is an isolated point of dom F then the right side in (18) is ∞ for all $x \in U$, $x \neq \bar{x}$, and then (18) holds automatically. On the other hand, for $x = \bar{x}$ the right side of (18) is always finite (since by assumption $\bar{x} \in$ dom F), and then $F^{-1}(y) \neq \emptyset$ for $y \in V$. This also follows from 3E.1 via 3E.6.

Exercise 3E.7 (equivalent formulation). *Prove that a mapping F is metrically regular at $\bar{x}$ for $\bar{y}$ with constant κ if and only if there are neighborhoods U of $\bar{x}$ and V of $\bar{y}$ such that*

$$(24) \quad d(x, F^{-1}(y)) \leq \kappa d(y, F(x)) \quad \text{for all } x \in U \text{ having } F(x) \cap V \neq \emptyset \text{ and all } y \in V.$$

Guide. First, note that (18) implies (24). Let (24) hold with constant κ and neighborhoods $B_a(\bar{x})$ and $B_b(\bar{y})$ having $b < a/\kappa$. Choose $y, y' \in B_b(\bar{y})$. As in the proof of 3E.6 show first that $F^{-1}(y) \cap B_a(\bar{x}) \neq \emptyset$ by noting that $F(\bar{x}) \cap B_b(\bar{y}) \neq \emptyset$ and hence the inequality in (24) holds for $\bar{x}$ and y. Then for any $x \in F^{-1}(y) \cap B_a(\bar{x})$ we have that $y \in F(x) \cap B_b(\bar{y})$, that is, $F(x) \cap B_b(\bar{y}) \neq \emptyset$. Thus, the inequality in (24) holds with y' and any $x \in F^{-1}(y) \cap B_a(\bar{x})$, which leads to (21) and hence to (19), in the same way as in the proof of 3E.6. The rest follows from the equivalence of (18) and (19) established in 3E.6. □

There is a third property, which we introduced for functions in Section 1F, and which is closely related to both metric regularity and the Aubin property.

Openness. *A mapping $F : \mathbb{R}^n \rightrightarrows \mathbb{R}^m$ is said to be open at $\bar{x}$ for $\bar{y}$ if $\bar{y} \in F(\bar{x})$ and for every neighborhood U of $\bar{x}$, $F(U)$ is a neighborhood of $\bar{y}$.*

From the equivalence of metric regularity of F at $\bar{x}$ for $\bar{y}$ and the Aubin property of F^{-1} at $\bar{y}$ for $\bar{x}$, and Proposition 3E.1, we obtain that if a mapping F is metrically regular at $\bar{x}$ for $\bar{y}$, then F is open at $\bar{x}$ for $\bar{y}$. Metric regularity is actually equivalent to the following stronger version of the openness property:

Linear openness. *A mapping $F : \mathbb{R}^n \rightrightarrows \mathbb{R}^m$ is said to be linearly open at $\bar{x}$ for $\bar{y}$ when $\bar{y} \in F(\bar{x})$ and there is a constant $\kappa \geq 0$ together with neighborhoods U of $\bar{x}$ and V of $\bar{y}$ such that*

$$(25) \quad F(x + \kappa r \,\mathrm{int}\, B) \supset [F(x) + r \,\mathrm{int}\, B] \cap V \quad \text{for all } x \in U \text{ and all } r > 0.$$

Openness is a particular case of linear openness and follows from (25) for $x = \bar{x}$. Linear openness postulates openness *around* the reference point with balls having proportional radii.

Theorem 3E.8 (equivalence of linear openness and metric regularity). *A set-valued mapping $F : \mathbb{R}^n \rightrightarrows \mathbb{R}^m$ is linearly open at $\bar{x}$ for $\bar{y}$ if and only if F is metrically regular at $\bar{x}$ for $\bar{y}$. In this case the infimum of κ for which (25) holds is equal to $\mathrm{reg}(F; \bar{x}|\bar{y})$.*

Proof. Let (25) hold. Choose $y \in V$ and $x' \in U$. Let $y' \in F(x')$ (if there is no such y' there is nothing to prove). Since $y = y' + |y - y'|w$ for some $w \in B$, denoting

$r = |y - y'|$, for every $\varepsilon > 0$ we have $y \in (F(x') + r(1+\varepsilon)\operatorname{int}\!B) \cap V$. From (25), there exists $x \in F^{-1}(y)$ with $|x - x'| \leq \kappa(1+\varepsilon)r = \kappa(1+\varepsilon)|y' - y|$. Then $d(x', F^{-1}(y)) \leq \kappa(1+\varepsilon)|y' - y|$. Taking infimum with respect to $y' \in F(x')$ on the right and passing to zero with ε (since the left side does not depend on ε), we obtain that F is metrically regular at $\bar{x}$ for $\bar{y}$ with constant κ.

For the converse implication, we use the characterization of the Aubin property given in Proposition 3E.4. Let $x \in U$, $r > 0$, and let $y' \in (F(x) + r\operatorname{int}\!B) \cap V$. Then there exists $y \in F(x)$ such that $|y - y'| < r$. Let $\varepsilon > 0$ be so small that $(\kappa + \varepsilon)|y - y'| < \kappa r$. From (10) we obtain $d(x, F^{-1}(y')) \leq \kappa|y - y'| \leq (\kappa + \varepsilon)|y - y'|$. Then there exists $x' \in F^{-1}(y')$ such that $|x - x'| \leq (\kappa + \varepsilon)|y - y'|$. But then $y' \in F(x') \subset F(x + (\kappa + \varepsilon)|y - y'|B) \subset F(x + \kappa r \operatorname{int}\!B)$, which yields (25) with constant κ. □

In the classical setting, of course, the equation $f(p, x) = 0$ is solved for x in terms of p, and the goal is to determine when this reduces to x being a function of p through a localization, moreover one with some kind of property of differentiability, or at least Lipschitz continuity. Relinquishing single-valuedness entirely, we can look at "solving" the relation

$$(26) \qquad G(p, x) \ni 0 \text{ for a mapping } G : \mathbb{R}^d \times \mathbb{R}^n \rightrightarrows \mathbb{R}^m,$$

or in other words studying the solution mapping $S : \mathbb{R}^d \rightrightarrows \mathbb{R}^n$ defined by

$$(27) \qquad S(p) = \{ x \mid G(p, x) \ni 0 \}.$$

Fixing a pair $(\bar{p}, \bar{x})$ such that $\bar{x} \in S(\bar{p})$, we can raise questions about local behavior of S as might be deduced from assumptions on G.

We will concentrate here on the extent to which S can be guaranteed to have the Aubin property at $\bar{p}$ for $\bar{x}$. This turns out to be true when G has the Aubin property with respect to p and a weakened metric regularity property with respect to x, but we have to formulate exactly what we need about this in a local sense.

Partial Aubin property. *The mapping $G : \mathbb{R}^d \times \mathbb{R}^n \rightrightarrows \mathbb{R}^m$ is said to have the partial Aubin property with respect to p uniformly in x at $(\bar{p}, \bar{x})$ for $\bar{y}$ if $\bar{y} \in G(\bar{p}, \bar{x})$ and there is a constant $\kappa \geq 0$ together with neighborhoods Q for $\bar{p}$, U of $\bar{x}$ and V of $\bar{y}$ such that*

$$(28) \qquad e(G(p, x) \cap V, G(p', x)) \leq \kappa |p - p'| \quad \text{for all } p, p' \in Q \text{ and } x \in U,$$

or equivalently, there exist κ, Q, U and V, as described, such that

$$G(p, x) \cap V \subset G(p', x) + \kappa |p - p'| B \quad \text{for all } p, p' \in Q \text{ and } x \in U.$$

The infimum of κ over all such combinations of κ, Q, U and V is called the partial Lipschitz modulus of G with respect to p uniformly in x at $(\bar{p}, \bar{x})$ for $\bar{y}$ and denoted by $\widehat{\operatorname{lip}}_p(G; \bar{p}, \bar{x} \mid \bar{y})$. The absence of this property is signaled by $\widehat{\operatorname{lip}}_p(G; \bar{p}, \bar{x} \mid \bar{y}) = \infty$.

The basic result we are able now to state about the solution mapping in (27) could be viewed as an "implicit function" complement to the "inverse function" result in

Theorem 3E.6, rather than as a result in the pattern of the implicit function theorem (which features approximations of one kind or another).

Theorem 3E.9 (Aubin property of general solution mappings). *In (26), let $G : \mathbb{R}^d \times \mathbb{R}^n \rightrightarrows \mathbb{R}^m$, with $G(\bar{p}, \bar{x}) \ni 0$, have the partial Aubin property with respect to p uniformly in x at $(\bar{p}, \bar{x})$ for 0 with constant κ. Furthermore, in the notation (27), let G enjoy the existence of a constant λ such that*

(29) $\qquad d(x, S(p)) \leq \lambda\, d(0, G(p,x)) \quad \text{for all } (p,x) \text{ close to } (\bar{p}, \bar{x}).$

Then the solution mapping S in (27) has the Aubin property at $\bar{p}$ for $\bar{x}$ with constant $\lambda\kappa$.

Proof. Take $p, p' \in Q$ and $x \in S(p) \cap U$ so that (28) holds for a neighborhood V of 0. From (29) and then (28) we have

$$d(x, S(p')) \leq \lambda\, d(0, G(p',x)) \leq \lambda\, e(G(p,x) \cap V, G(p',x)) \leq \lambda\kappa |p - p'|.$$

Taking the supremum of the left side with respect to $x \in S(p) \cap U$, we obtain that S has the Aubin property with constant $\lambda\kappa$. □

Example 3E.10. Theorem 3E.9 cannot be extended to a two-way characterization parallel to Theorem 3E.6. Indeed, consider the "saddle" function of two real variables $f(p,x) = x^2 - p^2$. In this case f does not satisfy (29) at the origin of $\mathbb{R}^2$, yet the solution mapping $S(p) = \{x \mid f(p,x) = 0\} = \{-p, p\}$ has the Aubin property at 0 for 0.

At the end of this section we will take a closer look at the following question. If a mapping F is simultaneous metrically regular and has the Aubin property, both at $\bar{x}$ for $\bar{y}$ for some $(\bar{x}, \bar{y}) \in \text{gph } F$, then what is the relation, if any, between $\text{reg}(F; \bar{x} | \bar{y})$ and $\text{lip}(F; \bar{x} | \bar{y})$? Having in mind 3E.6, it is the same as asking what is the relation between $\text{reg}(F; \bar{x} | \bar{y})$ and $\text{reg}(F^{-1}; \bar{y} | \bar{x})$ or between $\text{lip}(F; \bar{x} | \bar{y})$ and $\text{lip}(F^{-1}; \bar{y} | \bar{x})$. When we exclude the trivial case when $(\bar{x}, \bar{y}) \in \text{int gph } F$, in which case both moduli would be zero, an answer to this question is stated in the following exercise.

Exercise 3E.11. *Consider a mapping $F : \mathbb{R}^n \rightrightarrows \mathbb{R}^m$ with closed graph and a point $(\bar{x}, \bar{y}) \in \text{gph } F \setminus \text{int gph } F$. Then*

$$\text{reg}(F; \bar{x} | \bar{y}) \cdot \text{lip}(F; \bar{x} | \bar{y}) \geq 1,$$

including the limit cases when either of these moduli is 0 and then the other is ∞ under the convention $0\infty = \infty$.

Guide. Let $\kappa > \text{reg}(F; \bar{x} | \bar{y})$ and $\gamma > \text{lip}(F; \bar{x} | \bar{y})$. Then there are neighborhoods U of $\bar{x}$ and V of $\bar{y}$ corresponding to metric regularity and the Aubin property of F with constants κ and γ, respectively. Let $(x,y) \in U \times V$ be such that $d(x, F^{-1}(y)) > 0$ (why does such a point exist?). Then there exists $x' \in F^{-1}(y)$ such that $0 < |x - x'| = d(x, F^{-1}(y))$. We have

$$|x-x'| = d(x,F^{-1}(y)) \leq \kappa d(y,F(x)) \leq \kappa e(F(x') \cap U, F(x)) \leq \kappa\gamma|x-x'|.$$

Hence, $\kappa\gamma \geq 1$. □

For a solution mapping $S = F^{-1}$ of an inclusion $F(x) \ni y$ with a parameter y, the quantity lip$(S;\bar{y}|\bar{x})$ measures how "stable" solutions near $\bar{x}$ are under changes of the parameter near $\bar{y}$. In this context, the smaller this modulus is, the "better" stability we have. In view of 3E.11, better stability means larger values of the regularity modulus reg$(S;\bar{y}|\bar{x})$. In the limit case, when S is a constant function near $\bar{y}$, that is, when the solution is not sensitive at all with respect to small changes of the parameter y near $\bar{y}$, then lip$(S;\bar{y}|S(\bar{y})) = 0$ while the metric regularity modulus of S there is infinity. In Section 6A we will see that the "larger" the regularity modulus of a mapping is, the "easier" it is to perturb the mapping so that it looses its metric regularity. In this connection, E3.11 provides a good candidate for a condition number of a general mapping, but we shall not go into this here.

3F. Implicit Mapping Theorems with Metric Regularity

In the paradigm of the implicit function theorem, as applied to a generalized equation $f(p,x) + F(x) \ni 0$ with solutions $x \in S(p)$, the focus is on some $\bar{p}$ and $\bar{x} \in S(\bar{p})$, and on some kind of approximation of the mapping $x \mapsto f(\bar{p},x) + F(x)$. Assumptions about this approximation lead to conclusions about the solution mapping S relative to $\bar{p}$ and $\bar{x}$. Stability properties of the approximation are the key to progress in this direction. Our aim now is to study such stability with respect to metric regularity and to show that this leads to implicit function theorem type results which apply to set-valued solution mappings S beyond any framework of single-valued localization.

We start this section with a particular case of a fundamental result in variational analysis and beyond, stated below as Theorem 3F.1, which goes back to works by Lyusternik and Graves. We will devote most of Chapter 5 to the full theory behind this result—in an infinite-dimensional setting. In the statement of the theorem the following concept is required.

Locally closed sets. *A set C is said to be* locally closed *at $x \in C$ if there exists a neighborhood U of x such that the intersection $C \cap U$ is closed.*

Local closedness of a set C at $x \in C$ can be equivalently defined as the existence of a scalar $a > 0$ such that the set $C \cap \mathbb{B}_a(x)$ is closed.

Theorem 3F.1 (inverse mapping theorem with metric regularity). *Consider a mapping $F : \mathbb{R}^n \rightrightarrows \mathbb{R}^m$ and any $(\bar{x},\bar{y}) \in \text{gph } F$ at which gph F is locally closed and let κ and μ be nonnegative constants such that*

$$\text{reg}(F;\bar{x}|\bar{y}) \leq \kappa \quad \text{and} \quad \kappa\mu < 1.$$

Then for any function $g : \mathbb{R}^n \to \mathbb{R}^m$ with $\bar{x} \in \text{int dom } g$ and $\text{lip}(g;\bar{x}) \leq \mu$, one has

(1) $$\text{reg}(g+F;\bar{x}|g(\bar{x})+\bar{y}) \leq \frac{\kappa}{1-\kappa\mu}.$$

Although formally there is no inversion of a mapping in Theorem 3F.1, if this result is stated equivalently in terms of the Aubin property of the inverse mapping F^{-1}, it fits then into the pattern of the inverse function theorem paradigm. It can also be viewed as a result concerning stability of metric regularity under perturbations by functions with small Lipschitz constants. We can actually deduce the classical inverse function theorem 1A.1 from 3F.1. Indeed, let $f : \mathbb{R}^n \to \mathbb{R}^n$ be a smooth function around $\bar{x}$ and let $\nabla f(\bar{x})$ be nonsingular. Then $F = Df(\bar{x})$ is metrically regular everywhere and from 3F.1 for the function $g(x) = f(x) - Df(\bar{x})(x-\bar{x})$ with $\text{lip}(g;\bar{x}) = 0$ we obtain that $g+F = f$ is metrically regular at $\bar{x}$ for $f(\bar{x})$. But then f must be open (cf. 3E.8). Establishing this fact is the main part of all proofs of 1A.1 presented so far.

We will postpone proving Theorem 3F.1 to Chapter 5, for a mapping F acting from a complete metric space to a linear metric space. In this section we focus on some consequences of this result and its implicit function version. Several corollaries of 3F.1 will lead the way.

Corollary 3F.2 (detailed estimates). *Consider a mapping $F : \mathbb{R}^n \rightrightarrows \mathbb{R}^m$ and any pair $(\bar{x}, \bar{y}) \in \text{gph } F$ at which $\text{gph } F$ is locally closed. If $\text{reg}(F;\bar{x}|\bar{y}) > 0$, then for any $g : \mathbb{R}^n \to \mathbb{R}^m$ such that $\text{reg}(F;\bar{x}|\bar{y}) \cdot \text{lip}(g;\bar{x}) < 1$, one has*

(2) $$\text{reg}(g+F;\bar{x}|g(\bar{x})+\bar{y}) \leq (\text{reg}(F;\bar{x}|\bar{y})^{-1} - \text{lip}(g;\bar{x}))^{-1}.$$

If $\text{reg}(F;\bar{x}|\bar{y}) = 0$, then $\text{reg}(g+F;\bar{x}|g(\bar{x})+\bar{y}) = 0$ for any $g : \mathbb{R}^n \to \mathbb{R}^m$ with $\text{lip}(g;\bar{x}) < \infty$. If $\text{reg}(F;\bar{x}|\bar{y}) = \infty$, then $\text{reg}(g+F;\bar{x}|g(\bar{x})+\bar{y}) = \infty$ for any $g : \mathbb{R}^n \to \mathbb{R}^m$ with $\text{lip}(g;\bar{x}) = 0$.

Proof. If $\text{reg}(F;\bar{x}|\bar{y}) < \infty$, then by choosing κ and μ appropriately and passing to limits in (1) we obtain the claimed inequality (2) also for the case where $\text{reg}(F;\bar{x}|\bar{y}) = 0$. Let $\text{reg}(F;\bar{x}|\bar{y}) = \infty$, and suppose that $\text{reg}(g+F;\bar{x}|g(\bar{x})+\bar{y}) < \kappa$ for some κ and a function g with $\text{lip}(g;\bar{x}) = 0$. Note that g is Lipschitz continuous around $\bar{x}$, hence the graph of g is locally closed around $(\bar{x}, g(\bar{x}))$. Then $g+F$ has locally closed graph at $(\bar{x}, g(\bar{x})+\bar{y})$. Applying Theorem 3F.1 to the mapping $g+F$ with perturbation $-g$, and noting that $\text{lip}(-g;\bar{x}) = 0$, we get $\text{reg}(F;\bar{x}|\bar{y}) \leq \kappa$, which constitutes a contradiction. $\square$

When the perturbation g has zero Lipschitz modulus at the reference point, we obtain another interesting fact.

Corollary 3F.3 (perturbations with Lipschitz modulus 0). *Consider a mapping $F : \mathbb{R}^n \rightrightarrows \mathbb{R}^m$ and any pair $(\bar{x}, \bar{y}) \in \text{gph } F$ at which $\text{gph } F$ is locally closed. Then for any $g : \mathbb{R}^n \to \mathbb{R}^m$ with $\text{lip}(g;\bar{x}) = 0$ one has*

$$\text{reg}(g+F;\bar{x}|g(\bar{x})+\bar{y}) = \text{reg}(F;\bar{x}|\bar{y}).$$

3 Regularity Properties of Set-valued Solution Mappings 171

Proof. The cases with $\operatorname{reg}(F;\bar{x}|\bar{y}) = 0$ or $\operatorname{reg}(F;\bar{x}|\bar{y}) = \infty$ are already covered by Corollary 3F.2. If $0 < \operatorname{reg}(F;\bar{x}|\bar{y}) < \infty$, we get from (2) that

$$\operatorname{reg}(g+F;\bar{x}|g(\bar{x})+\bar{y}) \leq \operatorname{reg}(F;\bar{x}|\bar{y}).$$

By exchanging the roles of F and $g+F$, we also get

$$\operatorname{reg}(F;\bar{x}|\bar{y}) \leq \operatorname{reg}(g+F;\bar{x}|g(\bar{x})+\bar{y}),$$

and in that way arrive at the claimed equality. □

An elaboration of Corollary 3F.3 employs first-order approximations of a function as were introduced in Section 1E.

Corollary 3F.4 (utilization of strict first-order approximations). *Consider a mapping $F: \mathbb{R}^n \rightrightarrows \mathbb{R}^m$ and any pair $(\bar{x},\bar{y}) \in \operatorname{gph} F$ at which $\operatorname{gph} F$ is locally closed. Let $f: \mathbb{R}^n \to \mathbb{R}^m$ be continuous in a neighborhood of $\bar{x}$. Then, for any $h: \mathbb{R}^n \to \mathbb{R}^m$ which is a strict first-order approximation to f at $\bar{x}$, one has*

$$\operatorname{reg}(f+F;\bar{x}|f(\bar{x})+\bar{y}) = \operatorname{reg}(h+F;\bar{x}|h(\bar{x})+\bar{y}).$$

In particular, when the strict first-order approximation is represented by the linearization coming from strict differentiability, we get something even stronger.

Corollary 3F.5 (utilization of strict differentiability). *Consider $M = f+F$ for a function $f: \mathbb{R}^n \to \mathbb{R}^m$ and a mapping $F: \mathbb{R}^n \rightrightarrows \mathbb{R}^m$, and let $\bar{y} \in M(\bar{x})$. Suppose that $f: \mathbb{R}^n \to \mathbb{R}^m$ is strictly differentiable at $\bar{x}$ and that $\operatorname{gph} M$ is locally closed at $(\bar{x},\bar{y})$ (or in other words that $\operatorname{gph} F$ is locally closed at $(\bar{x},\bar{y}-f(\bar{x}))$). Then, for the linearization*

$$M_0(x) = f(\bar{x}) + \nabla f(\bar{x})(x-\bar{x}) + F(x)$$

one has

$$\operatorname{reg}(M;\bar{x}|\bar{y}) = \operatorname{reg}(M_0;\bar{x}|\bar{y}).$$

In the case when $m = n$ and the mapping F is the normal cone mapping to a polyhedral convex set, we can likewise employ a "first-order approximation" of F. When f is linear the corresponding result parallels 2E.6.

Corollary 3F.6 (affine-polyhedral variational inequalities). *For an $n \times n$ matrix A and a polyhedral convex set $C \subset \mathbb{R}^n$, consider the variational inequality*

$$Ax + N_C(x) \ni y.$$

Let $\bar{x}$ be a solution for $\bar{y}$, let $\bar{v} = \bar{y} - A\bar{x}$, so that $\bar{v} \in N_C(\bar{x})$, and let $K = K_C(\bar{x},\bar{v})$ be the critical cone to C at $\bar{x}$ for $\bar{v}$. Then, for the mappings

$$G(x) = Ax + N_C(x) \text{ with } G(\bar{x}) \ni \bar{y},$$
$$G_0(w) = Aw + N_K(w) \text{ with } G_0(0) \ni 0,$$

we have
$$\operatorname{reg}(G;\bar{x}|\bar{y}) = \operatorname{reg}(G_0;0|0).$$

Proof. From reduction lemma 2E.4, for (w,u) in a neighborhood of $(0,0)$, we have that $\bar{v}+u \in N_C(\bar{x}+w)$ if and only if $u \in N_K(w)$. Then, for (w,v) in a neighborhood of $(0,0)$, we obtain $\bar{y}+v \in G(\bar{x}+w)$ if and only if $v \in G_0(w)$. Thus, metric regularity of $A+N_C$ at $\bar{x}$ for $\bar{y}$ with a constant κ implies metric regularity of $A+N_K$ at 0 for 0 with the same constant κ, and conversely. □

Combining 3F.5 and 3F.6 we obtain the following corollary:

Corollary 3F.7 (strict differentiability and polyhedral convexity). *Consider $M = f+N_C$ for a function $f : \mathbb{R}^n \to \mathbb{R}^n$ and a polyhedral convex set $C \subset \mathbb{R}^n$, let $\bar{y} \in M(\bar{x})$ and let f be strictly differentiable at $\bar{x}$. For $\bar{v} = \bar{y} - f(\bar{x})$, let $K = K_C(\bar{v},\bar{x})$ be the critical cone to the set C at $\bar{x}$ for $\bar{v}$. Then, for $M_0(x) = \nabla f(\bar{x})x + N_K(x)$ one has*
$$\operatorname{reg}(M;\bar{x}|\bar{y}) = \operatorname{reg}(M_0;0|0).$$

We are ready now to take up once more the study of a generalized equation having the form

$$(3) \qquad f(p,x) + F(x) \ni 0$$

for $f : \mathbb{R}^d \times \mathbb{R}^n \to \mathbb{R}^m$ and $F : \mathbb{R}^n \rightrightarrows \mathbb{R}^m$, and its solution mapping $S : \mathbb{R}^d \to \mathbb{R}^n$ defined by

$$(4) \qquad S(p) = \{x \mid f(p,x) + F(x) \ni 0\}.$$

This time, however, we are not looking for single-valued localizations of S but aiming at a better understanding of situations in which S may not have any such localization, as in the example of parameterized constraint systems. Recall from Chapter 1 that, for $f : \mathbb{R}^d \times \mathbb{R}^n \to \mathbb{R}^m$ and a point $(\bar{p},\bar{x}) \in \operatorname{int} \operatorname{dom} f$, a function $h : \mathbb{R}^n \to \mathbb{R}^m$ is said to be a strict estimator of f with respect to x uniformly in p at $(\bar{p},\bar{x})$ with constant μ if $h(\bar{x}) = f(\bar{x},\bar{p})$ and

$$\widehat{\operatorname{lip}}_x(e;(\bar{p},\bar{x})) \leq \mu < \infty \quad \text{for } e(p,x) = f(p,x) - h(x).$$

Theorem 3F.8 (implicit mapping theorem with metric regularity). *For the generalized equation (3) and its solution mapping S in (4), and a pair $(\bar{p},\bar{x})$ with $\bar{x} \in S(\bar{p})$, let $h : \mathbb{R}^n \to \mathbb{R}^m$ be a strict estimator of f with respect to x uniformly in p at $(\bar{p},\bar{x})$ with constant μ, let $\operatorname{gph}(h+F)$ be locally closed at $(\bar{x},0)$ and let $h+F$ be metrically regular at $\bar{x}$ for 0 with $\operatorname{reg}(h+F;\bar{x}|0) \leq \kappa$. Suppose that*

$$(5) \qquad \kappa\mu < 1 \quad \text{and} \quad \widehat{\operatorname{lip}}_p(f;(\bar{p},\bar{x})) \leq \lambda < \infty.$$

Then S has the Aubin property at $\bar{p}$ for $\bar{x}$, and moreover

3 Regularity Properties of Set-valued Solution Mappings

(6) $$\mathrm{lip}(S;\bar{p}|\bar{x}) \leq \frac{\kappa\lambda}{1-\kappa\mu}.$$

This theorem will be established in an infinite-dimensional setting in Section 5E, so we will not prove it separately here. An immediate consequence is obtained by specializing the function h in Theorem 3F.8 to a linearization of f with respect to x. We add to this the effect of ample parameterization, in parallel to the case of single-valued localization in Theorem 2C.2.

Theorem 3F.9 (using strict differentiability and ample parameterization). *For the generalized equation (3) and its solution mapping S in (4), and a pair $(\bar{p},\bar{x})$ with $\bar{x} \in S(\bar{p})$, suppose that f is strictly differentiable at $(\bar{p},\bar{x})$ and that gph F is locally closed at $(\bar{x}, -f(\bar{p},\bar{x}))$. If the mapping*

$$h + F \text{ for } h(x) = f(\bar{p},\bar{x}) + \nabla_x f(\bar{p},\bar{x})(x-\bar{x})$$

is metrically regular at $\bar{x}$ for 0, then S has the Aubin property at $\bar{p}$ for $\bar{x}$ with

(7) $$\mathrm{lip}(S;\bar{p}|\bar{x}) \leq \mathrm{reg}(h+F;\bar{x}|0) \cdot |\nabla_p f(\bar{p},\bar{x})|.$$

Furthermore, when f satisfies the ample parameterization condition

(8) $$\mathrm{rank}\, \nabla_p f(\bar{p},\bar{x}) = m,$$

then the converse implication holds as well: the mapping $h + F$ is metrically regular at $\bar{x}$ for 0 provided that S has the Aubin property at $\bar{p}$ for $\bar{x}$.

Proof of 3F.9, initial part. In these circumstances with this choice of h, the conditions in (5) are satisfied because, for $e = f - h$,

$$\widehat{\mathrm{lip}}_x(e;(\bar{p},\bar{x})) = 0 \text{ and } \widehat{\mathrm{lip}}_p(f;(\bar{p},\bar{x})) = |\nabla_p f(\bar{p},\bar{x})|.$$

Thus, (7) follows from (6) with $\mu = 0$. In the remainder of the proof, regarding ample parameterization, we will make use of the following fact.

Proposition 3F.10 (Aubin property in composition). *For a mapping $M : \mathbb{R}^d \rightrightarrows \mathbb{R}^n$ and a function $\psi : \mathbb{R}^n \times \mathbb{R}^m \to \mathbb{R}^d$ consider the composite mapping $N : \mathbb{R}^m \rightrightarrows \mathbb{R}^n$ of the form*

$$y \mapsto N(y) = \{x \mid x \in M(\psi(x,y))\} \text{ for } y \in \mathbb{R}^m.$$

Let ψ satisfy

(9) $$\widehat{\mathrm{lip}}_x(\psi;(\bar{x},0)) = 0 \text{ and } \widehat{\mathrm{lip}}_y(\psi;(\bar{x},0)) < \infty,$$

and, for $\bar{p} = \psi(\bar{x},0)$, let $(\bar{p},\bar{x}) \in \mathrm{gph}\, M$ at which gph M is locally closed. Under these conditions, if M has the Aubin property at $\bar{p}$ for $\bar{x}$, then N has the Aubin property at 0 for $\bar{x}$.

Proof. Let the mapping M have the Aubin property at $\bar{p}$ for $\bar{x}$ with neighborhoods Q of $\bar{p}$ and U of $\bar{x}$ and constant $\kappa > \text{lip}(M;\bar{p}|\bar{x})$. Choose $\lambda > 0$ with $\lambda < 1/\kappa$ and let $\gamma > \widehat{\text{lip}}_y(\psi;(\bar{x},0))$. By (9) there exist positive constants a and b such that for any $y \in B_a(0)$ the function $\psi(\cdot,y)$ is Lipschitz continuous on $B_b(\bar{x})$ with Lipschitz constant λ and for every $x \in B_b(\bar{x})$ the function $\psi(x,\cdot)$ is Lipschitz continuous on $B_a(0)$ with Lipschitz constant γ. Pick a positive constant c and make a and b smaller if necessary so that:
- (a) $B_c(\bar{p}) \subset Q$ and $B_b(\bar{x}) \subset U$,
- (b) the set $\text{gph } M \cap (B_c(\bar{p}) \times B_b(\bar{x}))$ is closed, and
- (c) the following inequalities are satisfied:

$$(10) \quad \frac{4\kappa\gamma a}{1-\kappa\lambda} \leq b \text{ and } \gamma a + \lambda b \leq c.$$

Let $y', y \in B_a(0)$ and let $x' \in N(y') \cap B_{b/2}(\bar{x})$. Then $x' \in M(\psi(x',y')) \cap B_{b/2}(\bar{x})$. Further, we have

$$|\psi(x',y') - \bar{p}| \leq |\psi(x',y') - \psi(x',0)| + |\psi(x',0) - \psi(\bar{x},0)| \leq \gamma a + \lambda b/2 \leq c$$

and the same for $\psi(x',y)$. From the Aubin property of M we obtain the existence of $x^1 \in M(\psi(x',y))$ such that

$$|x^1 - x'| \leq \kappa|\psi(x',y') - \psi(x',y)| \leq \kappa\gamma|y' - y|.$$

Thus, through the first inequality in (10),

$$|x^1 - \bar{x}| \leq |x^1 - x'| + |x' - \bar{x}| \leq \kappa\gamma|y' - y| + |x' - \bar{x}| \leq \kappa\gamma(2a) + \frac{b}{2} \leq b,$$

and consequently

$$|\psi(x^1,y) - \bar{p}| = |\psi(x^1,y) - \psi(\bar{x},0)| \leq \lambda b + \gamma a \leq c,$$

utilizing the second inequality in (10). Hence again, from the Aubin property of M applied to $x^1 \in M(\psi(x',y)) \cap B_b(\bar{x})$, there exists $x^2 \in M(\psi(x^1,y))$ such that

$$|x^2 - x^1| \leq \kappa|\psi(x^1,y) - \psi(x',y)| \leq \kappa\lambda|x^1 - x'| \leq (\kappa\lambda)\kappa\gamma|y' - y|.$$

Employing induction, assume that we have a sequence $\{x^j\}$ with

$$x^j \in M(\psi(x^{j-1},y)) \text{ and } |x^j - x^{j-1}| \leq (\kappa\lambda)^{j-1}\kappa\gamma|y' - y| \text{ for } j = 1,\ldots,k.$$

Setting $x^0 = x'$, for $k = 1, 2, \ldots$ we get

$$|x^k - \bar{x}| \leq |x^0 - \bar{x}| + \sum_{j=1}^{k} |x^j - x^{j-1}|$$

$$\leq \frac{b}{2} + \sum_{j=0}^{k-1}(\kappa\lambda)^j \kappa\gamma |y' - y| \leq \frac{b}{2} + \frac{2a\kappa\gamma}{1-\kappa\lambda} \leq b,$$

where we use the first inequality in (10). Hence $|\psi(x^k,y) - \bar{p}| \leq \lambda b + \gamma a \leq c$. Then there exists $x^{k+1} \in M(\psi(x^k,y))$ such that

$$|x^{k+1} - x^k| \leq \kappa |\psi(x^k,y) - \psi(x^{k-1},y)| \leq \kappa\lambda |x^k - x^{k-1}| \leq (\kappa\lambda)^k \kappa\gamma |y' - y|,$$

and the induction step is complete.

The sequence $\{x^k\}$ is Cauchy, hence convergent to some $x \in {I\!B}_b(\bar{x}) \subset U$. From the local closedness of gph M and the continuity of ψ we deduce that $x \in M(\psi(x,y))$, hence $x \in N(y)$. Furthermore, using the estimate

$$|x^k - x^0| \leq \sum_{j=1}^{k}|x^j - x^{j-1}| \leq \sum_{j=0}^{k-1}(\kappa\lambda)^j \kappa\gamma |y' - y| \leq \frac{\kappa\gamma}{1-\kappa\lambda}|y' - y|$$

we obtain, for any $\kappa' \geq (\kappa\gamma)/(1-\kappa\lambda)$, and on passing to the limit with respect to $k \to \infty$, that $|x - x'| \leq \kappa'|y' - y|$. Thus, N has the Aubin property at 0 for $\bar{x}$ with constant κ'. □

Proof of 3F.9, final part. Under the ample parameterization condition (8), Lemma 2C.1 guarantees the existence of neighborhoods U of $\bar{x}$, V of 0, and Q of $\bar{p}$, as well as a local selection $\psi : U \times V \to Q$ around $(\bar{x}, 0)$ for $\bar{p}$ of the mapping

$$(x,y) \mapsto \{p \,|\, y + f(p,x) = h(x)\}$$

for $h(x) = f(\bar{p},\bar{x}) + \nabla_x f(\bar{p},\bar{x})(x-\bar{x})$ which satisfies the conditions in (9). Hence,

$$y + f(\psi(x,y),x) = h(x) \quad \text{and} \quad \psi(x,y) \in Q \quad \text{for } x \in U, y \in V.$$

Fix $y \in V$. If $x \in (h+F)^{-1}(y) \cap U$ and $p = \psi(x,y)$, then $p \in Q$ and $y + f(p,x) = h(x)$, hence $x \in S(p) \cap U$. Conversely, if $x \in S(\psi(x,y)) \cap U$, then clearly $x \in (h+F)^{-1}(y) \cap U$. Thus,

(11) $$(h+F)^{-1}(y) \cap U = \{x \,|\, x \in S(\psi(x,y)) \cap U\}.$$

Since the Aubin property of S at $\bar{p}$ for $\bar{x}$ is a local property of the graph of S relative to the point $(\bar{p},\bar{x})$, it holds if and only if the same holds for the truncated mapping $S_U : p \mapsto S(p) \cap U$ (see Exercise 3F.11 below). That equivalence is valid for $(h+F)^{-1}$ as well. Thus, if the mapping S_U has the Aubin property at $\bar{p}$ for $\bar{x}$, from Proposition 3F.10 in the context of (11), we obtain that $(h+F)^{-1}$ has the Aubin property at 0 for $\bar{x}$, hence, by 3E.6, $h+F$ is metrically regular at $\bar{x}$ for 0 as desired. □

Exercise 3F.11. Let $S : {I\!R}^m \rightrightarrows {I\!R}^n$ have the Aubin property at $\bar{y}$ for $\bar{x}$ with constant κ. Show that for any neighborhood U of $\bar{x}$ the mapping $S_U : y \mapsto S(y) \cap U$ also has the Aubin property at $\bar{y}$ for $\bar{x}$ with constant κ.

Guide. Choose sufficiently small $a > 0$ and $b > 0$ such that $\mathbb{B}_a(\bar{x}) \subset U$ and $b \leq a/(4\kappa)$. Then for every $y, y' \in \mathbb{B}_b(\bar{y})$ and every $x \in S(y) \cap \mathbb{B}_{a/2}(\bar{x})$ there exists $x' \in S(y')$ with $|x' - x| \leq \kappa |y' - y| \leq 2\kappa b \leq a/2$. Then both x and x' are from U. □

Let us now look at the case of 3F.9 in which F is a constant mapping, $F(x) \equiv K$, which was featured at the beginning of this chapter as a motivation for investigating real set-valuedness in solution mappings. Solving $f(p,x) + F(x) \ni 0$ for a given p then means finding an x such that $-f(p,x) \in K$. For particular choices of K this amounts to solving some mixed system of equations and inequalities, for example.

Example 3F.12 (application to general constraint systems). For $f : \mathbb{R}^d \times \mathbb{R}^n \to \mathbb{R}^m$ and a closed set $K \subset \mathbb{R}^m$, let

$$S(p) = \{x \mid f(p,x) \in K\}.$$

Fix $\bar{p}$ and $\bar{x} \in S(\bar{p})$. Suppose that f is continuously differentiable on a neighborhood of $(\bar{p}, \bar{x})$, and consider the solution mapping for an associated linearized system:

$$\bar{S}(y) = \{x \mid y - f(\bar{p}, \bar{x}) - \nabla_x f(\bar{p}, \bar{x})(x - \bar{x}) \in K\}.$$

If $\bar{S}$ has the Aubin property at 0 for $\bar{x}$, then S has the Aubin property at $\bar{p}$ for $\bar{x}$. The converse implication holds under the ample parameterization condition (8).

The key to applying this result, of course, is being able to ascertain when the linearized system does have the Aubin property in question. In the important case of $K = \mathbb{R}_-^s \times \{0\}^{m-s}$, a necessary and sufficient condition will emerge in the so-called Mangasarian–Fromovitz constraint qualification. This will be seen in Section 4D.

Example 3F.13 (application to polyhedral variational inequalities). For $f : \mathbb{R}^d \times \mathbb{R}^n \to \mathbb{R}^n$ and a convex polyhedral set $C \subset \mathbb{R}^n$, let

$$S(p) = \{x \mid f(p,x) + N_C(x) \ni 0\}.$$

Fix $\bar{p}$ and $\bar{x} \in S(\bar{p})$ and for $\bar{v} = -f(\bar{p}, \bar{x})$ let $K = K_C(\bar{x}, \bar{v})$ be the associated critical cone to C. Suppose that f is continuously differentiable on a neighborhood of $(\bar{p}, \bar{x})$, and consider the solution mapping for an associated reduced system:

$$\bar{S}(y) = \{x \mid \nabla_x f(\bar{p}, \bar{x})x + N_K(x) \ni y\}.$$

If $\bar{S}$ has the Aubin property at 0 for 0, then S has the Aubin property at $\bar{p}$ for $\bar{x}$. The converse implication holds under the ample parameterization condition (8).

If a mapping $F : \mathbb{R}^n \rightrightarrows \mathbb{R}^m$ has the Aubin property at $\bar{x}$ for $\bar{y}$, then for any function $f : \mathbb{R}^n \to \mathbb{R}^m$ with $\text{lip}(f; \bar{x}) < \infty$, the mapping $f + F$ has the Aubin property at $\bar{x}$ for $f(\bar{x}) + \bar{y}$ as well. This is a particular case of the following observation which utilizes ample parameterization.

Exercise 3F.14. Consider a mapping $F : \mathbb{R}^n \rightrightarrows \mathbb{R}^m$ with $(\bar{x}, \bar{y}) \in \text{gph } F$ and a function $f : \mathbb{R}^d \times \mathbb{R}^n \to \mathbb{R}^m$ having $\bar{y} = -f(\bar{p}, \bar{x})$ and which is strictly differentiable at $(\bar{p}, \bar{x})$ and satisfies the ample parameterization condition (8). Prove that the mapping

$$x \mapsto P(x) = \{p \mid 0 \in f(p,x) + F(x)\}$$

has the Aubin property at $\bar{x}$ for $\bar{p}$ if and only if F has the Aubin property at $\bar{x}$ for $\bar{y}$.

Guide. First, apply 3F.9 to show that, under the ample parameterization condition (8), the mapping

$$(x,y) \mapsto \Omega(x,y) = \{p \mid y + f(p,x) = 0\}$$

has the Aubin property at $(\bar{x},\bar{y})$ for $\bar{p}$. Let F have the Aubin property at $\bar{x}$ for $\bar{y}$ with neighborhoods U of $\bar{x}$ of V for $\bar{y}$ and constant κ. Choose a neighborhood Q of $\bar{p}$ and adjust U and V accordingly so that Ω has the Aubin property with constant λ and neighborhoods $U \times V$ and Q. Let $b > 0$ be such that $\mathbb{B}_b(\bar{y}) \subset V$, then choose $a > 0$ and adjust Q such that $\mathbb{B}_a(\bar{x}) \subset U$, $a \leq b/(4\kappa)$ and also $-f(p,x) \in \mathbb{B}_{b/2}(\bar{y})$ for $x \in \mathbb{B}_a(\bar{x})$ and $p \in Q$. Let $x, x' \in \mathbb{B}_a(\bar{x})$ and $p \in P(x) \cap Q$. Then $y = -f(p,x) \in F(x) \cap V$ and by the Aubin property of F there exists $y' \in F(x')$ such that $|y - y'| \leq \kappa|x - x'|$. But then $|y' - \bar{y}| \leq \kappa(2a) + b/2 \leq b$. Thus $y' \in V$ and hence, by the Aubin property of Ω, there exists p' satisfying $y' + f(p',x') = 0$ and

$$|p' - p| \leq \lambda(|y' - y| + |x' - x|) \leq \lambda(\kappa + 1)|x' - x|.$$

Noting that $p' \in P(x')$ we get that P has the Aubin property at $\bar{x}$ for $\bar{p}$.

Conversely, let P have the Aubin property at $\bar{x}$ for $\bar{p}$ with associated constant κ and neighborhoods U and Q of $\bar{x}$ and $\bar{p}$, respectively. Let f be Lipschitz continuous on $Q \times U$ with constant μ. We already know that the mapping Ω has the Aubin property at $(\bar{x}, \bar{y})$ for $\bar{p}$; let λ be the associated constant and $U \times V$ and Q the neighborhoods of $(\bar{x}, \bar{y})$ and of $\bar{p}$, respectively. Choose $c > 0$ such that $\mathbb{B}_c(\bar{p}) \subset Q$ and let $a > 0$ satisfy

$$\mathbb{B}_a(\bar{x}) \subset U, \quad \mathbb{B}_a(\bar{y}) \subset V \text{ and } a \max\{\kappa, \lambda\} \leq c/4.$$

Let $x, x' \in \mathbb{B}_a(\bar{x})$ and $y \in F(x) \cap \mathbb{B}_a(\bar{y})$. Since Ω has the Aubin property and $\bar{p} \in \Omega(\bar{x},\bar{y}) \cap \mathbb{B}_c(\bar{p})$, there exists $p \in \Omega(x,y)$ such that $|p - \bar{p}| \leq \lambda(2a) \leq c/2$. This means that $p \in P(x) \cap \mathbb{B}_{c/2}(\bar{p})$ and from the Aubin property of P there exists $p' \in P(x')$ so that $|p' - p| \leq \kappa|x' - x|$. Thus, $|p' - \bar{p}| \leq \kappa(2a) + c/2 \leq c$. Let $y' = -f(p',x')$. Then $y' \in F(x')$ because $p' \in P(x')$ and the Lipschitz continuity of f gives us

$$|y - y'| = |f(p,x) - f(p',x')| \leq \mu(|p - p'| + |x - x'|) \leq \mu(\kappa + 1)|x - x'|.$$

Hence, F has the Aubin property at $\bar{x}$ for $\bar{y}$. □

In none of the directions of the statement of 3F.14 we can replace the Aubin property by metric regularity. Indeed, the function $f(p,x) = x + p$ satisfies the assumptions, but if we add to it the zero mapping $F \equiv 0$, which is not metrically regular (anywhere), we get the mapping $P(x) = -x$ which is metrically regular (everywhere). Taking the same f and $F(x) = -x$ contradicts the other direction.

3G. Strong Metric Regularity

Although our chief goal in this chapter has been the treatment of solution mappings for which Lipschitz continuous single-valued localizations need not exist or even be a topic of interest, the concepts and results we have built up can shed new light on our earlier work with such localizations through their connection with metric regularity.

Proposition 3G.1 (single-valued localizations and metric regularity). *For a mapping $F : \mathbb{R}^n \rightrightarrows \mathbb{R}^m$ and a pair $(\bar{x}, \bar{y}) \in \text{gph } F$, the following properties are equivalent:*

(a) *F^{-1} has a Lipschitz continuous single-valued localization s around $\bar{y}$ for $\bar{x}$.*

(b) *F is metrically regular at $\bar{x}$ for $\bar{y}$ and F^{-1} has a localization at $\bar{y}$ for $\bar{x}$ that is nowhere multi-valued.*

Indeed, in the circumstances of (b) the localization s in (a) has $\text{lip}(s; \bar{y}) = \text{reg}(F; \bar{x} | \bar{y})$.

Proof. According to 3E.2 as applied to $S = F^{-1}$, condition (a) is equivalent to F^{-1} having the Aubin property at $\bar{y}$ for $\bar{x}$ and a localization around $\bar{y}$ for $\bar{x}$ that is nowhere multi-valued. When F^{-1} has the Aubin property at $\bar{y}$ for $\bar{x}$, by 3E.1 the domain of F^{-1} contains a neighborhood of $\bar{y}$, hence any localization of F^{-1} at $\bar{y}$ for $\bar{x}$ is actually a localization around $\bar{y}$ for $\bar{x}$. On the other hand, we know from 3E.6 that F^{-1} has the Aubin property at $\bar{y}$ for $\bar{x}$ if and only if F is metrically regular at $\bar{x}$ for $\bar{y}$. That result also relates the constants κ in the two properties and yields for us the final statement. □

Proposition 3G.2 (stability of single-valuedness under perturbation). *For a mapping $F : \mathbb{R}^n \rightrightarrows \mathbb{R}^m$ and a pair $(\bar{x}, \bar{y}) \in \text{gph } F$, let F^{-1} have a Lipschitz continuous single-valued localization s around $\bar{y}$ for $\bar{x}$ and let $\lambda > \text{lip}(s; \bar{y})$. Then for any positive $\nu < \lambda^{-1}$ and for any function $g : \mathbb{R}^n \to \mathbb{R}^m$ with $\bar{x} \in \text{int dom } g$ and $\text{lip}(g; \bar{x}) < \nu$ the mapping $(g + F)^{-1}$ has a localization at $g(\bar{x}) + \bar{y}$ for $\bar{x}$ which is nowhere multi-valued.*

Proof. Our hypothesis says that there are neighborhoods U of $\bar{x}$ and V of $\bar{y}$ such that for any $y \in V$ the set $F^{-1}(y) \cap U$ consists of exactly one point, $s(y)$, and that the function $s : y \mapsto F^{-1}(y) \cap U$ is Lipschitz continuous on V with Lipschitz constant λ. Let $0 < \nu < \lambda^{-1}$ and choose a function $g : \mathbb{R}^n \to \mathbb{R}^m$ and a neighborhood U' of $\bar{x}$ on which g is Lipschitz continuous with constant ν. We can find neighborhoods $U_0 = \mathbb{B}_\tau(\bar{x}) \subset U \cap U'$ and $V_0 = \mathbb{B}_\varepsilon(g(\bar{x}) + \bar{y}) \subset (g(\bar{x}) + V)$ such that

(1) $$x \in U_0, \, y \in V_0 \implies y - g(x) \in V.$$

Consider now the graphical localization of $(g + F)^{-1}$ corresponding to U_0 and V_0. Each $y \in V_0$ maps to the set $(g + F)^{-1}(y) \cap U_0$; it will be demonstrated that this set can have at most one element, and that will finish the proof.

Suppose to the contrary that $y \in V_0$ and $x, x' \in U_0$, $x \neq x'$, are such that both x and x' belong to $(g+F)^{-1}(y)$. Clearly $x \in (g+F)^{-1}(y) \cap U_0$ if and only if $x \in U_0$ and $y \in g(x) + F(x)$, or equivalently $y - g(x) \in F(x)$. The latter, in turn, is the same as having $x \in F^{-1}(y - g(x)) \cap U_0 \subset F^{-1}(y - g(x)) \cap U = s(y - g(x))$, where $y - g(x) \in V$ by (1). Then

$$0 < |x - x'| = |s(y - g(x)) - s(y - g(x'))| \leq \lambda |g(x) - g(x')| \leq \lambda \nu |x - x'| < |x - x'|,$$

which is absurd. □

The observation in 3G.1 leads to a definition.

Strong metric regularity. *A mapping $F : \mathbb{R}^n \rightrightarrows \mathbb{R}^m$ having the equivalent properties in 3G.1 will be called strongly metrically regular at $\bar{x}$ for $\bar{y}$.*

For a linear mapping represented by an $m \times n$ matrix A, strong metric regularity comes out as the nonsingularity of A and thus requires that $m = n$. Moreover, for any single-valued function $f : \mathbb{R}^n \to \mathbb{R}^m$, strong metric regularity requires $m = n$ by Theorem 1F.1 on the invariance of domain. This property can be seen therefore as corresponding closely to the one in the classical implicit function theorem, except for its focus on Lipschitz continuity instead of continuous differentiability. It was the central property in fact, if not in name, in Robinson's implicit function theorem 2B.1.

The terminology of *strong* metric regularity offers a way of gaining new perspectives on earlier results by translating them into the language of metric regularity. Indeed, strong metric regularity is just metric regularity plus the existence of a single-valued localization of the inverse. According to Theorem 3F.1, metric regularity of a mapping F with a locally closed graph is stable under addition of a function g with a "small" Lipschitz constant, and so too is local single-valuedness, according to 3G.2 above. Thus, strong metric regularity must be stable under perturbation in the same way as metric regularity. The corresponding result is a version of the inverse function result in 2B.10 corresponding to the extended form of Robinson's implicit function theorem in 2B.5.

Theorem 3G.3 (inverse function theorem with strong metric regularity). *Consider a mapping $F : \mathbb{R}^n \rightrightarrows \mathbb{R}^m$ and any $(\bar{x}, \bar{y}) \in \text{gph } F$ such that F is strongly metrically regular at $\bar{x}$ for $\bar{y}$. Let κ and μ be nonnegative constants such that*

$$\text{reg}(F; \bar{x} | \bar{y}) \leq \kappa \quad \text{and} \quad \kappa \mu < 1.$$

Then for any function $g : \mathbb{R}^n \to \mathbb{R}^m$ with $\bar{x} \in \text{int dom } g$ and $\text{lip}(g; \bar{x}) \leq \mu$, the mapping $g + F$ is strongly metrically regular at $\bar{x}$ for $g(\bar{x}) + \bar{y}$. Moreover,

$$\text{reg}\left(g + F; \bar{x} | g(\bar{x}) + \bar{y}\right) \leq \frac{\kappa}{1 - \kappa \mu}.$$

Proof. Our hypothesis that F is strongly metrically regular at $\bar{x}$ for $\bar{y}$ implies that a graphical localization of F^{-1} around $(\bar{y}, \bar{x})$ is single-valued and continuous near

$\bar{y}$ and therefore that gph F is locally closed at $(\bar{x},\bar{y})$. Further, by fixing $\lambda > \kappa$ and using Proposition 3G.1, we can get neighborhoods U of $\bar{x}$ and V of $\bar{y}$ such that for any $y \in V$ the set $F^{-1}(y) \cap U$ consists of exactly one point, which we may denote by $s(y)$ and know that the function $s : y \mapsto F^{-1}(y) \cap U$ is Lipschitz continuous on V with Lipschitz constant λ. Let $\mu < \nu < \lambda^{-1}$ and choose a function $g : \mathbb{R}^n \to \mathbb{R}^m$ and a neighborhood $U' \subset U$ of $\bar{x}$ on which g is Lipschitz continuous with constant ν. Applying Proposition 3G.2 we obtain that the mapping $(g+F)^{-1}$ has a localization at $g(\bar{x}) + \bar{y}$ for $\bar{x}$ which is nowhere multi-valued. On the other hand, since F has locally closed graph at $(\bar{x},\bar{y})$, we know from Theorem 3F.1 that for such g the mapping $g + F$ is metrically regular at $g(\bar{x}) + \bar{y}$ for $\bar{x}$. Applying Proposition 3G.1 once more, we complete the proof. □

In much the same way we can state in terms of strong metric regularity an implicit function result paralleling Theorem 2B.5.

Theorem 3G.4 (implicit function theorem with strong metric regularity). *For the generalized equation $f(p,x) + F(x) \ni 0$ with $f : \mathbb{R}^n \to \mathbb{R}^m$ and $F : \mathbb{R}^n \rightrightarrows \mathbb{R}^m$ and its solution mapping*

$$S : p \mapsto \big\{ x \,\big|\, f(p,x) + F(x) \ni 0 \big\},$$

consider a pair $(\bar{p},\bar{x})$ with $\bar{x} \in S(\bar{p})$. Let $h : \mathbb{R}^n \to \mathbb{R}^m$ be a strict estimator of f with respect to x uniformly in p at $(\bar{p},\bar{x})$ with constant μ and let $h + F$ be strongly metrically regular at $\bar{x}$ for 0 with $\operatorname{reg}(h+F;\bar{x}|0) \leq \kappa$. Suppose that

$$\kappa\mu < 1 \quad \text{and} \quad \widehat{\operatorname{lip}}_p(f;(\bar{p},\bar{x})) \leq \lambda < \infty.$$

Then S has a Lipschitz continuous single-valued localization s around $\bar{p}$ for $\bar{x}$, moreover with

$$\operatorname{lip}(s;\bar{p}) \leq \frac{\kappa\lambda}{1 - \kappa\mu}.$$

Many corollaries of this theorem could be stated in a mode similar to that in Section 3F, but the territory has already been covered essentially in Chapter 2. We will get back to this result in Section 5F.

In some situations, metric regularity automatically entails strong metric regularity. That is the case, for instance, for a linear mapping from $\mathbb{R}^n$ to itself represented by an $n \times n$ matrix A. Such a mapping is metrically regular if and only if it is surjective, which means that A has full rank, but then A is nonsingular, so that we have strong metric regularity. More generally, for any mapping which describes the Karush-Kuhn-Tucker optimality system in a nonlinear programming problem, metric regularity implies strong metric regularity. We will prove this fact in Section 4F.

We will describe now another class of mappings for which metric regularity and strong metric regularity come out to be the same thing. This class depends on a localized, set-valued form of the monotonicity concept which appeared in Section 2F in the context of variational inequalities.

3 Regularity Properties of Set-valued Solution Mappings

Locally monotone mappings. A mapping $F : \mathbb{R}^n \rightrightarrows \mathbb{R}^n$ is said to be *locally monotone* at $\bar{x}$ for $\bar{y}$ if $(\bar{x}, \bar{y}) \in \operatorname{gph} F$ and for some neighborhood W of $(\bar{x}, \bar{y})$, one has

$$\langle y' - y, x' - x \rangle \geq 0 \quad \text{whenever } (x', y'), (x, y) \in \operatorname{gph} F \cap W.$$

Theorem 3G.5 (strong metric regularity of locally monotone mappings). *If a mapping $F : \mathbb{R}^n \rightrightarrows \mathbb{R}^n$ that is locally monotone at $\bar{x}$ for $\bar{y}$ is metrically regular at $\bar{x}$ for $\bar{y}$, then it must be strongly metrically regular at $\bar{x}$ for $\bar{y}$.*

Proof. According to 3G.1, all we need to show is that a mapping F which is locally monotone and metrically regular at $\bar{x}$ for $\bar{y}$ must have a localization around $\bar{y}$ for $\bar{x}$ which is nowhere multi-valued. Suppose to the contrary that every graphical localization of F^{-1} at $\bar{y}$ for $\bar{x}$ is multi-valued. Then there are infinite sequences $y_k \to \bar{y}$ and $x_k, z_k \in F^{-1}(y_k)$, $x_k \to \bar{x}$, $z_k \to \bar{x}$ such that $x_k \neq z_k$ for all k. Let $b_k = |z_k - x_k| > 0$ and $h_k = (z_k - x_k)/b_k$. Then we have

$$(2) \qquad \langle z_k, h_k \rangle = b_k + \langle x_k, h_k \rangle \quad \text{for all } k = 1, 2, \ldots.$$

Since the metric regularity of F implies through 3E.6 the Aubin property of F^{-1} at $\bar{y}$ for $\bar{x}$, there exist $\kappa > 0$ and $a > 0$ such that

$$F^{-1}(y) \cap \mathbb{B}_a(\bar{x}) \subset F^{-1}(y') + \kappa |y - y'| \mathbb{B} \quad \text{for all } y, y' \in \mathbb{B}_a(\bar{y}).$$

Choose a sequence of positive numbers τ_k satisfying

$$(3) \qquad \tau_k \searrow 0 \quad \text{and} \quad \tau_k < b_k / 2\kappa.$$

Then for k large, we have $y_k, y_k + \tau_k h_k \in \mathbb{B}_a(\bar{y})$ and $x_k \in F^{-1}(y_k) \cap \mathbb{B}_a(\bar{x})$, and hence there exists $u_k \in F^{-1}(y_k + \tau_k h_k)$ satisfying

$$(4) \qquad |u_k - x_k| \leq \kappa \tau_k.$$

By the local monotonicity of F at $(\bar{x}, \bar{y})$ we have

$$\langle u_k - z_k, y_k + \tau_k h_k - y_k \rangle \geq 0.$$

This, combined with (2), yields

$$(5) \qquad \langle u_k, h_k \rangle \geq \langle z_k, h_k \rangle \geq b_k + \langle x_k, h_k \rangle.$$

We get from (3), (4) and (5) that

$$b_k + \langle x_k, h_k \rangle \leq \langle u_k, h_k \rangle \leq \langle x_k, h_k \rangle + \kappa \tau_k < \langle x_k, h_k \rangle + (b_k/2),$$

which is impossible. Therefore, F^{-1} must indeed have a localization around $\bar{y}$ for $\bar{x}$ which is not multi-valued. □

3H. Calmness and Metric Subregularity

A "one-point" variant of the Aubin property can be defined for set-valued mappings in the same way as calmness of functions, and this leads to another natural topic of investigation.

Calmness. *A mapping $S : \mathbb{R}^m \rightrightarrows \mathbb{R}^n$ is said be calm at $\bar{y}$ for $\bar{x}$ if $(\bar{y},\bar{x}) \in \operatorname{gph} S$ and there is a constant $\kappa \geq 0$ along with neighborhoods U of $\bar{x}$ and V of $\bar{y}$ such that*

(1) $$e(S(y) \cap U, S(\bar{y})) \leq \kappa |y - \bar{y}| \quad \text{for all } y \in V.$$

Equivalently, the property in (1) can be also written as

(2) $$S(y) \cap U \subset S(\bar{y}) + \kappa |y - \bar{y}| \mathbb{B} \quad \text{for all } y \in V$$

although perhaps with larger constant κ. The infimum of κ over all such combinations of κ, U and V is called the calmness modulus *of S at $\bar{y}$ for $\bar{x}$ and denoted by $\operatorname{clm}(S; \bar{y}|\bar{x})$. The absence of this property is signaled by $\operatorname{clm}(S; \bar{y}|\bar{x}) = \infty$.*

As in the case of the Lipschitz modulus $\operatorname{lip}(S; \bar{y}|\bar{x})$ in 3E, it is not claimed that (1) and (2) are themselves equivalent, although that is true when $S(y)$ is closed for every $y \in V$. But anyway, the infimum furnishing $\operatorname{clm}(S; \bar{y}|\bar{x})$ is the same with respect to (2) as with respect to (1).

In the case when S is not multi-valued, the definition above reduces to the definition of calmness of a function in Section 1C relative to a neighborhood V of $\bar{y}$; $\operatorname{clm}(S; \bar{y}, S(\bar{y})) = \operatorname{clm}(S; \bar{y})$. Indeed, for any $y \in V \setminus \operatorname{dom} S$ the inequality (1) holds automatically.

Clearly, for mappings with closed values, outer Lipschitz continuity implies calmness. In particular, we get the following fact from Theorem 3D.1.

Proposition 3H.1 (calmness of polyhedral mappings). *Any mapping $S : \mathbb{R}^m \rightrightarrows \mathbb{R}^n$ whose graph is the union of finitely many polyhedral convex sets is calm with the same constant κ at any $\bar{y}$ for any $\bar{x}$ provided that $(\bar{y},\bar{x}) \in \operatorname{gph} S$.*

In particular, any linear mapping is calm at any point of its graph, and this is also true for its inverse. For comparison, the inverse of a linear mapping has the Aubin property at some point if and only if the mapping is surjective.

Exercise 3H.2 (local outer Lipschitz continuity under truncation). *Show that a mapping $S : \mathbb{R}^m \rightrightarrows \mathbb{R}^n$ with $(\bar{y},\bar{x}) \in \operatorname{gph} S$ and with $S(\bar{y})$ convex is calm at $\bar{y}$ for $\bar{x}$ if and only if there is a neighborhood U of $\bar{x}$ such that the truncated mapping $y \mapsto S(y) \cap U$ is outer Lipschitz continuous at $\bar{y}$.*

Guide. Mimic the proof of 3E.3 with $y = \bar{y}$. □

Is there a "one-point" variant of the metric regularity which would characterize calmness of the inverse, in the way metric regularity characterizes the Aubin property of the inverse? Yes, as we explore next.

3 Regularity Properties of Set-valued Solution Mappings

Metric subregularity. A mapping $F: \mathbb{R}^n \rightrightarrows \mathbb{R}^m$ is called *metrically subregular* at $\bar{x}$ for $\bar{y}$ if $(\bar{x}, \bar{y}) \in \text{gph } F$ and there exists $\kappa \geq 0$ along with neighborhoods U of $\bar{x}$ and V of $\bar{y}$ such that

$$(3) \qquad d(x, F^{-1}(\bar{y})) \leq \kappa d(\bar{y}, F(x) \cap V) \quad \text{for all } x \in U.$$

The infimum of all κ for which this holds is the *modulus of metric subregularity*, denoted by $\text{subreg}(F; \bar{x} | \bar{y})$. The absence of metric subregularity is signaled by $\text{subreg}(F; \bar{x} | \bar{y}) = \infty$.

The main difference between metric subregularity and metric regularity is that the data input $\bar{y}$ is now fixed and not perturbed to a nearby y. Since $d(\bar{y}, F(x)) \leq \kappa d(\bar{y}, F(x) \cap V)$, it is clear that subregularity is a weaker condition than metric regularity, and

$$\text{subreg}(F; \bar{x} | \bar{y}) \leq \text{reg}(F; \bar{x} | \bar{y}).$$

The following result reveals the equivalence of metric subregularity of a mapping with calmness of its inverse:

Theorem 3H.3 (characterization by inverse calmness). *For a mapping $F: \mathbb{R}^n \rightrightarrows \mathbb{R}^m$, let $F(\bar{x}) \ni \bar{y}$. Then F is metrically subregular at $\bar{x}$ for $\bar{y}$ with a constant κ if and only if its inverse $F^{-1}: \mathbb{R}^m \rightrightarrows \mathbb{R}^n$ is calm at $\bar{y}$ for $\bar{x}$ with the same constant κ, i.e., there exist neighborhoods U of $\bar{x}$ and V of $\bar{y}$ such that*

$$(4) \qquad F^{-1}(y) \cap U \subset F^{-1}(\bar{y}) + \kappa |y - \bar{y}| \mathbb{B} \quad \text{for all } y \in V.$$

Moreover, $\text{clm}(F^{-1}; \bar{y} | \bar{x}) = \text{subreg}(F; \bar{x} | \bar{y})$.

Proof. Assume first that (4) holds. Let $x \in U$. If $F(x) = \emptyset$, then the right side of (3) is ∞ and we are done. If not, having $x \in U$ and $y \in F(x) \cap V$ is the same as having $x \in F^{-1}(y) \cap U$ and $y \in V$. For such x and y, the inclusion in (4) requires the ball $x + \kappa |y - \bar{y}| \mathbb{B}$ to have nonempty intersection with $F^{-1}(\bar{y})$. Then $d(x, F^{-1}(\bar{y})) \leq \kappa |y - \bar{y}|$. Thus, for any $x \in U$, we must have $d(x, F^{-1}(\bar{y})) \leq \inf_y \{ \kappa |y - \bar{y}| \, | \, y \in F(x) \cap V \}$, which is (3). This shows that (4) implies (3) and that

$$\inf \{ \kappa \, | \, U, V, \kappa \text{ satisfying (4)} \} \geq \inf \{ \kappa \, | \, U, V, \kappa \text{ satisfying (3)} \},$$

the latter being by definition $\text{subreg}(F; \bar{x} | \bar{y})$.

For the opposite direction, we have to demonstrate that if $\text{subreg}(F; \bar{x} | \bar{y}) < \kappa < \infty$, then (4) holds for some choice of neighborhoods U and V. Consider any κ' with $\text{subreg}(F; \bar{x} | \bar{y}) < \kappa' < \kappa$. For this κ', there exist U and V such that $d(x, F^{-1}(\bar{y})) \leq \kappa' d(\bar{y}, F(x) \cap V)$ for all $x \in U$. Then we have $d(x, F^{-1}(\bar{y})) \leq \kappa' |y - \bar{y}|$ when $x \in U$ and $y \in F(x) \cap V$, or equivalently $y \in V$ and $x \in F^{-1}(y) \cap U$. Fix $y \in V$. If $y = \bar{y}$ there is nothing to prove; let $y \neq \bar{y}$. If $x \in F^{-1}(y) \cap U$, then $d(x, F^{-1}(\bar{y})) \leq \kappa' |y - \bar{y}| < \kappa |y - \bar{y}|$. Then there must be a point of $x' \in F^{-1}(\bar{y})$ having $|x' - x| \leq \kappa |y - \bar{y}|$. Hence we have (4), as required. $\square$

As we will see next, there is no need at all to mention a neighborhood V of $\bar{y}$ in the description of calmness and subregularity in (2) and (3).

Exercise 3H.4 (equivalent formulations). *For a mapping $F : \mathbb{R}^n \rightrightarrows \mathbb{R}^m$ and a point $(\bar{x},\bar{y}) \in \mathrm{gph}\, F$ metric subregularity of F at $\bar{x}$ for $\bar{y}$ with constant κ is equivalent simply to the existence of a neighborhood U of $\bar{x}$ such that*

(5) $$d(x, F^{-1}(\bar{y})) \leq \kappa d(\bar{y}, F(x)) \quad \text{for all } x \in U,$$

whereas the calmness of F^{-1} at $\bar{y}$ for $\bar{x}$ with constant κ can be identified with the existence of a neighborhood U of $\bar{x}$ such that

(6) $$F^{-1}(y) \cap U \subset F^{-1}(\bar{y}) + \kappa |y - \bar{y}| \mathbb{B} \quad \text{for all } y \in \mathbb{R}^m.$$

Guide. Assume that (3) holds with $\kappa > 0$ and associated neighborhoods U and V. We can choose within V a neighborhood of the form $V' = \mathbb{B}_\varepsilon(\bar{y})$ for some $\varepsilon > 0$. Let $U' := U \cap (\bar{x} + \varepsilon \kappa \mathbb{B})$ and pick $x \in U'$. If $F(x) \cap V' \neq \emptyset$ then $d(\bar{y}, F(x) \cap V') = d(\bar{y}, F(x))$ and (3) becomes (5) for this x. Otherwise, $F(x) \cap V' = \emptyset$ and then

$$d(\bar{y}, F(x)) \geq \varepsilon \geq \frac{1}{\kappa}|x - \bar{x}| \geq \frac{1}{\kappa} d(x, F^{-1}(\bar{y})),$$

which is (5).

Similarly, (6) entails the calmness in (4), so attention can be concentrated on showing that we can pass from (4) to (6) under an adjustment in the size of U. We already know from 3H.3 that the calmness condition in (4) leads to the metric subregularity in (3), and further, from the argument just given, that such subregularity yields the condition in (5). But that condition can be plugged into the argument in the proof of 3H.3, by taking $V = \mathbb{R}^m$, to get the corresponding calmness property with $V = \mathbb{R}^m$ but with U replaced by a smaller neighborhood of $\bar{x}$. ☐

Although we could take (5) as a redefinition of metric subregularity, we prefer to retain the neighborhood V in (3) in order to underscore the parallel with metric regularity; similarly for calmness.

Does metric subregularity enjoy stability properties under perturbation resembling those of metric regularity and strong metric regularity? In other words, does metric subregularity obey the general paradigm of the implicit function theorem? The answer to this question turns out to be *no* even for simple functions. Indeed, the function $f(x) = x^2$ is clearly not metrically subregular at 0 for 0, but its derivative $Df(0)$, which is the zero mapping, is metrically subregular.

More generally, every linear mapping $A : \mathbb{R}^n \to \mathbb{R}^m$ is metrically subregular, and hence the derivative mapping of any smooth function is metrically subregular. But of course, not every smooth function is subregular. *For this reason, there cannot be an implicit mapping theorem in the vein of 3F.8 in which metric regularity is replaced by metric subregularity, even for the classical case of an equation with smooth f and no set-valued F.*

An illuminating but more intricate counterexample of instability of metric subregularity of set-valued mappings is as follows. In $\mathbb{R} \times \mathbb{R}$, let $\mathrm{gph}\, F$ be the set of all (x, y) such that $x \geq 0$, $y \geq 0$ and $yx = 0$. Then $F^{-1}(0) = [0, \infty) \supset F^{-1}(y)$ for all

3 Regularity Properties of Set-valued Solution Mappings

y, so F is metrically subregular at $\bar{y} = 0$ for $\bar{x} = 0$, even "globally" with $\kappa = 0$. By Theorem 3H.3, $\mathrm{subreg}(F;0|0) = 0$.

Consider, however the function $f(x) = -x^2$ for which $f(0) = 0$ and $\nabla f(0) = 0$. The perturbed mapping $f + F$ has $(f+F)^{-1}$ single-valued everywhere: $(f+F)^{-1}(y) = 0$ when $y \geq 0$, and $(f+F)^{-1}(y) = \sqrt{|y|}$ when $y \leq 0$. This mapping is not calm at 0 for 0. Then, from Theorem 3H.3 again, $f + F$ is not metrically subregular; we have $\mathrm{subreg}(f+F;0|0) = \infty$.

To conclude this section, we point out some other properties which are, in a sense, derived from metric regularity but, like subregularity, lack such kind of stability. One of these properties is the openness which we introduced in Section 1F: a function $f : \mathbb{R}^n \to \mathbb{R}^m$ is said to be open at $\bar{x} \in \mathrm{dom}\, f$ when for any $a > 0$ there exists $b > 0$ such that

$$(7) \qquad f(\bar{x} + a\,\mathrm{int}\mathbb{B}) \supset f(\bar{x}) + b\,\mathrm{int}\mathbb{B}.$$

It turns out that this property likewise fails to be preserved when f is perturbed to $f + g$ by a function g with $\mathrm{lip}(g;\bar{x}) = 0$. This is demonstrated by the following example. Define $f : \mathbb{R}^2 \to \mathbb{R}^2$ by taking $f(0,0) = (0,0)$ and, for $x = (x_1, x_2) \neq (0,0)$,

$$f(x_1, x_2) = \frac{1}{\sqrt{x_1^2 + x_2^2}} \begin{pmatrix} x_1^2 - x_2^2 \\ 2|x_1||x_2| \end{pmatrix}.$$

Then f satisfies (7) at $\bar{x} = 0$ with $b = a$, since $|f(x)| = |x|$. The function $g(x_1, x_2) = (0, x_2^3)$ has $g(0,0) = (0,0)$ and $\mathrm{lip}(g;(0,0)) = 0$, but $(f+g)^{-1}(c,0) = \emptyset$ when $c < 0$.

A "metric regularity variant" of the openness property (7), equally failing to be preserved under small Lipschitz continuous perturbations, as shown by this same example, is the requirement that $d(\bar{x}, F^{-1}(y)) \leq \kappa|y - \bar{y}|$ for y close to $\bar{y}$. The same trouble comes up also for an "inner semicontinuity variant," namely the condition that there exist neighborhoods U of $\bar{x}$ and V of $\bar{y}$ such that $F^{-1}(y) \cap U \neq \emptyset$ for all $y \in V$.

If we consider calmness as a local version of the outer Lipschitz continuity, then it might seem to be worthwhile to define a local version of inner Lipschitz continuity, introduced in Section 3D. For a mapping $S: \mathbb{R}^m \to \mathbb{R}^n$ with $(\bar{y},\bar{x}) \in \mathrm{gph}\, S$, this would refer to the existence of neighborhoods U of $\bar{x}$ and V of $\bar{y}$ such that

$$(8) \qquad S(\bar{y}) \cap U \subset S(y) + \kappa|y - \bar{y}|\mathbb{B} \quad \text{for all } y \in V.$$

We will not give a name to this property here, or a name to the associated property of the inverse of a mapping satisfying (8). We will only demonstrate, by an example, that the property of the inverse associated to (8), similar to metric subregularity, is not stable under perturbation, in the sense we have been exploring, and hence does not support the implicit function theorem paradigm.

Consider the mapping $S: \mathbb{R} \rightrightarrows \mathbb{R}$ whose values are the set of three points $\{-\sqrt{|y|}, 0, \sqrt{|y|}\}$ for all $y \geq 0$ and the empty set for $y < 0$. This mapping has the property in (8) at $\bar{y} = 0$ for $\bar{x} = 0$. Now consider the inverse S^{-1} and add to it the function $g(x) = -x^2$, which has zero derivative at $\bar{x} = 0$. The sum $S^{-1} + g$ is the mapping whose value at $x = 0$ is the interval $[0, \infty)$ but is just zero for $x \neq 0$. The inverse $(S^{-1} + g)^{-1}$ has $(-\infty, \infty)$ as its value for $y = 0$, but 0 for $y > 0$ and the empty set for $y < 0$. Clearly, this inverse does not have the property displayed in (8) at $\bar{y} = 0$ for $\bar{x} = 0$.

In should be noted that for special cases of mappings with particular perturbations one might still obtain stability of metric subregularity, or the property associated to (8), but we shall not go into this further.

3I. Strong Metric Subregularity

The handicap of serious instability of calmness and metric subregularity can be obviated by passing to strengthened forms of these properties.

Isolated calmness. *A mapping $S : \mathbb{R}^m \rightrightarrows \mathbb{R}^n$ is said to have the* isolated calmness *property if it is calm at $\bar{y}$ for $\bar{x}$ and, in addition, S has a graphical localization at $\bar{y}$ for $\bar{x}$ that is single-valued at $\bar{y}$ itself (with value $\bar{x}$). Specifically, this refers to the existence of a constant $\kappa \geq 0$ and neighborhoods U of $\bar{x}$ and V of $\bar{y}$ such that*

(1) $$|x - \bar{x}| \leq \kappa |y - \bar{y}| \quad \text{when } x \in S(y) \cap U \text{ and } y \in V.$$

Observe that in this definition $S(\bar{y}) \cap U$ is a singleton, namely the point $\bar{x}$, so $\bar{x}$ is an isolated point in $S(\bar{y})$, hence the terminology. Isolated calmness can equivalently be defined as the existence of a (possibly slightly larger) constant κ and neighborhoods U of $\bar{x}$ and V of $\bar{y}$ such that

(2) $$S(y) \cap U \subset \bar{x} + \kappa |y - \bar{y}| \mathbb{B} \quad \text{when } y \in V.$$

For a linear mapping A, isolated calmness holds at every point, whereas isolated calmness of A^{-1} holds at some point of $\operatorname{dom} A^{-1}$ if and only if A is nonsingular. More generally we have the following fact through Theorem 3D.1 for polyhedral mappings, as defined there.

Proposition 3I.1 (isolated calmness of polyhedral mappings). *A polyhedral mapping $S : \mathbb{R}^m \rightrightarrows \mathbb{R}^n$ has the isolated calmness property at $\bar{y}$ for $\bar{x}$ if and only if $\bar{x}$ is an isolated point of $S(\bar{y})$.*

Once again we can ask whether there is a property of a mapping that corresponds to isolated calmness of its inverse. Such a property exists and, in order to unify the terminology, we name it as follows.

Strong metric subregularity. *A mapping $F : \mathbb{R}^n \rightrightarrows \mathbb{R}^m$ is said to be* strongly metrically subregular *at $\bar{x}$ for $\bar{y}$ if $(\bar{x}, \bar{y}) \in \operatorname{gph} F$ and there is a constant $\kappa \geq 0$ along with neighborhoods U of $\bar{x}$ and V of $\bar{y}$ such that*

3 Regularity Properties of Set-valued Solution Mappings

(3) $\qquad |x - \bar{x}| \leq \kappa d(\bar{y}, F(x) \cap V) \quad \text{for all } x \in U.$

Equivalently, F is strongly metrically subregular at $\bar{x}$ for $\bar{y}$ if it is metrically subregular at $\bar{x}$ for $\bar{y}$ and there exist neighborhoods U of $\bar{x}$ and V of $\bar{y}$ such that the graphical localization $V \ni y \mapsto F^{-1}(y) \cap U$ is single-valued at $\bar{y}$ with value $\bar{x}$. Clearly, (3) implies that $\bar{x}$ is an isolated point of $F^{-1}(\bar{y})$ and that F is metrically subregular there; hence the infimum of κ for which (3) holds is equal to $\text{subreg}(F; \bar{x}|\bar{y})$.

Note however that, in general, the condition $\text{subreg}(F; \bar{x}|\bar{y}) < \infty$ is *not* a characterization of strong metric subregularity, but becomes such a criterion under the isolatedness assumption. As an example, observe that, for a linear mapping A is always metrically subregular at 0 for 0, but it is strongly metrically subregular at 0 for 0 if and only if $\ker A$ consists of just 0, which corresponds to A being injective. The equivalence of strong metric subregularity and isolated calmness of the inverse is shown next:

Theorem 3I.2 (characterization by inverse isolated calmness). *A mapping* $F : \mathbb{R}^n \rightrightarrows \mathbb{R}^m$ *is strongly metrically subregular at $\bar{x}$ for $\bar{y}$ with constant κ if and only if its inverse F^{-1} has the isolated calmness property at $\bar{y}$ for $\bar{x}$ with the same constant κ, i.e., there exist neighborhoods U of $\bar{x}$ and V of $\bar{y}$ such that*

(4) $\qquad F^{-1}(y) \cap U \subset \bar{x} + \kappa |y - \bar{y}| \mathbb{B} \quad \text{when } y \in V.$

Then infimum of all κ such that the inclusion holds for some U and V equals $\text{subreg}(F; \bar{x}|\bar{y})$.

Proof. Assume first that F is strongly subregular at $\bar{x}$ for $\bar{y}$. Let $\kappa > \text{subreg}(F; \bar{x}|\bar{y})$. Then there are neighborhoods U for $\bar{x}$ and V for $\bar{y}$ such that (3) holds with the indicated κ. Consider any $y \in V$. If $F^{-1}(y) \cap U = \emptyset$, then (4) holds trivially. If not, let $x \in F^{-1}(y) \cap U$. This entails $y \in F(x) \cap V$, hence $d(\bar{y}, F(x) \cap V) \leq |y - \bar{y}|$ and consequently $|x - \bar{x}| \leq \kappa |y - \bar{y}|$ by (3). Thus, $x \in \bar{x} + \kappa |y - \bar{y}| \mathbb{B}$, and we conclude that (4) holds. Also, we see that $\text{subreg}(F; \bar{x}|\bar{y})$ is not less than the infimum of all κ such that (4) holds for some choice of U and V.

For the converse, suppose (4) holds for some κ and neighborhoods U and V. Consider any $x \in U$. If $F(x) \cap V = \emptyset$ the right side of (3) is ∞ and there is nothing more to prove. If not, for an arbitrary $y \in F(x) \cap V$ we have $x \in F^{-1}(y) \cap U$, and therefore $x \in \bar{x} + \kappa |y - \bar{y}| \mathbb{B}$ by (4), which means $|x - \bar{x}| \leq \kappa |y - \bar{y}|$. This being true for all $y \in F(x) \cap V$, we must have $|x - \bar{x}| \leq \kappa d(\bar{y}, F(x) \cap V)$. Thus, (3) holds, and in particular we have $\kappa \geq \text{subreg}(F; \bar{x}|\bar{y})$. Therefore, the infimum of κ in (4) equals $\text{subreg}(F; \bar{x}|\bar{y})$. □

Observe also, through 3H.4, that the neighborhood V in (2) and (3) can be chosen to be the entire space $\mathbb{R}^m$, by adjusting the size of U; that is, strong metric subregularity as in (3) with constant κ is equivalent to the existence of a neighborhood U' of $\bar{x}$ such that

(5) $\qquad |x - \bar{x}| \leq \kappa d(\bar{y}, F(x)) \quad \text{for all } x \in U'.$

Accordingly, the associated isolated calmness of the inverse is equivalent to the existence of a neighborhood U' of $\bar{x}$ such that

(6) $\quad F^{-1}(y) \cap U' \subset \bar{x} + \kappa|y - \bar{y}|\mathbb{B} \quad \text{when } y \in \mathbb{R}^m.$

Exercise 3I.3. *Provide direct proofs of the equivalence of (3) and (5), and (4) and (6), respectively.*

Guide. Use the argument in the proof of 3H.4. □

Similarly to the distance function characterization in Theorem 3E.8 for the Aubin property, the isolated calmness property is characterized by uniform calmness of the distance function associated with the inverse mapping:

Theorem 3I.4 (distance function characterization of strong metric subregularity). *For a mapping $F : \mathbb{R}^n \rightrightarrows \mathbb{R}^m$ and a point $(\bar{x}, \bar{y}) \in \operatorname{gph} F$, suppose that $\bar{x}$ is an isolated point in $F^{-1}(\bar{y})$ and moreover*

(7) $\quad \bar{x} \in \liminf_{y \to \bar{y}} F^{-1}(y).$

Consider the function $s(y, x) = d(x, F^{-1}(y))$. Then the mapping F is strongly metrically subregular at $\bar{x}$ for $\bar{y}$ if and only if s is calm with respect to y uniformly in x at $(\bar{y}, \bar{x})$, in which case

$$\widehat{\operatorname{clm}}_y(s; (\bar{y}, \bar{x})) = \operatorname{subreg}(F; \bar{x}|\bar{y}).$$

Proof. Let F be strongly metrically subregular at $\bar{x}$ for $\bar{y}$ and let $\kappa > \operatorname{subreg}(F; \bar{x}|\bar{y})$. Let (5) and (6) hold with $U' = \mathbb{B}_a(\bar{x})$ and also $F^{-1}(\bar{y}) \cap \mathbb{B}_a(\bar{x}) = \bar{x}$. Let $b > 0$ be such that, according to (7), $F^{-1}(y) \cap \mathbb{B}_a(\bar{x}) \neq \emptyset$ for all $y \in \mathbb{B}_b(\bar{y})$. Make b smaller if necessary so that $b \leq a/(10\kappa)$. Choose $y \in \mathbb{B}_b(\bar{y})$ and $x \in \mathbb{B}_{a/4}(\bar{x})$; then from (6) we have

(8) $\quad d(\bar{x}, F^{-1}(y) \cap \mathbb{B}_a(\bar{x})) \leq \kappa|y - \bar{y}|.$

Since all points in $F^{-1}(\bar{y})$ except $\bar{x}$ are at distance from x more than $a/4$ we obtain

(9) $\quad d(x, F^{-1}(\bar{y})) = |x - \bar{x}|.$

Using the triangle inequality and (8), we get

(10) $\quad \begin{aligned} d(x, F^{-1}(y)) &\leq |x - \bar{x}| + d(\bar{x}, F^{-1}(y)) \\ &= |x - \bar{x}| + d(\bar{x}, F^{-1}(y) \cap \mathbb{B}_a(\bar{x})) \leq |x - \bar{x}| + \kappa|y - \bar{y}|. \end{aligned}$

Then, taking (9) into account, we have

(11) $\quad s(y, x) - s(\bar{y}, x) = d(x, F^{-1}(y)) - d(x, F^{-1}(\bar{y})) \leq \kappa|y - \bar{y}|.$

Let $\tilde{x}$ be a projection of x on cl $F^{-1}(y)$. Using (10) we obtain

$$|x - \tilde{x}| = d(x, F^{-1}(y)) \leq |x - \bar{x}| + \kappa|y - \bar{y}| \leq a/4 + \kappa b \leq a/4 + \kappa a/(10\kappa) \leq a/2,$$

and consequently

$$|\bar{x} - \tilde{x}| \leq |\bar{x} - x| + |x - \tilde{x}| \leq a/4 + a/2 = 3a/4 < a.$$

Therefore,

(12) $$d(x, F^{-1}(y) \cap \mathbb{B}_a(\bar{x})) = d(x, F^{-1}(y)).$$

According to (6),

$$d(x, \bar{x} + \kappa|y - \bar{y}|\mathbb{B}) \leq d(x, F^{-1}(y) \cap \mathbb{B}_a(\bar{x}))$$

and then, by (12)

(13) $$|x - \bar{x}| - \kappa|y - \bar{y}| \leq d(x, F^{-1}(y) \cap \mathbb{B}_a(\bar{x})) = d(x, F^{-1}(y)).$$

Plugging (9) into (13), we conclude that

(14) $$s(\bar{y}, x) - s(y, x) = d(x, F^{-1}(\bar{y})) - d(x, F^{-1}(y)) \leq \kappa|y - \bar{y}|.$$

Since x and y were arbitrarily chosen in dom s and close to $\bar{x}$ and $\bar{y}$, respectively, we obtain by combining (11) and (14) that $\widehat{\text{clm}}_y(s;(\bar{y},\bar{x})) \leq \kappa$, hence

(15) $$\widehat{\text{clm}}_y(s;(\bar{y},\bar{x})) \leq \text{subreg}(F;\bar{x}|\bar{y}).$$

To show the converse inequality, let $\kappa > \widehat{\text{clm}}_y(s;(\bar{y},\bar{x}))$; then there exists $a > 0$ such that $s(\cdot, x)$ is calm on $\mathbb{B}_a(\bar{y})$ with constant κ uniformly in $x \in \mathbb{B}_a(\bar{x})$. Adjust a so that $F^{-1}(\bar{y}) \cap \mathbb{B}_a(\bar{x}) = \bar{x}$. Pick any $x \in \mathbb{B}_{a/3}(\bar{x})$. If $F(x) = \emptyset$, (5) holds automatically. If not, choose any $y \in \mathbb{R}^n$ such that $(x,y) \in \text{gph } F$. Since $s(y,x) = 0$, we have

$$|x - \bar{x}| = d(x, F^{-1}(\bar{y})) = s(\bar{y}, x) \leq s(y, x) + \kappa|y - \bar{y}| = \kappa|y - \bar{y}|.$$

Since y is arbitrarily chosen in $F(x)$, this gives us (5). This means that F is strongly subregular at $\bar{x}$ for $\bar{y}$ with constant κ and hence

$$\widehat{\text{clm}}_y(s;(\bar{y},\bar{x})) \geq \text{subreg}(F;\bar{x}|\bar{y}).$$

Combining this with (15) brings the proof to a finish. □

Exercise 3I.5 (counterexample). *Show that the mapping* $F : \mathbb{R} \rightrightarrows \mathbb{R}$ *given by*

$$F(x) = \begin{cases} x & \text{if } 0 \leq x < 1, \\ \mathbb{R} & \text{if } x \geq 1, \\ \emptyset & \text{if } x < 0 \end{cases}$$

does not satisfy condition (7) and has $\operatorname{subreg}(F;0|0) = 1$ while $\widehat{\operatorname{clm}}_y(s;(0,0)) = \infty$.

We look next at perturbations of F by single-valued mappings g in the pattern that was followed for the other regularity properties considered in the preceding sections.

Theorem 3I.6 (inverse mapping theorem for strong metric subregularity). *Consider a mapping $F : \mathbb{R}^n \rightrightarrows \mathbb{R}^m$ and a point $(\bar{x}, \bar{y}) \in \operatorname{gph} F$ such that F is strongly metrically subregular at $\bar{x}$ for $\bar{y}$ and let κ and μ be nonnegative constants such that*

$$\operatorname{subreg}(F;\bar{x}|\bar{y}) \leq \kappa \quad \text{and} \quad \kappa\mu < 1.$$

Then for any function $g : \mathbb{R}^n \to \mathbb{R}^m$ with $\bar{x} \in \operatorname{dom} g$ and $\operatorname{clm}(g;\bar{x}) \leq \mu$, one has

$$\operatorname{subreg}(g + F;\bar{x}|g(\bar{x}) + \bar{y}) \leq \frac{\kappa}{1 - \kappa\mu}.$$

Proof. Choose κ and μ as in the statement of the theorem and let $\lambda > \kappa$, $\nu > \mu$ be such that $\lambda\nu < 1$. Pick $g : \mathbb{R}^n \to \mathbb{R}^m$ with $\operatorname{clm}(g;\bar{x}) < \nu$. Without loss of generality, let $g(\bar{x}) = 0$; then there exists $a > 0$ such that

(16) $\qquad |g(x)| \leq \nu|x - \bar{x}|$ when $x \in \mathbb{B}_a(\bar{x})$.

Since $\operatorname{subreg}(F;\bar{x}|\bar{y}) < \lambda$, we can arrange, by taking a smaller if necessary, that

(17) $\qquad |x - \bar{x}| \leq \lambda|y - \bar{y}|$ when $(x,y) \in \operatorname{gph} F \cap (\mathbb{B}_a(\bar{x}) \times \mathbb{B}_a(\bar{y}))$.

Let $\nu' = \max\{1, \nu\}$ and consider any

(18) $\qquad z \in \mathbb{B}_{a/2}(\bar{y})$ with $x \in (g + F)^{-1}(z) \cap \mathbb{B}_{a/2\nu'}(\bar{x})$.

These relations entail $z \in g(x) + F(x)$, hence $z = y + g(x)$ for some $y \in F(x)$. From (16) and since $x \in \mathbb{B}_{a/2\nu'}(\bar{x})$, we have $|g(x)| \leq \mu(a/2\nu') \leq a/2$ (inasmuch as $\nu' \geq \nu$). Using the equality $y - \bar{y} = z - g(x) - \bar{y}$ we get $|y - \bar{y}| \leq |z - \bar{y}| + |g(x)| \leq (a/2) + (a/2) = a$. However, because $(x, y) \in \operatorname{gph} F \cap (\mathbb{B}_a(\bar{x}) \times \mathbb{B}_a(\bar{y}))$, through (17),

$$|x - \bar{x}| \leq \lambda|(z - g(x)) - \bar{y}| \leq \lambda|z - \bar{y}| + \lambda|g(x)| \leq \lambda|z - \bar{y}| + \lambda\nu|x - \bar{x}|,$$

hence $|x - \bar{x}| \leq \lambda/(1 - \lambda\nu)|z - \bar{y}|$. Since x and z are chosen as in (18) and λ and ν could be arbitrarily close to κ and μ, respectively, the proof is complete. □

Corollaries that parallel those for metric regularity given in Section 3F can immediately be derived.

Corollary 3I.7 (detailed estimate). *Consider a mapping $F : \mathbb{R}^n \rightrightarrows \mathbb{R}^m$ which is strongly metrically subregular at $\bar{x}$ for $\bar{y}$ and a function $g : \mathbb{R}^n \to \mathbb{R}^m$ such that $\operatorname{subreg}(F;\bar{x}|\bar{y}) \cdot \operatorname{clm}(g;\bar{x}) < 1$. Then the mapping $g + F$ is strongly metrically subregular at $\bar{x}$ for $\bar{y}$, and one has*

3 Regularity Properties of Set-valued Solution Mappings

$$\text{subreg}\,(g+F;\bar{x}|g(\bar{x})+\bar{y}) \leq \left(\text{subreg}\,(F;\bar{x}|\bar{y})^{-1} - \text{clm}\,(g;\bar{x})\right)^{-1}.$$

If $\text{subreg}\,(F;\bar{x}|\bar{y}) = 0$, then $\text{subreg}\,(g+F;\bar{x}|\bar{y}) = 0$ for any $g : \mathbb{R}^n \to \mathbb{R}^m$ with $\text{clm}\,(g;\bar{x}) < \infty$. If $\text{subreg}\,(F;\bar{x}|\bar{y}) = \infty$, then $\text{subreg}\,(g+F;\bar{x}|g(\bar{x})+\bar{y}) = \infty$ for any $g : \mathbb{R}^n \to \mathbb{R}^m$ with $\text{clm}\,(g;\bar{x}) = 0$.

This result implies in particular that the property of strong metric subregularity is preserved under perturbations with zero calmness moduli. The only difference with the corresponding results for metric regularity in Section 3F is that now a larger class of perturbation is allowed with first-order approximations replacing the strict first-order approximations.

Corollary 3I.8 (utilizing first-order approximations). *Consider $F : \mathbb{R}^n \rightrightarrows \mathbb{R}^m$, a point $(\bar{x},\bar{y}) \in \text{gph}\,F$ and two functions $f : \mathbb{R}^n \to \mathbb{R}^m$ and $g : \mathbb{R}^n \to \mathbb{R}^m$ with $\bar{x} \in \text{int dom}\,f \cap \text{int dom}\,g$ which are first-order approximations to each other at $\bar{x}$. Then the mapping $f + F$ is strongly metrically subregular at $\bar{x}$ for $f(\bar{x}) + \bar{y}$ if and only if $g + F$ is strongly metrically subregular at $\bar{x}$ for $g(\bar{x}) + \bar{y}$, in which case*

$$\text{subreg}\,(f+F;\bar{x}|f(\bar{x})+\bar{y}) = \text{subreg}\,(g+F;\bar{x}|g(\bar{x})+\bar{y}).$$

This corollary takes a more concrete form when the first-order approximation is represented by a linearization:

Corollary 3I.9 (linearization). *Let $M = f + F$ for mappings $f : \mathbb{R}^n \to \mathbb{R}^m$ and $F : \mathbb{R}^n \rightrightarrows \mathbb{R}^m$, and let $\bar{y} \in M(\bar{x})$. Suppose f is differentiable at $\bar{x}$, and let*

$$M_0 = h + F \text{ for } h(x) = f(\bar{x}) + \nabla f(\bar{x})(x - \bar{x}).$$

Then M is strongly metrically subregular at $\bar{x}$ for $\bar{y}$ if and only if M_0 has this property. Moreover $\text{subreg}\,(M;\bar{x}|\bar{y}) = \text{subreg}\,(M_0;\bar{x}|\bar{y})$.

Through 3I.1, the result in Corollary 3I.9 could equally well be stated in terms of the isolated calmness property of M^{-1} in relation to that of M_0^{-1}. We can specialize that result in the following way.

Corollary 3I.10 (linearization with polyhedrality). *Let $M : \mathbb{R}^n \rightrightarrows \mathbb{R}^m$ with $\bar{y} \in M(\bar{x})$ be of the form $M = f + F$ for $f : \mathbb{R}^n \to \mathbb{R}^m$ and $F : \mathbb{R}^n \rightrightarrows \mathbb{R}^m$ such that f is differentiable at $\bar{x}$ and F is polyhedral. Let $M_0(x) = f(\bar{x}) + \nabla f(\bar{x})(x - \bar{x}) + F(x)$. Then M^{-1} has the isolated calmness property at $\bar{y}$ for $\bar{x}$ if and only if $\bar{x}$ is an isolated point of $M_0^{-1}(\bar{y})$.*

Proof. This applies 3I.1 in the framework of the isolated calmness restatement of 3I.9 in terms of the inverses. □

Applying Corollary 3I.9 to the case where F is the zero mapping, we obtain yet another inverse function theorem in the classical setting:

Corollary 3I.11 (an inverse function result). *Let $f : \mathbb{R}^n \to \mathbb{R}^m$ be differentiable at $\bar{x}$ and such that $\ker \nabla f(\bar{x}) = \{0\}$. Then there exist $\kappa > 0$ and a neighborhood U of*

$\bar{x}$ such that
$$|x-\bar{x}| \leq \kappa |f(x) - f(\bar{x})| \quad \text{for every } x \in U.$$

Proof. This comes from (5). □

Next, we state and prove an implicit function theorem for strong metric subregularity:

Theorem 3I.12 (implicit mapping theorem with strong metric subregularity). *For the generalized equation $f(p,x) + F(x) \ni 0$ and its solution mapping*
$$S : p \mapsto \{x \mid f(p,x) + F(x) \ni 0\},$$
consider a pair $(\bar{p},\bar{x})$ with $\bar{x} \in S(\bar{p})$. Let $h : \mathbb{R}^n \to \mathbb{R}^m$ be an estimator of f with respect to x at $(\bar{p},\bar{x})$ with constant μ and let $h + F$ be strongly metrically subregular at $\bar{x}$ for 0 with $\operatorname{subreg}(h+F;\bar{x}|0) \leq \kappa$. Suppose that

(19) $$\kappa\mu < 1 \quad \text{and} \quad \widehat{\operatorname{clm}}_p(f;(\bar{p},\bar{x})) \leq \lambda < \infty.$$

Then S has the isolated calmness property at $\bar{p}$ for $\bar{x}$, moreover with
$$\operatorname{clm}(S;\bar{p}|\bar{x}) \leq \frac{\kappa\lambda}{1 - \kappa\mu}.$$

Proof. The proof goes along the lines of the proof of Theorem 3I.6 with different choice of constants. Let κ, μ and λ be as required and let $\delta > \kappa$ and $\nu > \mu$ be such that $\delta\nu < 1$. Let $\gamma > \lambda$. By the assumptions for the mapping $h+F$ and the functions f and h, there exist positive scalars a and r such that

(20) $$|x - \bar{x}| \leq \delta|y| \quad \text{for all } x \in (h+F)^{-1}(y) \cap \mathbb{B}_a(\bar{x}) \text{ and } y \in \mathbb{B}_{\nu a + \gamma r}(0),$$

(21) $$|f(p,x) - f(\bar{p},x)| \leq \gamma|p - \bar{p}| \quad \text{for all } p \in \mathbb{B}_r(\bar{p}) \text{ and } x \in \mathbb{B}_a(\bar{x}),$$

and also, for $e = f - h$,

(22) $$|e(p,x) - e(p,\bar{x})| \leq \nu|x - \bar{x}| \quad \text{for all } x \in \mathbb{B}_a(\bar{x}) \text{ and } p \in \mathbb{B}_r(\bar{p}).$$

Let $x \in S(p) \cap \mathbb{B}_a(\bar{x})$ for some $p \in \mathbb{B}_r(\bar{p})$. Then, since $h(\bar{x}) = f(\bar{p},\bar{x})$, we obtain from (21) and (22) that

(23) $$\begin{aligned} |e(p,x)| &\leq |e(p,x) - e(p,\bar{x})| + |f(p,\bar{x}) - f(\bar{p},\bar{x})| \\ &\leq \nu|x - \bar{x}| + \gamma|p - \bar{p}| \leq \nu a + \gamma r. \end{aligned}$$

Observe that $x \in (h+F)^{-1}(-f(p,x) + h(x)) \cap \mathbb{B}_a(\bar{x})$, and then from (20) and (23) we have
$$|x - \bar{x}| \leq \delta|-f(p,x) + h(x)| \leq \delta\nu|x - \bar{x}| + \delta\gamma|p - \bar{p}|.$$

In consequence,

3 Regularity Properties of Set-valued Solution Mappings

$$|x - \bar{x}| \leq \frac{\delta \gamma}{1 - \delta \nu} |p - \bar{p}|.$$

Since δ is arbitrarily close to κ, ν is arbitrarily close to μ and γ is arbitrarily close to λ, we arrive at the desired result. □

In the theorem we state next, we can get away with a property of f at $(\bar{p},\bar{x})$ which is weaker than local continuous differentiability, namely a kind of uniform differentiability we introduced in Section 1C. We say that $f(p,x)$ is differentiable in x *uniformly with respect to* p at $(\bar{p},\bar{x})$ if f is differentiable with respect to (p,x) at $(\bar{p},\bar{x})$ and for every $\varepsilon > 0$ there is a (p,x)-neighborhood of $(\bar{p},\bar{x})$ in which

$$|f(p,x) - f(p,\bar{x}) - \nabla_x f(\bar{p},\bar{x})(x - \bar{x})| \leq \varepsilon |x - \bar{x}|.$$

Symmetrically, we define what it means for $f(p,x)$ to be differentiable in p *uniformly with respect to* x at $(\bar{p},\bar{x})$. Note that the combination of these two properties is implied by, yet weaker than, the continuous differentiability of f at $(\bar{p},\bar{x})$. For instance, the two uniformity properties hold when $f(p,x) = f_1(p) + f_2(x)$ and we simply have f_1 differentiable at $\bar{p}$ and f_2 differentiable at $\bar{x}$.

Theorem 3I.13 (utilizing differentiability and ample parameterization). *For the generalized equation in Theorem 3I.12 and its solution mapping S, and a pair $(\bar{p},\bar{x})$ with $\bar{x} \in S(\bar{p})$, suppose that f is differentiable in x uniformly with respect to p at $(\bar{p},\bar{x})$, and at the same time differentiable in p uniformly with respect to x at $(\bar{p},\bar{x})$. If the mapping*

$$h + F \text{ for } h(x) = f(\bar{p},\bar{x}) + \nabla_x f(\bar{p},\bar{x})(x - \bar{x})$$

is strongly metrically subregular at $\bar{x}$ for 0, then S has the isolated calmness property at $\bar{p}$ for $\bar{x}$ with

(24) $$\text{clm}(S;\bar{p}|\bar{x}) \leq \text{subreg}(h + F; \bar{x}|0) \cdot |\nabla_p f(\bar{p},\bar{x})|.$$

Furthermore, when f is continuously differentiable on a neighborhood of $(\bar{p},\bar{x})$ and satisfies the ample parameterization condition

$$\text{rank } \nabla_p f(\bar{p},\bar{x}) = m,$$

then the converse implication holds as well: the mapping $h + F$ is strongly metrically subregular at $\bar{x}$ for 0 provided that S has the isolated calmness property at $\bar{p}$ for $\bar{x}$.

Proof. With this choice of h, the assumption (19) of 3I.12 holds and then (24) follows from the conclusion of this theorem. To handle the ample parameterization we employ Lemma 2C.1 by repeating the argument in the proof of 3F.10, simply replacing the composition rule there with the one in the following proposition. □

Proposition 3I.14 (isolated calmness in composition). *For a mapping $M : \mathbb{R}^d \rightrightarrows \mathbb{R}^n$ and a function $\psi : \mathbb{R}^n \times \mathbb{R}^m \to \mathbb{R}^d$ consider the composite mapping $N : \mathbb{R}^m \rightrightarrows \mathbb{R}^n$ of the form*

$$y \mapsto N(y) = \{x \mid x \in M(\psi(x,y))\} \quad \text{for } y \in \mathbb{R}^m.$$

Let ψ satisfy

(25) $$\widehat{\text{clm}}_x(\psi;(\bar{x},0)) = 0 \quad \text{and} \quad \widehat{\text{clm}}_y(\psi;(\bar{x},0)) < \infty,$$

and let $(\psi(\bar{x},0),\bar{x}) \in \text{gph } M$. If M has the isolated calmness property at $\psi(\bar{x},0)$ for $\bar{x}$, then N has the isolated calmness property at 0 for $\bar{x}$.

Proof. Let M have the isolated calmness property with neighborhoods $\mathbb{B}_b(\bar{x})$, $\mathbb{B}_c(\bar{p})$ and constant $\kappa > \text{clm}(M;\bar{p}|\bar{x})$, where $\bar{p} = \psi(\bar{x},0)$. Choose $\lambda > 0$ with $\lambda < 1/\kappa$ and $a > 0$ such that for any $y \in \mathbb{B}_a(0)$ the function $\psi(\cdot,y)$ is calm on $\mathbb{B}_b(\bar{x})$ with calmness constant λ. Pick $\gamma > \widehat{\text{clm}}_y(\psi;(\bar{x},0))$ and make a and b smaller if necessary so that the function $\psi(x,\cdot)$ is calm on $\mathbb{B}_a(0)$ with constant γ and also

(26) $$\lambda b + \gamma a \le c.$$

Let $y \in \mathbb{B}_a(0)$ and $x \in N(y) \cap \mathbb{B}_b(\bar{x})$. Then $x \in M(\psi(x,y)) \cap \mathbb{B}_b(\bar{x})$. Using the assumed calmness properties (25) of ψ and utilizing (26) we see that

$$|\psi(x,y) - \bar{p}| = |\psi(x,y) - \psi(\bar{x},0)| \le \lambda b + \gamma a \le c.$$

From the isolated calmness of M we then have

$$|x - \bar{x}| \le \kappa |\psi(x,y) - \psi(\bar{x},0)| \le \kappa \lambda |x - \bar{x}| + \kappa \gamma |y|,$$

hence

$$|x - \bar{x}| \le \frac{\kappa \gamma}{1 - \kappa \lambda} |y|.$$

This establishes that the mapping N has the isolated calmness property at 0 for $\bar{x}$ with constant $\kappa\gamma/(1-\kappa\lambda)$. □

Example 3I.15 (application to complementarity problems). For $f : \mathbb{R}^d \times \mathbb{R}^n \to \mathbb{R}^n$, consider the complementarity problem of finding for given p an x such that

(27) $$x \ge 0, \quad f(p,x) \ge 0, \quad x \perp f(p,x).$$

This corresponds to solving $f(p,x) + N_{\mathbb{R}^n_+} \ni 0$, as seen in 2A. Let $\bar{x}$ be a solution for $\bar{p}$, and suppose that f is continuously differentiable in a neighborhood of $(\bar{p},\bar{x})$. Consider now the linearized problem

(28) $$x \ge 0, \quad Ax + y \ge 0, \quad x \perp Ax + y, \quad \text{with } A = \nabla_x f(\bar{p},\bar{x}).$$

Then, from 3I.13 we obtain that if the solution mapping for (28) has the isolated calmness property at $\bar{y} = f(\bar{p},\bar{x}) - A\bar{x}$ for $\bar{x}$, then the solution mapping for (27) has the isolated calmness property at $\bar{p}$ for $\bar{x}$. Under the ample parameterization condition, rank $\nabla_p f(\bar{p},\bar{x}) = n$, the converse implication holds as well.

3 Regularity Properties of Set-valued Solution Mappings

Commentary

The inner and outer limits of sequences of sets were introduced by Painlevé in his lecture notes as early as 1902 and later popularized by Hausdorff [1927] and Kuratowski [1933]. The definition of excess was first given by Pompeiu [1905], who also defined the distance between sets C and D as $e(C,D) + e(D,C)$. Hausdorff [1927] gave the definition we use here. These two definitions are equivalent in the sense that they induce the same convergence of sets. The reader can find much more about set-convergence and continuity properties of set-valued mappings together with extended historical commentary in Rockafellar and Wets [1998]. This includes the reason why we prefer "inner and outer" in contrast to the more common terms "lower and upper," so as to avoid certain conflicts in definition that unfortunately pervade the literature.

Theorem 3B.4 is a particular case of a result sometimes referred to as the Berge theorem; see Section 8.1 in Dontchev and Zolezzi [1993] for a general statement. Theorem 3C.3 comes from Walkup and Wets [1969], while the Hoffman lemma, 3C.4, is due to Hoffman [1952].

The concept of outer Lipschitz continuity was introduced by Robinson [1981] under the name "upper Lipschitz continuity" and adjusted to "outer Lipschitz continuity" later in Robinson [2007]. Theorem 3D.1 is due to Robinson [1981] while 3D.3 is a version, given in Robinson [2007], of a result due to Wu Li [1994].

The Aubin property of set-valued mappings was introduced by J.-P. Aubin [1984], who called it "pseudo-Lipschitz continuity"; it was renamed after Aubin in Dontchev and Rockafellar [1996]. In the literature one can also find it termed "Aubin continuity," but we do not use that here since the Aubin property does not imply continuity. Theorem 3E.3 is from Bessis, Ledyaev and Vinter [2001]. The name "metric regularity" was coined by J. M. Borwein [1986a], but the origins of this concept go back to the Banach open mapping theorem and even earlier. In the literature, metric regularity is defined in various ways, for example in Schirotzek [2007] the property expressed in 3E.6 is called weak metric regularity; see e.g. Mordukhovich [2006] for other names. Theorem 3E.4 is from Rockafellar [1985]. Theorem 3E.9 comes from Ledyaev and Zhu [1999]. For historical remarks regarding inverse and implicit mapping theorems with metric regularity, see the commentary to Chapter 5.

As we mentioned earlier in Chapter 2, the term "strong regularity" comes from Robinson [1980], who used it in the framework of variational inequalities. Theorem 3F.5 is a particular case of a more general result due to Kenderov [1975]; see also Levy and Poliquin [1997].

Calmness and metric subregularity, as well as isolated calmness and metric subregularity, have been considered in various contexts and under various names in the literature; here we follow the terminology of Dontchev and Rockafellar [2004]. Isolated calmness was formally introduced in Dontchev [1995a], where its stability (Theorem 3I.6) was first proved. The equivalent property of strong metric subregularity was considered earlier, without giving it a name, by Rockafellar [1989]; see also the commentary to Chapter 4.

Chapter 4
Regularity Properties Through Generalized Derivatives

In the wide-ranging generalizations we have been developing of the inverse function theorem and implicit function theorem, we have followed the idea that conclusions about a solution mapping, concerning the Aubin property, say, or the existence of a single-valued localization, can be drawn by confirming that some auxiliary solution mapping, obtained from a kind of approximation, has the property in question. In the classical framework, we can appeal to a condition like the invertibility of a Jacobian matrix and thus tie in with standard calculus. Now, though, we are far away in another world where even a concept of differentiability seems to be lacking. However, substitutes for classical differentiability can very well be introduced and put to work. In this chapter we show the way to that and explain numerous consequences.

First, graphical differentiation of a set-valued mapping is defined through the variational geometry of the mapping's graph. A characterization of the Aubin property is derived and applied to the case of a solution mapping. Strong metric subregularity is characterized next. Applications are made to parameterized constraint systems and special features of solution mappings for variational inequalities. There is a review then of some other derivative concepts and the associated inverse function theorems of Clarke and Kummer. Finally, alternative results using coderivatives are described.

4A. Graphical Differentiation

The concept of the tangent cone $T_C(x)$ to a set C in $\mathbb{R}^n$ at a point $x \in C$ was introduced in 2A, but it was only utilized there in the case of C being closed and convex. In 2E, the geometry of tangent cones to polyhedral convex sets received special attention and led to significant insights in the study of variational inequalities. Now, tangent cones to possibly nonconvex sets will come strongly onto the stage as well, serving as a tool for a kind of generalized differentiation. The definition of the tangent cone is the same as before.

Tangent cones. *A vector $v \in \mathbb{R}^n$ is said to be* tangent *to a set $C \subset \mathbb{R}^n$ at a point $x \in C$ if*
$$\frac{1}{\tau^k}(x^k - x) \to v \text{ for some } x^k \to x, \; x^k \in C, \; \tau^k \searrow 0.$$
The set of all such vectors v is called the tangent cone *to C at x and is denoted $T_C(x)$. The* tangent cone mapping *is defined as*
$$T_C : x \mapsto \begin{cases} T_C(x) & \text{for } x \in C, \\ \emptyset & \text{otherwise.} \end{cases}$$

A description equivalent to this definition is that $v \in T_C(x)$ if and only if there are sequences $v^k \to v$ and $\tau^k \searrow 0$ with $x + \tau^k v^k \in C$, or equivalently, if there are sequences $x_k \in C$, $x_k \to x$ and $\tau_k \searrow 0$ such that $v_k := (x_k - x)/\tau_k \to v$ as $k \to \infty$.

Note that $T_C(x)$ is indeed a *cone*: it contains $v = 0$ (as seen from taking $x^k \equiv x$), and contains along with any vector v all positive multiples of v. The definition can also be recast in the notation of set convergence:

(1) $$T_C(x) = \limsup_{\tau \searrow 0} \tau^{-1}(C - x).$$

Described as an outer limit in this way, it is clear in particular that $T_C(x)$ is always a closed set. When C is a "smooth manifold" in $\mathbb{R}^n$, $T_C(x)$ is the usual tangent subspace, but in general, of course, $T_C(x)$ need not even be convex. The tangent cone mapping T_C has dom $T_C = C$ but gph T_C is not necessarily a closed subset of $\mathbb{R}^n \times \mathbb{R}^n$ even when C is closed.

As noted in 2A.4, when the set C is convex, the tangent cone $T_C(x)$ is also convex for any $x \in C$. In this case the limsup in (1) can be replaced by lim, as shown in the following proposition.

Proposition 4A.1 (tangent cones to convex sets). *For a convex set $C \subset \mathbb{R}^n$ and a point $x \in C$,*

(2) $$T_C(x) = \lim_{\tau \searrow 0} \tau^{-1}(C - x).$$

Proof. Consider the set

$$K_C(x) = \{v \mid \exists \tau > 0 \text{ with } x + \tau v \in C\}.$$

Let $v \in T_C(x)$. Then there exist sequences $\tau_k \searrow 0$ and $x_k \in C$, $x_k \to x$, such that $v_k := (x_k - x)/\tau_k \to v$. Hence $v_k \in K_C$ for all k and therefore $v \in \operatorname{cl} K$. Thus, we obtain

(3) $$T_C(x) \subset \operatorname{cl} K_C(x).$$

Now let $v \in K_C(x)$. Then $v = (\tilde{x} - x)/\tau$ for some $\tau > 0$ and $\tilde{x} \in C$. Take an arbitrary sequence $\tau_k \searrow 0$ as $k \to \infty$. Since C is convex, we have

$$x + \tau_k v = (1 - \frac{\tau_k}{\tau})x + \frac{\tau_k}{\tau}\tilde{x} \in C \text{ for all } k.$$

But then $v \in (C - x)/\tau_k$ for all k and hence $v \in \liminf_k \tau_k^{-1}(C - x)$. Since τ_k was arbitrarily chosen, we conclude that

$$K_C(x) \subset \liminf_{\tau \searrow 0} \tau^{-1}(C - x) \subset \limsup_{\tau \searrow 0} \tau^{-1}(C - x) = T_C(x).$$

This combined with (3) gives us (2). □

We should note that, in order to have the equality (2), the set C does not need to be convex. Generally, sets C for which (2) is satisfied are called *geometrically derivable*. Proposition 4A.1 simply says that all convex sets are geometrically derivable.

Starting in elementary calculus, students are taught to view differentiation in terms of tangents to the graph of a function. This can be formulated in the notation of tangent cones as follows. Let $f: \mathbb{R}^n \to \mathbb{R}^m$ be a function which is differentiable at x with derivative mapping $Df(x): \mathbb{R}^n \to \mathbb{R}^m$. Then

$$(u,v) \in \operatorname{gph} Df(x) \iff (u,v) \in T_{\operatorname{gph} f}(x, f(x)).$$

In other words, the derivative is completely represented geometrically by the tangent cone to the set $\operatorname{gph} f$ at the point $(x, f(x))$. In fact, differentiability is more or less equivalent to having $T_{\operatorname{gph} f}(x, f(x))$ turn out to be the graph of a linear mapping.

By adopting such a geometric characterization as a definition, while not insisting on linearity, we can introduce derivatives for an arbitrary set-valued mapping $F: \mathbb{R}^n \rightrightarrows \mathbb{R}^m$. However, because $F(x)$ may have more than one element y, it is essential for the derivative mapping to depend not just on x but also on a choice of $y \in F(x)$.

Graphical derivatives. For a mapping $F: \mathbb{R}^n \rightrightarrows \mathbb{R}^m$ and a pair (x,y) with $y \in F(x)$, the *graphical derivative* of F at x for y is the mapping $DF(x|y): \mathbb{R}^n \rightrightarrows \mathbb{R}^m$ whose graph is the tangent cone $T_{\operatorname{gph} F}(x,y)$ to $\operatorname{gph} F$ at (x,y):

$$v \in DF(x|y)(u) \iff (u,v) \in T_{\operatorname{gph} F}(x,y).$$

Thus, $v \in DF(x|y)(u)$ if and only if there exist sequences $u^k \to u$, $v^k \to v$ and $\tau^k \searrow 0$ such that $y + \tau^k v^k \in F(x + \tau^k u^k)$ for all k.

On this level, derivative mappings may no longer even be single-valued. But because their graphs are cones, they do always belong to the following class of mappings, at least.

Positively homogeneous mappings. A mapping $H : \mathbb{R}^n \rightrightarrows \mathbb{R}^m$ is called *positively homogeneous* when gph H is a cone, which is equivalent to H satisfying

$$0 \in H(0) \quad \text{and} \quad H(\lambda x) = \lambda H(x) \text{ for } \lambda > 0.$$

Clearly, the inverse of a positively homogeneous mapping is another positively homogeneous mapping. Linear mappings are positively homogeneous as a special case, their graphs being not just cones but linear subspaces.

Since the graphical differentiation comes from an operation on graphs, and the graph of a mapping F can be converted to the graph of its inverse F^{-1} just by interchanging variables, we immediately have the rule that

$$D(F^{-1})(y|x) = DF(x|y)^{-1}.$$

Another useful relation is available for sums.

Proposition 4A.2 (sum rule). *For a function $f : \mathbb{R}^n \to \mathbb{R}^m$ which is differentiable at x, a set-valued mapping $F : \mathbb{R}^n \rightrightarrows \mathbb{R}^m$ and any $y \in F(x)$, one has*

$$D(f+F)(x|f(x)+y) = Df(x) + DF(x|y).$$

Proof. If $v \in D(f+F)(x|f(x)+y)(u)$ there exist sequences $\tau^k \searrow 0$ and $u^k \to u$ and $v^k \to v$ such that

$$f(x) + y - f(x + \tau^k u^k) + \tau^k v^k \in F(x + \tau^k u^k) \quad \text{for every } k.$$

By using the definition of the derivative for f we get

$$y + \tau^k(-Df(x)u + v^k) + o(\tau^k) \in F(x + \tau^k u^k).$$

Hence, by the definition of the graphical derivative, $v \in Df(x)u + DF(x|y)(u)$.

Conversely, if $v - Df(x)u \in DF(x|y)(u)$ then there exist sequences $\tau^k \searrow 0$, and $u^k \to u$ and $w^k \to v - Df(x)u$ such that $y + \tau^k w^k \in F(x + \tau^k u^k)$. Again, by the differentiability of f,

$$y + f(x) + \tau^k v^k + o(\tau^k) \in (f+F)(x + \tau^k u^k) \quad \text{for } v^k = w^k + Df(x)u^k,$$

which yields $v \in D(f+F)(x|f(x)+y)(u)$. □

Example 4A.3 (graphical derivative for a constraint system). Consider a general constraint system of the form

(4) $$f(x) - K \ni y,$$

for a function $f : \mathbb{R}^n \to \mathbb{R}^m$, a set $K \subset \mathbb{R}^m$ and a parameter vector y, and let x be a solution of (4) for y at which f is differentiable. Then for the mapping

$$G : x \mapsto f(x) - K, \quad \text{with } y \in G(x),$$

one has

(5) $$DG(x|y)(u) = Df(x)u - T_K(f(x) - y).$$

Detail. This applies the sum rule to the case of a constant mapping $F \equiv -K$, for which the definition of the graphical derivative gives $DF(x|z) = T_{-K}(z) = -T_K(-z)$. □

In the special but important case of Example 4A.3 in which $K = \mathbb{R}_-^s \times \{0\}^{m-s}$ with $f = (f_1, \ldots, f_m)$, the constraint system (4) with respect to $y = (y_1, \ldots, y_m)$ takes the form

$$f_i(x) \begin{cases} \leq y_i & \text{for } i = 1, \ldots, s, \\ = y_i & \text{for } i = s+1, \ldots, m. \end{cases}$$

The graphical derivative formula (5) says then that a vector $v = (v_1, \ldots, v_m)$ is in $DG(x|y)(u)$ if and only if

$$Df_i(x)u \begin{cases} \leq v_i & \text{for } i \in [1,s] \text{ with } f_i(x) = y_i, \\ = v_i & \text{for } i = s+1, \ldots, m. \end{cases}$$

Example 4A.4 (graphical derivative for a variational inequality). For a function $f : \mathbb{R}^n \to \mathbb{R}^n$ and a convex set $C \subset \mathbb{R}^n$ that is polyhedral, consider the variational inequality

(6) $$f(x) + N_C(x) \ni y$$

in which y is a parameter. Let x be a solution of (6) at which f is differentiable. Let $v = y - f(x) \in N_C(x)$ and let $K_C(x,v)$ be the corresponding critical cone, this being the polyhedral convex cone $T_C(x) \cap [v]^\perp$. Then for the mapping

$$G : x \mapsto f(x) + N_C(x), \quad \text{with } y \in G(x),$$

one has

(7) $$DG(x|y)(u) = Df(x)u + N_{K_C(x,v)}(u).$$

Detail. From the sum rule in 4A.2 we have $DG(x|y)u = Df(x)u + DN_C(x|v)(u)$. According to Lemma 2E.4 (the reduction lemma for normal cone mappings to polyhedral convex sets), for any $(x,v) \in \text{gph } N_C$ there exists a neighborhood O of the origin in $\mathbb{R}^n \times \mathbb{R}^n$ such that for $(x',v') \in O$ one has

$$v + v' \in N_C(x + x') \iff v' \in N_{K_C(x,v)}(x').$$

This reveals in particular that the tangent cone to gph N_C at (x,v) is just gph $N_{K_C(x,v)}$, or in other words, that $DN_C(x|v)$ is the normal cone mapping $N_{K_C(x,v)}$. Thus we have (7). □

Because graphical derivative mappings are positively homogeneous, general properties of positively homogeneous mappings can be applied to them. Norm concepts are available in particular for capturing quantitative characteristics.

Outer and inner norms. *For any positively homogeneous mapping $H : \mathbb{R}^n \rightrightarrows \mathbb{R}^m$, the* outer norm *and the* inner norm *are defined, respectively, by*

$$(8) \qquad |H|^+ = \sup_{|x|\leq 1} \sup_{y\in H(x)} |y| \quad \text{and} \quad |H|^- = \sup_{|x|\leq 1} \inf_{y\in H(x)} |y|$$

with the convention $\inf_{y\in \emptyset} |y| = \infty$ *and* $\sup_{y\in \emptyset} |y| = -\infty$.

When H is a linear mapping, both $|H|^+$ and $|H|^-$ reduce to the operator (matrix) norm $|H|$ associated with the Euclidean norm. However, it must be noted that neither $|H|^+$ nor $|H|^-$ satisfies the conditions in the definition of a true "norm," inasmuch as set-valued mappings do not even form a vector space.

The inner and outer norms have simple interpretations when $H = A^{-1}$ for a linear mapping $A : \mathbb{R}^n \to \mathbb{R}^m$. Let the $m \times n$ matrix for this linear mapping be denoted likewise by A, for simplicity. If $m < n$, we have A surjective (the associated matrix being of rank m) if and only if $|A^{-1}|^-$ is finite, this expression being the norm of the right inverse of A: $|A^{-1}|^- = |A^T(AA^T)^{-1}|$. Then $|A^{-1}|^+ = \infty$. On the other hand, if $m > n$, we have $|A^{-1}|^+ < \infty$ if and only if A is injective (the associated matrix has rank n), and then $|A^{-1}|^+ = |(A^T A)^{-1} A^T|$ but $|A^{-1}|^- = \infty$. For $m = n$, of course, both norms agree with the usual matrix norm $|A^{-1}|$, and the finiteness of this quantity is equivalent to nonsingularity of A.

Proposition 4A.5 (domains of positively homogeneous mappings). *For a positively homogeneous mapping $H : \mathbb{R}^n \rightrightarrows \mathbb{R}^m$,*

$$(9) \qquad \text{dom}\, H = \mathbb{R}^n \quad \Longrightarrow \quad |H|^+ \geq |H|^-.$$

Moreover,

$$(10) \qquad |H|^- < \infty \quad \Longrightarrow \quad \text{dom}\, H = \mathbb{R}^n;$$

thus, if $|H^{-1}|^- < \infty$ then H must be surjective.

Proof. The implications (9) and (10) are immediate from the definition (8) and its conventions concerning the empty set. □

In cases where dom H is not all of $\mathbb{R}^n$, it is possible for the inequality in (9) to fail. As an illustration, this occurs for the positively homogeneous mapping $H : \mathbb{R} \rightrightarrows \mathbb{R}$ defined by

$$H(x) = \begin{cases} 0 & \text{for } x \geq 0, \\ \emptyset & \text{for } x < 0, \end{cases}$$

for which $|H|^+ = 0$ while $|H|^- = \infty$.

Proposition 4A.6 (norm characterizations). *The inner norm of a positively homogeneous mapping $H : \mathbb{R}^n \rightrightarrows \mathbb{R}^m$ satisfies*

$$(11) \qquad |H|^- = \inf\Big\{ \kappa > 0 \,\Big|\, H(x) \cap \kappa \mathbb{B} \neq \emptyset \text{ for all } x \in \mathbb{B} \Big\}.$$

In parallel, the outer norm satisfies

$$(12) \quad |H|^+ = \inf\Big\{ \kappa \in (0,\infty) \,\Big|\, y \in H(x) \Rightarrow |y| \leq \kappa|x| \Big\} = \sup_{|y|=1} \frac{1}{d(0, H^{-1}(y))}.$$

If H has closed graph, then furthermore

$$(13) \qquad |H|^+ < \infty \iff H(0) = \{0\}.$$

If H has closed and convex graph, then the implication (10) becomes equivalence:

$$(14) \qquad |H|^- < \infty \iff \operatorname{dom} H = \mathbb{R}^n$$

and in that case $|H^{-1}|^- < \infty$ if and only if H is surjective.

Proof. We get (11) and the first part of (12) simply by rewriting the formulas in terms of $\mathbb{B} = \{x \mid |x| \leq 1\}$ and utilizing the positive homogeneity. The infimum so obtained in (12) is unchanged when y is restricted to have $|y| = 1$, and in this way it can be identified with the infimum of all $\kappa \in (0, \infty)$ such that $\kappa \geq 1/|x|$ whenever $x \in H^{-1}(y)$ and $|y| = 1$. (It is correct in this to interpret $1/|x| = \infty$ when $x = 0$.) This shows that the middle expression in (12) agrees with the final one.

Moving on to (13), we observe that when $|H|^+ < \infty$ the middle expression in (12) implies that if $(0, y) \in \operatorname{gph} H$ then y must be 0. To prove the converse implication we will need the assumption that $\operatorname{gph} H$ is closed. Suppose that $H(0) = \{0\}$. If $|H|^+ = \infty$, there has to be a sequence of points $(x_k, y_k) \in \operatorname{gph} H$ such that $0 < |y_k| \to \infty$ but x_k is bounded. Consider then the sequence of pairs (w_k, u_k) in which $w_k = x_k/|y_k|$ and $u_k = y_k/|y_k|$. We have $w_k \to 0$, while u_k has a cluster point $\bar{u}$ with $|\bar{u}| = 1$. Moreover, $(w_k, u_k) \in \operatorname{gph} H$ by the positive homogeneity and hence, through the closedness of $\operatorname{gph} H$, we must have $H(0) \ni \bar{u}$. This contradicts our assumption that $H(0) = \{0\}$ and terminates the proof of (13). The equivalence (14) follows from (9) and a general result (Robinson-Ursescu theorem) which we will prove in Section 5B. □

Corollary 4A.7 (norms of linear-constraint-type mappings). *Suppose*

$$H(x) = Ax - K$$

for a linear mapping $A : \mathbb{R}^n \to \mathbb{R}^m$ and a closed convex cone $K \subset \mathbb{R}^m$. Then H is positively homogeneous with closed and convex graph. Moreover

(15) $$|H^{-1}|^- = \sup_{|y|\leq 1} d(0, A^{-1}(y+K))$$

and

(16) $$|H^{-1}|^- < \infty \iff \operatorname{rge} A - K = \mathbb{R}^m.$$

On the other hand,

(17) $$|H^{-1}|^+ = \sup_{|x|=1} \frac{1}{d(Ax, K)}$$

and

(18) $$|H^{-1}|^+ < \infty \iff \left[Ax - K \ni 0 \implies x = 0 \right].$$

Proof. Formula (15) follows from the definition (8) while (16) comes from (14) applied to this case. Formula (17) follows from (12) while (18) is the specification of (13). □

We will come back to the general theory of positively homogeneous mappings and their norms in Section 5A. In the meantime there will be applications to the case of derivative mappings.

Some properties of the graphical derivatives of convex-valued mappings under Lipschitz continuity are displayed in the following exercise.

Exercise 4A.8. *Consider a mapping $F : \mathbb{R}^n \rightrightarrows \mathbb{R}^m$ that is convex-valued and Lipschitz continuous in its domain and let $(x,y) \in \operatorname{gph} F$. Prove that in this case $DF(x|y)$ is convex-valued and*

(19) $$DF(x|y)(u) = \lim_{\tau \searrow 0} \tau^{-1}(F(x+\tau u) - y),$$

and in particular,

(20) $$DF(x|y)(0) = T_{F(x)}(y).$$

Guide. Observe that, by definition
$$DF(x|y)(u) = \liminf_{\tau \searrow 0, u' \to u} \tau^{-1}(F(x+\tau u') - y).$$

Since F is Lipschitz continuous, this equality reduces to

(21) $$DF(x|y)(u) = \liminf_{\tau \searrow 0} \tau^{-1}(F(x+\tau u) - y).$$

Then use the convexity of the values of F as in the proof of Proposition 4A.1 to show that liminf in (21) can be replaced by lim and use this to obtain convexity of

$DF(x|y)(u)$ from the convexity of $F(x+\tau u)$. Lastly, to show (20) apply 4A.1 to (19) in the case $u = 0$. □

Exercise 4A.9. For a positively homogeneous mapping $H : \mathbb{R}^n \rightrightarrows \mathbb{R}^m$, show that

$$|H|^- = 0 \iff \operatorname{cl} H(x) \ni 0 \text{ for all } x \in \mathbb{R}^n,$$

$$|H|^+ = 0 \iff \operatorname{rge} H = \{0\}.$$

Guide. Apply the norm characterizations in (11) and (12). □

4B. Derivative Criteria for the Aubin Property

Conditions will next be developed which characterize metric regularity and the Aubin property in terms of graphical derivatives. From these conditions, new forms of implicit mapping theorems will be obtained. First, we state a fundamental fact.

Theorem 4B.1 (derivative criterion for metric regularity). *For a mapping $F : \mathbb{R}^n \rightrightarrows \mathbb{R}^m$ and a point $(\bar{x}, \bar{y}) \in \operatorname{gph} F$ at which the gph F is locally closed, one has*

(1) $$\operatorname{reg}(F; \bar{x}|\bar{y}) = \limsup_{\substack{(x,y) \to (\bar{x}, \bar{y}) \\ (x,y) \in \operatorname{gph} F}} |DF(x|y)^{-1}|^-.$$

Thus, F is metrically regular at $\bar{x}$ for $\bar{y}$ if and only if the right side of (1) is finite.

The proof of Theorem 4B.1 will be furnished later in this section. Note that in the case when $m \leq n$ and F is a function f which is differentiable on a neighborhood of $\bar{x}$, the representation of the regularity modulus in (1) says that f is metrically regular precisely when the Jacobians $\nabla f(x)$ for x near $\bar{x}$ are of full rank and the inner norms of their inverses $\nabla f(x)^{-1}$ are uniformly bounded. This holds automatically when f is continuously differentiable around $\bar{x}$ with $\nabla f(\bar{x})$ of full rank, in which case we get not only metric regularity but also existence of a continuously differentiable local selection of f^{-1}, as in 1F.3. When $m = n$ this becomes nonsingularity and we come to the classical inverse function theorem.

Also to be kept in mind here is the connection between metric regularity and the Aubin property in 3E.6. This allows Theorem 4B.1 to be formulated equivalently as a statement about that property of the inverse mapping.

Theorem 4B.2 (derivative criterion for the Aubin property). *For a mapping $F : \mathbb{R}^n \rightrightarrows \mathbb{R}^m$, consider the inverse mapping $S = F^{-1}$ or, equivalently, the solution mapping $S : \mathbb{R}^m \rightrightarrows \mathbb{R}^n$ for the generalized equation $F(x) \ni y$:*

$$S(y) = \{x \mid F(x) \ni y\}.$$

Fix $\bar{y}$ and any $\bar{x} \in S(\bar{y})$, and suppose that gph F or, equivalently, gph S, is locally closed at $(\bar{x}, \bar{y})$. Then

$$\text{lip}\,(S;\bar{y}|\bar{x}) = \limsup_{\substack{(y,x) \to (\bar{y},\bar{x}) \\ (y,x) \in \text{gph}\,S}} |DS(y|x)|^-. \tag{2}$$

Thus, S has the Aubin property at $\bar{y}$ for $\bar{x}$ if and only if the right side of (2) is finite.

Note that 4B.2 can be stated for the mapping S alone, without referring to it as an inverse or a solution mapping.

Solution mappings of much greater generality can also be handled with these ideas. For this, we return to the framework introduced briefly at the end of Section 3E and delve into it much further. We consider the parameterized relation

$$G(p,x) \ni 0 \quad \text{for a mapping} \quad G: \mathbb{R}^d \times \mathbb{R}^n \rightrightarrows \mathbb{R}^m \tag{3}$$

and its solution mapping $S: \mathbb{R}^d \rightrightarrows \mathbb{R}^n$ defined by

$$S(p) = \{x \,|\, G(p,x) \ni 0\}. \tag{4}$$

In Theorem 3E.9, a result was presented in which a partial Aubin property of G with respect to p, combined with other assumptions, led to a conclusion that S has the Aubin property. We are looking now toward finding derivative criteria for these Aubin properties, so as to obtain a different type of statement about the "implicit mapping" S.

The following theorem will be our stepping stone to progress and will have many other interesting consequences as well. It makes use of the *partial* graphical derivative of $G(p,x)$ with respect to x, which is defined as the graphical derivative of the mapping $x \mapsto G(p,x)$ with p fixed and denoted by $D_x G(p,x|y)$. Of course, $D_p G(p,x)$ has a similar meaning.

Theorem 4B.3 (solution mapping estimate). *For the generalized equation (3) and its solution mapping S in (4), let $\bar{x} \in S(\bar{p})$, so that $(\bar{p},\bar{x},0) \in \text{gph}\,G$. Suppose that gph G is locally closed at $(\bar{p},\bar{x},0)$ and that the distance mapping $p \mapsto d(0,G(p,\bar{x}))$ is upper semicontinuous at $\bar{p}$. Then for every $c \in (0,\infty)$ satisfying*

$$\limsup_{\substack{(p,x,y) \to (\bar{p},\bar{x},0) \\ (p,x,y) \in \text{gph}\,G}} |D_x G(p,x|y)^{-1}|^- < c \tag{5}$$

there are neighborhoods V of $\bar{p}$ and U of $\bar{x}$ such that

$$d(x, S(p)) \leq c\, d(0, G(p,x)) \quad \text{for } x \in U \text{ and } p \in V. \tag{6}$$

Proof. Let c satisfy (5). Then there exists $\eta > 0$ such that

$$\begin{cases} \text{for every } (p,x,y) \in \text{gph}\,G \text{ with } |p - \bar{p}| + \max\{|x - \bar{x}|, c|y|\} \leq 2\eta, \\ \text{and for every } v \in \mathbb{R}^m, \text{ there exists } u \in D_x G(p,x|y)^{-1}(v) \text{ with } |u| \leq c|v|. \end{cases} \tag{7}$$

4 Regularity Properties Through Generalized Derivatives

We can always choose η smaller so that the intersection

(8) $\quad \text{gph } G \cap \{(p,x,y) \mid |p-\bar{p}| + \max\{|x-\bar{x}|, c|y|\} \leq 2\eta\}$ is closed.

The next part of the proof is developed as a lemma.

Lemma 4B.4 (intermediate estimate). *For c and η as above, let $\varepsilon > 0$ and $s > 0$ be such that*

(9) $\quad\quad\quad\quad\quad\quad\quad c\varepsilon < 1 \quad \text{and} \quad s < \varepsilon\eta$

and let $(p, \omega, v) \in \text{gph } G$ satisfy

(10) $\quad\quad\quad\quad\quad |p-\bar{p}| + \max\{|\omega-\bar{x}|, c|v|\} \leq \eta.$

Then for every $y' \in \mathbb{B}_s(v)$ there exists $\hat{x}$ with $y' \in G(p, \hat{x})$ such that

(11) $\quad\quad\quad\quad\quad\quad |\hat{x} - \omega| \leq \frac{1}{\varepsilon}|y' - v|.$

In the proof of the lemma we apply a fundamental result in variational analysis, which is stated next:

Theorem 4B.5 (Ekeland variational principle). *Let (X, ρ) be a complete metric space and let $f : X \to (-\infty, \infty]$ be a lower semicontinuous function on X which is bounded from below. Let $\bar{u} \in \text{dom } f$. Then for every $\delta > 0$ there exists u_δ such that*

$$f(u_\delta) + \delta\rho(u_\delta, \bar{u}) \leq f(\bar{u}),$$

and

$$f(u_\delta) < f(u) + \delta\rho(u, u_\delta) \quad \text{for every } u \in X, \ u \neq u_\delta.$$

Proof of Lemma 4B.4. On the product space $Z := \mathbb{R}^n \times \mathbb{R}^m$ we introduce the norm

$$\|(x,y)\| := \max\{|x|, c|y|\},$$

which is equivalent to the Euclidean norm. Pick ε, s and $(p, \omega, v) \in \text{gph } G$ as required in (9) and (10) and let $y' \in \mathbb{B}_s(v)$. By (8) the set

$$E_p := \{(x,y) \mid (p,x,y) \in \text{gph } G, |p-\bar{p}| + \|(x,y) - (\bar{x},0)\| \leq 2\eta\} \subset \mathbb{R}^n \times \mathbb{R}^m$$

is closed, hence, equipped with the metric induced by the norm in question, it is a complete metric space. The function $V_p : E_p \to \mathbb{R}$ defined by

(12) $\quad\quad\quad\quad V_p : (x,y) \mapsto |y' - y| \quad \text{for } (x,y) \in E_p$

is continuous on its domain E_p. Also, $(\omega, v) \in \text{dom } V_p$. We apply Ekeland's variational principle 4B.5 to V_p with $\bar{u} = (\omega, v)$ and the indicated ε to obtain the existence of $(\hat{x}, \hat{y}) \in E_p$ such that

(13) $$V_p(\hat{x},\hat{y}) + \varepsilon\|(\omega,v) - (\hat{x},\hat{y})\| \leq V_p(\omega,v)$$

and

(14) $$V_p(\hat{x},\hat{y}) \leq V_p(x,y) + \varepsilon\|(x,y) - (\hat{x},\hat{y})\| \quad \text{for every } (x,y) \in E_p.$$

With V_p as in (12), the inequalities (13) and (14) come down to

(15) $$|y' - \hat{y}| + \varepsilon\|(\omega,v) - (\hat{x},\hat{y})\| \leq |y' - v|$$

and

(16) $$|y' - \hat{y}| \leq |y' - y| + \varepsilon\|(x,y) - (\hat{x},\hat{y})\| \quad \text{for every } (x,y) \in E_p.$$

Through (15) we obtain in particular that

(17) $$\|(\omega,v) - (\hat{x},\hat{y})\| \leq \frac{1}{\varepsilon}|y' - v|.$$

Since $y' \in \mathbb{B}_s(v)$, we then have

$$\|(\omega,v) - (\hat{x},\hat{y})\| \leq \frac{s}{\varepsilon}$$

and consequently, from the choice of (p,ω,v) in (10) and s in (9),

(18) $$\begin{aligned}|p - \bar{p}| + \|(\hat{x},\hat{y}) - (\bar{x},0)\| \\ \leq |p - \bar{p}| + \|(\omega,v) - (\bar{x},0)\| + \|(\omega,v) - (\hat{x},\hat{y})\| \leq \eta + \frac{s}{\varepsilon} < 2\eta.\end{aligned}$$

Thus, $(p,\hat{x},\hat{y})$ satisfies the condition in (7), so there exists $u \in \mathbb{R}^n$ for which

(19) $$y' - \hat{y} \in D_x G(p,\hat{x}|\hat{y})(u) \quad \text{and} \quad |u| \leq c|y' - \hat{y}|.$$

By the definition of the partial graphical derivative, there exist sequences $\tau^k \searrow 0$, $u^k \to u$, and $v^k \to y' - \hat{y}$ such that

$$\hat{y} + \tau^k v^k \in G(p,\hat{x} + \tau^k u^k) \quad \text{for all } k.$$

Also, from (18) we know that, for sufficiently large k,

$$|p - \bar{p}| + \|(\hat{x} + \tau^k u^k, \hat{y} + \tau^k v^k) - (\bar{x},0)\| \leq 2\eta,$$

implying $(\hat{x} + \tau^k u^k, \hat{y} + \tau^k v^k) \in E_p$. If we now plug the point $(\hat{x} + \tau^k u^k, \hat{y} + \tau^k v^k)$ into (16) in place of (x,y), we get

$$|y' - \hat{y}| \leq |y' - (\hat{y} + \tau^k v^k)| + \varepsilon\|(\hat{x} + \tau^k u^k, \hat{y} + \tau^k v^k) - (\hat{x},\hat{y})\|.$$

This gives us

4 Regularity Properties Through Generalized Derivatives

$$|y' - \hat{y}| \leq (1 - \tau^k)|y' - \hat{y}| + \tau^k|v^k - (y' - \hat{y})| + \varepsilon \tau^k \|(u^k, v^k)\|,$$

that is,

$$|y' - \hat{y}| \leq |v^k - (y' - \hat{y})| + \varepsilon \|(u^k, v^k)\|.$$

Passing to the limit with $k \to \infty$ leads to $|y' - \hat{y}| \leq \varepsilon \|(u, y' - \hat{y})\|$ and then, taking into account the second relation in (19), we conclude that $|y' - \hat{y}| \leq \varepsilon c |y' - \hat{y}|$. Since $\varepsilon c < 1$ by (9), the only possibility here is that $y' = \hat{y}$. But then $y' \in G(p, \hat{x})$ and (17) yields (11). This proves the lemma. □

We continue now with the proof of Theorem 4B.3. Let $\tau = \eta/(4c)$. Since the function $p \to d(0, G(p, \bar{x}))$ is upper semicontinuous at $\bar{p}$, there exists a positive $\delta \leq c\tau$ such that $d(0, G(p, \bar{x})) \leq \tau/2$ for all p with $|p - \bar{p}| < \delta$. Set $V := B_\delta(\bar{p})$, $U := B_{c\tau}(\bar{x})$ and pick any $p \in V$ and $x \in U$. We can find y such that $y \in G(p, \bar{x})$ with $|y| \leq d(0, G(p, \bar{x})) + \tau/3 < \tau$. Note that

(20) $$|p - \bar{p}| + \|(\bar{x}, y) - (\bar{x}, 0)\| = |p - \bar{p}| + c|y| \leq \delta + c\tau \leq \eta.$$

Choose $\varepsilon > 0$ such that $1/2 < \varepsilon c < 1$ and let $s = \varepsilon \eta$. Then $s > \tau$. We apply Lemma 4B.4 with the indicated ε and s, and with $(p, \omega, v) = (p, \bar{x}, y)$ which, as seen in (20), satisfies (10), and with $y' = 0$, since $0 \in B_s(y)$. Thus, there exists $\hat{x}$ such that $0 \in G(p, \hat{x})$, that is, $\hat{x} \in S(p)$, and also, from (11), $|\hat{x} - \bar{x}| \leq |y|/\varepsilon$. Therefore, in view of the choice of y, we have $\hat{x} \in B_{\tau/\varepsilon}(\bar{x})$. We now consider two cases.

CASE 1. $d(0, G(p, x)) \geq 2\tau$. We just proved that there exists $\hat{x} \in S(p)$ with $\hat{x} \in B_{\tau/\varepsilon}(\bar{x})$; then

(21)
$$d(x, S(p)) \leq d(\bar{x}, S(p)) + |x - \bar{x}|$$
$$\leq |\bar{x} - \hat{x}| + |x - \bar{x}| \leq \frac{\tau}{\varepsilon} + c\tau \leq \frac{2\tau}{\varepsilon} \leq \frac{1}{\varepsilon} d(0, G(p, x)).$$

CASE 2. $d(0, G(p, x)) < 2\tau$. In this case, for any y with $|y| \leq 2\tau$ we have

$$|p - \bar{p}| + \max\{|x - \bar{x}|, c|y|\} \leq \delta + \max\{c\tau, 2c\tau\} \leq 3c\tau \leq \eta$$

and then, by (8), the nonempty set $G(p, x) \cap 2\tau B$ is closed. Hence, there exists $\tilde{y} \in G(p, x)$ such that $|\tilde{y}| = d(0, G(p, x)) < 2\tau$ and therefore

$$c|\tilde{y}| < 2c\tau = \frac{\eta}{2}.$$

We conclude that the point $(p, x, \tilde{y}) \in \text{gph } G$ satisfies

$$|p - \bar{p}| + \max\{|x - \bar{x}|, c|\tilde{y}|\} \leq \delta + \max\{c\tau, \frac{\eta}{2}\} \leq \eta.$$

Thus, the assumptions of Lemma 4B.4 hold for $(p, \omega, v) = (p, x, \tilde{y})$, $s = 2\tau$, and $y' = 0$. Hence there exists $\tilde{x} \in S(p)$ such that

$$|\tilde{x} - x| \leq \frac{1}{\varepsilon}|\tilde{y}|.$$

Then, by the choice of $\tilde{y}$,

$$d(x, S(p)) \leq |x - \tilde{x}| \leq \frac{1}{\varepsilon}|\tilde{y}| = \frac{1}{\varepsilon}d(0, G(p,x)).$$

Hence, by (21), for both cases 1 and 2, and therefore for any p in V and $x \in U$, we have

$$d(x, S(p)) \leq \frac{1}{\varepsilon}d(0, G(p,x)).$$

Since U and V do not depend on ε, and $1/\varepsilon$ can be arbitrarily close to c, this gives us (6). □

With this result in hand, we can confirm the criterion for metric regularity presented at the beginning of this section.

Proof of Theorem 4B.1. For short, let d_{DF} denote the right side of (1). We will start by showing that $\text{reg}(F; \bar{x}|\bar{y}) \leq d_{DF}$. If $d_{DF} = \infty$ there is nothing to prove. Let $d_{DF} < c < \infty$. Applying Theorem 4B.3 to $G(p,x) = F(x) - p$ and this c, letting y take the place of p, we have $S(y) = F^{-1}(y)$ and $d(0, G(y,x)) = d(y, F(x))$. Condition (6) becomes the definition of metric regularity of F at $\bar{x}$ for $\bar{y} = \bar{p}$, and therefore $\text{reg}(F; \bar{x}|\bar{y}) \leq c$. Since c can be arbitrarily close to d_{DF} we conclude that $\text{reg}(F; \bar{x}|\bar{y}) \leq d_{DF}$.

We turn now to demonstrating the opposite inequality,

(22) $$\text{reg}(F; \bar{x}|\bar{y}) \geq d_{DF}.$$

If $\text{reg}(F; \bar{x}|\bar{y}) = \infty$ we are done. Suppose therefore that F is metrically regular at $\bar{x}$ for $\bar{y}$ with respect to a constant κ and neighborhoods U for $\bar{x}$ and V for $\bar{y}$. Then

(23) $$d(x', F^{-1}(y)) \leq \kappa|y - y'| \quad \text{whenever } (x', y') \in \text{gph } F, x' \in U, y \in V.$$

We know from 3E.1 that V can be chosen so small that $F^{-1}(y) \cap U \neq \emptyset$ for every $y \in V$. Pick any $y' \in V$ and $x' \in F^{-1}(y') \cap U$, and let $v \in \mathbb{B}$. Take a sequence $\tau^k \searrow 0$ such that $y^k := y' + \tau^k v \in V$ for all k. By (23) and the local closedness of gph F at $(\bar{x}, \bar{y})$ there exists $x^k \in F^{-1}(y' + \tau^k v)$ such that

$$|x' - x^k| = d(x', F^{-1}(y^k)) \leq \kappa|y^k - y'| = \kappa\tau^k|v|.$$

For $u^k := (x^k - x')/\tau^k$ we obtain

(24) $$|u^k| \leq \kappa|v|.$$

Thus, u^k is bounded, so $u^{k_i} \to u$ for a subsequence $k_i \to \infty$. Since $(x^{k_i}, y' + \tau^{k_i}v) \in$ gph F, we obtain $(u, v) \in T_{\text{gph } F}(x', y')$. Hence, by the definition of the graphical derivative, we have $u \in DF^{-1}(y'|x')(v) = DF(x'|y')^{-1}(v)$. The bound (24) guarantees that

… Regularity Properties Through Generalized Derivatives

$$|DF(x|y)^{-1}|^- \leq \kappa.$$

Since $(y,x) \in \text{gph } S$ is arbitrarily chosen near $(\bar{x},\bar{y})$, and κ is independent of this choice, we conclude that (22) holds and hence we have (1). □

We apply Theorem 4B.3 now to obtain for the implicit mapping result in Theorem 3E.9 an elaboration in which graphical derivatives provide estimates. Recall here the definition of $\widehat{\text{lip}}_p(G;\bar{p},\bar{x}|\bar{y})$, the modulus of the partial Aubin property introduced just before 3E.9.

Theorem 4B.6 (implicit mapping theorem with graphical derivatives). *For the general inclusion (3) and its solution mapping S in (4), let $\bar{x} \in S(\bar{p})$, so that $(\bar{p},\bar{x},0) \in \text{gph } G$. Suppose that $\text{gph } G$ is locally closed at $(\bar{p},\bar{x},0)$ and that the distance $d(0,G(p,\bar{x}))$ depends upper semicontinuously on p at $\bar{p}$. Assume further that G has the partial Aubin property with respect to p uniformly in x at $(\bar{p},\bar{x})$, and that*

(25) $$\limsup_{\substack{(p,x,y)\to(\bar{p},\bar{x},0) \\ (p,x,y)\in\text{gph } G}} |D_x G(p,x|y)^{-1}|^- \leq \lambda < \infty.$$

Then S has the Aubin property at $\bar{p}$ for $\bar{x}$ with

(26) $$\text{lip}(S;\bar{p}|\bar{x}) \leq \lambda \widehat{\text{lip}}_p(G;\bar{p},\bar{x}|0).$$

Proof. This just combines Theorem 3E.9 with the estimate now available from Theorem 4B.3. □

Note from Proposition 4A.5 that finiteness in condition (25) necessitates, in particular, having the range of $D_x G(p,x|y)$ be all of $\mathbb{R}^m$ when (p,x,y) is sufficiently close to $(\bar{p},\bar{x},0)$ in gph G.

Next we specialize Theorem 4B.6 to the generalized equations we studied in detail in Chapters 2 and 3, or in other words, to a solution mapping of the type

(27) $$S(p) = \{x \mid f(p,x) + F(x) \ni 0\},$$

where $f : \mathbb{R}^d \times \mathbb{R}^n \to \mathbb{R}^m$ and $F : \mathbb{R}^n \rightrightarrows \mathbb{R}^m$. In the next two corollaries we take a closer look at the Aubin property of the solution mapping (27).

Corollary 4B.7 (derivative criterion for generalized equations). *For the solution mapping S in (27), and a pair $(\bar{p},\bar{x})$ with $\bar{x} \in S(\bar{p})$, suppose that $\widehat{\text{lip}}_p(f;(\bar{p},\bar{x})) < \infty$. Then the mapping $G(p,x) := f(p,x) + F(x)$ has the partial Aubin property with respect to p uniformly in x at $(\bar{p},\bar{x})$ with*

(28) $$\widehat{\text{lip}}_p(G;\bar{p},\bar{x}|0) \leq \widehat{\text{lip}}_p(f;(\bar{p},\bar{x})).$$

In addition, if f is differentiable in a neighborhood of $(\bar{p},\bar{x})$, gph F is locally closed at $(\bar{x},-f(\bar{p},\bar{x}))$ and

(29) $$\limsup_{\substack{(p,x,y)\to(\bar{p},\bar{x},0)\\ y\in f(p,x)+F(x)}} |(D_x f(p,x)+DF(x|y-f(p,x)))^{-1}|^- \leq \lambda < \infty,$$

then S has the Aubin property at $\bar{p}$ for $\bar{x}$ with

(30) $$\text{lip}(S;\bar{p}|\bar{x}) \leq \lambda \widehat{\text{lip}}_p(f;\bar{p}|\bar{x}).$$

Proof. By definition, the mapping G has $(\bar{p},\bar{x},0) \in \text{gph } G$. Let $\mu > \widehat{\text{lip}}_p(f;(\bar{p},\bar{x}))$ and let Q and U be neighborhoods of $\bar{p}$ and $\bar{x}$ such that f is Lipschitz continuous with respect to $p \in Q$ uniformly in $x \in U$ with Lipschitz constant μ. Let $p, p' \in Q$, $x \in U$ and $y \in G(p,x)$; then $y - f(p,x) \in F(x)$ and we have

$$d(y, G(p',x)) = d(y - f(p',x), F(x)) \leq |f(p,x) - f(p',x)| \leq \mu |p - p'|.$$

Thus,
$$e(G(p,x), G(p',x)) \leq \mu |p' - p|$$

and hence G has the partial Aubin (actually, Lipschitz) property with respect to p uniformly in x at $(\bar{p},\bar{x})$ with modulus satisfying (28). The assumptions that f is differentiable near $(\bar{p},\bar{x})$ and gph F is locally closed at $(\bar{x},-f(\bar{p},\bar{x}))$ yield that gph G is locally closed at $(\bar{p},\bar{x},0)$ as well. Further, observe that the function $p \mapsto d(0, G(p,\bar{x})) = d(-f(p,\bar{x}), F(\bar{x}))$ is Lipschitz continuous near $\bar{p}$ and therefore upper semicontinuous at $\bar{p}$. Then we can apply Theorem 4B.6 where, by using the sum rule 4A.2, the condition (25) comes down to (29) while (26) yields (30). □

From Section 3F we know that when the function f is continuously differentiable, the Aubin property of the solution mapping in (27) can be obtained by passing to the linearized generalized equation, in which case we can also utilize the ample parameterization condition. Specifically, we have the following result:

Corollary 4B.8 (derivative criterion with continuous differentiability and ample parameterization). *For the solution mapping S in (27), and a pair $(\bar{p},\bar{x})$ with $\bar{x} \in S(\bar{p})$, suppose that f is continuously differentiable on a neighborhood of $(\bar{p},\bar{x})$ and that gph F is locally closed at $(\bar{x},-f(\bar{p},\bar{x}))$. If*

(31) $$\limsup_{\substack{(x,y)\to(\bar{x},-f(\bar{p},\bar{x}))\\ y\in D_x f(\bar{p},\bar{x})(x-\bar{x})+F(x)}} |(D_x f(\bar{p},\bar{x})+DF(x|y-D_x f(\bar{p},\bar{x})(x-\bar{x})))^{-1}|^- \leq \lambda < \infty,$$

then S has the Aubin property at $\bar{p}$ for $\bar{x}$, moreover with

(32) $$\text{lip}(S;\bar{p}|\bar{x}) \leq \lambda |\nabla_p f(\bar{p},\bar{x})|.$$

Furthermore, when f satisfies the ample parameterization condition

(33) $$\text{rank } \nabla_p f(\bar{p},\bar{x}) = m,$$

then the converse implication holds as well; that is, S has the Aubin property at $\bar{p}$ for $\bar{x}$ if and only if condition (31) is satisfied.

4 Regularity Properties Through Generalized Derivatives

Proof. According to Theorem 3F.9, the mapping S has the Aubin property at $\bar{p}$ for $\bar{x}$ provided that the linearized mapping

(34) $$h + F \quad \text{for} \quad h(x) = f(\bar{p},\bar{x}) + D_x f(\bar{p},\bar{x})(x - \bar{x})$$

is metrically regular at $\bar{x}$ for 0, and the converse implication holds under the ample parameterization condition (33). Further, according to the derivative criterion for metric regularity 4B.1, metric regularity of the mapping $h + F$ in (34) is equivalent to condition (31) and its regularity modulus is bounded by λ. Then the estimate (32) follows from formula 3F(7) in the statement of 3F.9. □

The purpose of the next exercise is to understand what condition (29) means in the setting of the classical implicit function theorem.

Exercise 4B.9 (application to classical implicit functions). *For a function $f : \mathbb{R}^d \times \mathbb{R}^n \to \mathbb{R}^m$, consider the solution mapping*

$$S : p \mapsto \{x \mid f(p,x) = 0\}$$

and a pair $(\bar{p},\bar{x})$ with $\bar{x} \in S(\bar{p})$. Suppose that f is differentiable in a neighborhood of $(\bar{p},\bar{x})$ with Jacobians satisfying

$$\limsup_{(p,x) \to (\bar{p},\bar{x})} |D_x f(p,x)^{-1}|^- < \lambda \quad \text{and} \quad \limsup_{(p,x) \to (\bar{p},\bar{x})} |D_p f(p,x)| < \kappa.$$

Show that then S has the Aubin property at $\bar{p}$ for $\bar{x}$ with constant $\lambda \kappa$.

When f is continuously differentiable we can apply Corollary 4B.8, and the assumptions in 4B.9 can in that case be captured by conditions on the Jacobian $\nabla f(\bar{p},\bar{x})$. Then 4B.8 goes a long way toward the classical implicit function theorem, 1A.1. But Steps 2 and 3 of Proof I of that theorem would afterward need to be carried out to reach the conclusion that S has a single-valued localization that is smooth around $\bar{p}$.

Applications of Theorem 4B.6 and its corollaries to constraint systems and variational inequalities will be worked out in Sections 4D and 4E. We conclude the present section with a variant of the graphical derivative formula for the modulus of metric regularity in Theorem 4B.1, which will be put to use in the numerical variational analysis of Chapter 6.

Recall that the *closed convex hull* of a set $C \subset \mathbb{R}^n$, which will be denoted by cl co C, is the smallest closed convex set that contains C.

Convexified graphical derivative. *For a mapping $F : \mathbb{R}^n \rightrightarrows \mathbb{R}^m$ and a pair (x,y) with $y \in F(x)$, the convexified graphical derivative of F at x for y is the mapping $\tilde{D}F(x|y) : \mathbb{R}^n \rightrightarrows \mathbb{R}^m$ whose graph is the closed convex hull of the tangent cone $T_{\text{gph } F}(x,y)$ to gph F at (x,y):*

$$v \in \tilde{D}F(x|y)(u) \iff (u,v) \in \text{cl co } T_{\text{gph } F}(x,y).$$

Theorem 4B.10 (alternative characterization of regularity modulus). *For a mapping $F : \mathbb{R}^n \rightrightarrows \mathbb{R}^m$ and a point $(\bar{x}, \bar{y}) \in \operatorname{gph} F$ at which the graph of F is locally closed, one has*

$$\operatorname{reg}(F; \bar{x}|\bar{y}) = \limsup_{\substack{(x,y) \to (\bar{x},\bar{y}) \\ (x,y) \in \operatorname{gph} F}} |\tilde{D}F(x|y)^{-1}|^{-}.$$

Proof. Since $\tilde{D}F(x|y)^{-1}(v) \supset DF(x|y)^{-1}(v)$ for any $v \in \mathbb{R}^n$, we have

$$\inf_{u \in \tilde{D}F(x|y)^{-1}(v)} |u| \leq \inf_{u \in DF(x|y)^{-1}(v)} |u|,$$

so that $|\tilde{D}F(x|y)^{-1}|^{-} \leq |DF(x|y)^{-1}|^{-}$. Thus,

$$\limsup_{\substack{(x,y) \to (\bar{x},\bar{y}) \\ (x,y) \in \operatorname{gph} F}} |\tilde{D}F(x|y)^{-1}|^{-} \leq \operatorname{reg}(F; \bar{x}|\bar{y}),$$

and to complete the proof we only need validate the opposite inequality. Choose λ such that

$$\limsup_{\substack{(x,y) \to (\bar{x},\bar{y}) \\ (x,y) \in \operatorname{gph} F}} |\tilde{D}F(x|y)^{-1}|^{-} \leq \lambda < \infty.$$

Let $r > 0$ be small enough that

(35) $\quad \sup\limits_{v \in \mathbb{B}} \inf\limits_{u \in \tilde{D}F(x|y)^{-1}(v)} |u| \leq \lambda \quad$ for all $(x,y) \in \operatorname{gph} F \cap \mathbb{B}_r(\bar{x},\bar{y})$,

and that the set $\operatorname{gph} F \cap \mathbb{B}_r(\bar{x},\bar{y})$ is closed. We will now demonstrate that

(36) $\quad \sup\limits_{v \in \mathbb{B}} \inf\limits_{u \in DF(x|y)^{-1}(v)} |u| \leq \lambda \quad$ for all $(x,y) \in \operatorname{gph} F \cap \operatorname{int} \mathbb{B}_r(\bar{x},\bar{y})$,

which will be enough to complete the proof.

Fix $v \in \mathbb{B}$. For any sets A and B let $d_-(A,B) := \inf\{|a-b| \,|\, a \in A, b \in B\}$. Fix $(x,y) \in \operatorname{gph} F \cap \operatorname{int} \mathbb{B}_r(\bar{x},\bar{y})$, and let $(u^*, v^*) \in \operatorname{gph} DF(x|y)$ and $w \in \lambda \mathbb{B}$ be such that

$$|(w,v) - (u^*, v^*)| = d_-(\lambda \mathbb{B} \times \{v\}, \operatorname{gph} DF(x|y)).$$

Observe that the point (u^*, v^*) is the unique projection of any point in the open segment $((u^*, v^*), (w, v))$ on $\operatorname{gph} DF(x|y)$. We will show that $(u^*, v^*) = (w, v)$, thereby confirming (36).

By the definition of the graphical derivative $DF(x|y)$, there exist sequences $\tau^k \searrow 0$, $u^k \to u^*$ and $v^k \to v^*$, such that $y + \tau^k v^k \in F(x + \tau^k u^k)$ for all k. Let (x^k, y^k) be a point in $\operatorname{cl} \operatorname{gph} F$ which is closest to $(x,y) + \frac{\tau^k}{2}(u^* + w, v^* + v)$. Since $(x,y) \in \operatorname{gph} F$ we have

$$\left|(x,y) + \frac{\tau^k}{2}(u^* + w, v^* + v) - (x^k, y^k)\right| \leq \frac{\tau^k}{2}|(u^* + w, v^* + v)|,$$

and consequently

4 Regularity Properties Through Generalized Derivatives

$$\left|(x,y) - (x^k, y^k)\right| \leq \left|(x,y) + \frac{\tau^k}{2}(u^* + w, v^* + v) - (x^k, y^k)\right|$$
$$+ \frac{\tau^k}{2}\left|(u^* + w, v^* + v)\right| \leq \tau^k \left|(u^* + w, v^* + v)\right|.$$

Thus, for k sufficiently large, we have $(x^k, y^k) \in \text{int}B_r(\bar{x}, \bar{y})$ and hence $(x^k, y^k) \in \text{gph } F \cap \text{int}B_r(\bar{x}, \bar{y})$. Setting $(\bar{u}^k, \bar{v}^k) = \frac{1}{\tau^k}(x^k - x, y^k - y)$, we get, from the basic properties of projections, that

$$\frac{1}{2}(u^* + w, v^* + v) - (\bar{u}^k, \bar{v}^k) \in [T_{\text{gph } F}(x^k, y^k)]^* = \text{gph}[\tilde{D}F(x^k|y^k)]^*.$$

Then, by (35), there exists $w^k \in \lambda B$ such that $v \in \tilde{D}F(x^k|y^k)(w^k)$ and also

(37) $\qquad \left\langle \dfrac{u^* + w}{2} - \bar{u}^k, w^k \right\rangle + \left\langle \dfrac{v^* + v}{2} - \bar{v}^k, v \right\rangle \leq 0.$

We will show now that $(\bar{v}^k, \bar{u}^k)$ converges to (v^*, u^*) as $k \to \infty$. First observe that

$$\left|\left(\frac{u^* + w}{2}, \frac{v^* + v}{2}\right) - (\bar{u}^k, \bar{v}^k)\right| = \frac{1}{\tau^k}\left|(x,y) + \tau^k\left(\frac{u^* + w}{2}, \frac{v^* + v}{2}\right) - (x^k, y^k)\right|$$
$$\leq \frac{1}{\tau^k}\left|(x,y) + \tau^k\left(\frac{u^* + w}{2}, \frac{v^* + v}{2}\right) - (x,y) - \tau^k(u^k, v^k)\right|$$
$$= \left|\left(\frac{u^* + w}{2}, \frac{v^* + v}{2}\right) - (u^k, v^k)\right|.$$

Therefore, since (u^k, v^k) is a bounded sequence, the sequence $\{(\bar{u}^k, \bar{v}^k)\}$ is bounded too and has a cluster point $(\bar{u}, \bar{v})$ which, since $y^k = y + \tau^k \bar{v}^k \in F(x^k) = F(x + \tau^k \bar{u}^k)$, belongs to gph $DF(x|y)$. Moreover, by the last estimation, the limit $(\bar{u}, \bar{v})$ satisfies

$$\left|\left(\frac{u^* + w}{2}, \frac{v^* + v}{2}\right) - (\bar{u}, \bar{v})\right| \leq \left|\left(\frac{u^* + w}{2}, \frac{v^* + v}{2}\right) - (u^*, v^*)\right|.$$

This inequality, together with the fact that (u^*, v^*) is the unique closest point to $\frac{1}{2}(u^* + w, v^* + v)$ in gph $DF(x, y)$ implies that $(\bar{u}, \bar{v}) = (u^*, v^*)$.

Up to a subsequence, the sequence of points w^k satisfying (37) converges to some $\bar{w} \in \lambda B$. Passing to the limit in (37) we obtain

(38) $\qquad \langle w - u^*, \bar{w} \rangle + \langle v - v^*, v \rangle \leq 0.$

Since (w, v) is the unique projection of (u^*, v^*) on the closed convex set $\lambda B \times \{v\}$, we have

(39) $\qquad \langle w - u^*, w - \bar{w} \rangle \leq 0.$

Finally, since (u^*, v^*) is the unique projection of $\frac{1}{2}(u^* + w, v^* + v)$ on $\operatorname{gph} DF(x|y)$ which is a closed cone, we get

$$\text{(40)} \qquad \langle w - u^*, u^* \rangle + \langle v - v^*, v^* \rangle = 0.$$

In view of (38), (39) and (40), we have

$$\begin{aligned}|(w,v) - (u^*, v^*)|^2 &= \langle w - u^*, w - \bar{w} \rangle \\ &\quad + (\langle w - u^*, \bar{w} \rangle + \langle v - v^*, v \rangle) \\ &\quad - (\langle w - u^*, u^* \rangle + \langle v - v^*, v^* \rangle) \le 0.\end{aligned}$$

Hence $w = u^*$ and $v = v^*$, and we are done. $\qquad \square$

Exercise 4B.11 (sum rule for convexified derivatives). *For a function $f : \mathbb{R}^n \to \mathbb{R}^m$ which is differentiable at x and a mapping $F : \mathbb{R}^n \rightrightarrows \mathbb{R}^m$, prove that*

$$\tilde{D}(f + F)(x | f(x) + y)(u) = Df(x)u + \tilde{D}F(x|y)(u).$$

Guide. Let $v \in \tilde{D}(f+F)(x|f(x)+y)(u)$. By Carathéodory's theorem on convex hull representation, there are sequences $\{u_i^k\}$, $\{v_i^k\}$ and $\{\lambda_i^k\}$ for $i = 0, 1, \ldots, n+m$ and $k = 1, 2, \ldots$ with $\lambda_i^k \ge 0$, $\sum_{i=0}^{n+m} \lambda_i^k = 1$, such that $v_i^k \in D(f+F)(x|f(x)+y)(u_i^k)$ for all i and k and $\sum_{i=0}^{n+m} \lambda_i^k (u_i^k, v_i^k) \to (u,v)$ as $k \to \infty$. From 4A.2, get $v_i^k \in Df(x)u_i^k + DF(x|y)(u_i^k)$ for all i and k. Hence $\sum_{i=0}^{n+m} \lambda_i^k (u_i^k, v_i^k - Df(x)u_i^k) \in \operatorname{cl co} \operatorname{gph} DF(x|y)$. Then pass to the limit. $\qquad \square$

We end this section with yet another proof of the classical inverse function theorem 1A.1. This time it is based on the Ekeland principle given in 4B.5.

Proof of Theorem 1A.1. Without loss of generality, let $\bar{x} = 0$, $f(\bar{x}) = 0$. Let $A = \nabla f(0)$ and let $\delta = |A^{-1}|^{-1}$. Choose $a > 0$ such that

$$\text{(41)} \qquad |f(x) - f(x') - A(x - x')| \le \frac{\delta}{2}|x - x'| \quad \text{for all } x, x' \in a\mathbb{B},$$

and let $b = a\delta/2$. We now redo Step 1 in Proof I that the localization s of f^{-1} with respect to the neighborhoods $b\mathbb{B}$ and $a\mathbb{B}$ is nonempty-valued. The other two steps remain the same as in Proof I.

Fix $y \in b\mathbb{B}$ and consider the function $|f(x) - y|$ with domain containing the closed ball $a\mathbb{B}$, which we view as a complete metric space equipped with the Euclidean metric. This function is continuous and bounded below, hence, by Ekeland principle 4B.5 with the indicated δ and $\bar{u} = 0$ there exists $x_\delta \in a\mathbb{B}$ such that

$$\text{(42)} \qquad |y - f(x_\delta)| < |y - f(x)| + \frac{\delta}{2}|x - x_\delta| \quad \text{for all } x \in a\mathbb{B},\ x \ne x_\delta.$$

Let us assume that $y \ne f(x_\delta)$. Then $\tilde{x} := A^{-1}(y - f(x_\delta)) + x_\delta \ne x_\delta$. Moreover, from (41) with $x = x_\delta$ and $x' = 0$ and the choice of δ and b we get

4 Regularity Properties Through Generalized Derivatives

$$|\tilde{x}| \leq |A^{-1}|(|y| + |-f(x_\delta) + Ax_\delta|) \leq |A^{-1}|\left(b + \frac{a\delta}{2}\right) = |A^{-1}|a\delta = a.$$

Hence we can set $x = \tilde{x}$ in (42), obtaining

(43) $$|y - f(x_\delta)| < |y - f(\tilde{x})| + \frac{\delta}{2}|\tilde{x} - x_\delta|.$$

Using (41), we have

(44) $$\begin{aligned} |y - f(\tilde{x})| &= |f(A^{-1}(y - f(x_\delta))) + x_\delta) - y| \\ &= |f(A^{-1}(y - f(x_\delta))) + x_\delta) - f(x_\delta) - A(A^{-1}(y - f(x_\delta)))| \\ &\leq \tfrac{\delta}{2}|A^{-1}(y - f(x_\delta))| \end{aligned}$$

and also

(45) $$|\tilde{x} - x_\delta| = |A^{-1}(y - f(x_\delta))|.$$

Plugging (44) and (45) into (43), we arrive at

$$|y - f(x_\delta)| < \left(\frac{\delta}{2} + \frac{\delta}{2}\right)|A^{-1}(y - f(x_\delta))| \leq \delta|A^{-1}||y - f(x_\delta)| = |y - f(x_\delta)|$$

which furnishes a contradiction. Thus, our assumption that $y \neq f(x_\delta)$ is voided, and we have $x_\delta \in f^{-1}(y) \cap (a\mathbb{B})$. This means that s is nonempty-valued, and the proof is complete. □

4C. Characterization of Strong Metric Subregularity

Strong metric subregularity of a mapping $F : \mathbb{R}^n \rightrightarrows \mathbb{R}^m$ at $\bar{x}$ for $\bar{y}$, where $\bar{y} \in F(\bar{x})$, was defined in Section 3I to mean the existence of a constant κ along with neighborhoods U of $\bar{x}$ and V of $\bar{y}$ such that

(1) $$|x - \bar{x}| \leq \kappa d(\bar{y}, F(x) \cap V) \quad \text{for all } x \in U.$$

This property is equivalent to the combination of two other properties: that F is *metrically subregular* at $\bar{x}$ for $\bar{y}$, and $\bar{x}$ is an *isolated point* of $F^{-1}(\bar{y})$. The associated modulus, the infimum of all $\kappa > 0$ for which this holds for some U and V, is thus the same as the modulus of subregularity, subreg$(F; \bar{x}|\bar{y})$.

It was shown in 3I.2 that F is strongly metrically subregular at $\bar{x}$ for $\bar{y}$ if and only if F^{-1} has the isolated calmness property at $\bar{y}$ for $\bar{x}$. As an illustration, a linear mapping A is strongly metrically subregular at $\bar{x}$ for $\bar{y} = A\bar{x}$ if and only if $A^{-1}(\bar{y})$ consists only of $\bar{x}$, i.e., A is injective. More generally, a mapping F that is polyhedral, as defined in 3D, is strongly metrically subregular at $\bar{x}$ for $\bar{y}$ if and only if $\bar{x}$ is an isolated point of $F^{-1}(\bar{y})$; this follows from 3I.1.

What makes the strong metric subregularity attractive, along the same lines as metric regularity and strong metric regularity, is its stability with respect to approximation as established in 3I.6. In particular, a function $f : \mathbb{R}^n \to \mathbb{R}^m$ which is differentiable at $\bar{x}$ is strongly metrically subregular at $\bar{x}$ for $f(\bar{x})$ if and only if its Jacobian $\nabla f(\bar{x})$ has rank n, so that $\nabla f(\bar{x})u = 0$ implies $u = 0$. According to 4A.6, this is also characterized by $|Df(\bar{x})^{-1}|^+ < \infty$. It turns out that such an outer norm characterization can be provided also for set-valued mappings by letting graphical derivatives take over the role of ordinary derivatives.

Theorem 4C.1 (derivative criterion for strong metric subregularity). *A mapping $F : \mathbb{R}^n \rightrightarrows \mathbb{R}^m$ whose graph is locally closed at $(\bar{x},\bar{y}) \in \mathrm{gph}\, F$ is strongly metrically subregular at $\bar{x}$ for $\bar{y}$ if and only if*

(2) $$DF(\bar{x}|\bar{y})^{-1}(0) = \{0\},$$

this being equivalent to

(3) $$|DF(\bar{x}|\bar{y})^{-1}|^+ < \infty,$$

and in that case

(4) $$\mathrm{subreg}\,(F;\bar{x}|\bar{y}) = |DF(\bar{x}|\bar{y})^{-1}|^+.$$

Proof. The equivalence between (2) and (3) comes from 4A.6. To get the equivalence of these conditions with strong metric subregularity, suppose first that $\kappa > \mathrm{subreg}\,(F;\bar{x}|\bar{y})$ so that F is strongly metrically subregular at $\bar{x}$ for $\bar{y}$ and (1) holds for some neighborhoods U and V. By definition, having $v \in DF(\bar{x}|\bar{y})(u)$ refers to the existence of sequences $u^k \to u$, $v^k \to v$ and $\tau^k \searrow 0$ such that $\bar{y} + \tau^k v^k \in F(\bar{x} + \tau^k u^k)$. Then $\bar{x} + \tau^k u^k \in U$ and $\bar{y} + \tau^k v^k \in V$ eventually, so that (1) yields $|(\bar{x} + \tau^k u^k) - \bar{x}| \leq \kappa |(\bar{y} + \tau^k v^k) - \bar{y}|$, which is the same as $|u^k| \leq \kappa |v^k|$. In the limit, this implies $|u| \leq \kappa |v|$. But then, by 4A.6, $|DF(\bar{x}|\bar{y})^{-1}|^+ \leq \kappa$ and hence

(5) $$\mathrm{subreg}\,(F;\bar{x}|\bar{y}) \geq |DF(\bar{x}|\bar{y})^{-1}|^+.$$

In the other direction, (3) implies the existence of a $\kappa > 0$ such that

$$\sup_{v \in B} \sup_{u \in DF(\bar{x}|\bar{y})^{-1}(v)} |u| < \kappa.$$

This in turn implies that $|x - \bar{x}| \leq \kappa |y - \bar{y}|$ for all $(x,y) \in \mathrm{gph}\, F$ close to $(\bar{x},\bar{y})$. That description fits with (1). Further, κ can be chosen arbitrarily close to $|DF(\bar{x}|\bar{y})^{-1}|^+$, and therefore $|DF(\bar{x}|\bar{y})^{-1}|^+ \geq \mathrm{subreg}\,(F;\bar{x}|\bar{y})$. This, combined with (5), finishes the argument. □

4 Regularity Properties Through Generalized Derivatives   219

Corollary 4C.2 (derivative criterion for isolated calmness). *For a mapping $F : \mathbb{R}^n \rightrightarrows \mathbb{R}^m$, consider the inverse $S = F^{-1}$, which can also be viewed as the solution mapping $S : \mathbb{R}^m \rightrightarrows \mathbb{R}^n$ for the generalized equation $F(x) \ni y$:*

$$S(y) = \{x \,|\, F(x) \ni y\}.$$

Fix $\bar{y}$ and any $\bar{x} \in S(\bar{y})$, and suppose that $\text{gph } F$ is locally closed at $(\bar{x},\bar{y})$, which is the same as $\text{gph } S$ being locally closed at $(\bar{y},\bar{x})$. Then

$$\text{clm}(S;\bar{y}|\bar{x}) = |DS(\bar{y}|\bar{x})|^+.$$

Thus, S has the isolated calmness property at $\bar{y}$ for $\bar{x}$ if and only if $|DS(\bar{y}|\bar{x})|^+ < \infty$.

Theorem 4C.1 immediately gives us the linearization result in Corollary 3I.9 by using the sum rule in 4A.2.

Implicit function theorems could be developed for the isolated calmness of solution mappings to general inclusions $G(p,x) \ni 0$ in parallel to the results in 4B, but we shall not do this here. We limit ourselves to an application of the derivative criterion 4C.1 to the solution mapping of a generalized equation

(6) $$S(p) = \{x \,|\, f(p,x) + F(x) \ni 0\},$$

where $f : \mathbb{R}^d \times \mathbb{R}^n \to \mathbb{R}^m$ and $F : \mathbb{R}^n \rightrightarrows \mathbb{R}^m$. In the following corollary we utilize Theorem 3I.13 and the ample parameterization condition.

Corollary 4C.3 (derivative rule for isolated calmness of solution mappings). *For the solution mapping S in (6) and a pair $(\bar{p},\bar{x})$ with $\bar{x} \in S(\bar{p})$, suppose that f is differentiable with respect to x uniformly in p at $(\bar{p},\bar{x})$ and also differentiable with respect to p uniformly in x at $(\bar{p},\bar{x})$. Also, suppose that $\text{gph } F$ is locally closed at $(\bar{x}, -f(\bar{p},\bar{x}))$. If*

(7) $$|(D_x f(\bar{p},\bar{x}) + DF(\bar{x}| - f(\bar{p},\bar{x})))^{-1}|^+ \leq \lambda < \infty,$$

then S has the isolated calmness property at $\bar{p}$ for $\bar{x}$, moreover with

(8) $$\text{clm}(S;\bar{p}|\bar{x}) \leq \lambda |\nabla_p f(\bar{p},\bar{x})|.$$

Furthermore, when f is continuously differentiable in a neighborhood of $(\bar{p},\bar{x})$ and satisfies the ample parameterization condition

(9) $$\text{rank } \nabla_p f(\bar{p},\bar{x}) = m,$$

then the converse implication holds as well; that is, S has isolated calmness property at $\bar{p}$ for $\bar{x}$ if and only if (7) is satisfied.

Proof. We apply Theorem 3I.13 according to which the mapping S has the isolated calmness property at $\bar{p}$ for $\bar{x}$ if the mapping

$$h + F \quad \text{for} \quad h(x) = f(\bar{p},\bar{x}) + D_x f(\bar{p},\bar{x})(x - \bar{x})$$

is strongly metrically subregular at $\bar{x}$ for 0, and the converse implication holds under the ample parameterization condition (9). Then, it is sufficient to apply 4C.1 together with the sum rule 4A.2 to the mapping $h + F$. The estimate (8) follows from formula 3I(24). □

The following simple example in terms of graphical derivatives illustrates further the distinction between metric regularity and strong metric subregularity.

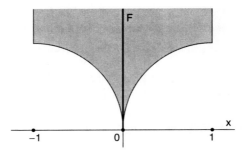

Fig. 4.1 Graph of the mapping in Example 4C.4.

Example 4C.4 (strong metric subregularity without metric regularity). Let $F : \mathbb{R} \rightrightarrows \mathbb{R}$ be defined by

$$F(x) = \begin{cases} [\sqrt{1-(x-1)^2}, \infty) & \text{for } 0 \leq x \leq 1, \\ [\sqrt{1-(x+1)^2}, \infty) & \text{for } -1 \leq x \leq 0, \\ \emptyset & \text{elsewhere,} \end{cases}$$

as shown in Figure 4.1. Then

$$DF(0|0)(u) = \begin{cases} [0, \infty) & \text{for } u = 0, \\ \emptyset & \text{for } u \neq 0, \end{cases}$$

and therefore

$$|DF(0|0)^{-1}|^+ = 0, \qquad |DF(0|0)^{-1}|^- = \infty.$$

This fits with F being strongly metrically subregular, but not metrically regular at 0 for 0.

4D. Applications to Parameterized Constraint Systems

Next we look at what the graphical derivative results in the preceding sections have to say about a constraint system

(1) $$f(p,x) - K \ni 0, \text{ or equivalently } f(p,x) \in K,$$

and its solution mapping

(2) $$S: p \mapsto \{x \mid f(p,x) \in K\}$$

for a function $f : \mathbb{R}^d \times \mathbb{R}^n \to \mathbb{R}^m$ and a set $K \subset \mathbb{R}^m$.

Theorem 4D.1 (implicit mapping theorem for a constraint system). *Let $\bar{x} \in S(\bar{p})$ for solution mapping S of the constraint system (1) and suppose that f is differentiable in a neighborhood of $(\bar{p},\bar{x})$ and satisfies $\widehat{\text{lip}}_p(f;(\bar{p},\bar{x})) < \infty$, and that the set K is closed. If*

(3) $$\limsup_{\substack{(p,x,y) \to (\bar{p},\bar{x},0) \\ f(p,x)-y \in K}} \sup_{|v| \leq 1} d\Big(0, D_x f(p,x)^{-1}(v + T_K(f(p,x) - y))\Big) \leq \lambda < \infty,$$

then S has the Aubin property at $\bar{p}$ for $\bar{x}$, with

(4) $$\text{lip}(S;\bar{p},\bar{x}) \leq \lambda \, \widehat{\text{lip}}_p(f;(\bar{p},\bar{x})).$$

Proof. The assumed closedness of K and continuous differentiability of f around $(\bar{p},\bar{x})$ allow us to apply Corollary 4B.8 to the case of $F(x) \equiv -K$. Further, according to 4A.3 we have

$$D_x G(p,x \mid y) = D_x f(p,x) - T_K(f(p,x) - y).$$

Next, we use the definition of inner norm in 4A(8) to write 4B(29) as (3) and apply 4B.7 to obtain that S has the Aubin property at $\bar{p}$ for $\bar{x}$. The estimate (4) follows immediately from 4B(30). □

A much sharper result can be obtained when f is continuously differentiable and the set K in the system (1) is polyhedral convex.

Theorem 4D.2 (constraint systems with polyhedral convexity). *Let $\bar{x} \in S(\bar{p})$ for the solution mapping S of the constraint system (1) in the case of a polyhedral convex set K. Suppose that f is continuously differentiable in a neighborhood of $(\bar{p},\bar{x})$. Then for S to have the Aubin property at $\bar{x}$ for $\bar{p}$, it is sufficient that*

(5) $$\text{rge } D_x f(\bar{p},\bar{x}) - T_K(f(\bar{p},\bar{x})) = \mathbb{R}^m,$$

in which case the corresponding modulus satisfies $\text{lip}(S;\bar{p} \mid \bar{x}) \leq \lambda |\nabla_p f(\bar{p},\bar{x})|$ for

(6) $$\lambda = \sup_{|v|\leq 1} d\Big(0, D_x f(\bar{p},\bar{x})^{-1}(v + T_K(f(\bar{p},\bar{x})))\Big).$$

Moreover (5) is necessary for S to have the Aubin property at $\bar{p}$ for $\bar{x}$ under the ample parameterization condition

(7) $$\operatorname{rank} \nabla_p f(\bar{p},\bar{x}) = m.$$

Proof. We invoke Theorem 4D.1 but make special use of the fact that K is polyhedral. That property implies that $T_K(w) \supset T_K(\bar{w})$ for all w sufficiently near to $\bar{w}$, as seen in 2E.3; we apply this to $w = f(p,x) - y$ and $\bar{w} = f(\bar{p},\bar{x})$ in the formulas (3) and (4) of 4D.1. The distances in question are greatest when the cone is as small as possible; this, combined with the continuous differentiability of f, allows us to drop the limit in (3). Further, from the equivalence relation 4A(16) in Corollary 4A.7, we obtain that the finiteness of λ in (6) is equivalent to (5).

For the necessity, we bring in a further argument which makes use of the ample parameterization condition (7). According to Theorem 3F.9, under (7) the Aubin property of S at $\bar{p}$ for $\bar{x}$ implies metric regularity of the linearized mapping $h - K$ for $h(x) = f(\bar{p},\bar{x}) + \nabla_x f(\bar{p},\bar{x})(x - \bar{x})$. The derivative criterion for metric regularity 4B.1 tells us then that

(8) $$\limsup_{\substack{(x,y)\to(\bar{x},0)\\ f(\bar{p},\bar{x})+D_x f(\bar{p},\bar{x})(x-\bar{x})-y\in K}} \sup_{|v|\leq 1} d\Big(0, D_x f(\bar{p},\bar{x})^{-1}(v + T_K(f(\bar{p},\bar{x}) + D_x f(\bar{p},\bar{x})(x - \bar{x}) - y))\Big) < \infty.$$

Taking $x = \bar{x}$ and $y = 0$ instead of limsup in (8) gives us the expression for λ in (6) and may only decrease the left side of this inequality. We already know that the finiteness of λ in (6) yields (5), and so we are done. □

Example 4D.3 (application to systems of inequalities and equalities). For $K = \mathbb{R}_-^s \times \{0\}^{m-s}$, the solution mapping $S(p)$ in (2) consists, in terms of $f(p,x) = (f_1(p,x),\ldots,f_m(p,x))$ of all solutions x to

$$f_i(p,x) \begin{cases} \leq 0 & \text{for } i \in [1,s], \\ = 0 & \text{for } i \in [s+1,m]. \end{cases}$$

Let $\bar{x}$ solve this for $\bar{p}$ and let each f_i be continuously differentiable around $(\bar{p},\bar{x})$. Then a sufficient condition for S to have the Aubin property for $\bar{p}$ for $\bar{x}$ is the *Mangasarian–Fromovitz condition*:

(9) $$\exists\, w \in \mathbb{R}^n \text{ with } \begin{cases} \nabla_x f_i(\bar{p},\bar{x})w < 0 & \text{for } i \in [1,s] \text{ with } f_i(\bar{p},\bar{x}) = 0, \\ \nabla_x f_i(\bar{p},\bar{x})w = 0 & \text{for } i \in [s+1,m], \end{cases}$$

and

(10) the vectors $\nabla_x f_i(\bar{p},\bar{x})$ for $i \in [s+1,m]$ are linearly independent.

4 Regularity Properties Through Generalized Derivatives

Moreover, the combination of (9) and (10) is also necessary for S to have the Aubin property under the ample parameterization condition (7). In particular, when f is independent of p and then $0 \in f(\bar{x}) - K$, the Mangasarian–Fromovitz condition (9)–(10) is a necessary and sufficient condition for metric regularity of the mapping $f - K$ at $\bar{x}$ for 0.

Detail. According to 4D.2, it is enough to show that (5) is equivalent to the combination of (9) and (10) in the case of $K = \mathbb{R}_-^s \times \{0\}^{m-s}$. Observe that the tangent cone to the set K at $f(\bar{p}, \bar{x})$ has the following form:

$$(11) \quad v \in T_K(f(\bar{p},\bar{x})) \iff v_i \begin{cases} \leq 0 & \text{for } i \in [1,s] \text{ with } f_i(\bar{p},\bar{x}) = 0, \\ = 0 & \text{for } i \in [s+1,m]. \end{cases}$$

Let (5) hold. Then, using (11), we obtain that the matrix with rows the vectors $\nabla_x f_{s+1}(\bar{p},\bar{x}), \ldots, \nabla_x f_m(\bar{p},\bar{x})$ must be of full rank, hence (10) holds. If (9) is violated, then for every $w \in \mathbb{R}^n$ either $\nabla_x f_i(\bar{p},\bar{x})w \geq 0$ for some $i \in [1,s]$ with $f_i(\bar{p},\bar{x}) = 0$, or $\nabla_x f_i(\bar{p},\bar{x})w \neq 0$ for some $i \in [s+1,m]$, which contradicts (5) in an obvious way.

The combination of (9) and (10) implies that for every $y \in \mathbb{R}^m$ there exist $w, v \in \mathbb{R}^n$ and $z \in \mathbb{R}^m$ with $z_i \leq 0$ for $i \in [1,s]$ with $f_i(\bar{p},\bar{x}) = 0$ such that

$$\begin{cases} \nabla_x f_i(\bar{p},\bar{x})w - z_i = y_i & \text{for } i \in [1,s] \text{ with } f_i(\bar{p},\bar{x}) = 0, \\ \nabla_x f_i(\bar{p},\bar{x})(w+v) = y_i & \text{for } i \in [s+1,m]. \end{cases}$$

But then (5) follows directly from the form (11) of the tangent cone.

If f is independent of p, by 3E.6 the metric regularity of $-f + K$ is equivalent to the Aubin property of the inverse $(-f + K)^{-1}$, which is the same as the solution mapping

$$S(p) = \{x \mid p + f(x) \in K\}$$

for which the ample parameterization condition (7) holds automatically. Then, from 4D.2, for $\bar{x} \in S(\bar{p})$, the Aubin property of S at $\bar{p}$ for $\bar{x}$ and hence metric regularity of $f - K$ at $\bar{x}$ for $\bar{p}$ is equivalent to (5) and therefore to (9)–(10). □

Exercise 4D.4. Consider the constraint system in 4D.3 with $f(p,x) = g(x) - p$, $\bar{p} = 0$ and g continuously differentiable near $\bar{x}$. Show that the existence of a Lipschitz continuous local selection of the solution mapping S at 0 for $\bar{x}$ implies the Mangasarian–Fromovitz condition. In other words, the existence of a Lipschitz continuous local selection of S at 0 for $\bar{x}$ implies metric regularity of the mapping $g - K$ at $\bar{x}$ for 0.

Guide. Utilizing 2B.11, from the existence of a local selection of S at 0 for $\bar{x}$ we obtain that the inverse F_0^{-1} of the linearization $F_0(x) := g(\bar{x}) + \nabla g(\bar{x})(x - \bar{x}) - K$ has a Lipschitz continuous local selection at 0 for $\bar{x}$. Then, in particular, for every $v \in \mathbb{R}^m$ there exists $w \in \mathbb{R}^n$ such that

$$\begin{cases} \nabla g_i(\bar{x})w \leq v_i & \text{for } i \in [1,s] \text{ with } g_i(\bar{x}) = 0, \\ \nabla g_i(\bar{x})w = v_i & \text{for } i \in [s+1,m]. \end{cases}$$

This is the same as (5). □

A result parallel to 4D.2 can be formulated also for isolated calmness instead of the Aubin property.

Proposition 4D.5 (isolated calmness of constraint systems). *In the setting of Theorem 4D.1, for S to have the isolated calmness property at $\bar{p}$ for $\bar{x}$ it is sufficient that*

$$\nabla_x f(\bar{p},\bar{x})u \in T_K(f(\bar{p},\bar{x})) \implies u = 0.$$

Moreover, this condition is necessary for S to have the isolated calmness property at $\bar{p}$ for $\bar{x}$ under the ample parameterization condition (8).

Proof. This is a special case of Corollary 4C.3 in which we utilize 4A.2. □

We note that the isolated calmness property offers little of interest in the case of solution mappings for constraint systems beyond equations, inasmuch as it necessitates $\bar{x}$ being an isolated point of the solution set $S(\bar{p})$; this restricts significantly the class of constraint systems for which such a property may occur. In the following section we will consider mappings associated with variational inequalities for which the isolated calmness is a more natural property.

4E. Isolated Calmness for Variational Inequalities

Now we take up once more the topic of variational inequalities, to which serious attention was already devoted in Chapter 2. This revolves around a generalized equation of the form

(1) $$f(p,x) + N_C(x) \ni 0, \quad \text{or} \quad -f(p,x) \in N_C(x),$$

for a function $f: \mathbb{R}^d \times \mathbb{R}^n \to \mathbb{R}^n$ and the normal cone mapping N_C associated with a nonempty, closed, convex set $C \subset \mathbb{R}^n$, and the solution mapping $S: \mathbb{R}^d \rightrightarrows \mathbb{R}^n$ defined by

(2) $$S(p) = \{x \mid f(p,x) + N_C(x) \ni 0\}.$$

Especially strong results were obtained in 2E for the case in which C is a *polyhedral* convex set, and that will also persist here. Of special importance in that setting is the critical cone associated with C at a point x with respect to a vector $v \in N_C(x)$, defined by

(3) $$K_C(x,v) = T_C(x) \cap [v]^\perp,$$

which is always polyhedral convex as well. Recall here that for any vector $v \in \mathbb{R}^n$ we denote $[v] = \{\tau v \mid \tau \in \mathbb{R}\}$; then $[v]$ is a subspace of dimension 1 if $v \neq 0$ and 0 if $v = 0$. Accordingly, $[v]^\perp$ is a hyperplane through the origin if $v \neq 0$ and the whole $\mathbb{R}^n$ when $v = 0$.

In this section we examine the isolated calmness property, which is inverse to strong metric subregularity by 3I.2.

Theorem 4E.1 (isolated calmness for variational inequalities). *For the variational inequality (1) and its solution mapping (2) under the assumption that the convex set C is polyhedral, let $\bar{x} \in S(\bar{p})$ and suppose that f is continuously differentiable around $(\bar{p}, \bar{x})$. Let $A = \nabla_x f(\bar{p}, \bar{x})$ and let $K = K_C(\bar{x}, \bar{v})$ be the corresponding critical cone in (3) for $\bar{v} = -f(\bar{p}, \bar{x})$. If*

(4) $$(A + N_K)^{-1}(0) = \{0\},$$

then the solution mapping S has the isolated calmness property at $\bar{p}$ for $\bar{x}$ with

$$\mathrm{clm}(S; \bar{p} | \bar{x}) \leq |(A + N_K)^{-1}|^+ \cdot |\nabla_p f(\bar{p}, \bar{x})|.$$

Moreover, under the ample parameterization condition rank $\nabla_p f(\bar{p}, \bar{x}) = n$, the property in (4) is not just sufficient but also necessary for S to have the isolated calmness property at $\bar{p}$ for $\bar{x}$.

Proof. Utilizing the specific form of the graphical derivative established in 4A.4 and the equivalence relation 4A(13) in 4A.6, we see that (4) is equivalent to the condition 4C(7) in Corollary 4C.3. Everything then follows from the claim of that corollary. □

Exercise 4E.2 (alternative cone condition). *In terms of the cone K^* that is polar to K, show that the condition in (4) is equivalent to*

(5) $$w \in K, \; -Aw \in K^*, \; w \perp Aw \;\Longrightarrow\; w = 0.$$

Guide. Make use of 2A.3. □

In the important case when $C = \mathbb{R}^n_+$, the variational inequality (1) turns into the *complementarity* relation

(6) $$x \geq 0, \quad f(p, x) \geq 0, \quad x \perp f(p, x).$$

This will serve to illustrate the result in Theorem 4E.1. Using the notation introduced in Section 2E for the analysis of a complementarity problem, we associate with the reference point $(\bar{x}, \bar{v}) \in \mathrm{gph}\, N_{\mathbb{R}^n_+}$ the index sets J_1, J_2 and J_3 in $\{1, \ldots, n\}$ given by

$$J_1 = \{j \mid \bar{x}_j > 0, \bar{v}_j = 0\}, \quad J_2 = \{j \mid \bar{x}_j = 0, \bar{v}_j = 0\}, \quad J_3 = \{j \mid \bar{x}_j = 0, \bar{v}_j < 0\}.$$

Then, by 2E.5, the critical cone $K = K_C(\bar{x}, \bar{v}) = T_{\mathbb{R}^n_+}(\bar{x}) \cap [f(\bar{p}, \bar{x})]^\perp$ is described by

(7) $\quad w \in K \iff \begin{cases} w_j \text{ free} & \text{for } i \in J_1, \\ w_j \geq 0 & \text{for } i \in J_2, \\ w_j = 0 & \text{for } i \in J_3. \end{cases}$

Example 4E.3 (isolated calmness in complementarity problems). In the case of $C = \mathbb{R}^n_+$ in which the variational inequality (1) reduces to the complementarity relation (6) and the critical cone K is given by (7), the condition (4) in Theorem 4E.1 reduces through (5) to having the following hold for the entries a_{ij} of the matrix A. If w_j for $j \in J_1 \cup J_2$ are real numbers satisfying

$$w_j \geq 0 \text{ for } j \in J_2 \quad \text{and} \quad \sum_{j \in J_1 \cup J_2} a_{ij} w_j \begin{cases} = 0 & \text{for } i \in J_1 \text{ and for } i \in J_2 \text{ with } w_i > 0, \\ \geq 0 & \text{for } i \in J_2 \text{ with } w_i = 0, \end{cases}$$

then $w_j = 0$ for all $j \in J_1 \cup J_2$.

In the particular case when $J_1 = J_3 = \emptyset$, the matrices satisfying the condition in 4E.3 are called R_0-matrices[1].

As another application of Theorem 4E.1, consider the tilted minimization problem from Section 2G:

(8) $\qquad$ minimize $g(x) - \langle v, x \rangle$ over $x \in C$,

where C is a nonempty *polyhedral* convex subset of $\mathbb{R}^n$, $v \in \mathbb{R}^n$ is a parameter, and the function $g : \mathbb{R}^n \to \mathbb{R}$ is twice continuously differentiable everywhere. We first give a brief summary of the optimality conditions from 2G.

If x is a local optimal solution of (8) for v then x satisfies the basic first-order necessary optimality condition

(9) $\qquad \nabla g(x) + N_C(x) \ni v.$

Any solution x of (9) is a stationary point for problem (8), denoted $S(v)$, and the associated stationary point mapping is $v \mapsto S(v) = (Dg + N_C)^{-1}(v)$. The set of local minimizers of (8) for v is a subset of $S(v)$. If the function g is convex, every stationary point is not only local but also a global minimizer. For the variational inequality (9), the critical cone to C associated with a solution x for v has the form

$$K_C(x, v - \nabla g(x)) = T_C(x) \cap [v - \nabla g(x)]^\perp.$$

If x furnishes a local minimum of (8) for v, then, according to 2G.1(a), x must satisfy the second-order necessary condition

(10) $\qquad \langle u, \nabla^2 g(x) u \rangle \geq 0 \quad \text{for all } u \in K_C(x, v - \nabla g(x)).$

In addition, from 2G.1(b), when $x \in S(v)$ satisfies the second-order sufficient condition

---

[1] For a detailed description of the classes of matrices appearing in the theory of linear complementarity problems, see the book Cottle, Pang and Stone [1992].

4 Regularity Properties Through Generalized Derivatives

(11) $\quad \langle u, \nabla^2 g(x) u \rangle > 0 \quad$ for all nonzero $u \in K_C(x, v - \nabla g(x))$,

then x is a local optimal solution of (8) for v. Having x to satisfy (9) and (11) is equivalent to the existence of $\varepsilon > 0$ and $\delta > 0$ such that

(12) $\quad g(y) - \langle v, y \rangle \geq g(x) - \langle v, x \rangle + \varepsilon |y - x|^2 \quad$ for all $y \in C$ with $|y - x| \leq \delta$,

meaning by definition that x furnishes a strong local minimum in (8).

We know from 2G.3 that the stationary point mapping S has a Lipschitz localization s around v for x with the property that $s(u)$ furnishes a strong local minimum for u near v if and only if the following stronger form of the second-order sufficient optimality condition holds:

$$\langle w, \nabla^2 g(\bar{x}) w \rangle > 0 \quad \text{for all nonzero } w \in K_C^+(x, v),$$

where $K_C^+(x, v) = K_C(x, v - \nabla g(x)) - K_C(x, v - \nabla g(x))$ is the critical subspace associated with x and v. We now complement this result with a necessary and sufficient condition for isolated calmness of S combined with local optimality at the reference point.

Theorem 4E.4 (role of second-order sufficiency). *Consider the stationary point mapping S for problem (8), that is, the solution mapping for (9), and let $\bar{x} \in S(\bar{v})$. Then the following are equivalent:*

(i) the second-order sufficient condition (11) holds at $\bar{x}$ for $\bar{v}$;

(ii) the point $\bar{x}$ is a local minimizer of (7) for $\bar{v}$ and the mapping S has the isolated calmness property at $\bar{v}$ for $\bar{x}$.

Moreover, in either case, $\bar{x}$ is actually a strong local minimizer of (7) for $\bar{v}$.

Proof. Denote $A := \nabla^2 g(\bar{x})$. According to Theorem 4E.1 complemented with 4E.2, the mapping S has the isolated calmness property at $\bar{v}$ for $\bar{x}$ if and only if

(13) $\quad u \in K, \quad -Au \in K^*, \quad u \perp Au \quad \Longrightarrow \quad u = 0,$

where $K = K_C(\bar{x}, \bar{v} - \nabla g(\bar{x}))$. Let (i) hold. Then of course $\bar{x}$ is a local optimal solution as described. If (ii) doesn't hold, there must exist some $u \neq 0$ satisfying the conditions in the left side of (13), and that would contradict the inequality $\langle u, Au \rangle > 0$ in (11).

Conversely, assume that (ii) is satisfied. Then the second-order necessary condition (10) must hold; this can be written as

$$u \in K \quad \Longrightarrow \quad -Au \in K^*.$$

The isolated calmness property of S at $\bar{v}$ for $\bar{x}$ is identified with (13), which in turn eliminates the possibility of there being a nonzero $u \in K$ such that the inequality in (10) fails to be strict. Thus, the necessary condition (10) turns into the sufficient condition (11). We already know that (11) implies (12), so the proof is complete. □

4F. Single-valued Localizations for Variational Inequalities

We now investigate applications of graphical derivatives to the issue of whether the solution mapping S to the variational inequality

(1) $$f(p,x) + N_C(x) \ni 0$$

has a Lipschitz continuous single-valued localization. As in the preceding section, we have a function $f : \mathbb{R}^d \times \mathbb{R}^n \to \mathbb{R}^n$ and a normal cone mapping N_C associated with a *polyhedral* convex set $C \subset \mathbb{R}^n$. Our starting point is Theorem 2E.8 which we state again for completeness.

Theorem 4F.1 (localization criterion under polyhedral convexity). *For the solution mapping S of (1) under the assumption that the convex set C is polyhedral, let $\bar{x} \in S(\bar{p})$ and suppose that f is continuously differentiable near $(\bar{p}, \bar{x})$. Let*

(2) $$A = \nabla_x f(\bar{p}, \bar{x}) \text{ and } K = \{ w \in T_C(\bar{x}) \, | \, w \perp f(\bar{p}, \bar{x}) \},$$

noting that the critical cone K is likewise polyhedral convex. Suppose the mapping $A + N_K$ has the property that

(3) $$\bar{s} := (A + N_K)^{-1} \text{ is everywhere single-valued,}$$

in which case $\bar{s}$ is Lipschitz continuous globally (this being equivalent to strong metric regularity of $A + N_K$ at 0 for 0). Then the solution mapping S has a Lipschitz continuous single-valued localization s around $\bar{p}$ for $\bar{x}$ which is semidifferentiable at $\bar{p}$ with

$$Ds(\bar{p})(q) = \bar{s}(-\nabla_p f(\bar{p}, \bar{x})q)$$

and moreover

(4) $$\text{lip}(s; \bar{p}) \leq \text{lip}(\bar{s}; 0) \cdot |\nabla_p f(\bar{p}, \bar{x})|.$$

In addition, under the ample parameterization condition rank $\nabla_p f(\bar{p}, \bar{x}) = n$ *the condition in (3) is not just sufficient but also necessary for S to have a Lipschitz continuous single-valued localization around $\bar{p}$ for $\bar{x}$.*

Through graphical derivatives, we will actually be able to show that strong metric regularity of $A + N_K$ is implied simply by metric regularity, or in other words, that the invertibility condition in (3) follows already from $(A + N_K)^{-1}$ having the Aubin property at 0 for 0, due to the special structure of the mapping $A + N_K$.

Our tactic for bringing this out will involve applying Theorem 4B.1 to $A + N_K$. Before that can be done, however, we put some effort into a better understanding of the normal cone mapping N_K.

Faces of a cone. *For a polyhedral convex cone K, a face is a set F of the form*

4 Regularity Properties Through Generalized Derivatives

$$F = K \cap [v]^\perp \text{ for some } v \in K^*.$$

The collection of all faces of K will be denoted by $\mathscr{F}_K$.

Since K is polyhedral, we obtain from the Minkowski-Weyl theorem, 2E.2, and its surroundings that $\mathscr{F}_K$ is a finite collection of polyhedral convex cones. This collection contains K itself and the zero cone, in particular.

Lemma 4F.2 (critical face lemma). *Let C be a convex polyhedral set, let $v \in N_C(x)$ and let $K = K_C(x,v)$ be the critical cone for C at (x,v),*

$$K = T_C(x) \cap [v]^\perp.$$

Then there exists a neighborhood O of (x,v) such that for every choice of $(x',v') \in \text{gph } N_C \cap O$ the corresponding critical cone $K_C(x',v')$ has the form

$$K_C(x',v') = F_1 - F_2$$

for some faces F_1, F_2 in $\mathscr{F}_K$ with $F_2 \subset F_1$. In particular, $K_C(x',v') \subset K - K$ for every $(x',v') \in \text{gph } N_C \cap O$. Conversely, for every two faces F_1, F_2 in $\mathscr{F}_K$ with $F_2 \subset F_1$ and every neighborhood O of (x,v) there exists $(x',v') \in \text{gph } N_C \cap O$ such that $K_C(x',v') = F_1 - F_2$.

Proof. Because C is polyhedral, all vectors of the form $x'' = x' - x$ with $x' \in C$ close to x are the vectors $x'' \in T_C(x)$ having sufficiently small norm. Also, for such x'',

(5) $$T_C(x') = T_C(x) + [x''] \supset T_C(x)$$

and

(6) $$N_C(x') = N_C(x) \cap [x'']^\perp \subset N_C(x).$$

Now, let $(x',v') \in \text{gph } N_C$ be close to (x,v) and let $x'' = x' - x$. Then from (5) we have

$$K_C(x',v') = T_C(x') \cap [v']^\perp = \left(T_C(x) + [x'']\right) \cap [v']^\perp.$$

Further, from (6) it follows that $v' \perp x''$ and then we obtain

(7) $$K_C(x',v') = T_C(x) \cap [v']^\perp + [x''] = K_C(x,v') + [x''].$$

We will next show that $K_C(x,v') \subset K$ for v' sufficiently close to v. If this were not so, there would be a sequence $v_k \to v$ and another sequence $w_k \in K_C(x,v_k)$ such that $w_k \notin K$ for all k. Each set $K_C(x,v_k)$ is a face of $T_C(x)$, but since $T_C(x)$ is polyhedral, the set of its faces is finite, hence for some face F of $T_C(x)$ we have $K_C(x,v_k) = F$ for infinitely many k. Note that the set $\text{gph } K_C(x,\cdot)$ is closed, hence for any $w \in F$, since (v_k,w) is in this graph, the limit (v,w) belongs to it as well. But then $w \in K$ and since $w \in F$ is arbitrarily chosen, we have $F \subset K$. Thus the sequence $w_k \in K$ for infinitely many k, which is a contradiction. Hence $K_C(x,v') \subset K$.

Let $(x',v') \in \mathrm{gph}\, N_C$ be close to (x,v). Relation (7) tells us that $K_C(x',v') = K_C(x,v') + [x'']$ for $x'' = x' - x$. Let $F_1 = T_C(x) \cap [v']^\perp$, this being a face of $T_C(x)$. The critical cone $K = K_C(x,v) = T_C(x) \cap [v]^\perp$ is itself a face of $T_C(x)$, and any face of $T_C(x)$ within K is also a face of K. Then F_1 is a face of the polyhedral cone K. Let F_2 be the face of F_1 having x'' in its relative interior. Then F_2 is also a face of K and therefore $K_C(x',v') = F_1 - F_2$, furnishing the desired representation.

Conversely, let F_1 be a face of K. Then there exists $v' \in K^* = N_K(0)$ such that $F_1 = K \cap [v']^\perp$. The size of v' does not matter; hence we may assume that $v + v' \in N_C(x)$ by the reduction lemma 2E.4. By repeating the above argument we have $F_1 = T_C(x) \cap [v'']^\perp$ for $v'' := v + v'$. Now let F_2 be a face of F_1. Let x' be in the relative interior of F_2. In particular, $x' \in T_C(x)$, so by taking the norm of x' sufficiently small we can arrange that the point $x'' = x + x'$ lies in C. We have $x' \perp v'$ and, as in (7),

$$F_1 - F_2 = T_C(x) \cap [v'']^\perp + [x'] = \Big(T_C(x) + [x']\Big) \cap [v'']^\perp = T_C(x'') \cap [v'']^\perp = K_C(x'', v'').$$

This gives us the form required. □

Our next step is to specify the derivative criterion for metric regularity in 4B.1 for the reduced mapping $A + N_K$ in 4F.1.

Lemma 4F.3 (regularity modulus from derivative criterion). *For $A + N_K$ with A and K as in (2), we have*

(8) $$\mathrm{reg}(A + N_K; 0|0) = \max_{\substack{F_1, F_2 \in \mathscr{F}_K \\ F_1 \supset F_2}} |(A + N_{F_1 - F_2})^{-1}|^-.$$

Thus, $A + N_K$ is metrically regular at 0 for 0 if and only if $|(A + N_{F_1 - F_2})^{-1}|^- < \infty$ for every $F_1, F_2 \in \mathscr{F}_K$ with $F_1 \supset F_2$.

Proof. From Theorem 4B.1, combined with Example 4A.4, we have that

$$\mathrm{reg}(A + N_K; 0|0) = \limsup_{\substack{(x,y) \to (0,0) \\ (x,y) \in \mathrm{gph}(A + N_K)}} |(A + N_{T_K(x) \cap [y - Ax]^\perp})^{-1}|^-.$$

Lemma 4F.2 with $(x,v) = (0,0)$ gives us the desired representation $N_{T_K(x) \cap [y - Ax]^\perp} = N_{F_1 - F_2}$ for (x,y) near zero and hence (8). □

Example 4F.4 (critical faces for complementarity problems). Consider the complementarity problem

$$f(p,x) + N_{\mathbb{R}^n_+}(x) \ni 0,$$

with a solution $\bar{x}$ for $\bar{p}$, with K and A as in (2) (with $C = \mathbb{R}^n_+$) and index sets

$$J_1 = \{j \mid \bar{x}_j > 0, \bar{v}_j = 0\}, \quad J_2 = \{j \mid \bar{x}_j = 0, \bar{v}_j = 0\}, \quad J_3 = \{j \mid \bar{x}_j = 0, \bar{v}_j < 0\}$$

for $\bar{v} = -f(\bar{p},\bar{x})$. Then the cones $F_1 - F_2$, where F_1 and F_2 are closed faces of K with $F_1 \supset F_2$, are the cones $\tilde{K}$ of the following form: There is a partition of $\{1,\ldots,n\}$ into index sets J'_1, J'_2, J'_3 with

4 Regularity Properties Through Generalized Derivatives

$$J_1 \subset J_1' \subset J_1 \cup J_2, \qquad J_3 \subset J_3' \subset J_2 \cup J_3,$$

such that

(9) $\qquad x' \in F_1 - F_2 \iff \begin{cases} x_i' \text{ free} & \text{for } i \in J_1', \\ x_i' \geq 0 & \text{for } i \in J_2', \\ x_i' = 0 & \text{for } i \in J_3'. \end{cases}$

Detail. Each face F of K has the form $K \cap [v']^\perp$ for some vector $v' \in K^*$. The vectors v' in question are those with

$$\begin{cases} v_i' = 0 & \text{for } i \in J_1, \\ v_i' \leq 0 & \text{for } i \in J_2, \\ v_i' \text{ free} & \text{for } i \in J_3. \end{cases}$$

The closed faces F of K correspond one-to-one therefore with the subsets of J_2: the face F corresponding to an index set J_2^F consists of the vectors x' such that

$$\begin{cases} x_i' \text{ free} & \text{for } i \in J_1, \\ x_i' \geq 0 & \text{for } i \in J_2 \setminus J_2^F, \\ x_i' = 0 & \text{for } i \in J_3 \cup J_2^F. \end{cases}$$

If F_1 and F_2 have $J_2^{F_1} \subset J_2^{F_2}$, so that $F_1 \supset F_2$, then $F_1 - F_2$ is given by (9) with $J_1' = J_1 \cup [J_2 \setminus J_2^{F_2}]$, $J_2' = J_2^{F_2} \setminus J_2^{F_1}$, $J_3' = J_3 \cup J_2^{F_1}$. □

Exercise 4F.5 (critical face criterion for metric regularity). *For a continuously differentiable function* $f : \mathbb{R}^n \to \mathbb{R}^n$ *and a polyhedral convex set* $C \subset \mathbb{R}^n$, *let* $f(\bar{x}) + N_C(\bar{x}) \ni 0$. *Show that the mapping* $f + N_C$ *is metrically regular at* $\bar{x}$ *for* 0 *if and only if, for all choices of faces* F_1 *and* F_2 *of the critical cone* K *to the set* C *at* $\bar{x}$ *for* $\bar{v} = -f(\bar{x})$, *with* $F_1 \supset F_2$, *the following condition holds with* $A = \nabla f(\bar{x})$:

$$\forall v \in \mathbb{R}^n \quad \exists u \in F_1 - F_2 \text{ such that } (v - Au) \in (F_1 - F_2)^* \text{ and } (v - Au) \perp u.$$

Guide. From 3F.7, metric regularity of $f + N_C$ at $\bar{x}$ for 0 is equivalent to metric regularity of $A + N_K$ at 0 for 0. Apply 4F.3 then with the characterization of the inner norm in 4A.6 and using the fact that $u \in N_K(w)$ whenever $w \in K$, $u \in K^*$, and $u \perp w$. □

Exercise 4F.6 (variational inequality over a subspace). *Show that when the critical cone K in 4F.5 is a subspace of $\mathbb{R}^n$ of dimension $m \leq n$, then the matrix BAB^T is nonsingular, where B is the matrix whose columns form an orthonormal basis in K.*

Using some of the results obtained so far in this section, we will now prove that, for a mapping appearing in particular in the Karush–Kuhn–Tucker (KKT) optimality conditions in nonlinear programming, *metric regularity and strong metric regularity are equivalent properties*. Specifically, consider the standard nonlinear programming problem

(10) $\quad$ minimize $g_0(x)$ over all x satisfying $g_i(x) \begin{cases} = 0 & \text{for } i \in [1,r], \\ \leq 0 & \text{for } i \in [r+1,m] \end{cases}$

with twice continuously differentiable functions $g_i : \mathbb{R}^n \to \mathbb{R}$, $i = 0, 1, \ldots, m$. The associated KKT optimality system has the form

(11) $\quad\quad\quad\quad f(x,y) + N_E(x,y) \ni (0,0),$

where

(12) $\quad\quad\quad\quad f(x,y) = \begin{pmatrix} \nabla g_0(x) + \sum_{i=1}^{m} \nabla g_i(x) y_i \\ -g_1(x) \\ \vdots \\ -g_m(x) \end{pmatrix}$

and

(13) $\quad\quad\quad\quad E = \mathbb{R}^n \times [\mathbb{R}^r \times \mathbb{R}_+^{m-r}].$

Theorem 2A.8 tells us that, under the constraint qualification condition 2A(14), for any local minimum x of (10) there exists a Lagrange multiplier y, with $y_i \geq 0$ for $i = r+1, \ldots, m$, such that (x,y) is a solution of (11). We will now establish an important fact for the mapping on the left side of (11).

Theorem 4F.7 (KKT metric regularity implies strong metric regularity). *Consider the mapping $F : \mathbb{R}^{n+m} \rightrightarrows \mathbb{R}^{n+m}$ defined as*

(14) $\quad\quad\quad\quad F : z \mapsto f(z) + N_E(z)$

with f as in (12) for $z = (x,y)$ and E as in (13), and let $\bar{z} = (\bar{x}, \bar{y})$ solve (11), that is, $F(\bar{z}) \ni 0$. If F is metrically regular at $\bar{z}$ for 0, then F is strongly metrically regular there.

We already showed in Theorem 3G.5 that this kind of equivalence holds for locally monotone mappings, but here F need not be monotone even locally, although it is a special kind of mapping in another way.

The claimed equivalence is readily apparent in a simple case of (10) when F is an affine mapping, which corresponds to problem (10) with no constraints and with g_0 being a quadratic function, $g_0(x) = \frac{1}{2}\langle x, Ax \rangle + \langle b, x \rangle$ for an $n \times n$ matrix A and a vector $b \in \mathbb{R}^n$. Then $F(x,y) = Ax + b$ and metric regularity of F (at any point) means that A has full rank. But then A must be nonsingular, so F is in fact strongly regular.

The general argument for $F = f + N_E$ is lengthy and proceeds through a series of reductions. First, since our analysis is local, we can assume without loss of generality that all inequality constraints are active at $\bar{x}$. Indeed, if for some index $i \in [r+1, m]$ we have $g_i(\bar{x}) < 0$, then $\bar{y}_i = 0$. For $q \in \mathbb{R}^{n+m}$ consider the solution set of the inclusion $F(z) \ni q$. Then for any q near zero and all x near $\bar{x}$ we will have $g_i(x) < q_i$, and hence any Lagrange multiplier y associated with such an x must have

4 Regularity Properties Through Generalized Derivatives

$y_i = 0$; thus, for q close to zero the solution set of $F(z) \ni q$ will not change if we drop the constraint with index i. Further, if there exists an index i such that $\bar{y}_i > 0$, then we can always rearrange the constraints so that $\bar{y}_i > 0$ for $i \in [r+1,s]$ for some $r < s \leq m$. Under these simplifying assumptions the critical cone $K = K_E(\bar{z}, \bar{v})$ to the set E in (13) at $\bar{z} = (\bar{x}, \bar{y})$ for $\bar{v} = -f(\bar{z})$ is the product $\mathbb{R}^n \times \mathbb{R}^s \times \mathbb{R}_+^{m-s}$. (Show that this form of the critical cone can be also derived by utilizing Example 2E.5.) The normal cone mapping N_K to the critical cone K has then the form $N_K = \{0\}^n \times \{0\}^s \times N_+^{m-s}$.

We next recall that metric regularity of F is equivalent to metric regularity of the mapping

$$L : z \mapsto \nabla f(\bar{z})z + N_K(z) \text{ for } z = (x,y) \in \mathbb{R}^{n+m}$$

at 0 for 0 and the same equivalence holds for strong metric regularity. This reduction to a simpler situation has already been highlighted several times in this book, e.g. in 2E.8 for strong metric regularity and 3F.7 for metric regularity. Thus, to achieve our goal of confirming the claimed equivalence between metric regularity and strong regularity for F, it is enough to focus on the mapping L which, in terms of the functions g_i in (10), has the form

$$(15) \qquad L = \begin{pmatrix} A & B^\mathsf{T} \\ -B & 0 \end{pmatrix} + N_K,$$

where

$$A = \nabla^2 g_0(\bar{x}) + \sum_{i=1}^m \nabla^2 g_i(\bar{x})\bar{y}_i \text{ and } B = \begin{pmatrix} \nabla_x g_1(\bar{x}) \\ \vdots \\ \nabla g_m(\bar{x}) \end{pmatrix}.$$

Taking into account the specific form of N_K, the inclusion $(v,w) \in L(x,y)$ becomes

$$(16) \qquad \begin{cases} v = Ax + B^\mathsf{T} y, \\ (w+Bx)_i = 0 & \text{for } i \in [1,s], \\ (w+Bx)_i \leq 0, \ y_i \geq 0, \ y_i(w+Bx)_i = 0 & \text{for } i \in [s+1,m]. \end{cases}$$

In further preparation for proving Theorem 4F.7, next we state and prove three lemmas. From now on any kind of regularity is at 0 for 0, unless specified otherwise.

Lemma 4F.8 (KKT metric regularity implies strong metric subregularity). *If the mapping L in (15) is metrically regular, then it is strongly subregular.*

Proof. Suppose that L is metrically regular. Then the critical face criterion displayed in 4F.5 with critical faces given in 4F.4 takes the following form: for every partition J_1', J_2', J_3' of $\{s+1,\ldots,m\}$ and for every $(v,w) \in \mathbb{R}^n \times \mathbb{R}^m$ there exists $(x,y) \in \mathbb{R}^n \times \mathbb{R}^m$ satisfying

(17) $$\begin{cases} v = Ax + B^\mathsf{T} y, \\ (w + Bx)_i = 0 & \text{for } i \in [1, s], \\ (w + Bx)_i = 0 & \text{for } i \in J_1', \\ (w + Bx)_i \leq 0, \ y_i \geq 0, \ y_i(w + Bx)_i = 0 & \text{for } i \in J_2', \\ y_i = 0 & \text{for } i \in J_3'. \end{cases}$$

In particular, denoting by B_0 the submatrix of B composed by the first s rows of B, for any index set $J \subset \{s+1, \ldots, m\}$, including the empty set, if $B(J)$ is the submatrix of B whose rows have indices in J, then the condition involving (17) implies that

(18) the matrix $N(J) = \begin{pmatrix} A & B_0^\mathsf{T} & B^\mathsf{T}(J) \\ -B_0 & 0 & 0 \\ -B(J) & 0 & 0 \end{pmatrix}$ is nonsingular.

Indeed, to reach such a conclusion it is enough to take $J = J_1'$ and $J_2' = \emptyset$ in (17). By 4E.1, the mapping L in (15) is strongly subregular if and only if

(19) the only solution of (17) with $(v, w) = 0$ is $(x, y) = 0$.

Now, suppose that L is not strongly subregular. Then, by (19), for some index set $J \subset \{s+1, \ldots, m\}$, possibly the empty set, there exists a nonzero vector $(x, y) \in \mathbb{R}^n \times \mathbb{R}^m$ satisfying (17) for $v = 0, w = 0$. Note that this y has $y_j = 0$ for $j \in \{s+1, \ldots, m\} \setminus J$. But then the nonzero vector $z = (x, y)$ with y having components in $\{1, \ldots, s\} \times J$ solves $N(J)z = 0$ where the matrix $N(J)$ is defined in (18). Hence, $N(J)$ is singular, and then the condition involving (17) is violated; thus, the mapping L is not metrically regular. This contradiction means that L must be strongly subregular. □

The next two lemmas present general facts that are separate from the specific circumstances of nonlinear programming problem (10) considered. The second lemma is a simple consequence of Brouwer's invariance of domain theorem 1F.1:

Lemma 4F.9 (single-valued localization from continuous local selection). *Let $f : \mathbb{R}^n \to \mathbb{R}^n$ be continuous and let there exist an open neighborhood V of $\bar{y} := f(\bar{x})$ and a continuous function $h : V \to \mathbb{R}^n$ such that $h(y) \in f^{-1}(y)$ for $y \in V$. Then f^{-1} has a single-valued graphical localization around $\bar{y}$ for $\bar{x}$.*

Proof. Since f is a function, we have $f^{-1}(y) \cap f^{-1}(y') = \emptyset$ for any $y, y' \in V$, $y \neq y'$. But then h is one-to-one and hence h^{-1} is a function defined in $U := h(V)$. Note that $x = h(h^{-1}(x)) \in f^{-1}(h^{-1}(x))$ implies $f(x) = h^{-1}(x)$ for all $x \in U$. Since h is a function, we have that $h^{-1}(x) \neq h^{-1}(x')$ for all $x, x' \in U$, $x \neq x'$. Thus, f is one-to-one on U, implying that the set $f^{-1}(y) \cap U$ consists of one point, $h(y)$. Hence, by Theorem 1F.1 applied to h, U is an open neighborhood of $\bar{x}$. Therefore, h is a single-valued graphical localization of f^{-1} around $\bar{y}$ for $\bar{x}$. □

Lemma 4F.10 (properties of optimal solutions). *Let $\varphi : \mathbb{R}^n \to \mathbb{R}$ be a continuous function and let $Q : \mathbb{R}^d \rightrightarrows \mathbb{R}^n$ have the Aubin property at $\bar{p}$ for $\bar{x}$. Then any graphical*

4 Regularity Properties Through Generalized Derivatives

localization around $\bar{p}$ for $\bar{x}$ of the solution mapping S_{opt} of the problem

$$\text{minimize } \varphi(x) \text{ subject to } x \in Q(p)$$

is either multi-valued or a continuous function on a neighborhood of $\bar{p}$.

Proof. Suppose that S_{opt} has a single-valued localization $\hat{x}(p) = S_{\text{opt}}(p) \cap U$ for all $p \in V$ for some neighborhoods U of $\bar{x}$ and V of $\bar{p}$ and, without loss of generality, that Q has the Aubin property at $\bar{p}$ for $\bar{x}$ with the same neighborhoods U and V. Let $p \in V$ and $V \ni p_k \to p$ as $k \to \infty$. Since $\hat{x}(p) \in Q(p) \cap U$ and $\hat{x}(p_k) \in Q(p_k) \cap U$, there exist $x_k \in Q(p_k)$ and $x'_k \in Q(p)$ such that $x_k \to \hat{x}(p)$ and also $|x'_k - \hat{x}(p_k)| \to 0$ as $k \to \infty$. From optimality,

$$\varphi(x_k) \geq \varphi(\hat{x}(p_k)) \quad \text{and} \quad \varphi(x'_k) \geq \varphi(\hat{x}(p)).$$

These two inequalities, combined with the continuity of φ, give us

$$\varphi(\hat{x}(p_k)) \to \varphi(\hat{x}(p)) \text{ as } k \to \infty.$$

Hence any limit of a sequence of minimizers $\hat{x}(p_k)$ is a minimizer for p, which implies that $\hat{x}$ is continuous at p. □

We are now ready to complete the proof of 4F.7.

Proof of Theorem 4F.7 (final part). We already know from the argument displayed after the statement of the theorem that metric regularity of the mapping F in (14) at $\bar{z} = (\bar{x}, \bar{y})$ for 0 is equivalent to the metric regularity of the mapping L in (15) at 0 for 0, and the same holds for the strong metric regularity. Our next step is to associate with the mapping L the function

(20) $$H(x, y) = \begin{pmatrix} Ax + \sum_{i=1}^{s} b_i y_i + \sum_{i=s+1}^{m} b_i y_i^+ \\ -\langle b_1, x \rangle + y_1 \\ \vdots \\ -\langle b_s, x \rangle + y_s \\ -\langle b_{s+1}, x \rangle + y_{s+1}^- \\ \vdots \\ -\langle b_m, x \rangle + y_m^- \end{pmatrix},$$

from $\mathbb{R}^n \times \mathbb{R}^m$ to itself, where b_i are the rows of the matrix B and where we let $y^+ = \max\{0, y\}$ and $y^- = y - y^+$.

For a given $(v, u) \in \mathbb{R}^n \times \mathbb{R}^m$, let $(x, y) \in H^{-1}(v, u)$. Then for $z_i = y_i^+$, $i = s+1, \ldots, q$, we have $(x, z) \in L^{-1}(v, u)$. Indeed, for each $i = s+1, \ldots, m$, if $y_i \leq 0$, then $u_i + \langle b_i, x \rangle = y_i^- \leq 0$ and $(u_i + \langle b_i, x \rangle) y_i^+ = 0$; otherwise $u_i + \langle b_i, x \rangle = y_i^- = 0$. Conversely, if $(x, z) \in L^{-1}(v, u)$ then for

(21) $$y_i = \begin{cases} z_i & \text{if } z_i > 0, \\ u_i + \langle b_i, x \rangle & \text{if } z_i = 0, \end{cases}$$

we obtain $(x,y) \in H^{-1}(v,u)$. Thus, in order to achieve our goal for the mapping L, we can focus on the same question for the equivalence between metric regularity and strong metric regularity for the *function H* in (20).

Suppose that H is metrically regular but not strongly metrically regular. Then, from 4F.8 and the equivalence between regularity properties of L and H, H is strongly subregular. Consequently, its inverse H^{-1} has both the Aubin property and the isolated calmness property, both at 0 for 0. In particular, since H is positively homogeneous and has closed graph, for each w sufficiently close to 0, $H^{-1}(w)$ is a compact set contained in an arbitrarily small ball around 0. Let $a > 0$. For any $w \in a\mathbb{B}$ the problem

$$(22) \qquad \text{minimize } y_m \text{ subject to } (x,y) \in H^{-1}(w)$$

has a solution $(x(w), y(w))$ which, from the property of H^{-1} mentioned just above (22), has a nonempty-valued graphical localization around 0 for 0. According to Lemma 4F.10, this localization is either a continuous function or a multi-valued mapping. If it is a continuous function, Lemma 4F.9 implies that H^{-1} has a continuous single-valued localization around 0 for 0. But then, since H^{-1} has the Aubin property at that point, we conclude that H must be strongly metrically regular, which contradicts the assumption made. Hence, any graphical localization of the solution mapping of (22) is multi-valued. Thus, there exists a sequence $z^k = (v^k, u^k) \to 0$ and two sequences $(x^k, y^k) \to 0$ and $(\xi^k, \eta^k) \to 0$, whose k-terms are both in $H^{-1}(z^k)$, such that the m-components of y^k and η^k are the same, $y_m^k = \eta_m^k$, but $(x^k, y^k) \neq (\xi^k, \eta^k)$ for all k. Remove from y^k the final component y_m^k and denote the remaining vector by y_{-m}^k. Do the same for η^k. Then (x^k, y_{-m}^k) and (ξ^k, η_{-m}^k) are both solutions of

$$v^k - b_m y_m^k = Ax^k + \sum_{i=1}^{s} b_i y_i + \sum_{i=s+1}^{m-1} b_i y_i^+$$

$$u_1^k = -\langle b_1, x \rangle + y_1$$

$$\vdots$$

$$u_s^k = -\langle b_s, x \rangle + y_s$$

$$\vdots$$

$$u_{s+1}^k = -\langle b_{s+1}, x \rangle + y_{s+1}^-$$

$$\vdots$$

$$u_{m-1}^k = -\langle b_{m-1}, x \rangle + y_{m-1}^-.$$

This relation concerns the reduced mapping H_{-m} with $m-1$ vectors b_i, and accordingly a vector y of dimension $m-1$:

$$H_{-m}(x,y) = \begin{pmatrix} Ax + \sum_{i=1}^{s} b_i y_i + \sum_{i=s+1}^{m-1} b_i y_i^+ \\ -\langle b_1, x \rangle + y_1 \\ \vdots \\ -\langle b_s, x \rangle + y_s \\ -\langle b_{s+1}, x \rangle + y_{s+1}^- \\ \vdots \\ -\langle b_{m-1}, x \rangle + y_{m-1}^- \end{pmatrix}.$$

We obtain that the mapping H_{-m} cannot be strongly metrically regular because for the same value $z^k = (v^k - b_m y_m^k, u_{-m}^k)$ of the parameter arbitrarily close to 0, we have two solutions (x^k, y_{-m}^k) and (ξ^k, η_{-m}^k). On the other hand, H_{-m} is metrically regular as a submapping of H; this follows e.g. from the characterization in (17) for metric regularity of the mapping L, which is equivalent to the metric regularity of H_{-m} if we choose J_3' in (17) always to include the index m.

Thus, our assumption for the mapping H leads to a submapping H_{-m}, of one less variable y associated with the "inequality" part of L, for which the same assumption is satisfied. By proceeding further with "deleting inequalities" we will end up with no inequalities at all, and then the mapping L becomes just the linear mapping represented by the square matrix

$$\begin{pmatrix} A & B_0^\mathsf{T} \\ B_0 & 0 \end{pmatrix}.$$

But this linear mapping cannot be simultaneously metrically regular and not strongly metrically regular, because a square matrix of full rank is automatically nonsingular. Hence, our assumption that the mapping H is metrically regular and not strongly regular is void. □

Exercise 4F.11. *Find a formula for the metric regularity modulus of the mapping F in (14).*

Guide. The regularity modulus of F at $\bar{z}$ for 0 equals the regularity modulus of the mapping L in (15) at 0 for 0. To find a formula for the latter, utilize Lemma 4F.3. □

4G. Special Nonsmooth Inverse Function Theorems

At the very beginning of this book, in Chapter 1, we were occupied with the classical idea of inverting a function $f : \mathbb{R}^n \to \mathbb{R}^n$ locally, in the pattern of solving $f(x) = y$ in a localized sense for x in terms of y. The notion of a single-valued localization of f^{-1} was introduced and studied under various assumptions about differentiability. After that, we moved on in Chapter 2 to single-valued localizations of solution mappings to generalized equations instead of just equations, and in Chapter 3 to results about solution mappings beyond single-valued localizations. Now, though, we can return to the original setting with new tools and describe two additional theorems in the inverse function mode. These theorems make use of other concepts of generalized differentiation, beyond the ones introduced in 4A.

In order to introduce the first of these alternative concepts, we have to rely on a theorem by Rademacher, according to which a function $f : \mathbb{R}^n \to \mathbb{R}^m$ that is Lipschitz continuous on an open set O is differentiable almost everywhere in O, hence at a set of points that is dense in O. If κ is a Lipschitz constant, then from the definition of Jacobian $\nabla f(x)$ we have $|\nabla f(x)| \leq \kappa$ at those points of differentiability. Recall that a set $C \subset \mathbb{R}^n$ is said to be dense in the closed set D when $\operatorname{cl} C = D$, or equivalently, when for any $x \in D$ any neighborhood U of x contains elements of C. One of the simplest examples of a dense set is the set of rational numbers relative to the set of real numbers.

Consider now any function $f : \mathbb{R}^n \to \mathbb{R}^m$ and any point $\bar{x} \in \operatorname{dom} f$ where $\operatorname{lip}(f;\bar{x}) < \infty$. For any $\kappa > \operatorname{lip}(f;\bar{x})$ we have f Lipschitz continuous with constant κ in some neighborhood U of $\bar{x}$, and hence, from Rademacher's theorem, there is a dense set of points x in U where f is differentiable with $|\nabla f(x)| \leq \kappa$. Hence there exist sequences $x_k \to \bar{x}$ such that f is differentiable at x_k, in which case the corresponding sequence of norms $|\nabla f(x_k)|$ is bounded by the Lipschitz constant κ and hence has at least one cluster point. This leads to the following definition.

Generalized Jacobian. *For $f : \mathbb{R}^n \to \mathbb{R}^m$ and any $\bar{x} \in \operatorname{dom} f$ where $\operatorname{lip}(f;\bar{x}) < \infty$, denote by $\bar{\nabla} f(\bar{x})$ the set consisting of all matrices $A \in \mathbb{R}^{m \times n}$ for which there is a sequence of points $x_k \to \bar{x}$ such that f is differentiable at x_k and $\nabla f(x_k) \to A$. The Clarke generalized Jacobian of f at $\bar{p}$ is the convex hull of this set: $\operatorname{co} \bar{\nabla} f(\bar{x})$.*

Note that $\bar{\nabla} f(\bar{x})$ is a nonempty, closed, bounded subset of $\mathbb{R}^{m \times n}$. This ensures that the convex set $\operatorname{co} \bar{\nabla} f(\bar{x})$ is nonempty, closed, and bounded as well. Strict differentiability of f at $\bar{x}$ is known to be characterized by having $\bar{\nabla} f(\bar{x})$ consist of a single matrix A (or equivalently by having $\operatorname{co} \bar{\nabla} f(\bar{x})$ consist of a single matrix A), in which case $A = \nabla f(\bar{x})$.

The inverse function theorem based on this notion, which we state next without proof[2], says roughly that a Lipschitz continuous function can be inverted when all elements of the generalized Jacobian are nonsingular. Compared with the classical inverse function theorem, the main difference is that the single-valued graphical localization so obtained can only be claimed to be Lipschitz continuous.

Theorem 4G.1 (Clarke's inverse function theorem). *Consider $f : \mathbb{R}^n \to \mathbb{R}^n$ and a point $\bar{x} \in \operatorname{int} \operatorname{dom} f$ where $\operatorname{lip}(f;\bar{x}) < \infty$. Let $\bar{y} = f(\bar{x})$. If all of the matrices in the generalized Jacobian $\operatorname{co} \bar{\nabla} f(\bar{x})$ are nonsingular, then f^{-1} has a Lipschitz continuous single-valued localization around $\bar{y}$ for $\bar{x}$.*

For illustration, we provide some elementary cases.

---

[2] Cf. Clarke [1976].

4 Regularity Properties Through Generalized Derivatives

Examples. The function $f: \mathbb{R} \to \mathbb{R}$ given by

$$f(x) = \begin{cases} x + x^3 & \text{for } x < 0, \\ 2x - x^2 & \text{for } x \geq 0 \end{cases}$$

has generalized Jacobian co $\bar{\nabla} f(0) = [1, 2]$, which does not contain 0. According to Theorem 4G.1, f^{-1} has a Lipschitz continuous single-valued localization around 0 for 0.

In contrast, the function $f: \mathbb{R} \to \mathbb{R}$ given by $f(x) = |x|$ has co $\bar{\nabla} f(0) = [-1, 1]$, which does contain 0. Although the theorem makes no claims about this case, there is no graphical localization of f^{-1} around 0 for 0 that is single-valued.

A simple 2-dimensional example[3] is with

$$f(x) = \begin{bmatrix} |x_1| + x_2 \\ 2x_1 + |x_2| \end{bmatrix},$$

for which

$$\text{co } \bar{\nabla} f(0,0) = \left\{ \begin{bmatrix} \lambda & 1 \\ 2 & \tau \end{bmatrix} \,\middle|\, -1 \leq \lambda \leq 1,\ -1 \leq \tau \leq 1 \right\}.$$

This set of matrices does not contain a singular matrix, and hence 4G.1 can be applied.

A better hold on the existence of single-valued Lipschitz continuous localizations can be gained through a different version of graphical differentiation.

Strict graphical derivative. *For a function $f: \mathbb{R}^n \to \mathbb{R}^m$ and any point $\bar{x} \in \text{dom } f$, the strict graphical derivative at $\bar{x}$ is the set-valued mapping $D_* f(\bar{x}): \mathbb{R}^n \rightrightarrows \mathbb{R}^m$ defined by*

$$D_* f(\bar{x})(u) = \left\{ w \,\middle|\, \exists (t^k, x^k, u^k) \to (0, \bar{x}, u) \text{ with } w = \lim_{k \to \infty} \frac{f(x^k + t^k u^k) - f(x^k)}{t^k} \right\}.$$

When $\text{lip}(f; \bar{x}) < \infty$, the set $D_* f(\bar{x})(u)$ is nonempty, closed and bounded in $\mathbb{R}^m$ for each $u \in \mathbb{R}^n$. Then too, the definition of $D_* f(\bar{x})(u)$ can be simplified by taking $u_k \equiv u$. In this Lipschitzian setting it can be shown that $D_* f(\bar{x})u = \{ Au \mid A \in \text{co } \bar{\nabla} f(\bar{x}) \}$ for all u if $m = 1$, but that fails for higher dimensions. In general, it is known[4] that for a function $f: \mathbb{R}^n \to \mathbb{R}^m$ with $\text{lip}(f; \bar{x}) < \infty$, one has

(1) $\qquad \text{co } \bar{\nabla} f(\bar{x})(u) \supset D_* f(\bar{x})(u) \quad \text{for all } u \in \mathbb{R}^n.$

Note that f is strictly differentiable at $\bar{x}$ if and only if $D_* f(\bar{x})$ is a linear mapping, with the matrix for that mapping then being $\nabla f(\bar{x})$. Anyway, strict graphical derivatives can be used without having to assume even that $\text{lip}(f; \bar{x}) < \infty$.

---

[3] This example is from Clarke [1983].
[4] Cf. Klatte and Kummer [2002], Section 6.3.

The computation of strict graphical derivatives will be illustrated now in a special case of nonsmoothness which has a basic role in various situations.

Example 4G.2. Consider the function
$$\theta^+ : x \mapsto x^+ := \max\{x, 0\}, \quad x \in \mathbb{R}.$$

Directly from the definition, we have, for any real u, that

(2) $$D_*\theta^+(\bar{x})(u) = \begin{cases} u & \text{for } \bar{x} > 0, \\ \{\lambda u \mid \lambda \in [0,1]\} & \text{for } \bar{x} = 0, \\ 0 & \text{for } \bar{x} < 0. \end{cases}$$

Similarly, the function
$$\theta^- : x \mapsto x^- := \min\{x, 0\}, \quad x \in \mathbb{R}$$

satisfies
$$\theta^-(x) = x - \theta^+(x).$$

Then, just by applying the definition, we get
$$v \in D_*\theta^+(\bar{x})(u) \iff (u - v) \in D_*\theta^-(\bar{x})(u) \text{ for any real } u.$$

Equipped with strict graphical derivatives, we are now able to present a generalization of the classical inverse function theorem which furnishes a complete characterization of the existence of a Lipschitz continuous localization of the inverse, and thus sharpens the theorem of Clarke.

Theorem 4G.3 (Kummer's inverse function theorem). *Let $f : \mathbb{R}^n \to \mathbb{R}^n$ be continuous around $\bar{x}$, with $f(\bar{x}) = \bar{y}$. Then f^{-1} has a Lipschitz continuous single-valued localization around $\bar{y}$ for $\bar{x}$ if and only if*

(3) $$0 \in D_*f(\bar{x})(u) \implies u = 0.$$

Proof. Recall Theorem 1F.2, which says that for a function $f : \mathbb{R}^n \to \mathbb{R}^n$ that is continuous around $\bar{x}$, the inverse f^{-1} has a Lipschitz continuous localization around $f(\bar{x})$ for $\bar{x}$ if and only if, in some neighborhood U of $\bar{x}$, there is a constant $c > 0$ such that

(4) $$c|x' - x| \le |f(x') - f(x)| \quad \text{for all } x', x \in O.$$

We will show first that (3) implies (4), from which the sufficiency of the condition will follow. With the aim of reaching a contradiction, let us assume there are sequences $c^k \to 0$, $x^k \to \bar{x}$ and $\tilde{x}^k \to \bar{x}$ such that
$$|f(x^k) - f(\tilde{x}^k)| \le c^k |x^k - \tilde{x}^k|.$$

Then the sequence of points

4 Regularity Properties Through Generalized Derivatives   241

$$u^k := \frac{\tilde{x}^k - x^k}{|x^k - \tilde{x}^k|}$$

satisfies $|u_k| = 1$ for all k, hence a subsequence u^{k_i} of it is convergent to some $u \neq 0$. Restricting ourselves to such a subsequence, we obtain for $t^{k_i} = |x^{k_i} - \tilde{x}^{k_i}|$ that

$$\lim_{k_i \to \infty} \frac{f(x^{k_i} + t^{k_i}u^{k_i}) - f(x^{k_i})}{t^{k_i}} = 0.$$

By definition, the limit on the left side belongs to $D_*f(\bar{x})(u)$, yet $u \neq 0$, which is contrary to (3). Hence (3) does imply (4).

For the converse, we argue that if (3) were violated, there would be sequences $t^k \to 0$, $x^k \to \bar{x}$, and $u^k \to u$ with $u \neq 0$, for which

(5)
$$\lim_{k \to \infty} \frac{f(x^k + t^k u^k) - f(x^k)}{t^k} = 0.$$

On the other hand, under (4) however, one has

$$\frac{|f(x^k + t^k u^k) - f(x^k)|}{t^k} \geq c|u^k|,$$

which combined with (5) and the assumption that u^k is away from 0 leads to an absurdity for large k. Thus (4) guarantees that (3) holds. □

The property recorded in (1) indicates clearly that Clarke's theorem follows from that of Kummer. However, although the characterization of Lipschitz invertibility in Kummer's theorem looks simple, the price to be paid still lies ahead: we have to be able to *calculate the strict graphical derivative* in every case of interest. This task could be quite hard without calculus rules.

A rule that immediately follows from the definitions, at least, is the following.

Exercise 4G.4 (strict graphical derivatives of a sum). *For a function $f_1 : \mathbb{R}^n \to \mathbb{R}^m$ that is strictly differentiable at $\bar{x}$ (or, in particular, as a special case, continuously differentiable in a neighborhood of $\bar{x}$), and a function $f_2 : \mathbb{R}^n \to \mathbb{R}^m$ that is Lipschitz continuous around $\bar{x}$, one has*

$$D_*(f_1 + f_2)(\bar{x})(u) = Df_1(\bar{x})u + D_*f_2(\bar{x})(u).$$

Guide. This can be deduced right from the definitions. □

Without going any further now into such rules, as developed in nonsmooth analysis, we devote the rest of this section to applying Kummer's theorem to a specific situation which we studied in Chapter 2.

Consider the following nonlinear programming problem with inequality constraints and special *canonical* perturbations

(6)   minimize $g_0(x) - \langle v, x \rangle$ over all x satisfying $g_i(x) \leq u_i$ for $i \in [1,m]$,

where the functions $g_i : \mathbb{R}^n \to \mathbb{R}$, $i = 0, 1, \ldots, m$ are twice continuously differentiable and $v \in \mathbb{R}^n$, $u = (u_1, \ldots, u_m)^T \in \mathbb{R}^m$ are parameters. According to the basic first-order optimality conditions established in Section 2A, if x is a solution to (6) and the constraint qualification condition 2A(13) holds, then there exists a multiplier vector $y = (y_1, \ldots, y_m)$ such that the pair (x, y) satisfies the Karush–Kuhn–Tucker conditions 2A(23), which are in the form of the variational inequality

$$(7) \qquad \begin{pmatrix} -v + \nabla g_0(x) + y \nabla g(x) \\ -u + g(x) \end{pmatrix} \in N_E(x, y) \text{ for } E = \mathbb{R}^n \times \mathbb{R}_+^m,$$

where we put

$$g(x) = \begin{pmatrix} g_1(x) \\ \vdots \\ g_m(x) \end{pmatrix} \quad \text{and} \quad y = \begin{pmatrix} y_1 \\ \vdots \\ y_m \end{pmatrix}.$$

More conveniently, for the mapping

$$(8) \qquad G : (x, y) \mapsto \begin{pmatrix} \nabla g_0(x) + y \nabla g(x) \\ -g(x) \end{pmatrix} + N_E(x, y),$$

the solution mapping of (7) is just G^{-1} (here, without changing anything, we take the negative of the first row since the normal cone to $\mathbb{R}^n$ is the zero mapping). We now focus on inverting the mapping G. Choose a reference value $(\bar{v}, \bar{u})$ of the parameters and let $(\bar{x}, \bar{y})$ solve (7) for $(\bar{v}, \bar{u})$, that is, $(\bar{v}, \bar{u}) \in G(\bar{x}, \bar{y})$.

To apply Kummer's theorem, we convert, as in the final part of the proof of 4F.7, the variational inequality (7) into an equation involving the function $H : \mathbb{R}^{n+m} \to \mathbb{R}^{n+m}$ defined as follows:

$$(9) \qquad H(x, y) = \begin{pmatrix} \nabla g_0(x) + \sum_{i=1}^m y_i^+ \nabla g_i(x) \\ -g_1(x) + y_1^- \\ \vdots \\ -g_m(x) + y_m^- \end{pmatrix}.$$

In Section 4F we showed in particular that strong metric regularity of the mapping in 4F(15) is equivalent to strong metric regularity of the associated mapping in 4F(20). The same argument works for the mappings in (8) and (9) where we now have non-linear functions; for completeness, we will repeat it here. If $((x, y), (v, u)) \in \text{gph } H$ then for $z_i = y_i^+$, $i = 1, \ldots, m$, we have that $((x, z), (v, u)) \in \text{gph } G$. Indeed, for each $i = 1, \ldots, m$, if $y_i \leq 0$, then $u_i + g_i(x) = y_i^- \leq 0$ and $(u_i + g_i(x)) y_i^+ = 0$; otherwise $u_i + g_i(x) = y_i^- = 0$. Conversely, if $((x, z), (v, u)) \in \text{gph } G$, then for

$$(10) \qquad y_i = \begin{cases} z_i & \text{if } z_i > 0, \\ u_i + g_i(x) & \text{if } z_i = 0, \end{cases}$$

we obtain $(x, y) \in H^{-1}(v, u)$. In particular, if H^{-1} has a Lipschitz continuous localization around $(\bar{v}, \bar{u})$ for $(\bar{x}, \bar{y})$ then G^{-1} has the same property at $(\bar{v}, \bar{u})$ for $(\bar{x}, \bar{z})$

where $\bar{z}_i = \bar{y}_i^+$, and if G^{-1} has a Lipschitz continuous localization around $(\bar{v}, \bar{u})$ for $(\bar{x}, \bar{z})$ then H^{-1} has the same property at $(\bar{v}, \bar{u})$ for $(\bar{x}, \bar{y})$, where $\bar{y}$ satisfies (10).

To invoke Kummer's theorem for the function H, we need to determine the strict graphical derivative of H. There is no trouble in differentiating the expressions $-g_i(x) + y_i^-$, inasmuch as we already know from 4G.2 the strict graphical derivative of y^-. A little bit more involved is the determination of the strict graphical derivative of $\varphi_i(x,y) := \nabla g_i(x) y_i^+$ for $i = 1, \ldots, m$. Adding and subtracting the same expressions, passing to the limit as in the definition, and using (2), we obtain

$$z \in D_* \varphi_i(\bar{x}, \bar{y})(u, v) \iff z = \bar{y}_i^+ \nabla^2 g_i(\bar{x}) u + \lambda_i v_i \nabla g_i(\bar{x}), \quad i = 1, \ldots, m,$$

where the coefficients λ_i for $i = 1, \ldots, m$ satisfy

(11)
$$\lambda_i \begin{cases} = 1 & \text{for } \bar{y}_i > 0, \\ \in [0,1] & \text{for } \bar{y}_i = 0, \\ = 0 & \text{for } \bar{y}_i < 0. \end{cases}$$

Taking into account 4G.4, the form thereby obtained for the strict graphical derivative of the function H in (9) at $(\bar{x}, \bar{y})$ is as follows:

$$(\xi, \eta) \in D_* H(\bar{x}, \bar{y})(u, v) \iff \begin{cases} \xi = Au + \sum_{i=1}^m \lambda_i v_i \nabla g_i(\bar{x}), \\ \eta_i = -\nabla g_i(\bar{x}) u + (1 - \lambda_i) v_i & \text{for } i = 1, \ldots, m, \end{cases}$$

where the λ_i's are as in (11), and

$$A = \nabla^2 g_0(\bar{x}) + \sum \bar{y}_i^+ \nabla^2 g_i(\bar{x}).$$

Denoting by Λ the $m \times m$ diagonal matrix with elements λ_i on the diagonal, by I_m the $m \times m$ identity matrix, and setting

$$B = \begin{pmatrix} \nabla g_1(\bar{x}) \\ \vdots \\ \nabla g_m(\bar{x}) \end{pmatrix},$$

we obtain that

(12) $$M(\Lambda) \in D_* H(\bar{x}, \bar{y}) \iff M(\Lambda) = \begin{pmatrix} A & B^T \Lambda \\ -B & I_m - \Lambda \end{pmatrix}.$$

This formula can be simplified by re-ordering the functions g_i according to the sign of $\bar{y}_i$. We first introduce some notation. Let $I = \{1, \ldots, m\}$ and, without loss of generality, suppose that

$$\{i \in I \mid \bar{y}_i > 0\} = \{1, \ldots, k\} \quad \text{and} \quad \{i \in I \mid \bar{y}_i = 0\} = \{k+1, \ldots, l\}.$$

Let

$$B_+ = \begin{pmatrix} \nabla g_1(\bar{x}) \\ \vdots \\ \nabla g_k(\bar{x}) \end{pmatrix} \text{ and } B_0 = \begin{pmatrix} \nabla g_{k+1}(\bar{x}) \\ \vdots \\ \nabla g_l(\bar{x}) \end{pmatrix},$$

let Λ_0 be the $(l-k) \times (l-k)$ diagonal matrix with diagonal elements $\lambda_i \in [0,1]$, let I_0 be the identity matrix for $\mathbb{R}^{l-k}$, and let I_{m-l} be the identity matrix for $\mathbb{R}^{m-l}$. Then, since $\lambda_i = 1$ for $i = 1, \ldots, k$ and $\lambda_i = 0$ for $i = l+1, \ldots, m$, the matrix $M(\Lambda)$ in (12) takes the form

$$M(\Lambda_0) = \begin{pmatrix} A & B_+^T & B_0^T \Lambda_0 & 0 \\ 0 & 0 & 0 & 0 \\ -B & 0 & I_0 - \Lambda_0 & 0 \\ 0 & 0 & 0 & I_{m-l} \end{pmatrix}.$$

Each column of $M(\Lambda_0)$ depends on at most one λ_i, hence there are numbers

$$a_{k+1}, b_{k+1}, \ldots, a_l, b_l$$

such that

$$\det M(\Lambda_0) = (a_{k+1} + \lambda_{k+1} b_{k+1}) \cdots (a_l + \lambda_l b_l).$$

Therefore, $\det M(\Lambda_0) \neq 0$ for all $\lambda_i \in [0,1]$, $i = k+1, \ldots, l$, if and only if the following condition holds:

$$a_i \neq 0, \ a_i + b_i \neq 0 \text{ and } [\text{sign}\, a_i = \text{sign}(a_i + b_i) \text{ or sign}\, b_i \neq \text{sign}(a_i + b_i)],$$
$$\text{for } i = k+1, \ldots, l.$$

Here we invoke the convention that

$$\text{sign}\, a = \begin{cases} 1 & \text{for } a > 0, \\ 0 & \text{for } a = 0, \\ -1 & \text{for } a < 0. \end{cases}$$

One can immediately note that it is not possible to have simultaneously $\text{sign}\, b_i \neq \text{sign}(a_i + b_i)$ and $\text{sign}\, a_i \neq \text{sign}(a_i + b_i)$ for some i. Therefore, it suffices to have

(13) $\quad a_i \neq 0, \ a_i + b_i \neq 0, \ \text{sign}\, a_i = \text{sign}(a_i + b_i) \text{ for all } i = k+1, \ldots, l.$

Now, let J be a subset of $\{k+1, \ldots, l\}$ and for $i = k+1, \ldots, l$, and let

$$\lambda_i^J = \begin{cases} 1 & \text{for } i \in J, \\ 0 & \text{otherwise.} \end{cases}$$

Let Λ^J be the diagonal matrix composed by these λ_i^J, and let $B_0(J) = \Lambda^J B_0$. Then we can write

$$M(J) := M(\Lambda^J) = \begin{pmatrix} A & B_+^T & B_0(J)^T & 0 \\ 0 & 0 & 0 & 0 \\ -B & 0 & I_0^J & 0 \\ 0 & 0 & 0 & I_l \end{pmatrix},$$

4 Regularity Properties Through Generalized Derivatives

where I_0^J is the diagonal matrix having 0 as $(i-k)$-th element if $i \in J$ and 1 otherwise. Clearly, all the matrices $M(J)$ are obtained from $M(\Lambda_0)$ by taking each λ_i either 0 or 1. The condition (13) can be then written equivalently as

(14) $\qquad \det M(J) \neq 0$ and $\operatorname{sign} \det M(J)$ is the same for all J.

Let the matrix $B(J)$ have as rows the row vectors $\nabla g_i(\bar{x})$ for $i \in J$. Reordering the last $m - k$ columns and rows of $M(J)$, if necessary, we obtain

$$M(J) = \begin{pmatrix} A & B_+^\mathsf{T} & B(J)^\mathsf{T} & 0 \\ -B & 0 & 0 & I \end{pmatrix},$$

where I is now the identity for $\mathbb{R}^{\{k+1,\ldots,m\}\setminus J}$. The particular form of the matrix $M(J)$ implies that $M(J)$ fulfills (14) if and only if (14) holds for just a part of it, namely for the matrix

(15) $$N(J) := \begin{pmatrix} H & B_+^\mathsf{T} & B(J)^\mathsf{T} \\ -B_+ & 0 & 0 \\ -B(J) & 0 & 0 \end{pmatrix}.$$

By applying Kummer's theorem, we arrive finally at the following result, which sharpens the first part of the statement of Theorem 2G.8.

Theorem 4G.5 (strong regularity characterization for KKT mappings). *The solution mapping of the Karush–Kuhn–Tucker variational inequality (7) has a Lipschitz continuous single-valued localization around $(\bar{v}, \bar{u})$ for $(\bar{x}, \bar{y})$ if and only if, for the matrix $N(J)$ in (15), $\det N(J)$ has the same nonzero sign for all $J \subset \{i \in I \,|\, \bar{y}_i = 0\}$.*

We should note that, in this specific case, the set of matrices $D_*H(\bar{x}, \bar{y})$ happens to be convex, and then $D_*H(\bar{x}, \bar{y})$ coincides with its generalized Jacobian $\operatorname{co} \bar{\nabla} H(\bar{x}, \bar{y})$.

4H. Results Utilizing Coderivatives

Graphical derivatives, defined in terms of tangent cones to graphs, have been the mainstay for most of the developments in this chapter so far, with the exception of the variants in 4G. However, an alternative approach can be made to many of the same issues in terms of graphical coderivatives, defined instead in terms of normal cones to graphs. This theory is readily available in other texts in variational analysis, so we will only lay out the principal ideas and facts here without going into their detailed development.

Normal cones $N_C(x)$ have already been prominent, of course, in our work with optimality conditions and variational inequalities, starting in Section 2A, but only in the case of convex sets C. To arrive at coderivatives for a mapping $F : \mathbb{R}^n \rightrightarrows \mathbb{R}^m$, we wish to make use of normal cones to gph F at points (x,y), but to keep the door open to significant applications we need to deal with graph sets that are not convex. The first task, therefore, is generalizing $N_C(x)$ to the case of nonconvex C.

General normal cones. *For a set $C \subset \mathbb{R}^n$ and a point $x \in C$ at which C is locally closed, a vector v is said to be a regular normal if $\langle v, x' - x \rangle \leq o(|x' - x|)$ for $x' \in C$. The set of all such vectors v is called the regular normal cone to C at x and is denoted by $\hat{N}_C(x)$. A vector v is said to be a general normal to C at x if there are sequences $\{x^k\}$ and $\{v^k\}$ with $x^k \in C$, such that*

$$x^k \to x \text{ and } v^k \to v \text{ with } v^k \in \hat{N}_C(x^k).$$

The set of all such vectors v is called the general normal cone to C at x and is denoted by $N_C(x)$. For $x \notin C$, $N_C(x)$ is taken to denote the empty set.

Very often, the limit process in the definition of the general normal cone $N_C(x)$ is superfluous: no additional vectors v are produced in that manner, and one merely has $N_C(x) = \hat{N}_C(x)$. This circumstance is termed the *Clarke regularity* of C at x. When C is convex, for instance, it is Clarke regular at every one of its points x, and the generalized normal cone $N_C(x)$ agrees with the normal cone defined earlier, in 2A. Anyway, $N_C(x)$ is always a closed cone.

Coderivatives of mappings. *For a mapping $F : \mathbb{R}^n \rightrightarrows \mathbb{R}^m$ and a pair $(x,y) \in \operatorname{gph} F$ at which $\operatorname{gph} F$ is locally closed, the coderivative of F at x for y is the mapping $D^*F(x|y) : \mathbb{R}^m \rightrightarrows \mathbb{R}^n$ defined by*

$$w \in D^*F(x|y)(z) \iff (w, -z) \in N_{\operatorname{gph} F}(x,y).$$

Obviously this is a "dual" sort of notion, but where does it fit in with classical differentiation? The answer can be seen by specializing to the case where F is single-valued, thus reducing to a function $f : \mathbb{R}^n \to \mathbb{R}^m$. Suppose f is strictly differentiable at x; then for $y = f(x)$, the graphical derivative $Df(x|y)$ is of course the linear mapping from $\mathbb{R}^n$ to $\mathbb{R}^m$ with matrix $\nabla f(x)$. In contrast, the coderivative $D^*f(x|y)$ comes out as the adjoint linear mapping from $\mathbb{R}^m$ to $\mathbb{R}^n$ with matrix $\nabla f(x)^\mathsf{T}$.

The most striking fact about coderivatives in our context is the following simple characterization of metric regularity.

Theorem 4H.1 (coderivative criterion for metric regularity). *For a mapping $F : \mathbb{R}^n \rightrightarrows \mathbb{R}^m$ and a pair $(\bar{x}, \bar{y}) \in \operatorname{gph} F$ at which $\operatorname{gph} F$ is locally closed, one has*

(1) $\operatorname{reg}(F; \bar{x}|\bar{y}) = |D^*F(\bar{x}|\bar{y})^{-1}|^+.$

Thus, F is metrically regular if and only if the right side of (1) is finite, which is equivalent to

(2) $D^*F(\bar{x}|\bar{y})(u) \ni 0 \implies u = 0.$

4 Regularity Properties Through Generalized Derivatives

If F is single-valued, that is, a function $f : \mathbb{R}^n \to \mathbb{R}^m$ which is strictly differentiable at $\bar{x}$, then the coderivative criterion means that the adjoint to the derivative mapping $Df(\bar{x})$ is injective, that is, $\ker \nabla f(x)^\mathsf{T} = \{0\}$. This is equivalent to surjectivity of $Df(\bar{x})$ which, as we know from 3F or 4B, is equivalent to metric regularity of f at $\bar{x}$ for $f(\bar{x})$.

We will now apply the coderivative criterion in 4H.1 to the variational inequality

$$(3) \qquad f(p,x) + N_C(x) \ni 0,$$

where $f : \mathbb{R}^d \times \mathbb{R}^n \to \mathbb{R}^n$ and C is a polyhedral convex set in $\mathbb{R}^n$. Let $\bar{x}$ be a solution of (2) for $\bar{p}$ and let f be continuously differentiable around $(\bar{p},\bar{x})$. From Theorem 4F.1 we know that a sufficient condition for the existence of a Lipschitz single-valued localization of the solution mapping S of (2) around $\bar{p}$ for $\bar{x}$ is the metric regularity at 0 for 0 of the reduced mapping

$$(4) \qquad A + N_K,$$

where the linear mapping A and the critical cone K are

$$(5) \qquad A = \nabla_x f(\bar{p},\bar{x}) \text{ and } K = K_C(\bar{x},\bar{v}) = \{w \in T_C(\bar{x}) \,|\, w \perp f(\bar{p},\bar{x})\}.$$

Based on earlier results in 2E we also noted there that in the case of ample parameterization, this sufficient condition is necessary as well. Further, in 4F.3 we obtained a characterization of metric regularity of (3) by means of the derivative criterion in 4B.1. We will now apply the coderivative criterion for metric regularity (2) to the mapping in (4); for that purpose we have to compute the coderivative of the mapping in (4). The first step to do that is easy and we will give it as an exercise.

Exercise 4H.2 (reduced coderivative formula). *Show that, for a linear mapping $A : \mathbb{R}^n \to \mathbb{R}^n$ and a closed convex cone $K \subset \mathbb{R}^n$ one has*

$$D^*(A + N_K)(\bar{x}\,|\,\bar{y}) = A^* + D^* N_K(\bar{x}\,|\,\bar{y} - A\bar{x}).$$

Guide. Apply the definition of the general normal cone above. □

Thus, everything hinges on determining the coderivative $D^* N_K(0\,|\,0)$ of the mapping N_K at the point $(0,0) \in G = \mathrm{gph}\, N_K$. By definition, the graph of the coderivative mapping consists of all pairs $(w, -z)$ such that $(w,z) \in N_G(0,0)$ where N_G is the general normal cone to the nonconvex set G. In these terms, for A and K as in (5), the coderivative criterion (2) becomes

$$(6) \qquad (u, A^\mathsf{T} u) \in N_G(0,0) \implies u = 0.$$

Everything depends then on determining $N_G(0,0)$.

We will next appeal to the known fact[5] that $N_G(0,0)$ is the limsup of polar cones $T_G(x,v)^*$ at $(x,v) \in G$ as $(x,v) \to (0,0)$. Because G is the union of finitely many

---
[5] See Proposition 6.5 in Rockafellar and Wets [1998].

polyhedral convex sets in $\mathbb{R}^{2n}$ (due to K being polyhedral), only finitely many cones can be manifested as $T_G(x,v)$ at points $(x,v) \in G$ near $(0,0)$. Thus, for a sufficiently small neighborhood O of the origin in $\mathbb{R}^{2n}$ we have that

$$(7) \qquad N_G(0,0) = \bigcup_{(x,v) \in O \cap G} T_G(x,v)^*.$$

It follows from reduction lemma 2E.4 that $T_G(x,v) = \text{gph } N_{K(x,v)}$, where $K(x,v) = \{x' \in T_K(x) \,|\, x' \perp v\}$. Therefore,

$$T_G(x,v) = \{(x',v') \,|\, x' \in K(x,v),\, v' \in K(x,v)^*, x' \perp v'\},$$

and we have

$$T_G(x,v)^* = \{(r,u) \,|\, \langle (r,u),(x',v') \rangle \leq 0 \text{ for all } (x',v') \in T_G(x,v)\}$$
$$= \{(r,u) \,|\, \langle r,x' \rangle + \langle u,v' \rangle \leq 0 \text{ for all}$$
$$x' \in K(x,v),\, v' \in K(x,v)^* \text{ with } x' \perp v'\}.$$

It is evident from this (first in considering $v' = 0$, then in considering $x' = 0$) that actually

$$(8) \qquad T_G(x,v)^* = K(x,v)^* \times K(x,v).$$

Hence $N_G(0,0)$ is the union of all product sets $\hat{K}^* \times \hat{K}$ associated with cones $\hat{K}$ such that $\hat{K} = K(x,v)$ for some $(x,v) \in G$ near enough to $(0,0)$.

It remains to observe that the form of the critical cones $\hat{K} = K(x,v)$ at points (x,v) close to $(0,0)$ is already derived in Lemma 4F.2, namely, for every choice of $(x,v) \in$ gph K near $(0,0)$ (this last requirement is actually not needed) the corresponding critical cone $\hat{K} = K(x,v)$ is given by

$$(9) \qquad \hat{K} = F_1 - F_2 \text{ for some faces } F_1, F_2 \in \mathscr{F}_K \text{ with } F_2 \subset F_1,$$

where $\mathscr{F}$ is the collection of all faces of K as defined in 4F. To see this all we need to do is to replace C by K and (x,v) by $(0,0)$ in the proof of 4F.2. Summarizing, from (6), (7), (8) and (9), and the coderivative criterion in 4H.1, we come to the following result:

Lemma 4H.3 (regularity modulus from coderivative criterion). *For the mapping in (4)(5) we have*

$$(10) \qquad \text{reg}\,(A + N_K; 0\,|\,0) = \max_{\substack{F_1, F_2 \in \mathscr{F}_K \\ F_1 \supset F_2}} \sup_{\substack{u \in F_1 - F_2 \\ |u|=1}} \frac{1}{d(A^\mathsf{T} u, (F_1 - F_2)^*)}.$$

Thus, $A + N_K$ is metrically regular at 0 for 0 if and only if for every choice of critical faces $F_1, F_2 \in \mathscr{F}_K$ with $F_2 \subset F_1$,

4 Regularity Properties Through Generalized Derivatives

$$u \in F_1 - F_2 \text{ and } A^\mathsf{T} u \in (F_1 - F_2)^* \implies u = 0.$$

Finally, relying on 4F.1, we arrive at the following characterization of the existence of Lipschitz single-valued localization of the solution mapping S of the variational inequality (3), which parallels the one in 4F.11:

Theorem 4H.4 (Lipschitz single-valued localization from a coderivative rule). *For the solution mapping S of (3) under the assumption that the convex set C is polyhedral, let $\bar{x} \in S(\bar{p})$, and suppose that f is continuously differentiable near $(\bar{p}, \bar{x})$. For the mapping in (4)(5), suppose that*

$$(11) \qquad \max_{\substack{F_1, F_2 \in \mathscr{F}_K \\ F_1 \supset F_2}} \sup_{\substack{u \in F_1 - F_2 \\ |u|=1}} \frac{1}{d(A^\mathsf{T} u, (F_1 - F_2)^*)} \leq \lambda < \infty.$$

Then the solution mapping S has a Lipschitz continuous single-valued localization s around $\bar{p}$ for $\bar{x}$ with

$$\mathrm{lip}\,(s; \bar{p}) \leq \lambda \, |\nabla_p f(\bar{p}, \bar{x})|.$$

Moreover, under the ample parameterization condition rank $\nabla_p f(\bar{p}, \bar{x}) = n$, *the condition in (11) is not just sufficient but also necessary for S to have a Lipschitz continuous single-valued localization around $\bar{p}$ for $\bar{x}$.*

Proof. We utilize 4F.1, taking into account that, because metric regularity of $A + N_K$ is equivalent to strong metric regularity, the Lipschitz modulus of $(A + N_K)^{-1}$ equals the regularity modulus of $A + N_K$. We then apply the formula in (10). □

Commentary

Graphical derivatives of set-valued mappings were introduced by Aubin [1981]; for more, see Aubin and Frankowska [1990]. The material in sections 4B and 4C is from Dontchev, Quincampoix and Zlateva [2006], where results of Aubin and Frankowska [1987, 1990] were used.

The statement 4B.5 of the Ekeland principle is from Ekeland [1990]. A detailed presentation of this principle along with various forms and extensions is given in Borwein and Zhu [2005]. The proof of the classical implicit function theorem 1A.1 given at the end of Section 4B is close, but not identical, to that in Ekeland [1990].

The derivative criterion for metric subregularity in 4C.1 was obtained by Rockafellar [1989], but the result itself was embedded in a proof of a statement requiring additional assumptions. The necessity without those assumptions was later noted in King and Rockafellar [1992] and in the case of sufficiency by Levy [1996]. The statement and the proof of 4C.1 are from Dontchev and Rockafellar [2004].

Sections 4E and 4F collect various results scattered in the literature. Theorem 4E.4 is from Dontchev and Rockafellar [2004] while the critical face lemma 4F.2 is a particular case of Lemma 3.5 in Robinson [1984]; see also Theorem 5.6 in Rockafellar [1989]. Theorem 4F.6 is a particular case of Theorem 3 in Dontchev and Rockafellar [1996] which in turn is based on a deeper result in Robinson [1992], see also Ralph [1993]. The presented proof uses a somewhat modified version of a reduction argument from the book Klatte and Kummer [2002], Section 7.5.

Clarke's inverse function theorem, 4G.1, was first published in Clarke [1976]; for more information regarding the generalized Jacobian see the book of Clarke [1983]. Theorem 4G.3 is from Kummer [1991]; see also Klatte and Kummer [2002] and Páles [1997]. It is interesting to note that a nonsmooth implicit function theorem which is a special case of both Clarke's theorem and Kummer's theorem, appeared as early as 1916 in a paper by Hedrick and Westfall [1916].

Theorem 4G.5 originates from Robinson [1980]; the proof given here uses some ideas from Kojima [1980] and Jongen et al. [1987].

The coderivative criterion in 4H.1 goes back to the early works of Ioffe [1981, 1984], Kruger [1982] and Mordukhovich [1984]. A broad review of the role of coderivatives in variational analysis is given in Mordukhovich [2006].

Chapter 5
Regularity in Infinite Dimensions

The theme of this chapter has origins in the early days of functional analysis and the Banach open mapping theorem, which concerns continuous linear mappings from one Banach space to another. The graphs of such mappings are subspaces of the product of the two Banach spaces, but remarkably much of the classical theory extends to set-valued mappings whose graphs are convex sets or cones instead of subspaces. Openness connects up then with metric regularity and interiority conditions on domains and ranges, as seen in the Robinson–Ursescu theorem. Infinite-dimensional inverse function theorems and implicit function theorems due to Lyusternik, Graves, and Bartle and Graves can be derived and extended. Banach spaces can even be replaced to some degree by more general metric spaces.

Before proceeding we review some notation and terminology. Already in the first section of Chapter 1 we stated the contraction mapping principle in metric spaces. Given a set X, a function $\rho : X \times X \to \mathbb{R}_+$ is said to be a *metric* in X when
  (i) $\rho(x,y) = 0$ if and only if $x = y$;
  (ii) $\rho(x,y) = \rho(y,x)$;
  (iii) $\rho(x,y) \leq \rho(x,z) + \rho(z,y)$ (triangle inequality).
A set X equipped with a metric ρ is called a *metric space* (X,ρ). In a metric space (X,ρ), a sequence $\{x_k\}$ is called a *Cauchy sequence* if for every $\varepsilon > 0$ there exists $n \in \mathbb{N}$ such that $\rho(x_k, x_j) < \varepsilon$ for all $k, j > n$. A metric space is *complete* if every Cauchy sequence converges to an element of the space. Any closed set in a Euclidean space is a complete metric space with the metric $\rho(x,y) = |x - y|$.

A *linear (vector) space* over the reals is a set X in which addition and scalar multiplication are defined obeying the standard algebraic laws of commutativity, associativity and distributivity. A linear space X with elements x is *normed* if it is furnished with a real-valued expression $\|x\|$, called the *norm* of x, having the properties
  (i) $\|x\| \geq 0$, $\|x\| = 0$ if and only if $x = 0$;
  (ii) $\|\alpha x\| = |\alpha| \, \|x\|$ for $\alpha \in \mathbb{R}$;
  (iii) $\|x + y\| \leq \|x\| + \|y\|$.
Any normed space is a metric space with the metric $\rho(x,y) = \|x - y\|$. A complete normed vector space is called a *Banach space*. On a finite-dimensional space, all

norms are equivalent, but when we refer specifically to $\mathbb{R}^n$ we ordinarily have in mind the Euclidean norm denoted by $|\cdot|$. Regardless of the particular norm being employed in a Banach space, the closed unit ball for that norm will be denoted by $\mathbb{B}$, and the distance from a point x to a set C will be denoted by $d(x,C)$, and so forth.

As in finite dimensions, a function A acting from a Banach space X into a Banach space Y is called a *linear mapping* if $\operatorname{dom} A = X$ and $A(\alpha x + \beta y) = \alpha A x + \beta A y$ for all $x, y \in X$ and all scalars α and β. The range of a linear mapping A from X to Y is always a subspace of Y, but it might not be a closed subspace, even if A is continuous. A linear mapping $A : X \to Y$ is *surjective* if $\operatorname{rge} A = Y$ and *injective* if $\ker A = \{0\}$.

Although in finite dimensions a linear mapping $A : X \to Y$ is automatically continuous, this fails in infinite dimensions; neither does surjectivity of A when $X = Y$ necessarily yield invertibility, in the sense that A^{-1} is single-valued. However, if A is continuous at any one point of X, then it is continuous at every point of X. That, moreover, is equivalent to A being *bounded*, in the sense that A carries bounded subsets of X into bounded subsets of Y, or what amounts to the same thing due to linearity, the image of the unit ball in X is included in some multiple of the unit ball in Y, i.e., the value

$$\|A\| = \sup_{\|x\| \leq 1} \|Ax\|$$

is finite. This expression defines the *operator norm* on the space $\mathscr{L}(X,Y)$, consisting of all continuous linear mappings $A : X \to Y$, which is then another Banach space.

Special and important in this respect is the Banach space $\mathscr{L}(X,\mathbb{R})$, consisting of all linear and continuous real-valued functions on X. It is the space *dual* to X, symbolized by X^*, and its elements are typically denoted by x^*; the value that an $x^* \in X^*$ assigns to an $x \in X$ is written as $\langle x^*, x \rangle$. The dual of the Banach space X^* is the bidual X^{**} of X; when every function $x^{**} \in X^{**}$ on X^* can be represented as $x^* \mapsto \langle x^*, x \rangle$ for some $x \in X$, the space X is called *reflexive*. This holds in particular when X is a *Hilbert* space with $\langle x, y \rangle$ as its *inner product*, and each $x^* \in X^*$ corresponds to a function $x \mapsto \langle x, y \rangle$ for some $y \in X$, so that X^* can be identified with X itself.

Another thing to be mentioned for a pair of Banach spaces X and Y and their duals X^* and Y^* is that any $A \in \mathscr{L}(X,Y)$ has an *adjoint* $A^* \in \mathscr{L}(Y^*, X^*)$ such that $\langle Ax, y^* \rangle = \langle x, A^* y^* \rangle$ for all $x \in X$ and $y^* \in Y^*$. Furthermore, $\|A^*\| = \|A\|$. A generalization of this to set-valued mappings having convex cones as their graphs will be seen later.

In fact most of the definitions, and even many of the results, in the preceding chapters will carry over with hardly any change, the major exception being results with proofs which truly depended on the compactness of $\mathbb{B}$. Our initial task, in Section 5A, will be to formulate various facts in this broader setting while coordinating them with classical theory. In the remainder of the chapter, we present inverse and implicit mapping theorems with metric regularity in abstract spaces. Parallel results for metric subregularity are not considered.

5A. Openness and Positively Homogeneous Mappings

At the end of Chapter 1 we introduced the concept of openness of a function and presented a Jacobian criterion for openness, but did not elaborate further. We now return to this property in the broader context of Banach spaces X and Y. A function $f : X \to Y$ is called *open* at $\bar{x}$ if $\bar{x} \in \text{int dom } f$ and, for every neighborhood U of $\bar{x}$ in X, the set $f(U)$ is a neighborhood of $f(\bar{x})$ in Y. This definition extends to set-valued mappings $F : X \rightrightarrows Y$; we say that F *is open at $\bar{x}$ for $\bar{y}$*, where $\bar{y} \in F(\bar{x})$, if

(1) $\qquad \bar{x} \in \text{int dom } F$ and for every neighborhood U of $\bar{x}$, the set $F(U) = \bigcup_{x \in U} F(x)$ is a neighborhood of $\bar{y}$.

We will also be concerned with another property, introduced for mappings $F : \mathbb{R}^n \rightrightarrows \mathbb{R}^m$ in 2D but likewise directly translatable to mappings $F : X \rightrightarrows Y$, namely that F is *metrically regular at $\bar{x}$ for $\bar{y}$*, where $\bar{y} \in F(\bar{x})$, if

(2) $\qquad$ there exists $\kappa > 0$ with neighborhoods U of $\bar{x}$ and V of $\bar{y}$ such that $d(x, F^{-1}(y)) \leq \kappa d(y, F(x))$ for all $(x, y) \in U \times V$.

As before, the infimum of all such κ associated with choices of U and V is denoted by $\text{reg}(F; \bar{x} | \bar{y})$ and called the modulus of metric regularity of F at $\bar{x}$ for $\bar{y}$.

The classical theorem about openness only addresses linear mappings. There are numerous versions of it available in the literature; we provide the following formulation:

Theorem 5A.1 (Banach open mapping theorem). *For any $A \in \mathscr{L}(X,Y)$ the following properties are equivalent:*
  (a) *A is surjective;*
  (b) *A is open (at every point);*
  (c) *$0 \in \text{int } A(\text{int }\mathbb{B})$;*
  (d) *there is a $\kappa > 0$ such that for all $y \in Y$ there exists $x \in X$ with $Ax = y$ and $\|x\| \leq \kappa \|y\|$.*

This theorem will be derived in Section 5B from a far more general result about set-valued mappings F than just linear mappings A. Our immediate interest lies in connecting it with the ideas in previous chapters, so as to shed light on where we have arrived and where we are going.

The first observation to make is that (d) of Theorem 5A.1 is the same as the existence of a $\kappa > 0$ such that $d(0, A^{-1}(y)) \leq \kappa \|y\|$ for all y. Clearly (d) does imply this, but the converse holds also by passing to a slightly higher κ if need be. But the linearity of A can also be brought in. For $x \in X$ and $y \in Y$ in general, we have $d(x, A^{-1}(y)) = d(0, A^{-1}(y) - x)$, and since $z \in A^{-1}(y) - x$ corresponds to $A(x+z) = y$, we have $d(0, A^{-1}(y) - x) = d(0, A^{-1}(y - Ax)) \leq \kappa \|y - Ax\|$. Thus, (d) of Theorem 5A.1 is actually equivalent to:

(3) $\qquad$ there exists $\kappa > 0$ such that $d(x, A^{-1}(y)) \leq \kappa d(y, Ax)$ for all $x \in X, y \in Y$.

Obviously this is the same as the metric regularity property in (2) as specialized to A, with the local character property becoming global through the arbitrary scaling made available because $A(\lambda x) = \lambda Ax$. In fact, due to linearity, metric regularity of A with respect to any pair $(\bar{x},\bar{y})$ in its graph is identical to metric regularity with respect to $(0,0)$, and the same modulus of metric regularity prevails everywhere. We can simply denote this modulus by reg A and use the formula that

$$\text{(4)} \qquad \text{reg } A = \sup_{\|y\| \leq 1} d(0, A^{-1}(y)) \quad \text{for } A \in \mathscr{L}(X,Y).$$

What we see then is that the condition

(e) A is metrically regular (everywhere): reg $A < \infty$

could be added to the equivalences in Theorem 5A.1 as a way of relating it to the broader picture we now have of the subject of openness.

Corollary 5A.2 (invertibility of linear mappings). *If a continuous linear mapping $A : X \to Y$ is both surjective and injective, then its inverse is a continuous linear mapping $A^{-1} : Y \to X$ with $\|A^{-1}\| = \text{reg } A$.*

Proof. When A is both surjective and injective, then A^{-1} is single-valued everywhere and linear. Observe that the right side of (4) reduces to $\|A^{-1}\|$. The finiteness of $\|A^{-1}\|$ corresponds to A^{-1} being bounded, hence continuous. □

It is worth noting also that if the range of $A \in \mathscr{L}(X,Y)$ is a *closed* subspace Y' of Y, then Y' is a Banach space in its own right, and the facts we have recorded can be applied to A as a surjective mapping from X to Y'.

Linear openness and the Aubin property. A result is available for set-valued mappings $F : X \rightrightarrows Y$ which has close parallels to the version of Theorem 5A.1 with (e) added, although it misses some aspects. This result corresponds in the case of $X = \mathbb{R}^n$ and $Y = \mathbb{R}^n$ to Theorems 3E.6 and 3E.8, where an equivalence was established between metric regularity, *linear* openness, and the inverse mapping having the Aubin property. The statements and proofs of the cited theorems carry over in the obvious manner to our present setting to yield a combined result, stated below as Theorem 5A.3. Linear openness of $F : X \to Y$ at $\bar{x}$ for $\bar{y}$, where $\bar{y} \in F(\bar{x})$, means

(5) there is a $\kappa > 0$ along with neighborhoods U of $\bar{x}$ and V of $\bar{y}$ such that
$F(x + \kappa r\,\text{int}\mathbb{B}) \supset [F(x) + r\,\text{int}\mathbb{B}] \cap V$ for all $x \in U, r > 0$,

whereas the Aubin property of F^{-1} at $\bar{y}$ for $\bar{x}$, where $\bar{x} \in F^{-1}(\bar{y})$, means

(6) there is a $\kappa > 0$ with neighborhoods U of $\bar{x}$ and V of $\bar{y}$ such that
$e(F^{-1}(y) \cap U, F^{-1}(y')) \leq \kappa \|y - y'\|$ for all $y, y' \in V$,

where the excess $e(C, D)$ is defined in Section 3A. The infimum of all κ in (6) over various choices of U and V is the modulus $\text{lip}(F^{-1}; \bar{y} | \bar{x})$.

5 Regularity in Infinite Dimensions

Theorem 5A.3 (metric regularity, linear openness and the inverse Aubin property). *For Banach spaces X and Y and a mapping $F : X \rightrightarrows Y$, the following properties with respect to a pair $(\bar{x}, \bar{y}) \in \mathrm{gph}\, F$ are equivalent:*
 (a) *F is linearly open at $\bar{x}$ for $\bar{y}$ with constant κ,*
 (b) *F is metrically regular at $\bar{x}$ for $\bar{y}$ with constant κ,*
 (c) *F^{-1} has the Aubin property at $\bar{y}$ for $\bar{x}$ with constant κ.*

Moreover $\mathrm{reg}(F; \bar{x} | \bar{y}) = \mathrm{lip}(F^{-1}; \bar{y} | \bar{x})$.

When F is taken to be a mapping $A \in \mathcal{L}(X,Y)$, how does the content of Theorem 5A.3 compare with that of Theorem 5A.1? With linearity, the openness in 5A.1(b) comes out the same as the linear openness in 5A.3(a) and is easily seen to reduce as well to the interiority condition in 5A.1(c). On the other hand, 5A.1(d) has already been shown to be equivalent to the subsequently added property (e), to which 5A.3(b) reduces when $F = A$. From 5A.3(c), though, we get yet another property which could be added to the equivalences in Theorem 5A.1 for $A \in \mathcal{L}(X,Y)$, specifically that

 (f) $A^{-1} : Y \rightrightarrows X$ *has the Aubin property at every $\bar{y} \in Y$ for every $\bar{x} \in A^{-1}(\bar{y})$*,

where $\mathrm{lip}(A^{-1}; \bar{y} | \bar{x}) = \mathrm{reg}\, A$ always. This goes farther than the observation in Corollary 5A.2, which covered only single-valued A^{-1}. In general, of course, the Aubin property in 5A.3(c) turns into local Lipschitz continuity when F^{-1} is single-valued.

An important feature of Theorem 5A.1, which is not represented at all in Theorem 5A.3, is the assertion that surjectivity is sufficient, as well as necessary, for all these properties to hold. An extension of that aspect to nonlinear F will be possible, in a local sense, under the restriction that $\mathrm{gph}\, F$ is closed and convex. This will emerge in the next section, in Theorem 5B.4.

Another result which we now wish to upgrade to infinite dimensions is the estimation for perturbed inversion which appeared in matrix form in Corollary 1E.7 with elaborations in 1E.8. It lies at the heart of the theory of implicit functions and will eventually be generalized in more than one way. We provide it here with a direct proof (compare with 1E.8(b)).

Lemma 5A.4 (estimation for perturbed inversion). *Let $A \in \mathcal{L}(X,Y)$ be invertible. Then for any $B \in \mathcal{L}(X,Y)$ with $\|A^{-1}\| \cdot \|B\| < 1$ one has*

$$(7) \qquad \|(A+B)^{-1}\| \leq \frac{\|A^{-1}\|}{1 - \|A^{-1}\| \|B\|}.$$

Proof. Let $C = BA^{-1}$; then $\|C\| < 1$ and hence $\|C^n\| \leq \|C\|^n \to 0$ as $n \to \infty$. Also, the elements

$$S_n = \sum_{i=0}^{n} C^i \quad \text{for } n = 0, 1, \ldots$$

form a Cauchy sequence in the Banach space $\mathcal{L}(X,Y)$ which therefore converges to some $S \in \mathcal{L}(X,Y)$. Observe that, for each n,

$$S_n(I - C) = I - C^{n+1} = (I - C)S_n,$$

and hence, through passing to the limit, one has $S = (I - C)^{-1}$. On the other hand

$$\|S_n\| \leq \sum_{i=0}^{n} \|C^i\| \leq \sum_{i=0}^{\infty} \|C\|^i = \frac{1}{1 - \|C\|}.$$

Thus, we obtain

$$\|(I - C)^{-1}\| \leq \frac{1}{1 - \|C\|}.$$

All that remains is to bring in the identity $(I - C)A = A - B$ and the inequality $\|C\| \leq \|A^{-1}\|\|B\|$, and to observe that the sign of B does not matter. □

Note that, with the conventions $\infty \cdot 0 = 0$, $1/0 = \infty$ and $1/\infty = 0$, Lemma 5A.4 also covers the cases $\|A^{-1}\| = \infty$ and $\|A^{-1}\| \cdot \|B\| = 1$.

Exercise 5A.5. *Derive Lemma 5A.4 from the contraction mapping principle 1A.2.*

Guide. Setting $a = \|A^{-1}\|$, choose $B \in \mathscr{L}(X,Y)$ with $\|B\| < \|A^{-1}\|^{-1}$ and $y \in Y$ with $\|y\| \leq 1 - a\|B\|$. Show that the mapping $\Phi : x \mapsto A^{-1}(y - Bx)$ satisfies the conditions in 1A.2 with $\lambda = a\|B\|$ and hence, there is a unique $x \in a\mathbb{B}$ such that $x = A^{-1}(y - Bx)$, that is $(A + B)x = y$. Thus, $A + B$ is invertible. Moreover $\|x\| = \|(A + B)^{-1}(y)\| \leq a$ for every $y \in (1 - a\|B\|)\mathbb{B}$, which implies that

$$\|(A + B)^{-1}z\| \leq \frac{\|A^{-1}\|}{1 - \|A^{-1}\|\|B\|} \quad \text{for every } z \in \mathbb{B}.$$

This yields (7). □

Exercise 5A.6. *Let $C \in \mathscr{L}(X,Y)$ satisfy $\|C\| < 1$. Prove that*

$$\|(I - C)^{-1} - I - C\| \leq \frac{\|C\|^2}{1 - \|C\|}.$$

Guide. Use the sequence of mappings S_n in the proof of 5A.4 and observe that

$$\|S_n - I - C\| = \|C^2 + C^3 + \cdots + C^n\| \leq \frac{\|C\|^2}{1 - \|C\|}.$$

□

Positive homogeneity. A mapping $H : X \rightrightarrows Y$ whose graph is a cone in $X \times Y$ is called *positively homogeneous*. In infinite dimensions such mappings have properties similar to those developed in finite dimensions in Section 4A, but with some complications. Outer and inner norms are defined for such mappings H as in Section 4A, but it is necessary to take into account the possible variety of underlying norms on X and Y in place of just the Euclidean norm earlier:

(8) $$\|H\|^+ = \sup_{\|x\| \leq 1} \sup_{y \in H(x)} \|y\|, \quad \|H\|^- = \sup_{\|x\| \leq 1} \inf_{y \in H(x)} \|y\|.$$

5 Regularity in Infinite Dimensions

When dom $H = X$ and H is single-valued, these two expressions agree. For $H = A \in \mathscr{L}(X,Y)$, they reduce to $\|A\|$.

The inverse H^{-1} of a positively homogeneous mapping H is another positively homogeneous mapping, and its outer and inner norms are therefore available also. The elementary relationships in Propositions 4A.5 and 4A.6 have the following update.

Proposition 5A.7 (outer and inner norms). *The inner norm of a positively homogeneous mapping $H : X \rightrightarrows Y$ satisfies*

$$\|H\|^- = \inf\{\,\kappa \in (0,\infty) \,|\, H(x) \cap \kappa \mathbb{B} \neq \emptyset \text{ for all } x \in \mathbb{B}\,\},$$

so that, in particular,

(9) $$\|H\|^- < \infty \implies \operatorname{dom} H = X.$$

In parallel, the outer norm satisfies

$$\|H\|^+ = \inf\{\,\kappa \in (0,\infty) \,|\, H(\mathbb{B}) \subset \kappa \mathbb{B}\,\} = \sup_{\|y\|=1} \frac{1}{d(0, H^{-1}(y))},$$

and we have

(10) $$\|H\|^+ < \infty \implies H(0) = \{0\},$$

with this implication becoming an equivalence when H has closed graph and $\dim X < \infty$.

The equivalence generally fails in (10) when $\dim X = \infty$ because of the lack of compactness then (with respect to the norm) of the ball $\mathbb{B}$ in X.

An extension of Lemma 5A.4 to possibly set-valued mappings that are positively homogeneous is now possible in terms of the outer norm. Recall that for a positively homogeneous $H : X \rightrightarrows Y$ and a linear $B : X \to Y$ we have $(H + B)(x) = H(x) + Bx$ for every $x \in X$.

Theorem 5A.8 (inversion estimate for the outer norm). *Let $H : X \rightrightarrows Y$ be positively homogeneous with $\|H^{-1}\|^+ < \infty$. Then for any $B \in \mathscr{L}(X,Y)$ with the property that $\|H^{-1}\|^+ \cdot \|B\| < 1$, one has*

(11) $$\|(H+B)^{-1}\|^+ \leq \frac{\|H^{-1}\|^+}{1 - \|H^{-1}\|^+ \|B\|}.$$

Proof. Having $\|H^{-1}\|^+ = 0$ is equivalent to having $\operatorname{dom} H = \{0\}$; in this case, since $\emptyset + y = \emptyset$ for any y, we get $\|(H+B)^{-1}\|^+ = 0$ for any $B \in \mathscr{L}(X,Y)$ as claimed. Suppose therefore instead that $0 < \|H^{-1}\|^+ < \infty$. If the estimate (11) is false, there is some $B \in \mathscr{L}(X,Y)$ with $\|B\| < [\|H^{-1}\|^+]^{-1}$ such that $\|(H+B)^{-1}\|^+ > ([\|H^{-1}\|^+]^{-1} - \|B\|)^{-1}$. In particular $B \neq 0$ then, and by definition there must exist

$y \in B$ and $x \in (H+B)^{-1}(y)$ such that $\|x\| > ([\|H^{-1}\|^+]^{-1} - \|B\|)^{-1}$, which is the same as

(12) $$\frac{1}{\|x\|^{-1} + \|B\|} > \|H^{-1}\|^+.$$

But then $y - Bx \in H(x)$ and

(13) $$\|y - Bx\| \leq \|y\| + \|B\|\|x\| \leq 1 + \|B\|\|x\|.$$

If $y = Bx$ then $0 \in H(x)$, so (10) yields $x = 0$, a contradiction. Hence $\alpha := \|y - Bx\|^{-1} > 0$, and due to the positive homogeneity of H we have $(\alpha x, \alpha(y - Bx)) \in$ gph H and $\alpha\|y - Bx\| = 1$, which implies, by definition,

$$\|H^{-1}\|^+ \geq \frac{\|x\|}{\|y - Bx\|}.$$

Combining this inequality with (12) and (13), we get

$$\|H^{-1}\|^+ \geq \frac{\|x\|}{\|y - Bx\|} \geq \frac{\|x\|}{1 + \|B\|\|x\|} = \frac{1}{\|x\|^{-1} + \|B\|} > \|H^{-1}\|^+.$$

This is impossible, and the proof is at its end. □

A corresponding extension of Lemma 5A.4 in terms of the inner norm will be possible later, in Section 5C.

Normal cones and polarity. For a closed, convex cone $K \subset X$, the *polar* of K is the subset K^* of the dual space X^* defined by

$$K^* = \{x^* \in X^* \mid \langle x, x^* \rangle \leq 0 \text{ for all } x \in K\}.$$

It is a closed convex cone in X^* from which K can be recovered as the polar $(K^*)^*$ of K^* in the sense that

$$K = \{x \in X \mid \langle x, x^* \rangle \leq 0 \text{ for all } x^* \in K^*\}.$$

For any set C in a Banach space X and any point $x \in C$, the *tangent cone* $T_C(x)$ at a point $x \in C$ is defined as in 2A to consist of all limits v of sequences $(1/\tau^k)(x^k - x)$ with $x^k \to x$ in C and $\tau^k \searrow 0$. When C is convex, $T_C(x)$ has an equivalent description as the closure of the convex cone consisting of all vectors $\lambda(x' - x)$ with $x' \in C$ and $\lambda > 0$.

In infinite dimensions, the *normal cone* $N_C(x)$ to C at x can be introduced in various ways that extend the general definition given for finite dimensions in 4H, but we will only be concerned with the case of convex sets C. For that case, the special definition in 2A suffices with only minor changes caused by the need to work with the dual space X^* and the pairing $\langle x, x^* \rangle$ between X and X^*. Namely, $N_C(x)$ consists of all $x^* \in X^*$ such that

5 Regularity in Infinite Dimensions

$$\langle x' - x, x^* \rangle \leq 0 \text{ for all } x' \in C.$$

Equivalently, through the alternative description of $T_C(x)$ for convex C, the normal cone $N_C(x)$ is the polar $T_C(x)^*$ of the tangent cone $T_C(x)$. It follows that $T_C(x)$ is in turn the polar cone $N_C(x)^*$.

As earlier, $N_C(x)$ is taken to be the empty set when $x \notin C$ so as to get a set-valued mapping N_C defined for all x, but this *normal cone mapping* now goes from X to X^* instead of from the underlying space into itself (except in the case of a Hilbert space, where X^* can be identified with X as recalled above). A generalized equation of the form

$$f(x) + N_C(x) \ni 0 \text{ for a function } f : X \to X^*$$

is again a *variational inequality*. Such generalized equations are central, for instance, to many applications involving differential or integral operators, especially in a Hilbert space framework.

Exercise 5A.9 (normals to cones). *Show that for a closed convex cone $K \subset X$ and its polar $K^* \subset X^*$, one has*

$$x^* \in N_K(x) \iff x \in K, \ x^* \in K, \ \langle x, x^* \rangle = 0.$$

Exercise 5A.10 (linear variational inequalities on cones). *Let $H(x) = Ax + N_K(x)$ for $A \in \mathcal{L}(X, X^*)$ and a closed, convex cone $K \subset X$. Show that H is positively homogeneous with closed graph, but this graph is not convex unless K is a subspace of X.*

5B. Mappings with Closed Convex Graphs

For any mapping $F : X \rightrightarrows Y$ with convex graph, the sets dom F and rge F, as the projections of gph F on the Banach spaces X and Y, are convex sets as well. When gph F is closed, these sets can fail to be closed (a famous example being the case where F is a "closed" linear mapping from X to Y with domain dense in X). However, if either of them has nonempty interior, or even nonempty "core" (in the sense about to be explained), there are highly significant consequences for the behavior of F. This section is dedicated to developing such consequences for properties like openness and metric regularity, but we begin with some facts that are more basic.

The *core* of a set $C \subset X$ is defined by

$$\text{core } C = \{x \mid \forall w \in X \ \exists \varepsilon > 0 \text{ such that } x + tw \in C \text{ when } 0 \leq t \leq \varepsilon\}.$$

A set C is called *absorbing* if $0 \in \text{core } C$. Obviously core $C \supset \text{int } C$ always, but there are circumstances where necessarily core $C = \text{int } C$. It is elementary that this holds when C is convex with $\text{int } C \neq \emptyset$, but more attractive is the potential of using the purely algebraic test of whether a point x belongs to core C to confirm that $x \in \text{int } C$ without first having to establish that $\text{int } C \neq \emptyset$. Most importantly for our purposes here,

(1) *for a closed convex subset C of a Banach space,* $\text{core } C = \text{int } C$.

An equivalent statement, corresponding to how this fact is often recorded in functional analysis, is that if C is a closed convex set which is absorbing, then C must be a neighborhood of 0. Through the observation already made about convexity, the confirmation of this comes down to establishing that C has nonempty interior. That can be deduced from the Baire category theorem, according to which the union of a sequence of nowhere dense subsets of a complete metric space cannot cover the whole space. If $\text{int } C$ were empty, the closed sets nC for $n = 1, 2, \ldots$ would be nowhere dense with the entire Banach space as their union, but this is impossible.

We demonstrate now that, for some of the convex sets central to the study of closed convex graphs on which we are embarking, the core and interior coincide even without closedness.

Theorem 5B.1 (interiority criteria for domains and ranges). *For any mapping $F : X \rightrightarrows Y$ with closed convex graph, one has*

(2) $\qquad \text{core rge } F = \text{int rge } F, \qquad \text{core dom } F = \text{int dom } F.$

In addition, $\text{core cl rge } F = \text{int cl rge } F$ *and* $\text{core cl dom } F = \text{int cl dom } F$, *where moreover*

(3) $\qquad \begin{array}{l} \text{int cl rge } F = \text{int rge } F \text{ when } \text{dom } F \text{ is bounded,} \\ \text{int cl dom } F = \text{int dom } F \text{ when } \text{rge } F \text{ is bounded.} \end{array}$

In particular, if $\text{dom } F$ is bounded and $\text{rge } F$ is dense in Y, and then $\text{rge } F = Y$. Likewise, if $\text{rge } F$ is bounded and $\text{dom } F$ is dense in X, and then $\text{dom } F = X$.

Proof. The equations in the line following (2) merely apply (1) to the closed convex sets $\text{cl rge } F$ and $\text{cl dom } F$. Through symmetry between F and F^{-1}, the two claims in (2) are equivalent to each other, as are the two claims in (3). Also, the claims after (3) are immediate from (3). Therefore, we only have to prove one of the claims in (2) and one of the claims in (3).

We start with the first claim in (3). Assuming that $\text{dom } F$ is bounded, we work toward verifying that $\text{int rge } F \supset \text{int cl rge } F$; this gives equality, inasmuch as the opposite inclusion is obvious. In fact, for this we only need to show that $\text{rge } F \supset \text{int cl rge } F$.

We choose $\tilde{y} \in \text{int cl rge } F$; then there exists $\delta > 0$ such that $\text{int} I\!\!B_{2\delta}(\tilde{y}) \subset \text{int cl rge } F$. We will find a point $\tilde{x}$ such that $(\tilde{x}, \tilde{y}) \in \text{gph } F$, so that $\tilde{y} \in \text{rge } F$. The point $\tilde{x}$ will be obtained by means of a sequence $\{(x^k, y^k)\}$ which we now construct by induction.

5 Regularity in Infinite Dimensions

Pick any $(x^0, y^0) \in \text{gph}\, F$. Suppose we have already determined $\{(x^j, y^j)\} \in \text{gph}\, F$ for $j = 0, 1, \ldots, k$. If $y^k = \tilde{y}$, then take $\tilde{x} = x^k$ and $(x^n, y^n) = (\tilde{x}, \tilde{y})$ for all $n = k, k+1, \ldots$; that is, after the index k the sequence is constant. Otherwise, with $\alpha^k = \delta / \|y^k - \tilde{y}\|$, we let $w^k := \tilde{y} + \alpha^k(\tilde{y} - y^k)$. Then $w^k \in \mathbb{B}_\delta(\tilde{y}) \subset \text{cl}\, \text{rge}\, F$. Hence there exists $v^k \in \text{rge}\, F$ such that $\|v^k - w^k\| \leq \|y^k - \tilde{y}\|/2$ and also u^k with $(u^k, v^k) \in \text{gph}\, F$. Having gotten this far, we pick

$$(x^{k+1}, y^{k+1}) = \frac{\alpha^k}{1+\alpha^k}(x^k, y^k) + \frac{1}{1+\alpha^k}(u^k, v^k).$$

Clearly, $(x^{k+1}, y^{k+1}) \in \text{gph}\, F$ by its convexity. Also, the sequence $\{y^k\}$ satisfies

$$\|y^{k+1} - \tilde{y}\| = \frac{\|v^k - w^k\|}{1+\alpha^k} \leq \frac{1}{2}\|y^k - \tilde{y}\|.$$

If $y^{k+1} = \tilde{y}$, we take $\tilde{x} = x^{k+1}$ and $(x^n, y^n) = (\tilde{x}, \tilde{y})$ for all $n = k+1, k+2, \ldots$. If not, we perform the induction step again. As a result, we generate an infinite sequence $\{(x^k, y^k)\}$, each element of which is equal to $(\tilde{x}, \tilde{y})$ after some k or has $y^k \neq \tilde{y}$ for all k and also

(4) $$\|y^k - \tilde{y}\| \leq \frac{1}{2^k}\|y^0 - \tilde{y}\| \quad \text{for all } k = 1, 2, \ldots.$$

In the latter case, we have $y^k \to \tilde{y}$. Further, for the associated sequence $\{x^k\}$ we obtain

$$\|x^{k+1} - x^k\| = \frac{\|x^k - u^k\|}{1+\alpha^k} \leq \frac{\|x^k\| + \|u^k\|}{\|y^k - \tilde{y}\| + \delta}\|y^k - \tilde{y}\|.$$

Both x^k and u^k are from $\text{dom}\, F$ and thus are bounded. Therefore, from (4), $\{x^k\}$ is a Cauchy sequence, hence (because X is a complete metric space) convergent to some $\tilde{x}$. Because $\text{gph}\, F$ is closed, we end up with $(\tilde{x}, \tilde{y}) \in \text{gph}\, F$, as required.

Next we address the second claim in (2), where the inclusion $\text{core}\, \text{dom}\, F \subset \text{int}\, \text{dom}\, F$ suffices for establishing equality. We must show that an arbitrarily chosen point of $\text{core}\, \text{dom}\, F$ belongs to $\text{int}\, \text{dom}\, F$, but through a translation of $\text{gph}\, F$ we can focus without loss of generality on that point in $\text{core}\, \text{dom}\, F$ being 0, with $F(0) \ni 0$. Let $F_0 : X \rightrightarrows Y$ be defined by $F_0(x) = F(x) \cap \mathbb{B}$. The graph of F_0, being $[X \times \mathbb{B}] \cap \text{gph}\, F$, is closed and convex, and we have $\text{dom}\, F_0 \subset \text{dom}\, F$ and $\text{rge}\, F_0 \subset \mathbb{B}$ (bounded). The relations already established in (3) tell us that $\text{int}\, \text{cl}\, \text{dom}\, F_0 = \text{int}\, \text{dom}\, F_0$, where $\text{cl}\, \text{dom}\, F_0$ is a closed convex set. By demonstrating that $\text{cl}\, \text{dom}\, F_0$ is absorbing, we will be able to conclude from (1) that $0 \in \text{int}\, \text{dom}\, F_0$, hence $0 \in \text{int}\, \text{dom}\, F$. It is enough actually to show that $\text{dom}\, F_0$ itself is absorbing.

Consider any $x \in X$. We have to show the existence of $\varepsilon > 0$ such that $tx \in \text{dom}\, F_0$ for $t \in [0, \varepsilon]$. We do know, because $\text{dom}\, F$ is absorbing, that $tx \in \text{dom}\, F$ for all $t > 0$ sufficiently small. Fix t_0 as such a t, and letting $y_0 \in F(t_0 x)$; let $y = y_0/t_0$, so that $t_0(x, y) \in \text{gph}\, F$. The pair $t(x, y) = (tx, ty)$ belongs then to $\text{gph}\, F$ for all $t \in [0, t_0]$ through the convexity of $\text{gph}\, F$ and our arrangement that $(0, 0) \in \text{gph}\, F$. Take $\varepsilon > 0$

small enough that $\varepsilon\|y\| \leq 1$. Then for $t \in [0,\varepsilon]$ we have $\|ty\| = t\|y\| \leq 1$, giving us $ty \in F(tx) \cap I\!B$ and therefore $tx \in \text{dom } F_0$, as required. □

Regularity properties will now be explored. The property of a mapping $F : X \rightrightarrows Y$ being open at $\bar{x}$ for $\bar{y}$, as extended to Banach spaces X and Y in 5A(1), can be restated equivalently in a manner that more closely resembles the linear openness property defined in 5A(5):

(5) for any $a > 0$ there exists $b > 0$ such that $F(\bar{x} + a\,\text{int}I\!B) \supset \bar{y} + b\,\text{int}I\!B$.

Linear openness requires a linear scaling relationship between a and b. Under positive homogeneity, such scaling is automatic. On the other hand, an intermediate type of property holds automatically without positive homogeneity when the graph of F is convex, and it will be a stepping stone toward other, stronger, consequences of convexity.

Proposition 5B.2 (openness of mappings with convex graph). *Consider a mapping $F : X \rightrightarrows Y$ with convex graph, and let $\bar{y} \in F(\bar{x})$. Then openness of F at $\bar{x}$ for $\bar{y}$ is equivalent to the simpler condition that*

(6) *there exists $c > 0$ with $F(\bar{x} + \text{int}I\!B) \supset \bar{y} + c\,\text{int}I\!B$.*

Proof. Clearly, (5) implies (6). For the converse, assume (6) and consider any $a > 0$. Take $b = \min\{1,a\}c$. If $a \geq 1$, the left side of (6) is contained in the left side of (5), and hence (5) holds. Suppose therefore that $a < 1$. Let $w \in \bar{y} + b\,\text{int}I\!B$. The point $v = (w/a) - (1-a)(\bar{y}/a)$ satisfies $\|v - \bar{y}\| = \|w - \bar{y}\|/a < b/a = c$, hence $v \in \bar{y} + c\,\text{int}I\!B$. Then from (6) there exists $u \in \bar{x} + \text{int}I\!B$ with $(u,v) \in \text{gph } F$. The convexity of gph F implies $a(u,v) + (1-a)(\bar{x},\bar{y}) \in \text{gph } F$ and yields $av + (1-a)\bar{y} \in F(au + (1-a)\bar{x}) \subset F(\bar{x} + a\,\text{int}I\!B)$. Substituting $v = (w/a) - (1-a)(\bar{y}/a)$ in this inclusion, we see that $w \in F(\bar{x} + a\,\text{int}I\!B)$, and since w was an arbitrary point in $\bar{y} + b\,\text{int}I\!B$, we get (5). □

The following fact bridges, for set-valued mappings with convex graphs, between condition (6) and metric regularity.

Lemma 5B.3 (metric regularity estimate). *Let $F : X \rightrightarrows Y$ have convex graph containing $(\bar{x},\bar{y})$, and suppose (6) is fulfilled. Then*

(7) $d(x, F^{-1}(y)) \leq \dfrac{1 + \|x - \bar{x}\|}{c - \|y - \bar{y}\|} d(y, F(x))$ *for all $x \in X$, $y \in \bar{y} + c\,\text{int}I\!B$.*

Proof. We may assume that $(\bar{x},\bar{y}) = (0,0)$, since this can be arranged by translating gph F to gph $F - (\bar{x},\bar{y})$. Then condition (6) has the simpler form

(8) there exists $c > 0$ with $F(\text{int}I\!B) \supset c\,\text{int}I\!B$.

Let $x \in X$ and $y \in c\,\text{int}I\!B$. Observe that (7) is automatically true when $x \notin \text{dom } F$ or $y \in F(x)$, so assume that $x \in \text{dom } F$ but $y \notin F(x)$. Let $\alpha := c - \|y\|$. Then $\alpha > 0$. Choose $\varepsilon \in (0,\alpha)$ and find $y' \in F(x)$ such that $\|y' - y\| \leq d(y, F(x)) + \varepsilon$. The point

5 Regularity in Infinite Dimensions   263

$\tilde{y} := y + (\alpha - \varepsilon)\|y' - y\|^{-1}(y - y')$ satisfies $\|\tilde{y}\| \leq \|y\| + \alpha - \varepsilon = c - \varepsilon < c$, hence $\tilde{y} \in c\,\text{int}B$. By (8) there exists $\tilde{x} \in \text{int}B$ with $\tilde{y} \in F(\tilde{x})$. Let $\beta := \|y - y'\|(\alpha - \varepsilon + \|y - y'\|)^{-1}$; then $\beta \in (0, 1)$. From the convexity of gph F we have

$$y = (1 - \beta)y' + \beta\tilde{y} \in (1 - \beta)F(x) + \beta F(\tilde{x}) \subset F((1 - \beta)x + \beta\tilde{x}).$$

Thus $x + \beta(\tilde{x} - x) \in F^{-1}(y)$, so $d(x, F^{-1}(y)) \leq \beta\|x - \tilde{x}\|$. Noting that $\|x - \tilde{x}\| \leq \|x\| + \|\tilde{x}\| < \|x\| + 1$ and $\beta \leq (\alpha - \varepsilon)^{-1}\|y - y'\|$, we obtain

$$d(x, F^{-1}(y)) < \frac{1 + \|x\|}{\alpha - \varepsilon}[d(y, F(x)) + \varepsilon].$$

Letting $\varepsilon \to 0$, we finish the proof. □

Condition (6) entails in particular having $\bar{y} \in \text{int rge } F$. It turns out that when the graph of F is not only convex but also closed, the converse implication holds as well, that is, $\bar{y} \in \text{int rge } F$ is equivalent to (6). This is a consequence of the following theorem, which furnishes a far-reaching generalization of the Banach open mapping theorem.

Theorem 5B.4 (Robinson–Ursescu). *Let $F : X \rightrightarrows Y$ have closed convex graph and let $\bar{y} \in F(\bar{x})$. Then the following are equivalent:*
(a) $\bar{y} \in \text{int rge } F$,
(b) F *is open at $\bar{x}$ for $\bar{y}$,*
(c) F *is metrically regular at $\bar{x}$ for $\bar{y}$.*

Proof. We first demonstrate that

(9) $\bar{y} \in \text{int } F(\bar{x} + \text{int}B)$ when $\bar{x} \in F^{-1}(\bar{y})$ and $\bar{y} \in \text{int rge } F$.

By a translation, we can reduce to the case of $(\bar{x}, \bar{y}) = (0, 0)$. To conclude (9) in this setting, where $F(0) \ni 0$ and $0 \in \text{int rge } F$, it will be enough to show, for an arbitrary $\delta \in (0, 1)$, that $0 \in \text{int } F(\delta B)$. Define the mapping $F_\delta : X \rightrightarrows Y$ by $F_\delta(x) = F(x)$ when $x \in \delta B$ but $F_\delta(x) = \emptyset$ otherwise. Then F_δ has closed convex graph given by $[\delta B \times Y] \cap \text{gph } F$. Also $F(\delta B) = \text{rge } F_\delta$ and $\text{dom } F_\delta \subset \delta B$. We want to show that $0 \in \text{int rge } F_\delta$, but have Theorem 5B.1 at our disposal, according to which we only need to show that $\text{rge } F_\delta$ is absorbing. For that purpose we use an argument which closely parallels one already presented in the proof of Theorem 5B.1. Consider any $y \in Y$. Because $0 \in \text{int rge } F$, there exists t_0 such that $ty \in \text{rge } F$ when $t \in [0, t_0]$. Then there exists x_0 such that $t_0 y \in F(x_0)$. Let $x = x_0/t_0$, so that $(t_0 x, t_0 y) \in \text{gph } F$. Since gph F is convex and contains $(0, 0)$, it also then contains (tx, ty) for all $t \in [0, t_0]$. Taking $\varepsilon > 0$ for which $\varepsilon\|x\| \leq \delta$, we get for all $t \in [0, \varepsilon]$ that $(tx, ty) \in \text{gph } F_\delta$, hence $ty \in \text{rge } F_\delta$, as desired.

Utilizing (9), we can put the argument for the equivalences in Theorem 5B.4 together. That (b) implies (a) is obvious. We work next on getting from (a) to (c). When (a) holds, we have from (9) that (6) holds for some c, in which case Lemma 5B.3 provides (7). By restricting x and y to small neighborhoods of $\bar{x}$ and $\bar{y}$ in (7), we deduce the metric regularity of F at $\bar{x}$ for $\bar{y}$ with any constant $\kappa > 1/c$. Thus, (c) holds. Finally, out of (c) and the equivalences in Theorem 5A.3 we may conclude that F is linearly open at $\bar{x}$ for $\bar{y}$, and this gets us back to (b). □

The preceding argument passed through linear openness as a fourth property which could be added to the equivalences in Theorem 5B.4, but which was left out of the theorem's statement for historical reasons. We now record this fact separately.

Theorem 5B.5 (linear openness from openness and convexity). *For a mapping $F : X \rightrightarrows Y$ with closed convex graph, openness at $\bar{x}$ for $\bar{y}$ always entails linear openness at $\bar{x}$ for $\bar{y}$.*

Another fact, going beyond the original versions of Theorem 5B.4, has come up as well.

Theorem 5B.6 (core criterion for regularity). *Condition (a) of Theorem 5B.4 can be replaced by the criterion that $\bar{y} \in \operatorname{core} \operatorname{rge} F$.*

Proof. This calls up the core property in Theorem 5B.1. □

We can finish tying up loose ends now by returning to the Banach open mapping theorem at the beginning of this chapter and tracing how it fits with the Robinson–Ursescu theorem.

Derivation of Theorem 5A.1 from Theorem 5B.4. It was already noted in the sequel to 5A.1 that condition (d) in that result was equivalent to the metric regularity of the linear mapping A, stated as condition (e). It remains only to observe that when Theorem 5B.4 is applied to $F = A \in \mathscr{L}(X,Y)$ with $\bar{x} = 0$ and $\bar{y} = 0$, the graph of A being a closed subspace of $X \times Y$ (in particular a convex set), and the positive homogeneity of A is brought in, we not only get (b) and (c) of Theorem 5A.1, but also (a). □

The argument for Theorem 5B.4, in obtaining metric regularity, also revealed a relationship between that property and the openness condition in 5B.2 which can be stated in the form

(10) $$\sup\{\, c \in (0,\infty) \,|\, (6) \text{ holds} \,\} \leq [\operatorname{reg}(F;\bar{x}|\bar{y})]^{-1}.$$

Exercise 5B.7 (a counterexample). *Show that for the mapping $F : \mathbb{R} \rightrightarrows \mathbb{R}$*

$$F(x) = \begin{cases} [-x, 0.5] & \text{if } x > 0.25, \\ [-x, 2x] & \text{if } x \in [0, 0.25], \\ \emptyset & \text{if } x < 0, \end{cases}$$

which is not positively homogeneous, and for $\bar{x} = \bar{y} = 0$ the inequality (10) is strict.

Exercise 5B.8 (effective domains of convex functions). *Let $g : X \to (-\infty, \infty]$ be convex and lower semicontinuous, and let $D = \{\, x \,|\, g(x) < \infty \,\}$. Show that D is a* convex set which, although not necessarily closed in X, is sure to have core $D =$ int D. Moreover, on that interior g is locally Lipschitz continuous.

Guide. Look at the mapping $F : X \rightrightarrows \mathbb{R}$ defined by $F(x) = \{y \in \mathbb{R} \,|\, y \geq g(x)\}$. Apply results in this section and also 5A.3. □

5C. Sublinear Mappings

An especially interesting class of positively homogeneous mappings $H : X \rightrightarrows Y$ acting between Banach spaces X and Y consists of the ones for which gph H is not just a cone, but a convex cone. Such mappings are called *sublinear*, because these geometric properties of gph H are equivalent to the rules that

(1) $\quad\quad\quad 0 \in H(0), \quad H(\lambda x) = \lambda H(x) \text{ for } \lambda > 0,$
$\quad\quad\quad\quad H(x+x') \supset H(x) + H(x') \text{ for all } x, x',$

which resemble linearity. Since the projection of a convex cone in $X \times Y$ into X or Y is another convex cone, it is clear for a sublinear mapping H that dom H is a convex cone in X and rge H is a convex cone in Y. The inverse H^{-1} of a sublinear mapping H is another sublinear mapping.

Although sublinearity has not been mentioned as a specific property before now, sublinear mappings have already appeared many times. Obviously, any linear mapping $A : X \to Y$ is sublinear (its graph being not just a convex cone but in fact a subspace of $X \times Y$). Sublinear also, though, is any mapping $H : X \rightrightarrows Y$ with $H(x) = Ax - K$ for a convex cone K in Y. Such mappings enter the study of constraint systems, with linear equations corresponding to $K = \{0\}$. When A is continuous and K is closed, their graphs are closed.

Sublinear mappings with closed graph enjoy the properties laid out in 5B along with those concerning outer and inner norms at the end of 5A. But their properties go a lot further, as in the result stated now about metric regularity.

Theorem 5C.1 (metric regularity of sublinear mappings). *For a sublinear mapping $H : X \rightrightarrows Y$ with closed graph, and any $(x,y) \in$ gph H we have*

(2) $\quad \text{reg}(H;x|y) \leq \text{reg}(H;0|0) = \inf\{\kappa > 0 \,|\, H(\kappa \text{int}\mathbb{B}) \supset \text{int}\mathbb{B}\} = \|H^{-1}\|^-.$

Moreover, $\text{reg}(H;0|0) < \infty$ *if and only if H is surjective, in which case H^{-1} is Lipschitz continuous on Y (in the sense of Pompeiu-Hausdorff distance as defined in 3A) and the infimum of the Lipschitz constant κ for this equals* $\|H^{-1}\|^-$.

Proof. Let $\kappa > \text{reg}(H;0|0)$. Then, from 5A.3, H is linearly open at 0 for 0 with constant κ, which reduces to $H(\kappa \text{int}\mathbb{B}) \supset \text{int}\mathbb{B}$. On the other hand, just from knowing that $H(\kappa \text{int}\mathbb{B}) \supset \text{int}\mathbb{B}$, we obtain for arbitrary $(x,y) \in$ gph H and $r > 0$ through the sublinearity of H that

$$H(x + \kappa r \,\text{int}\mathbb{B}) \supset H(x) + rH(\kappa \,\text{int}\mathbb{B}) \supset y + r\,\text{int}\mathbb{B}.$$

This establishes that $\text{reg}(H;x|y) \leq \text{reg}(H;0|0)$ for all $(x,y) \in \text{gph}\,H$. Appealing again to positive homogeneity, we get

(3) $$\text{reg}(H;0|0) = \inf\{\kappa > 0 \,|\, H(\kappa\,\text{int}\mathbb{B}) \supset \text{int}\mathbb{B}\}.$$

The right side of (3) does not change if we replace the open balls with their closures, hence, by 5A.7 or just by the definition of the inner norm, it equals $\|H^{-1}\|^-$. This confirms (2).

The finiteness of the right side of (3) corresponds to H being surjective, by virtue of positive homogeneity. We are left now with showing that H^{-1} is Lipschitz continuous on Y with $\|H^{-1}\|^-$ as the infimum of the available constants κ.

If H^{-1} is Lipschitz continuous on Y with constant κ, it must in particular have the Aubin property at 0 for 0 with this constant, and then $\kappa \geq \text{reg}(H;0|0)$ by 5A.3. We already know that this regularity modulus equals $\|H^{-1}\|^-$, so we are left with proving that, for every $\kappa > \text{reg}(H;0|0)$, H^{-1} is Lipschitz continuous on Y with constant κ.

Let $c < [\|H^{-1}\|^-]^{-1}$ and $\kappa > 1/c$. Taking (2) into account, we apply the inequality 5B(7) derived in 5B.3 with $x = \bar{x} = 0$ and $\bar{y} = 0$, obtaining the existence of $a > 0$ such that

$$d(0, H^{-1}(y)) \leq \kappa d(y, H(0)) \leq \kappa \|y\| \quad \text{for all } y \in a\mathbb{B}.$$

(Here, without loss of generality, we replace the open ball for y by its closure.) For any $y \in Y$, we have $ay/\|y\| \in a\mathbb{B}$, and from the positive homogeneity of H we get

(4) $$d(0, H^{-1}(y)) \leq \kappa \|y\| \quad \text{for all } y \in Y.$$

If $\|H^{-1}\|^- = 0$ then $0 \in H^{-1}(y)$ for all $y \in Y$ (see 4A.9), hence (4) follows automatically.

Let $y, y' \in Y$ and $x' \in H^{-1}(y')$. Through the surjectivity of H again, we can find for any $\delta > 0$ an $x_\delta \in H^{-1}(y - y')$ such that $\|x_\delta\| \leq d(0, H^{-1}(y - y')) + \delta$, and then from (4) we get

(5) $$\|x_\delta\| \leq \kappa \|y - y'\| + \delta.$$

Invoking the sublinearity of H yet once more, we obtain

$$x := x' + x_\delta \in H^{-1}(y') + H^{-1}(y - y') \subset H^{-1}(y' + y - y') = H^{-1}(y).$$

Hence $x' = x - x_\delta \in H^{-1}(y) + \|x_\delta\|\mathbb{B}$. Recalling (5), we arrive finally at the existence of $x \in H^{-1}(y)$ such that $\|x - x'\| \leq \kappa \|y - y'\| + \delta$. Since δ can be arbitrarily small, this yields Lipschitz continuity of H^{-1}, and we are done. □

5 Regularity in Infinite Dimensions

Corollary 5C.2 (finiteness of the inner norm). *Let $H : X \rightrightarrows Y$ be a sublinear mapping with closed graph. Then*

$$\mathrm{dom}\, H = X \iff \|H\|^- < \infty,$$
$$\mathrm{rge}\, H = Y \iff \|H^{-1}\|^- < \infty.$$

Proof. This comes from applying Theorem 5C.1 to both H and H^{-1}. □

Exercise 5C.3 (regularity modulus at zero). *For a sublinear mapping $H : X \rightrightarrows Y$ with closed graph, prove that*

$$\mathrm{reg}\,(H;0|0) = \inf\{\kappa \,|\, H(x + \kappa r \mathbb{B}) \supset H(x) + r\mathbb{B} \text{ for all } x \in X, r > 0\}.$$

Guide. Utilize the connections with openness properties. □

Example 5C.4 (application to linear constraints). *For $A \in \mathscr{L}(X,Y)$ and a closed, convex cone $K \subset Y$, define the solution mapping $S : Y \rightrightarrows X$ by*

$$S(y) = \{x \in X \,|\, Ax - y \in K\}.$$

Then S is a sublinear mapping with closed graph, and the following properties are equivalent:
(a) $S(y) \neq \emptyset$ for all $y \in Y$;
(b) there exists κ such that $d(x, S(y)) \leq \kappa d(Ax - y, K)$ for all $x \in X$, $y \in Y$;
(c) there exists κ such that $h(S(y), S(y')) \leq \kappa \|y - y'\|$ for all $y, y' \in Y$,

in which case the infimum of the constants κ that work in (b) coincides with the infimum of the constants κ that work in (c) and equals $\|S\|^-$.

Detail. Here $S = H^{-1}$ for $H(x) = Ax - K$, and the assertions of Theorem 5C.1 then translate into this form. □

Additional insights into the structure of sublinear mappings will emerge from applying a notion which comes out of the following fact.

Exercise 5C.5 (directions of unboundedness in convexity). *Let C be a closed, convex subset of X and let x_1 and x_2 belong to C. If $w \neq 0$ in X has the property that $x_1 + tw \in C$ for all $t \geq 0$, then it also has the property that $x_2 + tw \in C$ for all $t \geq 0$.*

Guide. Fixing any $t_2 > 0$, show that $x_2 + t_2 w$ can be approached arbitrarily closely by points on the line segment between x_2 and $x_1 + t_1 w$ by taking t_1 larger. □

On the basis of the property in 5C.5, the *recession cone* $\mathrm{rc}\, C$ of a closed, convex set $C \subset X$, defined by

(6) $$\mathrm{rc}\, C = \{w \in X \,|\, \forall x \in C, \forall t \geq 0 : x + tw \in C\},$$

can equally well be described by

(7) $$\operatorname{rc} C = \{ w \in X \mid \exists x \in C, \, \forall t \geq 0 : x + tw \in C \}.$$

It is easily seen that rc C is a closed, convex cone. In finite dimensions, C is bounded if and only if rc C is just $\{0\}$, but in infinite dimensions there are unbounded sets for which that holds. For a closed, convex cone K, one just has rc $K = K$, as seen from the equivalence between (6) and (7) by taking $x = 0$ in (7).

We will apply this now to the graph of a sublinear mapping H. It should be recalled that dom H is a convex cone, and for any convex cone K the set $K \cap [-K]$ is a subspace, in fact the largest subspace within K. On the other hand, $K - K$ is the smallest subspace that includes K.

Proposition 5C.6 (recession cones in sublinearity). *A sublinear mapping $H : X \rightrightarrows Y$ with closed graph has*

(8) $$\operatorname{rc} H(x) = H(0) \quad \text{for all } x \in \operatorname{dom} H,$$

and on the other hand,

(9) $$x \in \operatorname{dom} H \cap [-\operatorname{dom} H] \implies \begin{cases} H(x) + H(-x) \subset H(0), \\ H(x) - H(x) \subset H(0) - H(0). \end{cases}$$

Proof. Let $G = \operatorname{gph} H$, this being a closed, convex cone in $X \times Y$, therefore having rc $G = G$. For any $(x, y) \in G$, the recession cone rc $H(x)$ consists of the vectors w such that $y + tw \in H(x)$ for all $t \geq 0$, which are the same as the vectors w such that $(x, y) + t(0, w) \in G$ for all $t \geq 0$, i.e., the vectors w such that $(0, w) \in \operatorname{rc} G = G$. But these are the vectors $w \in H(0)$. That proves (8).

The first inclusion in (9) just reflects the rule that $H(x + [-x]) \supset H(x) + H(-x)$ by sublinearity. To obtain the second inclusion, let y_1 and y_2 belong to $H(x)$, which is the same as having $y_1 - y_2 \in H(x) - H(x)$, and let $y \in H(-x)$. Then by the first inclusion we have both $y_1 + y$ and $y_2 + y$ in $H(0)$, hence their difference lies in $H(0) - H(0)$. □

Theorem 5C.7 (single-valuedness of sublinear mappings). *For a sublinear mapping $H : X \rightrightarrows Y$ with closed graph and $\operatorname{dom} H = X$, the following conditions are equivalent:*
   (a) *H is a linear mapping from $\mathscr{L}(X, Y)$;*
   (b) *H is single-valued at some point $\{x\}$;*
   (c) *$\|H\|^+ < \infty$.*

Proof. Certainly (a) leads to (b). On the other hand, (c) necessitates $H(0) = \{0\}$ and hence (b) for $x = 0$. By 5C.6, specifically the second part of (9), if (b) holds for any x it must hold for all x. The first part of (9) reveals then that if $H(x)$ consists just of y, then $H(-x)$ consists just of $-y$. This property, along with the rules of sublinearity, implies linearity. The closedness of the graph of H implies that the linear mapping so obtained is continuous, so we have come back to (a). □

5 Regularity in Infinite Dimensions

Note that, without the closedness of the graph of H in Theorem 5C.7, there would be no assurance that (b) implies (a). We would still have a linear mapping, but it might not be continuous.

Corollary 5C.8 (single-valuedness of solution mappings). *In the context of Theorem 5C.1, it is impossible for H^{-1} to be single-valued at any point without actually turning out to be a continuous linear mapping from Y to X. The same holds for the solution mapping S for the linear constraint system in 5C.4 when a solution exists for every $y \in Y$.*

We next state the counterpart to Lemma 5A.4 which works for the inner norm of a positively homogeneous mapping. In contrast to the result presented in 5A.8 for the outer norm, convexity is now essential: we must limit ourselves to sublinear mappings.

Theorem 5C.9 (inversion estimate for the inner norm). *Let $H : X \rightrightarrows Y$ be sublinear with closed graph and have $\|H^{-1}\|^- < \infty$. Then for any $B \in \mathscr{L}(X,Y)$ such that $\|H^{-1}\|^- \cdot \|B\| < 1$, one has*

$$\|(H+B)^{-1}\|^- \leq \frac{\|H^{-1}\|^-}{1 - \|H^{-1}\|^- \|B\|}.$$

The proof of this is postponed until 5E, where it will be deduced from the connection between these properties and metric regularity in 5C.1. Perturbations of metric regularity will be a major theme, starting in Section 5D.

Duality. A special feature of sublinear mappings, with parallels linear mappings, is the availability of "adjoints" in the framework of the duals X^* and Y^* of the Banach spaces X and Y. For a sublinear mapping $H : X \rightrightarrows Y$, the *upper adjoint* $H^{*+} : Y^* \rightrightarrows X^*$ is defined by

(10) $\quad (y^*, x^*) \in \text{gph } H^{*+} \iff \langle x^*, x \rangle \leq \langle y^*, y \rangle$ for all $(x,y) \in \text{gph } H$,

whereas the *lower adjoint* $H^{*-} : Y^* \rightrightarrows X^*$ is defined by

(11) $\quad (y^*, x^*) \in \text{gph } H^{*-} \iff \langle x^*, x \rangle \geq \langle y^*, y \rangle$ for all $(x,y) \in \text{gph } H$.

These formulas correspond to modified polarity operations on the convex cone $\text{gph } H \subset X \times Y$ with polar $[\text{gph } H]^* \subset X^* \times Y^*$. They say that $\text{gph } H^{*+}$ consists of the pairs (y^*, x^*) such that $(x^*, -y^*) \in [\text{gph } H]^*$, while $\text{gph } H^{*-}$ consists of the pairs (y^*, x^*) such that $(-x^*, y^*) \in [\text{gph } H]^*$, and thus imply in particular that the graphs of these adjoints are closed, convex cones — so that both of these mappings from Y^* to X^* are sublinear with closed graph.

The switches of sign in (10) and (11) may seem a pointless distinction to make, but they are essential in capturing rules for recovering H from its adjoints through the fact that, when the spaces X and Y are reflexive, $[\text{gph } H]^{**} = \text{gph } H$ when the convex cone $\text{gph } H$ is closed, and then we get

(12) $\quad [H^{*+}]^{*-} = H$ and $[H^{*-}]^{*+} = H$ for sublinear H with closed graph.

When H reduces to a linear mapping $A \in \mathscr{L}(X,Y)$, both adjoints come out as the usual adjoint $A^* \in \mathscr{L}(Y^*,X^*)$. In that setting the graphs are subspaces instead of just cones, and the difference between (10) and (11) has no effect. The fact that $\|A^*\| = \|A\|$ in this case has the following generalization.

Theorem 5C.10 (duality of inner and outer norms). *For any sublinear mapping $H : X \rightrightarrows Y$ with closed graph, one has*

(13) $$\begin{aligned} \|H\|^+ &= \|H^{*-}\|^- = \|H^{*+}\|^-, \\ \|H\|^- &= \|H^{*-}\|^+ = \|H^{*+}\|^+. \end{aligned}$$

The proof requires some additional background. First, we need to update to Banach spaces the semicontinuity properties introduced in a finite-dimensional framework in Section 3B, but this only involves an extension of notation. A mapping $F : X \rightrightarrows Y$ is *inner semicontinuous* at $\bar{x} \in \text{dom}\, F$ if for every $y \in F(\bar{x})$ and every neighborhood V of y one can find a neighborhood U of $\bar{x}$ with $U \subset F^{-1}(V)$ or, equivalently, $F(x) \cap V \neq \emptyset$ for all $x \in U$ (this corresponds to 3B.2). Outer semicontinuity has a parallel extension. Next, we record a standard fact in functional analysis which will be called upon.

Theorem 5C.11 (Hahn–Banach). *Let M be a linear subspace of a Banach space X, and let $p : X \to \mathbb{R}$ satisfy*

(14) $\quad p(x+y) \leq p(x) + p(y)$ *and* $p(tx) = tp(x)$ *for all $x,y \in X$, $t \geq 0$.*

Let $f : M \to \mathbb{R}$ be a linear functional such that $f(x) \leq p(x)$ for all $x \in M$. Then there exists a linear functional $l : X \to \mathbb{R}$ such that $l(x) = f(x)$ for all $x \in M$, and $l(x) \leq p(x)$ for all $x \in X$.

In this formulation of the Hahn–Banach theorem, nothing is said about continuity, so X could really be any linear space — no topology is involved. But the main applications are ones in which p is continuous and it follows that l is continuous. Another standard fact in functional analysis, which can be derived from the Hahn–Banach theorem in that manner, is the following separation theorem.

Theorem 5C.12 (separation theorem). *Let C be a nonempty, closed, convex subset of a Banach space X, and let $x_0 \in X$. Then $x_0 \notin C$ if and only if there exists $x^* \in X^*$ such that*
$$\langle x^*, x_0 \rangle > \sup_{x \in C} \langle x^*, x \rangle.$$

Essentially, this says geometrically that a closed convex set is the intersection of all the "closed half-spaces" that include it.

5 Regularity in Infinite Dimensions

Proof of Theorem 5C.10. First, observe from (10) and (11) that

$$H^{*+}(y^*) = -H^{*-}(-y^*) \quad \text{for any } y^* \in Y^*,$$

so that

$$\|H^{*-}\|^- = \|H^{*+}\|^- \quad \text{and} \quad \|H^{*-}\|^+ = \|H^{*+}\|^+.$$

To prove that $\|H\|^+ = \|H^{*-}\|^-$ we fix any $y^* \in Y^*$ and show that

(15) $$\sup_{x \in B} \sup_{y \in H(x)} \langle y^*, y \rangle = \inf_{x^* \in H^{*-}(y^*)} \|x^*\| \quad \text{for all } y^* \in B.$$

If $\inf_{x^* \in H^{*-}(y^*)} \|x^*\| < r$ for some $r > 0$, then there exist $x^* \in H^{*-}(y^*)$ such that $\|x^*\| < r$. For any $\tilde{x} \in B$ and $\tilde{y} \in H(\tilde{x})$ we have

$$\langle y^*, \tilde{y} \rangle \leq \langle x^*, \tilde{x} \rangle \leq \sup_{x \in B} \langle x^*, x \rangle = \|x^*\| < r,$$

and then of course $\sup_{x \in B} \sup_{y \in H(x)} \langle y^*, y \rangle \leq r$. Hence

(16) $$\sup_{x \in B} \sup_{y \in H(x)} \langle y^*, y \rangle \leq \inf_{x^* \in H^{*-}(y^*)} \|x^*\|.$$

To prove the inequality opposite to (16) and hence the equality (15), assume that $\sup_{x \in B} \sup_{y \in H(x)} \langle y^*, y \rangle < r$ for some $r > 0$ and pick $0 < d < r$ such that

(17) $$\sup_{x \in B} \sup_{y \in H(x)} \langle y^*, y \rangle \leq d.$$

Define the mapping $G : X \rightrightarrows \mathbb{R}$ by

$$G : x \mapsto \{ z \mid z = \langle y^*, y \rangle, y \in H(x + B) \}.$$

First, observe that gph G is convex. Indeed, if $(x_1, z_1), (x_2, z_2) \in$ gph G and $0 < \lambda < 1$, then there exist $y_i \in Y$ and $w_i \in B$ with $z_i = \langle y^*, y_i \rangle$ and $y_i \in H(x_i + w_i)$, for $i = 1, 2$. Since H is sublinear, we get $\lambda y_1 + (1-\lambda)y_2 \in H(\lambda(x_1+w_1)+(1-\lambda)(x_2+w_2))$. Hence, $\lambda y_1 + (1-\lambda)y_2 \in H(\lambda x_1 + (1-\lambda)x_2 + B)$, and thus,

$$\lambda(x_1, z_1) + (1-\lambda)(x_2, z_2) = (\lambda x_1 + (1-\lambda)x_2, \langle y^*, \lambda y_1 + (1-\lambda)y_2 \rangle) \in \text{gph } G.$$

We will show next that G is inner semicontinuous at 0. Take $\tilde{z} \in G(0)$ and $\varepsilon > 0$. Let $\tilde{z} = \langle y^*, \tilde{y} \rangle$ for $\tilde{y} \in H(\tilde{w})$ and $\tilde{w} \in B$. Since $\langle y^*, \cdot \rangle$ is continuous, there is some $\gamma > 0$ such that $|\langle y^*, y \rangle - \tilde{z}| \leq \varepsilon$ when $\|y - \tilde{y}\| \leq \gamma$. Choose $\delta \in (0, 1)$ such that $\delta \|\tilde{y}\| \leq \gamma$. If $\|x\| \leq \delta$, we have

$$\|(1-\delta)\tilde{w} - x\| \leq \|(1-\delta)\tilde{w}\| + \|x\| \leq 1,$$

and hence $(1-\delta)\tilde{w} - x \in B$. Because H is sublinear,

$$(1-\delta)\tilde{y} \in H((1-\delta)\tilde{w}) = H(x + ((1-\delta)\tilde{w} - x)) \subset H(x + B) \quad \text{whenever } \|x\| \leq \delta.$$

Moreover, $\|(1-\delta)\tilde{y} - \tilde{y}\| = \delta\|\tilde{y}\| \leq \gamma$, and then $|\langle y^*, (1-\delta)\tilde{y}\rangle - \tilde{z}| \leq \varepsilon$. Therefore, for all $x \in \delta B$, we have $\langle y^*, (1-\delta)\tilde{y}\rangle \in G(x) \cap B_\varepsilon(\tilde{z})$, and hence G is inner semicontinuous at 0 as desired.

Let us now define a mapping $K : X \rightrightarrows \mathbb{R}$ whose graph is the *conical hull* of $\mathrm{gph}\,(d - G)$ where d is as in (17); that is, its graph is the set of points λh for $h \in \mathrm{gph}\,(d - G)$ and $\lambda \geq 0$. The conical hull of a convex set is again convex, so K is another sublinear mapping. Since G is inner semicontinuous at 0, there is some neighborhood U of 0 with $U \subset \mathrm{dom}\, G$, and therefore $\mathrm{dom}\, K = X$. Consider the functional

$$k : x \mapsto \inf\{z \mid z \in K(x)\} \quad \text{for } x \in X.$$

Because K is sublinear and $d - H(0) \subset \mathbb{R}_+$, we have

(18) $$K(x) + K(-x) \subset K(0) \subset \mathbb{R}_+.$$

This inclusion implies in particular that any point in $-K(-x)$ furnishes a lower bound in $\mathbb{R}$ for the set of values $K(x)$, for any $x \in X$. Indeed, let $x \in X$ and $y \in -K(-x)$. Then (18) yields $K(x) - y \subset \mathbb{R}_+$, and consequently $y \leq z$ for all $z \in K(x)$. Therefore $k(x)$ is finite for all $x \in X$; we have $\mathrm{dom}\, k = X$. Also, from the sublinearity of K and the properties of the infimum, we have

$$k(x+y) \leq k(x) + k(y) \quad \text{and} \quad k(\alpha x) = \alpha k(x) \quad \text{for all } x, y \in X \text{ and } \alpha \geq 0.$$

Consider the subspace $M = \{0\} \subset X$ and define $f : M \to \mathbb{R}$ simply by $f(0) := k(0) = 0$. Applying Hahn–Banach theorem 5C.11 to f, we get a linear functional $l : X \to \mathbb{R}$ such that $l(0) = 0$ and $l(x) \leq k(x)$ for all $x \in X$. We will show now that l is continuous at 0 and hence continuous on the whole X.

Continuity at 0 means that for any $\varepsilon > 0$ there is $\delta > 0$ such that $(l(x) + \mathbb{R}_+) \cap \varepsilon B \neq \emptyset$ whenever $x \in \delta B$. Let $z \in d - G(0)$ and take $0 < \lambda < 1$ along with a neighborhood V of z such that $\lambda V \subset \varepsilon B$. Since G is inner semicontinuous at 0, there is some $\delta > 0$ such that

$$(d - G(x)) \cap V \neq \emptyset, \quad \text{for all } x \in (\delta/\lambda) B.$$

Since $d - G(x) \subset k(x) + \mathbb{R}_+$ and $k(x) \geq l(x)$, we have $d - G(x) \subset l(x) + \mathbb{R}_+$ and $(l(x) + \mathbb{R}_+) \cap V \neq \emptyset$ for all $x \in (\delta/\lambda) B$, so that $(l(\lambda x) + \mathbb{R}_+) \cap \lambda V \neq \emptyset$ for all $x \in (\delta/\lambda) B$. This yields

$$(l(x) + \mathbb{R}_+) \cap \varepsilon B \neq \emptyset \quad \text{for all } x \in \delta B,$$

which means that for all $x \in \delta B$ there exists some $z \geq l(x)$ with $|z| \leq \varepsilon$. The linearity of l makes $l(x) = -l(-x)$, and therefore $|l(x)| \leq \varepsilon$ for all $x \in \delta B$. This confirms the continuity of l.

The inclusion $d - G(x) - l(x) \subset \mathbb{R}_+$ is by definition equivalent to having $d - \langle y^*, y \rangle - l(x) \geq 0$ whenever $x \in H^{-1}(y) - B$. Let $x^* \in X^*$ be such that $\langle x^*, x \rangle = -l(x)$ for all $x \in X$. Then

5 Regularity in Infinite Dimensions

$$\langle y^*, y\rangle - \langle x^*, x\rangle \leq d \text{ for all } y \in Y \text{ and all } x \in H^{-1}(y) - \mathbb{B}.$$

Pick any $y \in H(x)$ and $\lambda > 0$. Then $\lambda y \in H(\lambda x)$ and $\langle y^*, \lambda y\rangle - \langle x^*, \lambda x\rangle \leq d$, or equivalently,

$$\langle y^*, y\rangle - \langle x^*, x\rangle \leq d/\lambda.$$

Passing to the limit with $\lambda \to \infty$, we obtain $x^* \in H^{*-}(y^*)$. Let now $x \in \mathbb{B}$. Since $0 \in H(0)$, we have $0 \in H(-x + \mathbb{B})$ and hence $\langle y^*, 0\rangle - \langle x^*, -x\rangle \leq d$. Therefore $\|x^*\| \leq d < r$, so that $\inf_{x^* \in H^{*-}(y^*)} \|x^*\| < r$. This, combined with (16), gives us the equality in (15) and hence the equalities in the first line of (13).

We will now confirm the equality in the second line of (13). Suppose $\|H\|^- < r$ for some $r > 0$. Then for any $\tilde{x} \in \mathbb{B}$ there is some $\tilde{y} \in H(\tilde{x})$ such that $\|\tilde{y}\| < r$. Given $y^* \in \mathbb{B}$ and $x^* \in H^{*+}(y^*)$, we have

$$\langle x^*, \tilde{x}\rangle \leq \langle y^*, \tilde{y}\rangle \leq \|\tilde{y}\| < r.$$

This being valid for arbitrary $\tilde{x} \in \mathbb{B}$, we conclude that $\|x^*\| \leq r$, and therefore $\|H^{*+}\|^+ \leq r$.

Suppose now that $\|H^{*+}\|^+ < r$ and pick $s > 0$ with

$$\sup_{x^* \in H^{*+}(B)} \|x^*\| = \|H^{*+}\|^+ \leq s < r,$$

in which case $H^{*+}(\mathbb{B}) \subset s\mathbb{B}$. We will show that

(19) $\quad\quad \langle x^*, x\rangle \leq 1 \text{ for all } x \in H^{-1}(\mathbb{B}) \implies \|x^*\| \leq s.$

The condition on the left of (19) can be written as $\sup_{y \in B} \sup_{x \in H^{-1}(y)} \langle x^*, x\rangle \leq 1$, which in turn is completely analogous to (17), with $d = 1$ and H replaced by H^{-1} and with y and y^* replaced by x and x^*, respectively. By repeating the argument in the first part of the proof after (17), we obtain $y^* \in (H^{-1})^{*-}(x^*) = (H^{*+})^{-1}(x^*)$ with $\|y^*\| \leq 1$. But then $x^* \in H^{*+}(\mathbb{B})$, and since $H^{*+}(\mathbb{B}) \subset s\mathbb{B}$ we have (19).

Now we will show that (19) implies

(20) $\quad\quad s^{-1}\mathbb{B} \subset \text{cl } H^{-1}(\mathbb{B}).$

If $u \notin \text{cl } H^{-1}(\mathbb{B})$, then from 5C.12 there exists $\tilde{x}^* \in X^*$ with

$$\langle \tilde{x}^*, u\rangle > \sup_{x \in \text{cl } H^{-1}(B)} \langle \tilde{x}^*, x\rangle \geq \langle \tilde{x}^*, 0\rangle = 0.$$

Choose $\lambda > 0$ such that

$$\sup_{x \in \text{cl } H^{-1}(B)} \langle \tilde{x}^*, x\rangle < \lambda^{-1} < \langle \tilde{x}^*, u\rangle.$$

Then

$$\langle \lambda \tilde{x}^*, u\rangle > 1 > \sup_{x \in \text{cl } H^{-1}(B)} \langle \lambda \tilde{x}^*, x\rangle.$$

According to (19) this implies that $\lambda \tilde{x}^* \in s\mathbb{B}$. Thus,

$$s \geq \|\lambda \tilde{x}^*\| \geq \langle \lambda \tilde{x}^*, u/\|u\| \rangle > \frac{1}{\|u\|},$$

and therefore $u \notin s^{-1}\mathbb{B}$, so (20) holds.

Our next task is to demonstrate that

(21) $$\operatorname{int} s^{-1}\mathbb{B} \subset \operatorname{int} H^{-1}(\mathbb{B}).$$

Define the mapping

$$x \mapsto H_0(x) = \begin{cases} H(x) & \text{for } x \in \mathbb{B}, \\ \emptyset & \text{otherwise.} \end{cases}$$

Then gph $H_0 = $ gph $H \cap (X \times \mathbb{B})$ is a closed convex set and rge $H_0 \subset \mathbb{B}$. By 5B.1 we have

$$\operatorname{int} \operatorname{cl} H^{-1}(\mathbb{B}) = \operatorname{int} \operatorname{cl} \operatorname{dom} H_0 = \operatorname{int} \operatorname{dom} H_0 = \operatorname{int} H^{-1}(\mathbb{B}).$$

This equality combined with the inclusion (20) gives us (21). But then $r^{-1}\mathbb{B} \subset \operatorname{int} s^{-1}\mathbb{B} \subset H^{-1}(\mathbb{B})$, ensuring $\|H\|^- \leq r$. This completes the proof of the second line in (13). □

The above proof can be shortened considerably in the case when X and Y are reflexive Banach spaces, by utilizing the equality (12).

Exercise 5C.13 (more norm duality). *For a sublinear mapping $H : X \rightrightarrows Y$ with closed graph show that*

$$\|(H^{*+})^{-1}\|^+ = \|H^{-1}\|^-.$$

Exercise 5C.14 (adjoint of a sum). *For a sublinear mapping $G : X \rightrightarrows Y$ and $B \in \mathscr{L}(X,Y)$ prove that*

$$(H+B)^{*+} = H^{*+} + B^* \quad \text{and} \quad (H+B)^{*-} = H^{*-} + B^*.$$

5D. The Theorems of Lyusternik and Graves

We start with the observation that inequality 5A(7) in Lemma 5A.4, giving an estimate for inverting a perturbed linear mapping A, can also be written in the form

$$\operatorname{reg} A \cdot \|B\| \leq 1 \quad \Longrightarrow \quad \operatorname{reg}(A+B) \leq \frac{\operatorname{reg} A}{1 - \operatorname{reg} A \cdot \|B\|},$$

since for an invertible mapping $A \in \mathscr{L}(X,Y)$ one has $\text{reg}\, A = \|A^{-1}\|$. This alternative formulation opens the way to extending the estimate to nonlinear and even set-valued mappings.

First, we recall a basic definition of differentiability in infinite dimensions, which is just an update of the definition employed in the preceding chapters in finite dimensions. With differentiability as well as Lipschitz continuity and calmness, the only difference is that the Euclidean norm is now replaced by the norms of the Banach spaces X and Y that we work with.

Fréchet differentiability and strict differentiability. *A function $f : X \to Y$ is said to be Fréchet differentiable at $\bar{x}$ if $\bar{x} \in \text{int dom}\, f$ and there is a mapping $M \in \mathscr{L}(X,Y)$ such that $\text{clm}\,(f - M; \bar{x}) = 0$. When such a mapping M exists, it is unique; it is called the Fréchet derivative of f at $\bar{x}$ and denoted by $Df(\bar{x})$, so that*

$$\text{clm}\,(f - Df(\bar{x}); \bar{x}) = 0.$$

If actually

$$\text{lip}\,(f - Df(\bar{x}); \bar{x}) = 0,$$

then f is said to be strictly differentiable at $\bar{x}$.

Partial Fréchet differentiability and partial strict differentiability can be introduced as well on the basis of the partial Lipschitz moduli, by updating the definitions in Section 1D to infinite dimensions. Building on the formulas for the calmness and Lipschitz moduli, we could alternatively express these definitions in an epsilon-delta mode as at the beginning of Chapter 1. If a function f is Fréchet differentiable at every point x of an open set O and the mapping $x \mapsto Df(x)$ is continuous from O to the Banach space $\mathscr{L}(X,Y)$, then f is said to be *continuously* Fréchet differentiable on O. Most of the assertions in Section 1D about functions acting in finite dimensions remain valid in Banach spaces, e.g., continuous Fréchet differentiability around a point implies strict differentiability at this point.

The extension of the Banach open mapping theorem to nonlinear and set-valued mappings goes back to the works of Lyusternik and Graves. In 1934, L. A. Lyusternik published a result, saying that when a function $f : X \to Y$ is continuously Fréchet differentiable in a neighborhood of a point $\bar{x}$ where $f(\bar{x}) = 0$ and its derivative mapping $Df(\bar{x})$ is surjective, then the tangent manifold to $f^{-1}(0)$ at $\bar{x}$ is the set $\bar{x} + \ker Df(\bar{x})$. In the current setting we adopt the following statement[1] of Lyusternik theorem:

Theorem 5D.1 (Lyusternik). *Consider a function $f : X \to Y$ that is continuously Fréchet differentiable in a neighborhood of a point $\bar{x}$ with the derivative mapping $Df(\bar{x})$ surjective. Then, in terms of $\bar{y} := f(\bar{x})$, for every $\varepsilon > 0$ there exists $\delta > 0$ such that*

$$d(x, f^{-1}(\bar{y})) \leq \varepsilon \|x - \bar{x}\| \quad \text{whenever } x \in (\bar{x} + \ker Df(\bar{x})) \text{ and } \|x - \bar{x}\| \leq \delta.$$

---

[1] In his paper of 1934 Lyusternik did not state his result as a theorem; the statement in 5D.1 is from Dmitruk, Milyutin and Osmolovskiĭ [1980].

In 1950 L. M. Graves published a result whose formulation and proof we present here in full, up to some minor adjustments in notation:

Theorem 5D.2 (Graves). *Consider a function $f : X \to Y$ and a point $\bar{x} \in \operatorname{int} \operatorname{dom} f$ and let f be continuous in $B_\varepsilon(\bar{x})$ for some $\varepsilon > 0$. Let $A \in \mathscr{L}(X,Y)$ be surjective and let $\kappa \geq \operatorname{reg} A$. Suppose there is a nonnegative μ such that $\mu\kappa < 1$ and*

(1) $\quad \|f(x) - f(x') - A(x - x')\| \leq \mu\|x - x'\| \quad$ *whenever* $x, x' \in B_\varepsilon(\bar{x})$.

Then, in terms of $\bar{y} := f(\bar{x})$ and $c = \kappa^{-1} - \mu$, if y is such that $\|y - \bar{y}\| \leq c\varepsilon$, then the equation $y = f(x)$ has a solution $x \in B_\varepsilon(\bar{x})$.

Proof. Without loss of generality, let $\bar{x} = 0$ and $\bar{y} = f(\bar{x}) = 0$. Note that $\kappa > 0$, hence $0 < c < \infty$. Take $y \in Y$ with $\|y\| \leq c\varepsilon$. Starting from $x^0 = 0$ we use induction to construct an infinite sequence $\{x^k\}$, the elements of which satisfy for all $k = 1, 2, \ldots$ the following three conditions:

(2a) $\quad\quad\quad\quad\quad\quad A(x^k - x^{k-1}) = y - f(x^{k-1}),$

(2b) $\quad\quad\quad\quad\quad\quad \|x^k - x^{k-1}\| \leq \kappa(\kappa\mu)^{k-1}\|y\|$

and

(2c) $\quad\quad\quad\quad\quad\quad \|x^k\| \leq \|y\|/c.$

By (d) in the Banach open mapping theorem 5A.1 there exists $x^1 \in X$ such that

$$Ax^1 = y \quad \text{and} \quad \|x^1\| \leq \kappa\|y\| \leq \|y\|/c.$$

That is, x^1 satisfies all three conditions (2a), (2b) and (2c). In particular, by the choice of y and the constant c we have $\|x^1\| \leq \varepsilon$.

Suppose now that for some $j \geq 1$ we have obtained points x^k satisfying (2a), (2b) and (2c) for $k = 1, \ldots, j$. Then, since $\|y\|/c \leq \varepsilon$, we have from (2c) that all the points x^k satisfy $\|x^k\| \leq \varepsilon$. Again using (d) in 5A.1, we can find x^{j+1} such that

(3) $\quad A(x^{j+1} - x^j) = y - f(x^j) \quad \text{and} \quad \|x^{j+1} - x^j\| \leq \kappa\|y - f(x^j)\|.$

If we plug $y = Ax^j - Ax^{j-1} + f(x^{j-1})$ into the second relation in (3) and use (1) for $x = x^j$ and $x' = x^{j-1}$ which, as we already know, are from εB, we obtain

$$\|x^{j+1} - x^j\| \leq \kappa\|f(x^j) - f(x^{j-1}) - A(x^j - x^{j-1})\| \leq \kappa\mu\|x^j - x^{j-1}\|.$$

Then, by the induction hypothesis,

$$\|x^{j+1} - x^j\| \leq \kappa(\kappa\mu)^j\|y\|.$$

Furthermore,

5 Regularity in Infinite Dimensions

$$\|x^{j+1}\| \leq \|x^1\| + \sum_{i=1}^{j} \|x^{i+1} - x^i\| \leq \sum_{i=0}^{j} (\kappa\mu)^i \kappa\|y\| \leq \frac{\kappa\|y\|}{1-\kappa\mu} = \|y\|/c.$$

The induction step is complete: we obtain an infinite sequence of points x^k satisfying (2a), (2b) and (2c). For any k and j with $k > j > 1$ we have

$$\|x^k - x^j\| \leq \sum_{i=j}^{k-1} \|x^{i+1} - x^i\| \leq \sum_{i=j}^{k-1} (\kappa\mu)^i \kappa\|y\| \leq (\kappa\mu)^j \kappa\|y\| \sum_{i=0}^{\infty} (\kappa\mu)^i \leq \frac{\kappa\|y\|}{1-\kappa\mu}(\kappa\mu)^j.$$

Thus, $\{x^k\}$ is a Cauchy sequence, hence convergent to some x, and then, passing to the limit with $k \to \infty$ in (2a) and (2c), this x satisfies $y = f(x)$ and $\|x\| \leq \|y\|/c$. The final inequality gives us $\|x\| \leq \varepsilon$ and the proof is finished. □

Observe that in the Graves theorem no differentiability of the function f is required, but only "approximate differentiability" as in the theorem of Hildebrand and Graves; see the commentary to Chapter 1. If we suppose that *for every $\mu > 0$ there exists $\varepsilon > 0$ such that (1) holds for every $x, x' \in B_\varepsilon(\bar{x})$*, then A is, by definition, the strict derivative of f at $\bar{x}$, $A = Df(\bar{x})$. That is, the Graves theorem encompasses the following special case: if f is strictly differentiable at $\bar{x}$ and its derivative $Df(\bar{x})$ is onto, then there exist $\varepsilon > 0$ and $c > 0$ such that for every $y \in Y$ with $\|y - \bar{y}\| \leq c\varepsilon$ there is an $x \in X$ such that $\|x - \bar{x}\| \leq \varepsilon$ and $y = f(x)$.

The statement of the Graves theorem above does not reflect all the information that can be extracted from its proof. In particular, a solution x of $f(x) = y$ which not only is in the ball $B_\varepsilon(\bar{x})$ but also satisfies $\|x - \bar{x}\| \leq \|y - \bar{y}\|/c$. Taking into account that $x \in f^{-1}(y)$, which yields $d(\bar{x}, f^{-1}(y)) \leq \|x - \bar{x}\|$, along with the form of the constant c, we get

$$d(\bar{x}, f^{-1}(y)) \leq \frac{\kappa}{1-\kappa\mu}\|y - f(\bar{x})\|.$$

Furthermore, this inequality actually holds not only at $\bar{x}$ but also for all x close to $\bar{x}$, and this important extension is hidden in the proof of the theorem.

Indeed, let (1) hold for $x, x' \in B_\varepsilon(\bar{x})$ and choose a positive $\tau < \varepsilon$. Then there is a neighborhood U of $\bar{x}$ such that $B_\tau(x) \subset B_\varepsilon(\bar{x})$ for all $x \in U$. Make U smaller if necessary so that $\|f(x) - f(\bar{x})\| < c\tau$ for $x \in U$. Pick $x \in U$ and a neighborhood V of $\bar{y}$ such that $\|y - f(x)\| \leq c\tau$ for $y \in V$. Then, remembering that in the proof $\bar{x} = 0$, modify the first induction step in the following way: there exists $x^1 \in X$ such that

$$Ax^1 = y - f(x) + Ax \quad \text{and} \quad \|x^1 - x\| \leq \kappa\|y - f(x)\|.$$

Then, construct a sequence $\{x^k\}$ with $x^0 = x$ satisfying (3), thereby obtaining

$$\|x^k - x^{k-1}\| \leq \kappa(\kappa\mu)^{k-1}\|y - f(x)\|$$

and then

(4) $\quad \|x^k - x\| \leq \sum_{i=1}^{k} \|x^i - x^{i-1}\| \leq \kappa\|y - f(x)\| \sum_{i=1}^{i} (\kappa\mu)^{i-1} \leq \frac{\kappa}{1-\kappa\mu}\|y - f(x)\|.$

Thus,
$$\|x^k - x\| \leq \|y - f(x)\|/c \leq \tau.$$

The sequence $\{x^k\}$ is a Cauchy sequence, and therefore convergent to some $\tilde{x}$. In passing to the limit in (4) we get
$$\|\tilde{x} - x\| \leq \frac{\kappa}{1 - \kappa\mu}\|y - f(x)\|.$$

Since $\tilde{x} \in f^{-1}(y)$, we see that, under the conditions of the Graves theorem, there exist neighborhoods U of $\bar{x}$ and V of $f(\bar{x})$ such that

(5) $\qquad d(x, f^{-1}(y)) \leq \dfrac{\kappa}{1 - \kappa\mu}\|y - f(x)\|\qquad$ for $(x, y) \in U \times V$.

The property described in (5) is something we know from Chapter 3: this is metric regularity of the function f at $\bar{x}$ for $\bar{y}$. Noting that the μ in (1) satisfies $\mu \geq \text{lip}(f - A)(\bar{x})$ we arrive at the following result:

Theorem 5D.3 (updated Graves' theorem). *Let $f : X \to Y$ be continuous in a neighborhood of $\bar{x}$, let $A \in \mathscr{L}(X, Y)$ satisfy $\text{reg } A \leq \kappa < \infty$, and suppose $\text{lip}(f - A)(\bar{x}) \leq \mu$ for some μ with $\mu\kappa < 1$. Then*

(6) $\qquad \text{reg}(f; \bar{x}|\bar{y}) \leq \dfrac{\kappa}{1 - \kappa\mu}\qquad$ *for $\bar{y} = f(\bar{x})$.*

For $f = A + B$ we obtain from this result the estimation for perturbed inversion of linear mappings in 5A.4.

The version of the Lyusternik theorem[2] stated as Theorem 5D.1 can be derived from the updated Graves theorem 5D.3. Indeed, the assumptions of 5D.1 are clearly stronger than those of Theorem 5D.3. From (5) with $y = f(\bar{x})$ we get

(7) $\qquad d(x, f^{-1}(\bar{y})) \leq \dfrac{\kappa}{1 - \kappa\mu}\|f(x) - f(\bar{x})\|$

for all x sufficiently close to $\bar{x}$. Let $\varepsilon > 0$ and choose $\delta > 0$ such that

(8) $\qquad \|f(x) - f(\bar{x}) + Df(\bar{x})(x - \bar{x})\| \leq \dfrac{(1 - \kappa\mu)\varepsilon}{\kappa}\|x - \bar{x}\|\quad$ whenever $x \in \mathbb{B}_\delta(\bar{x})$.

But then for any $x \in (\bar{x} + \ker Df(\bar{x})) \cap \mathbb{B}_\delta(\bar{x})$, from (7) and (8) we obtain
$$d(x, f^{-1}(y)) \leq \frac{\kappa}{1 - \kappa\mu}\|f(x) - f(\bar{x})\| \leq \varepsilon\|x - \bar{x}\|,$$

which is the conclusion of 5D.1.

---

[2] The iteration (3), which is a key step in the proof of Graves, is also present in the original proof of Lyusternik [1934], see also Lyusternik and Sobolev [1965]. In the case when A is invertible, it goes back to Goursat [1903], see the commentary to Chapter 1.

Exercise 5D.4 (correction function version of Graves theorem). *Show that on the conditions of Theorem 5D.3, for $\bar{y} = f(\bar{x})$ there exist neighborhoods U of $\bar{x}$ and V of $\bar{y}$ such that for every $y \in V$ and $x \in U$ there exists ξ with the property*

$$f(\xi + x) = y \quad \text{and} \quad \|\xi\| \leq \frac{\kappa}{1 - \kappa\mu} \|f(x) - y\|.$$

Guide. From Theorem 5D.3 we see that there exist neighborhoods U of $\bar{x}$ and V of $\bar{y}$ such that for every $x \in U$ and $y \in V$

$$d(x, f^{-1}(y)) \leq \frac{\kappa}{1 - \kappa\mu} \|y - f(x)\|.$$

Without loss of generality, let $y \neq f(x)$; then we can slightly increase μ so that the latter inequality becomes strict. Then there exists $\eta \in f^{-1}(y)$ such that $\|x - \eta\| \leq \kappa/(1 - \kappa\mu) \|y - f(x)\|$. Next take $\xi = \eta - x$. □

If the function f in 5D.3 is strictly differentiable at $\bar{x}$, we can choose $A = Df(\bar{x})$, and then $\mu = 0$. In this case (6) reduces to

(9) $$\operatorname{reg}(f; \bar{x} | \bar{y}) \leq \operatorname{reg} Df(\bar{x}) \quad \text{for } \bar{y} = f(\bar{x}).$$

In the following section we will show that this inequality actually holds as equality:

Theorem 5D.5. *For a function $f : X \to Y$ which is strictly differentiable at $\bar{x}$, one has*

$$\operatorname{reg}(f; \bar{x} | \bar{y}) = \operatorname{reg} Df(\bar{x}).$$

We terminate this section with two major observations. For the first, assume that the mapping A in Theorem 5D.2 is not only surjective but also invertible. Then the iteration procedure used in the proof of the Graves theorem becomes the iteration used first by Goursat (see the commentary to Chapter 1), namely

$$x^{j+1} = x^j - A^{-1}(f(x^j) - y).$$

In that case, one obtains the existence of a single-valued graphical localization of the inverse f^{-1} around $f(\bar{x})$ for $\bar{x}$. If the derivative mapping $Df(\bar{x})$ is merely surjective, as assumed in the Graves theorem, the inverse f^{-1} may not have a single-valued graphical localization at $\bar{y}$ for $\bar{x}$ but, still, this inverse, being a set-valued mapping, has the Aubin property at $\bar{y}$ for $\bar{x}$.

Our second observation is that in the proof of Theorem 5D.2 we use the linearity of the mapping A only to apply the Banach open mapping theorem. But we can employ the regularity modulus for any, even set-valued, mapping. After this somewhat historical section, we will explore this idea further in the section which follows.

5E. Metric Regularity in Metric Spaces

In this section we show that the updated Graves theorem 5D.3, in company with the stability of metric regularity under perturbations, demonstrated in Theorem 3F.1, can be extended to a much broader framework of set-valued mappings acting in abstract spaces. Specifically, we consider a set-valued mapping F acting from a metric space (X, ρ) to another metric space (Y, σ). In such spaces the standard definitions, e.g. of the ball in X with center x and radius r and the distance from a point x to a set C in Y, need only be adapted to metric notation:

$$\mathbb{B}_r(\bar{x}) = \{x \in X \,|\, \rho(x,\bar{x}) \leq r\}, \qquad d(x,C) = \inf_{x' \in C} \rho(x,x').$$

Recall that a subset C of a complete metric space is closed when $d(x,C) = 0 \Rightarrow x \in C$. Also recall that a set C is locally closed at a point $x \in C$ if there is a neighborhood U of x such that the intersection $C \cap U$ is closed.

In metric spaces (X, ρ) and (Y, σ), the definition of the Lipschitz modulus of a function $g : X \to Y$ is extended in a natural way, with attention paid to the metric notation of distance:

$$\mathrm{lip}\,(g;\bar{x}) = \limsup_{\substack{x,x' \to \bar{x}, \\ x \neq x'}} \frac{\sigma(g(x), g(x'))}{\rho(x,x')}.$$

For set-valued mappings F acting in such spaces, the definitions of metric regularity and the Aubin property persist in the same manner. The equivalence of metric regularity with the Aubin property of the inverse (Theorem 3E.4) with the same constant remains valid as well.

It will be important for our efforts to take the metric space X to be *complete* and to suppose that Y is a *linear* space equipped with a *shift-invariant* metric σ. Shift invariance means that

$$\sigma(y+z, y'+z) = \sigma(y,y') \text{ for all } y, y', z \in Y.$$

Of course, any Banach space meets these requirements.

The result stated next is just a reformulation of Theorem 3F.1 for mappings acting in metric spaces:

Theorem 5E.1 (inverse mapping theorem for metric regularity in metric spaces). Let (X, ρ) be a complete metric space and let (Y, σ) be a linear space with shift-invariant metric σ. Consider a mapping $F : X \rightrightarrows Y$ and any $(\bar{x}, \bar{y}) \in \mathrm{gph}\, F$ at which $\mathrm{gph}\, F$ is locally closed, and let κ and μ be nonnegative constants such that

$$\mathrm{reg}\,(F; \bar{x}|\bar{y}) \leq \kappa \quad \text{and} \quad \kappa\mu < 1.$$

Then for any function $g : X \to Y$ with $\bar{x} \in \mathrm{int}\,\mathrm{dom}\, g$ and $\mathrm{lip}\,(g;\bar{x}) \leq \mu$, one has

5 Regularity in Infinite Dimensions

(1) $$\operatorname{reg}(g+F;\bar{x}|g(\bar{x})+\bar{y}) \leq \frac{\kappa}{1-\kappa\mu}.$$

Before arguing this, we note that it immediately allows us to supply 5C.9 and 5D.5 with proofs.

Proof of 5C.9. We apply 5E.1 with X and Y Banach spaces, $F = H$, $\bar{x} = 0$ and $\bar{y} = 0$. According to 5C.1, $\operatorname{reg}(H;0|0) = \|H^{-1}\|^{-}$, so 5E.1 tells us that for any $\kappa > \|H^{-1}\|^{-}$, any $B \in \mathscr{L}(X,Y)$ with $\|B\| < 1/\kappa$, and any μ with $\|B\| < \mu < 1/\kappa$ one has from (1) that $\|(H+B)^{-1}\|^{-} \leq \kappa/(1-\kappa\mu)$. It remains only to pass to the limit as $\kappa \to \|H^{-1}\|^{-}$ and $\mu \to \|B\|$. □

Proof of 5D.5. To obtain the inequality opposite to 5D(9), choose $F = f$ and $g = Df(\bar{x}) - f$ and apply 5E.1, in this case with $\mu = 0$. □

We proceed now with presenting two separate proofs of Theorem 5E.1, which echo on a more abstract level the way we proved the classical inverse function theorem 1A.1 in Chapter 1. The first proof uses an iteration in line with the original argument in the proof of the Graves theorem 5D.2, while the second proof is based on a contraction mapping principle for set-valued mappings.

Proof I of Theorem 5E.1. Let κ and μ be as in the statement of the theorem and choose a function $g : X \to Y$ with $\operatorname{lip}(g;\bar{x}) \leq \mu$. Without loss of generality, suppose $g(\bar{x}) = 0$. Let $\lambda > \kappa$ and $\nu > \mu$ satisfy $\lambda \nu < 1$. Let $\alpha > 0$ be small enough that the set $\operatorname{gph} F \cap (\mathbb{B}_\alpha(\bar{x}) \times \mathbb{B}_\alpha(\bar{y}))$ is closed, g is Lipschitz continuous with constant ν on $\mathbb{B}_\alpha(\bar{x})$, and

(2) $$d(x, F^{-1}(y)) \leq \lambda d(y, F(x)) \quad \text{for all } (x,y) \in \mathbb{B}_\alpha(\bar{x}) \times \mathbb{B}_\alpha(\bar{y}).$$

From (2) with $x = \bar{x}$, it follows that

(3) $$F^{-1}(y) \neq \emptyset \quad \text{for all } y \in \mathbb{B}_\alpha(\bar{y}).$$

Having fixed λ, α and ν, consider the following system of inequalities:

(4) $$\begin{cases} \lambda \nu + \varepsilon < 1, \\ \frac{1}{1-(\lambda\nu+\varepsilon)}[(1+\lambda\nu)a + \lambda b + \varepsilon] + a \leq \alpha, \\ b + \nu\left(\frac{1}{1-(\lambda\nu+\varepsilon)}[(1+\lambda\nu)a + \lambda b + \varepsilon] + a\right) \leq \alpha. \end{cases}$$

It is not difficult to see that there are positive a, b and ε that satisfy this system. Indeed, first fix ε such that these inequalities hold strictly for $a = b = 0$; then pick sufficiently small a and b so that both the second and the third inequality are not violated.

Let $x \in \mathbb{B}_a(\bar{x})$ and $y \in \mathbb{B}_b(\bar{y})$. We will show that

(5) $$d(x, (g+F)^{-1}(y)) \leq \frac{\lambda}{1-\lambda\nu} d(y, (g+F)(x)).$$

Since x and y are arbitrarily chosen in the corresponding balls around $\bar{x}$ and $\bar{y}$, and λ and ν are arbitrarily close to κ and μ, respectively, this gives us (1).

According to the choice of a and of b in (4), we have

(6) $$\sigma(y-g(x),\bar{y}) \leq \nu\rho(x,\bar{x}) + \sigma(y,\bar{y}) \leq \nu a + b \leq \alpha.$$

Through (3) and (6), there exists $z^1 \in F^{-1}(y-g(x))$ such that

(7) $$\rho(z^1,x) \leq d(x,F^{-1}(y-g(x))) + \varepsilon \leq \lambda d(y,(g+F)(x)) + \varepsilon.$$

If $z^1 = x$, then $x \in F^{-1}(y-g(x))$, which is the same as $x \in (g+F)^{-1}(y)$. Then (5) holds automatically, since its left side is 0. Let $z^1 \neq x$. In this case, using (2), we obtain

(8) $$\begin{aligned}\rho(z^1,x) &\leq \rho(x,\bar{x}) + d(\bar{x},F^{-1}(y-g(x))) + \varepsilon \\ &\leq \rho(x,\bar{x}) + \lambda d(y-g(x),F(\bar{x})) + \varepsilon \\ &\leq \rho(x,\bar{x}) + \lambda\sigma(y,\bar{y}) + \lambda\sigma(g(x),g(\bar{x})) + \varepsilon \\ &\leq \rho(x,\bar{x}) + \lambda\sigma(y,\bar{y}) + \lambda\nu\rho(x,\bar{x}) + \varepsilon \\ &\leq (1+\lambda\nu)a + \lambda b + \varepsilon.\end{aligned}$$

Hence, by (4),

(9) $$\rho(z^1,\bar{x}) \leq \rho(z^1,x) + \rho(x,\bar{x}) \leq (1+\lambda\nu)a + \lambda b + \varepsilon + a \leq \alpha.$$

By induction, we construct a sequence of vectors $z^k \in \mathbb{B}_\alpha(\bar{x})$, with $z^0 = x$, such that, for $k = 0,1,\ldots$,

(10) $$z^{k+1} \in F^{-1}(y-g(z^k)) \quad \text{and} \quad \rho(z^{k+1},z^k) \leq (\lambda\nu+\varepsilon)^k \rho(z^1,x).$$

We already found z^1 which gives us (10) for $k=0$. Suppose that for some $n \geq 1$ we have generated $z^1,z^2,\ldots,z^n$ satisfying (10). If $z^n = z^{n-1}$ then $z^n \in F^{-1}(y-g(z^n))$ and hence $z^n \in (g+F)^{-1}(y)$. Then, by using (2), (7) and (10), we get

$$\begin{aligned}d(x,(g+F)^{-1}(y)) \leq \rho(z^n,x) &\leq \sum_{i=0}^{n-1} \rho(z^{i+1},z^i) \\ &\leq \sum_{i=0}^{n-1}(\lambda\nu+\varepsilon)^i \rho(z^1,x) \leq \frac{1}{1-(\lambda\nu+\varepsilon)}\rho(z^1,x) \\ &\leq \frac{\lambda}{1-(\lambda\nu+\varepsilon)}\left[d(y,(g+F)(x)) + \frac{\varepsilon}{\lambda}\right].\end{aligned}$$

Since the left side of this inequality does not depend on the ε on the right, we are able to obtain (5) by letting ε go to 0.

Assume $z^n \neq z^{n-1}$. We will first show that $z^i \in \mathbb{B}_\alpha(\bar{x})$ for all $i = 2,3,\ldots,n$. Utilizing (10), for such an i we have

5 Regularity in Infinite Dimensions

$$\rho(z^i,x) \leq \sum_{j=0}^{i-1} \rho(z^{j+1},z^j) \leq \sum_{j=0}^{i-1}(\lambda v+\varepsilon)^j \rho(z^1,x) \leq \frac{1}{1-(\lambda v+\varepsilon)}\rho(z^1,x)$$

and therefore, through (8) and (4),

(11) $\quad \rho(z^i,\bar{x}) \leq \rho(z^i,x) + \rho(x,\bar{x}) \leq \frac{1}{1-(\lambda v+\varepsilon)}[(1+\lambda v)a+\lambda b+\varepsilon]+a \leq \alpha.$

Thus, we have $z^i \in \mathbb{B}_\alpha(\bar{x})$ for all $i=1,\ldots,n$.

Taking into account the estimate in (11) for $i=n$ and the third inequality in (4), we get

$$\sigma(y-g(z^n),\bar{y}) \leq \sigma(y,\bar{y})+v\rho(z^n,\bar{x})$$
$$\leq b+v\left(\frac{1}{1-(\lambda v+\varepsilon)}[(1+\lambda v)a+\lambda b+\varepsilon]+a\right) \leq \alpha.$$

Since $\rho(z^n,z^{n-1}) > 0$, from (3) there exists $z^{n+1} \in F^{-1}(y-g(z^n))$ such that

$$\rho(z^{n+1},z^n) \leq d(z^n,F^{-1}(y-g(z^n)))+\varepsilon\rho(z^n,z^{n-1}),$$

and then (2) yields

$$\rho(z^{n+1},z^n) \leq \lambda d(y-g(z^n),F(z^n))+\varepsilon\rho(z^n,z^{n-1}).$$

Since $z^n \in F^{-1}(y-g(z^{n-1}))$ and hence $y-g(z^{n-1}) \in F(z^n)$, by invoking the induction hypothesis, we obtain

$$\rho(z^{n+1},z^n) \leq \lambda \sigma(g(z^n),g(z^{n-1}))+\varepsilon\rho(z^n,z^{n-1})$$
$$\leq (\lambda v+\varepsilon)\rho(z^n,z^{n-1}) \leq (\lambda v+\varepsilon)^n \rho(z^1,x).$$

The induction is complete, and therefore (10) holds for all k.

Right after (10) we showed that when $z^k = z^{k-1}$ for some k then (5) holds. Suppose now that $z^{k+1} \neq z^k$ for all k. By virtue of the second condition in (10), we see for any natural n and m with $m < n$ that

$$\rho(z^n,z^m) \leq \sum_{k=m}^{n-1}\rho(z^{k+1},z^k) \leq \sum_{k=m}^{n-1}(\lambda v+\varepsilon)^k \rho(z^1,x) \leq \frac{\rho(z^1,x)}{1-(\lambda v+\varepsilon)}(\lambda v+\varepsilon)^m.$$

We conclude that the sequence $\{z^k\}$ satisfies the Cauchy condition, and all its elements are in $\mathbb{B}_\alpha(\bar{x})$. Hence this sequence converges to some $z \in \mathbb{B}_\alpha(\bar{x})$ which, from (10) and the local closedness of gph F, satisfies $z \in F^{-1}(y-g(z))$, that is, $z \in (g+F)^{-1}(y)$. Moreover,

$$d(x,(g+F)^{-1}(y)) \leq \rho(z,x) = \lim_{k\to\infty}\rho(z^k,x) \leq \lim_{k\to\infty}\sum_{i=0}^{k}\rho(z^{i+1},z^i)$$

$$\leq \lim_{k\to\infty} \sum_{i=0}^{k} (\lambda v + \varepsilon)^i \rho(z^1, x) \leq \frac{1}{1-(\lambda v + \varepsilon)} \rho(z^1, x)$$

$$\leq \frac{1}{1-(\lambda v + \varepsilon)} \Big[\lambda d(y, (g+F)(x)) + \varepsilon\Big],$$

the final inequality being obtained from (2) and (7). Taking the limit as $\varepsilon \to 0$ we obtain (5), and the proof is finished. $\square$

The second proof of Theorem 5E.1 uses the following extension of the contraction mapping principle 1A.2 for set-valued mappings, furnished with a proof the idea of which goes back to Banach [1922], if not earlier.

Theorem 5E.2 (contraction mapping principle for set-valued mappings). *Let (X, ρ) be a complete metric space, and consider a set-valued mapping $\Phi : X \rightrightarrows X$ and a point $\bar{x} \in X$. Suppose that there exist scalars $a > 0$ and $\lambda \in (0, 1)$ such that the set gph $\Phi \cap (\mathbb{B}_a(\bar{x}) \times \mathbb{B}_a(\bar{x}))$ is closed and*
  (a) $d(\bar{x}, \Phi(\bar{x})) < a(1-\lambda)$;
  (b) $e(\Phi(u) \cap \mathbb{B}_a(\bar{x}), \Phi(v)) \leq \lambda \rho(u, v)$ for all $u, v \in \mathbb{B}_a(\bar{x})$.
Then Φ has a fixed point in $\mathbb{B}_a(\bar{x})$; that is, there exists $x \in \mathbb{B}_a(\bar{x})$ such that $x \in \Phi(x)$.

Proof. By assumption (a) there exists $x^1 \in \Phi(\bar{x})$ such that $\rho(x^1, \bar{x}) < a(1-\lambda)$. Proceeding by induction, let $x^0 = \bar{x}$ and suppose that there exists $x^{k+1} \in \Phi(x^k) \cap \mathbb{B}_a(\bar{x})$ for $k = 0, 1, \ldots, j-1$ with

$$\rho(x^{k+1}, x^k) < a(1-\lambda)\lambda^k.$$

By assumption (b),

$$d(x^j, \Phi(x^j)) \leq e(\Phi(x^{j-1}) \cap \mathbb{B}_a(\bar{x}), \Phi(x^j)) \leq \lambda \rho(x^j, x^{j-1}) < a(1-\lambda)\lambda^j.$$

This implies there is an $x^{j+1} \in \Phi(x^j)$ such that

$$\rho(x^{j+1}, x^j) < a(1-\lambda)\lambda^j.$$

By the triangle inequality,

$$\rho(x^{j+1}, \bar{x}) \leq \sum_{i=0}^{j} \rho(x^{i+1}, x^i) < a(1-\lambda) \sum_{i=0}^{j} \lambda^i < a.$$

Hence $x^{j+1} \in \Phi(x^j) \cap \mathbb{B}_a(\bar{x})$ and the induction step is complete.

For any $k > m > 1$ we then have

$$\rho(x^k, x^m) \leq \sum_{i=m}^{k-1} \rho(x^{i+1}, x^i) < a(1-\lambda) \sum_{i=m}^{k-1} \lambda^i < a\lambda^m.$$

Thus, $\{x^k\}$ is a Cauchy sequence and consequently converges to some $x \in \mathbb{B}_a(\bar{x})$. Since $(x^{k-1}, x^k) \in \text{gph } \Phi \cap (\mathbb{B}_a(\bar{x}) \times \mathbb{B}_a(\bar{x}))$ which is a closed set, we conclude that $x \in \Phi(x)$. □

For completeness, we now supply with a proof the (standard) contraction mapping principle 1A.2.

Proof of Theorem 1A.2. Let Φ be a function which is Lipschitz continuous on $\mathbb{B}_a(\bar{x})$ with constant $\lambda \in [0,1)$ and let $\rho(\bar{x}, \Phi(\bar{x})) \leq a(1-\lambda)$. By repeating the argument in the proof of 5E.2 for the sequence of points x^k satisfying $x^{k+1} = \Phi(x^k)$, $k = 0, 1, \ldots, x^0 = \bar{x}$, with all strict inequalities replaced by non-strict ones, we obtain that Φ has a fixed point in $\mathbb{B}_a(\bar{x})$. Suppose that Φ has two fixed points in $\mathbb{B}_a(\bar{x})$, that is, there are $x, x' \in \mathbb{B}_a(\bar{x})$, $x \neq x'$, with $x = \Phi(x)$ and $x' = \Phi(x')$. Then we have

$$0 < \rho(x, x') = \rho(\Phi(x), \Phi(x')) \leq \lambda \rho(x, x') < \rho(x, x'),$$

which is absurd. Hence, in this case Φ has a unique fixed point. □

Proof II of Theorem 5E.1. Pick the constants κ, μ, ν and λ and a function g as in the beginning of Proof I. Then, utilizing 5A.3, there exist positive constants a and b such that

(12) $\quad e(F^{-1}(y') \cap \mathbb{B}_a(\bar{x}), F^{-1}(y)) \leq \lambda \sigma(y', y) \quad \text{for all } y', y \in \mathbb{B}_{b+\nu a}(\bar{y}),$

(13) $\quad$ the set $\text{gph } F \cap (\mathbb{B}_a(\bar{x}) \times \mathbb{B}_{b+\nu a}(\bar{y}))$ is closed

and also

$$\sigma(g(x'), g(x)) \leq \nu \rho(x', x) \quad \text{for all } x', x \in \mathbb{B}_a(\bar{x}).$$

Take any $\lambda^+ > \lambda$ and make $b > 0$ smaller if necessary so that

(14) $\quad \dfrac{\lambda^+ b}{1 - \lambda \nu} < a/4.$

For any $y \in \mathbb{B}_b(g(\bar{x}) + \bar{y})$ and $x \in \mathbb{B}_a(\bar{x})$, using the shift-invariance of the metric σ and the triangle inequality, we obtain

(15) $\quad \sigma(-g(x) + y, \bar{y}) = \sigma(y, g(x) + \bar{y}) \leq \sigma(y, g(\bar{x}) + \bar{y}) + \sigma(g(x), g(\bar{x})) \leq b + \nu a.$

Fix $y \in \mathbb{B}_b(g(\bar{x}) + \bar{y})$ and consider the mapping

$$\Phi_y : x \mapsto F^{-1}(-g(x) + y) \quad \text{for } x \in \mathbb{B}_a(\bar{x}).$$

Let $y, y' \in \mathbb{B}_b(g(\bar{x}) + \bar{y})$, $y \neq y'$, and let $x' \in (g+F)^{-1}(y') \cap \mathbb{B}_{a/2}(\bar{x})$. We will establish now that there is a fixed point $x \in \Phi_y(x)$ in the closed ball centered at x' with radius

$$\varepsilon := \dfrac{\lambda^+ \sigma(y, y')}{1 - \lambda \nu}.$$

Since $x' \in F^{-1}(-g(x')+y') \cap I\!B_a(\bar{x})$ and both (x',y') and (x',y) satisfy (15), from (12) we get

$$d(x', \Phi_y(x')) \leq e(F^{-1}(-g(x')+y') \cap I\!B_a(\bar{x}), F^{-1}(-g(x')+y))$$
$$\leq \lambda \sigma(y',y) < \lambda^+ \sigma(y',y) = \varepsilon(1-\lambda v).$$

By the triangle inequality and (14), $\varepsilon \leq \lambda^+(2b)/(1-\lambda v) < a/2$, so that $I\!B_\varepsilon(x') \subset I\!B_a(\bar{x})$. Then we have that for any $u, v \in I\!B_\varepsilon(x')$,

$$e(\Phi_y(u) \cap I\!B_\varepsilon(x'), \Phi_y(v)) \leq e(F^{-1}(-g(u)+y) \cap I\!B_a(\bar{x}), F^{-1}(-g(v)+y))$$
$$\leq \lambda \sigma(g(u), g(v)) \leq \lambda v \rho(u,v).$$

By (13) the set $\text{gph } \Phi_y \cap (I\!B_\varepsilon(\bar{x}) \times I\!B_\varepsilon(\bar{x}))$ is closed; hence we can apply the contraction mapping principle in Theorem 5E.2 to the mapping Φ_y, with constants $a = \varepsilon$ and the λ taken to be the λv here, to obtain the existence of a fixed point $x \in \Phi_y(x)$ within distance ε from x'. Since $x \in (g+F)^{-1}(y)$, we obtain

$$d(x', (g+F)^{-1}(y)) \leq \varepsilon = \frac{\lambda^+}{1-\lambda v} \sigma(y', y).$$

This tells us that $(g+F)^{-1}$ has the Aubin property at $g(\bar{x}) + \bar{y}$ for $\bar{x}$ with constant $\lambda^+/(1-\lambda v)$. Hence, by 5A.3, the mapping $g+F$ is metrically regular at $\bar{x}$ for $g(\bar{x}) + \bar{y}$ with the same constant. Since x and y are arbitrarily chosen in the corresponding balls around $\bar{x}$ and $\bar{y}$, λ^+ and λ are arbitrarily close to κ, and v is arbitrarily close to μ, this comes down to (1). □

Next, we put together an implicit function version of Theorem 5E.1.

Theorem 5E.3 (implicit mapping theorem for metric regularity in metric spaces). *Let (X, ρ) be a complete metric space and let (Y, σ) be a linear metric space with shift-invariant metric. Let (P, π) be a metric space. For $f : P \times X \to Y$ and $F : X \rightrightarrows Y$, consider the generalized equation $f(p,x) + F(x) \ni 0$ with solution mapping*

$$S(p) = \{x \mid f(p,x) + F(x) \ni 0\} \text{ having } \bar{x} \in S(\bar{p}).$$

Let $h : X \to Y$ be a strict estimator of f with respect to x uniformly in p at $(\bar{p}, \bar{x})$ with constant μ, let $\text{gph}(h+F)$ be locally closed at $(\bar{x}, 0)$ and suppose that $h+F$ is metrically regular at $\bar{x}$ for 0 with $\text{reg}(h+F; \bar{x}|0) \leq \kappa$. Assume

$$\kappa\mu < 1 \quad \text{and} \quad \widehat{\text{lip}}_p(f;(\bar{p},\bar{x})) \leq \gamma < \infty.$$

Then S has the Aubin property at $\bar{p}$ for $\bar{x}$, and moreover

(16) $$\text{lip}(S; \bar{p}|\bar{x}) \leq \frac{\kappa\gamma}{1-\kappa\mu}.$$

Proof. Choose $\lambda > \kappa$ and $v > \mu$ such that $\lambda v < 1$. Also, let $\beta > \gamma$. Then there exist positive scalars α and τ such that the set $\text{gph}(h+F) \cap (I\!B_\alpha(\bar{x}) \times I\!B_\alpha(0))$ is closed,

5 Regularity in Infinite Dimensions

(17) $\quad e\Big((h+F)^{-1}(y') \cap B_\alpha(\bar{x}), (h+F)^{-1}(y)\Big) \leq \lambda\, \sigma(y',y) \quad \text{for all } y', y \in B_\alpha(0),$

(18) $\quad \sigma(r(p,x'), r(p,x)) \leq \nu \rho(x',x) \quad \text{for all } x', x \in B_\alpha(\bar{x}) \text{ and } p \in B_\tau(\bar{p})$

where $r(p,x) = f(p,x) - h(x)$, and

(19) $\quad \sigma(f(p',x), f(p,x)) \leq \beta\, \pi(p',p) \quad \text{for all } p', p \in B_\tau(\bar{p}) \text{ and } x \in B_\alpha(\bar{x}).$

Let

(20) $\quad \dfrac{2\lambda\beta}{1-\lambda\nu} \geq \lambda^+ > \dfrac{\lambda\beta}{1-\lambda\nu}.$

Now, choose positive $a < \alpha$ and then positive $q \leq \tau$ such that

(21) $\quad \nu a + \beta q \leq \alpha \quad \text{and} \quad \dfrac{4\lambda\beta q}{1-\lambda\nu} + a \leq \alpha.$

Then, from (18) and (19), for every $x \in B_a(\bar{x})$ and $p \in B_q(\bar{p})$ we have

(22) $\quad \begin{aligned} \sigma(r(p,x), 0) &\leq \sigma(r(p,x), r(p,\bar{x})) + \sigma(r(p,\bar{x}), r(\bar{p},\bar{x})) \\ &\leq \nu\rho(x,\bar{x}) + \beta\pi(p,\bar{p}) \leq \nu a + \beta q \leq \alpha. \end{aligned}$

Fix $p \in B_q(\bar{p})$ and consider the mapping

$$\Phi_p : x \mapsto (h+F)^{-1}(-r(p,x)) \quad \text{for } x \in B_\alpha(\bar{x}).$$

Observe that for any $x \in B_a(\bar{x})$ and $p \in B_q(\bar{p})$, $x \in \Phi_p(x) \iff x \in S(p)$ and also that the set $\mathrm{gph}\,\Phi_p \cap (B_\alpha(\bar{x}) \times B_\alpha(\bar{x}))$ is closed. Let $p', p \in B_q(\bar{p})$ with $p \neq p'$ and let $x' \in S(p') \cap B_a(\bar{x})$. Let $\varepsilon := \lambda^+ \pi(p', p)$; then $\varepsilon \leq \lambda^+(2q)$. Thus, remembering that $x' \in \Phi_{p'}(x') \cap B_\alpha(\bar{x})$, from (17), where we use (22), and from (18), (19) and (20), we deduce that

$$d(x', \Phi_p(x')) \leq e\Big((h+F)^{-1}(-r(p',x')) \cap B_\alpha(x'), (h+F)^{-1}(-r(p,x'))\Big)$$
$$\leq \lambda\, \sigma(f(p',x'), f(p,x')) \leq \lambda\beta\, \pi(p',p) < \lambda^+ \pi(p',p)(1-\lambda\nu) = \varepsilon(1-\lambda\nu).$$

Since $x' \in B_a(\bar{x})$ and, by (20) and (21),

$$\varepsilon \leq 2\lambda^+ q \leq \dfrac{4\lambda\beta q}{1-\lambda\nu},$$

we get $B_\varepsilon(x') \subset B_\alpha(\bar{x})$. Then, for any $u, v \in B_\varepsilon(x')$ using again (17) (with (22)) and (18), we see that

$$e(\Phi_p(u) \cap I\!B_\varepsilon(x'), \Phi_p(v))$$
$$\leq e\left((h+F)^{-1}(-r(u,p)) \cap I\!B_\varepsilon(\bar{x}), (h+F)^{-1}(-r(v,p))\right)$$
$$\leq \lambda\,\sigma(r(p,u), r(p,v)) \leq \lambda v \rho(u,v).$$

Hence the contraction mapping principle in Theorem 5E.2 applies, with the λ there taken to be the λv here, and it follows that there exists $x \in \Phi_p(x) \cap I\!B_\varepsilon(x')$ and hence $x \in S(p) \cap I\!B_\varepsilon(x')$. Thus,

$$d(x', S(p)) \leq \rho(x', x) \leq \varepsilon = \lambda^+ \pi(p', p).$$

Since this inequality holds for any $x' \in S(p') \cap I\!B_a(\bar{x})$ and any λ^+ fulfilling (20), we arrive at

$$e(S(p') \cap I\!B_a(\bar{x}), S(p)) \leq \lambda^+ \pi(p', p).$$

That is, S has the Aubin property at $\bar{p}$ for $\bar{x}$ with modulus not greater than λ^+. Since λ^+ can be arbitrarily close to $\lambda/(1-\lambda v)$, and λ, v and β can be arbitrarily close to κ, μ and γ, respectively, we achieve the estimate (16). □

We can also state Theorem 3F.9 in Banach spaces with only minor adjustments in notation and terminology.

Theorem 5E.4 (using strict differentiability and ample parameterization). *Let X, Y and P be Banach spaces. For $f: P \times X \to Y$ and $F: X \rightrightarrows Y$, consider the generalized equation $f(p,x) + F(x) \ni 0$ with solution mapping S and a pair $(\bar{p}, \bar{x})$ with $\bar{x} \in S(\bar{p})$. Suppose that f is strictly differentiable at $(\bar{p}, \bar{x})$ and that gph F is locally closed at $(\bar{x}, -f(\bar{p}, \bar{x}))$. If the mapping*

$$h + F \text{ for } h(x) = f(\bar{p}, \bar{x}) + D_x f(\bar{p}, \bar{x})(x - \bar{x})$$

is metrically regular at $\bar{x}$ for 0, then S has the Aubin property at $\bar{p}$ for $\bar{x}$ with

$$\mathrm{lip}\,(S; \bar{p}|\bar{x}) \leq \mathrm{reg}\,(h+F; \bar{x}|0) \cdot \|D_p f(\bar{p}, \bar{x})\|.$$

Furthermore, when f satisfies the ample parameterization condition:

the mapping $D_p f(\bar{p}, \bar{x})$ is surjective,

then the converse implication holds as well: the mapping $h + F$ is metrically regular at $\bar{x}$ for 0 provided that S has the Aubin property at $\bar{p}$ for $\bar{x}$.

The following exercise, which we supply with a detailed guide, deals with a more general kind of perturbation and shows that not only the constant but also the neighborhoods of metric regularity of the perturbed mapping can be independent of the perturbation provided that its Lipschitz constant is "sufficiently small." For better transparency, we consider mappings with closed graphs acting in Banach spaces.

Exercise 5E.5. *Let X and Y be Banach spaces and consider a continuous function $f: X \to Y$, a mapping $F: X \rightrightarrows Y$ with closed graph and a point $\bar{x} \in X$ such that*

5 Regularity in Infinite Dimensions

$0 \in f(\bar{x}) + F(\bar{x})$. Let κ and μ be positive constants such that

$$\operatorname{reg}(F; \bar{x} | \bar{y}) < \kappa \quad \text{and} \quad \kappa\mu < 1.$$

Prove that for any κ' satisfying

(23) $$\frac{\kappa}{1 - \kappa\mu} < \kappa'$$

there exist positive constants α and β such that for every mapping $A : X \times X \to Y$ and every $\tilde{x} \in \mathbb{B}_\alpha(\bar{x})$ and $\tilde{y} \in \mathbb{B}_\beta(0)$ with the properties that

$$\tilde{y} \in A(\tilde{x}, \tilde{x}) + F(\tilde{x}), \quad \|A(\tilde{x}, u) - f(\tilde{x})\| \leq \beta \text{ for every } u \in \mathbb{B}_\alpha(\bar{x})$$

and

$$\|[A(x', u) - f(x')] - [A(x, u) - f(x)]\| \leq \mu \|x' - x\| \text{ for every } x', x, u \in \mathbb{B}_{\alpha + 5\kappa'\beta}(\bar{x}),$$

we have that for any $u \in \mathbb{B}_\alpha(\bar{x})$ the mapping $A(\cdot, u) + F(\cdot)$ is metrically regular at $\tilde{x}$ for $\tilde{y}$ with constant κ' and neighborhoods $\mathbb{B}_\alpha(\tilde{x})$ and $\mathbb{B}_\beta(\tilde{y})$.

Guide. Let a and b be positive constants such that $f + F$ is metrically regular with constant κ and neighborhoods $\mathbb{B}_a(\bar{x})$ and $\mathbb{B}_b(0)$. Choose κ' satisfying (23) and then positive α and β such that

$$\alpha \leq 2\kappa'\beta, \quad 2\alpha + 5\kappa'\beta \leq a \text{ and } 6\beta + \mu\alpha \leq b.$$

Pick $x, u \in \mathbb{B}_\alpha(\tilde{x})$ and $y \in \mathbb{B}_\beta(\tilde{y})$. First prove that

(24) $$d(x, (A(\cdot, u) + F(\cdot))^{-1}(y)) \leq \kappa' \|y - y'\|$$
$$\text{for any } y' \in (A(x, u) + F(x)) \cap \mathbb{B}_{4\beta}(\tilde{y}).$$

Let $y' \in (A(x, u) + F(x)) \cap \mathbb{B}_{4\beta}(\tilde{y})$, $y' \neq y$. Observe that

$$\| -A(x,u) + f(x) + y' \| \leq \|y' - \tilde{y}\| + \|\tilde{y}\| + \| -A(x,u) + f(x) + A(\tilde{x}, u) - f(\tilde{x}) \|$$
$$+ \|A(\tilde{x}, u) - A(\tilde{x}, \tilde{x})\| + \|A(\tilde{x}, \tilde{x}) - f(\tilde{x})\|$$
$$\leq 4\beta + \mu\alpha + 2\beta \leq 6\beta + \mu\alpha \leq b.$$

The same estimate holds of course with y' replaced by y; that is, both $-A(x,u) + f(x) + y'$ and $-A(x,u) + f(x) + y$ are in $\mathbb{B}_b(0)$. Consider the mapping

$$\Phi : v \mapsto (f + F)^{-1}(-A(v,u) + f(v)) + y) \text{ for } v \in \mathbb{B}_\alpha(\tilde{x}).$$

Since $\mathbb{B}_\alpha(\tilde{x}) \subset \mathbb{B}_a(\bar{x})$, utilizing the metric regularity of $f + F$ we obtain

$$d(x, \Phi(x)) = d(x, (f+F)^{-1}(-A(x,u) + f(x) + y))$$
$$\leq \kappa d(-A(x,u) + f(x) + y, (f+F)(x))$$

$$\leq \kappa\|-A(x,u)+f(x)+y-(y'-A(x,u)+f(x))\|$$
$$= \kappa\|y-y'\| < \kappa'\|y-y'\|(1-\kappa\mu) = r(1-\kappa\mu),$$

where $r := \kappa'\|y-y'\|$. Observe that $r \leq 5\kappa'\beta$ and $I\!B_r(x) \subset I\!B_{\alpha+5\kappa'\beta}(\tilde{x}) \subset I\!B_a(\tilde{x})$. Again, by utilizing the metric regularity of $f+F$ we get that for every $v,w \in I\!B_r(x)$,

$$e(\Phi(v) \cap I\!B_r(x), \Phi(w))$$
$$\leq \sup_{z \in (f+F)^{-1}(-A(v,u)+f(v)+y) \cap B_a(\tilde{x})} d(z, (f+F)^{-1}(-A(w,u)+f(w)+y))$$
$$\leq \sup_{z \in (f+F)^{-1}(-A(v,u)+f(v)+y) \cap B_a(\tilde{x})} \kappa d(-A(w,u)+f(w)+y, f(z)+F(z))$$
$$\leq \kappa\|-A(v,u)+f(v)-[-A(w,v)+f(w)]\| \leq \kappa\mu\|v-w\|.$$

Noting that the set $\mathrm{gph}\, \Phi \cap (I\!B_r(x) \times I\!B_r(x))$ is closed, Theorem 5E.2 then yields the existence of a fixed point $\hat{x} \in \Phi_y(\hat{x}) \cap I\!B_r(x)$; that is,

$$y \in A(\hat{x},u)+F(\hat{x}) \quad \text{and} \quad \|\hat{x}-x\| \leq \kappa'\|y-y'\|.$$

Since $\hat{x} \in (A(\cdot,u)+F(\cdot))^{-1}(y)$ we obtain (24).

Now we are ready to show the desired inequality

(25) $\qquad d(x, (A(\cdot,u)+F(\cdot))^{-1}(y)) \leq \kappa' d(y, A(x,u)+F(x)).$

If $A(x,u)+F(x) = \emptyset$ there is nothing to prove. If not, choose $\varepsilon > 0$ and $w \in A(x,y)+F(x)$ such that $\|w-y\| \leq d(y, A(x,u)+F(x))+\varepsilon$. If $w \in I\!B_{4\beta}(\tilde{y})$, then from (25) with $y' = w$ we get

$$d(x, (A(\cdot,u)+F(\cdot))^{-1})(y)) \leq \kappa'\|w-y\| \leq \kappa' d(y, A(x,u)+F(x))+\kappa'\varepsilon,$$

which yields (25) since the left side does not depend on ε. Otherwise, we have

$$\|y-w\| \geq \|w-\tilde{y}\| - \|y-\tilde{y}\| \geq 4\beta - \beta = 3\beta.$$

But then,

$$d(x, (A(\cdot,u)+F(\cdot))^{-1}(y)) \leq \alpha + d(\tilde{x}, (A(\cdot,u)+F(\cdot))^{-1}(y))$$
$$\leq \alpha + \kappa'\|y-\tilde{y}\| \leq \alpha + \kappa'\beta \leq 3\kappa'\beta \leq \kappa'\|y-w\|$$
$$\leq \kappa' d(y, A(x,u)+F(x)) + \kappa'\varepsilon,$$

which leads (25) by letting $\varepsilon \to 0$. $\square$

Example 5E.6 (counterexample for set-valued perturbations). Consider $F : \mathbb{R} \rightrightarrows \mathbb{R}$ and $G : \mathbb{R} \rightrightarrows \mathbb{R}$ specified by

$$F(x) = \{-2x, 1\} \quad \text{and} \quad G(x) = \{x^2, -1\} \text{ for } x \in \mathbb{R}.$$

Then F is metrically regular at 0 for 0 while G has the Aubin property at 0 for 0. Moreover, both F^{-1} and G are Lipschitz continuous (with respect to the Pompeiu-Hausdorff distance) on the whole of $\mathbb{R}$. We have $\text{reg}(F;0|0) = 1/2$ whereas the Lipschitz modulus of the single-valued localization of G around 0 for 0 is 0 and serves also as the infimum of all Aubin constants. The mapping

$$(F+G)(x) = \{x^2 - 2x, x^2 + 1, -2x - 1, 0\} \text{ for } x \in \mathbb{R}$$

is not metrically regular at 0 for 0. Indeed, $(F+G)^{-1}$ has a single-valued localization s around $(0,0)$ of the form $s(y) = 1 - \sqrt{1+y}$, so that $d(x, (F+G)^{-1}(y)) = |x - 1 + \sqrt{1+y}|$, but also $d(y, (F+G)(x)) = \min\{|x^2 - 2x - y|, y\}$. Take $x = \varepsilon > 0$ and $y = \varepsilon^2$. Then, since $(\varepsilon - 1 + \sqrt{1+\varepsilon^2})/\varepsilon^2 \to \infty$, the mapping $F + G$ is seen not to be metrically regular at 0 for 0.

Exercise 5E.7. *Prove that, under the conditions of Theorem 5E.1,*

$$\inf_{g:X \to Y} \left\{ \text{lip}(g; \bar{x}) \,\Big|\, F + g \text{ is not metrically regular at } \bar{x} \text{ for } \bar{y} + g(\bar{x}) \right\} \geq \frac{1}{\text{reg}(F; \bar{x}|\bar{y})}.$$

At the end of this section we will derive from Theorem 5E.2 the following fixed point theorem due to Nadler [1969]:

Theorem 5E.8 (Nadler). *Let (X, ρ) be a complete metric space and suppose that Φ maps X into the set of closed subsets of X and is Lipschitz continuous in the sense of Pompeiu-Hausdorff distance on X with Lipschitz constant $\lambda \in (0,1)$. Then Φ has a fixed point.*

Proof. We will first show that Φ has closed graph. Indeed, let $(x_k, y_k) \in \text{gph } \Phi$ and $(x_k, y_k) \to (x, y)$. Then

$$d(y, \Phi(x)) \leq \rho(y, y_k) + d(y_k, \Phi(x))$$
$$\leq \rho(y, y_k) + h(\Phi(x_k), \Phi(x))$$
$$\leq \rho(y, y_k) + \lambda \rho(x_k, x) \to 0 \text{ as } k \to \infty.$$

Hence $d(y, \Phi(x)) = 0$ and since $\Phi(x)$ is closed we have $(x, y) \in \text{gph } \Phi$, and therefore $\text{gph } \Phi$ is closed as claimed.

Let $\bar{x} \in X$ and choose $a > d(\bar{x}, \Phi(\bar{x}))/(1 - \lambda)$. Since $\text{gph } \Phi$ is closed, the set $\text{gph } \Phi(x) \cap (\mathbb{B}_a(\bar{x}) \times \mathbb{B}_a(\bar{x}))$ is closed as well. Furthermore, for every $u, v \in \mathbb{B}_a(\bar{x})$ we obtain

$$e(\Phi(u) \cap \mathbb{B}_a(\bar{x}), \Phi(v)) \leq e(\Phi(u), \Phi(v)) \leq h(\Phi(u), \Phi(v)) \leq \lambda \rho(u, v).$$

Hence, by Theorem 5E.2 there exists $x \in X$ such that $x \in \Phi(x)$. □

5F. Strong Metric Regularity and Implicit Function Theorems

Here we first present a strong regularity analogue of Theorem 5E.1 that provides a sharper view of the interplay among the constants and neighborhoods of a mapping and its perturbation. As in the preceding section, we consider mappings acting in metric spaces, to which the concept of strong regularity can be extended in an obvious way. The following result significantly generalizes Theorem 3G.3 (and moreover its previous version 2B.10).

Theorem 5F.1 (inverse function theorem with strong metric regularity in metric spaces). *Let (X,ρ) be a complete metric space and let (Y,σ) be a linear metric space with shift-invariant metric σ. Consider a mapping $F : X \rightrightarrows Y$ and any $(\bar{x},\bar{y}) \in$ gph F such that, for a nonnegative constant κ and neighborhoods U of $\bar{x}$ and V of $\bar{y}$, the mapping $y \mapsto F^{-1}(y) \cap U$ is a Lipschitz continuous function on V with Lipschitz constant κ.*

Then for every nonnegative constant μ with $\kappa\mu < 1$ there exist neighborhoods U' of $\bar{x}$ and V' of $\bar{y}$ such that, for every function $g : X \to Y$ which is Lipschitz continuous on U with Lipschitz constant μ, the mapping $y \mapsto (g+F)^{-1}(y) \cap U'$ is a Lipschitz continuous function on $g(\bar{x}) + V'$ with Lipschitz constant $\kappa/(1 - \kappa\mu)$.

Proof. We apply the standard (single-valued) version of the contracting mapping principle, 1A.2, as in Proof II of 5E.1 but with some adjustments in the argument. By assumption, for the function $s(y) = F^{-1}(y) \cap U$ for $y \in V$ we have

(1) $$\rho(s(y'),s(y)) \leq \kappa\sigma(y',y) \text{ for all } y',y \in V.$$

Pick $\mu > 0$ such that $\kappa\mu < 1$ and then choose positive constants a and b such that

(2) $$\mathbb{B}_a(\bar{x}) \subset U, \quad \mathbb{B}_{b+\mu a}(\bar{y}) \subset V \quad \text{and} \quad \kappa b \leq a(1 - \kappa\mu).$$

Choose any function $g : X \to Y$ such that

(3) $$\sigma(g(x'),g(x)) \leq \mu\rho(x',x) \text{ for all } x',x \in U.$$

For any $y \in \mathbb{B}_b(g(\bar{x}) + \bar{y})$ and any $x \in \mathbb{B}_a(\bar{x})$ we have

$$\sigma(-g(x)+y,\bar{y}) = \sigma(y, g(x)+\bar{y}) \leq \sigma(y, g(\bar{x})+\bar{y}) + \sigma(g(x), g(\bar{x})) \leq b + \mu a,$$

and hence, by (2), $-g(x)+y \in V \subset$ dom s. Fix $y \in \mathbb{B}_b(g(\bar{x})+\bar{y})$ and consider the mapping

$$\Phi_y : x \mapsto s(-g(x)+y) \text{ for } x \in \mathbb{B}_a(\bar{x}).$$

Then, by using (1), (2) and (3) we get

$$\rho(\bar{x}, \Phi_y(\bar{x})) = \rho(s(\bar{y}), s(y-g(\bar{x}))) \leq \kappa\sigma(y, \bar{y}+g(\bar{x})) \leq \kappa b \leq a(1-\kappa\mu).$$

Moreover, for any $u, v \in \mathbb{B}_a(\bar{x})$,

5 Regularity in Infinite Dimensions   293

$$\rho(\Phi_y(u), \Phi_y(v)) = \rho(s(y-g(u)), s(y-g(v))) \leq \kappa\sigma(g(u),g(v)) \leq \kappa\mu\rho(u,v).$$

Thus, by the contraction mapping principle 1A.2, there exists a fixed point $x = \Phi_y(x)$ in $\mathbb{B}_a(\bar{x})$, and there is no more than one such fixed point in $\mathbb{B}_a(\bar{x})$. The mapping from $y \in \mathbb{B}_b(g(\bar{x})+\bar{y})$ to the unique fixed point $x(y)$ of Φ_y in $\mathbb{B}_a(\bar{x})$ is a function which satisfies $x(y) = s(y - g(x(y)))$; therefore, for any $y, y' \in \mathbb{B}_b(g(\bar{x})+\bar{y})$ we have

$$\begin{aligned}\rho(x(y), x(y')) &= \rho(s(y-g(x(y))), s(y'-g(x(y')))) \\ &\leq \kappa(\sigma(y,y') + \sigma(g(x(y)), g(x(y')))) \\ &\leq \kappa\sigma(y,y') + \kappa\mu\rho(x(y),x(y')).\end{aligned}$$

Hence,

$$\rho(x(y),x(y')) \leq \frac{\kappa}{1-\kappa\mu}\sigma(y,y').$$

Choosing $U' = \mathbb{B}_a(\bar{x})$ and $V' = \mathbb{B}_b(\bar{y})$, and noting that $\mathbb{B}_b(g(\bar{x})+\bar{y}) = g(\bar{x}) + \mathbb{B}_b(\bar{y})$, we end the proof. □

Compared with 3G.3, the above theorem exposes the fact that not only the Lipschitz constant, but also the neighborhoods associated with the Lipschitz localization of $(g + F)^{-1}$ depend on the mapping F only, and *not* on the perturbation g, as long as its Lipschitz modulus is less than the reciprocal to the regularity modulus of F. We already stated such a result for metric regularity in 5E.5.

Exercise 5F.2. *Derive 5F.1 from a reformulation of 5E.5 in metric spaces.*

Exercise 5F.3. *In the framework of 5F.1, let $F : X \rightrightarrows Y$ be strongly metrically regular at $\bar{x}$ for $\bar{y}$ and and let $\kappa > \operatorname{reg}(F;\bar{x}|\bar{y})$. For a metric space (P,π) consider a function $r : P \times X \to Y$ having $(\bar{p},\bar{x}) \in \operatorname{int}\operatorname{dom} r$ and such that, for some $\mu \in (0, 1/\kappa)$,*

$$r(\cdot,\bar{x}) \text{ is continuous at } \bar{p} \text{ and } \widehat{\operatorname{lip}}_x(r;(\bar{p},\bar{x})) < \mu.$$

Prove that for each $\gamma \geq \kappa/(1 - \kappa\mu)$ there are neighborhoods U of $\bar{x}$, V of $\bar{y}$ and Q of $\bar{p}$ such that, for every $p \in Q$, the mapping $y \mapsto (r(p,\cdot)+F)^{-1}(y) \cap U$ is a Lipschitz continuous function on $r(\bar{p},\bar{x}) + V$ with Lipschitz constant γ.

Guide. Choose neighborhoods U' of $\bar{x}$ and Q of $\bar{p}$ such that for each $p \in Q$ the function $r(p,\cdot)$ is Lipschitz continuous on U' with Lipschitz constant μ. Applying Theorem 5F.1, we obtain a constant γ and neighborhoods U of $\bar{x}$, and V' of $\bar{y}$ such that for every $p \in Q$ the mapping $y \mapsto (r(p,\cdot)+F)^{-1}(y) \cap U$ is a Lipschitz continuous function on $r(p,\bar{x}) + V'$ with Lipschitz constant γ. Since V' is independent of $p \in Q$, by making Q small enough we can find a neighborhood V of $\bar{y}$ such that $r(\bar{p},\bar{x}) + V \subset r(p,\bar{x}) + V'$ for every $p \in Q$. □

We present next a strong regularity extension of Theorem 5E.3 such as has already appeared in various forms in the preceding chapters. We proved a weaker version of this result in 2B.5 via Lemma 2B.6 and stated it again in Theorem 3G.4, which we left unproved. Here we treat a general case which can be deduced from Theorem 5E.3 by taking into account that the strong metric regularity of $h+F$ automatically implies local closedness of its graph, and by adjoining to that the argument in the proof of 3G.2. Since the result is central in this book, with the risk of repeating ourselves, we supply it with an unabbreviated proof.

Theorem 5F.4 (implicit function theorem with strong metric regularity in metric spaces). *Let (X,ρ) be a complete metric space and let (Y,σ) be a linear metric space with shift-invariant metric. Let (P,π) be a metric space. For $f : P \times X \to Y$ and $F : X \rightrightarrows Y$, consider the generalized equation $f(p,x) + F(x) \ni 0$ with solution mapping*

$$S(p) = \{x \,|\, f(p,x) + F(x) \ni 0\} \quad \text{having } \bar{x} \in S(\bar{p}).$$

Let $f(\cdot,\bar{x})$ be continuous at $\bar{p}$ and let $h : X \to Y$ be a strict estimator of f with respect to x uniformly in p at $(\bar{p},\bar{x})$ with constant μ. Suppose that $h+F$ is strongly metrically regular at $\bar{x}$ for 0 or, equivalently, the inverse $(h+F)^{-1}$ has a Lipschitz continuous single-valued localization ω around 0 for $\bar{x}$ such that there exists $\kappa \geq \mathrm{reg}(h+F;\bar{x}|0) = \mathrm{lip}(\omega;0)$ with $\kappa\mu < 1$.

Then the solution mapping S has a single-valued localization s around $\bar{p}$ for $\bar{x}$. Moreover, for every $\varepsilon > 0$ there exists a neighborhood Q of $\bar{p}$ such that

$$(4) \quad \rho(s(p'),s(p)) \leq \frac{\kappa + \varepsilon}{1 - \kappa\mu} \sigma(f(p',s(p)), f(p,s(p))) \quad \text{for all } p', p \in Q.$$

In particular, s is continuous at $\bar{p}$. In addition, if

$$(5) \quad \mathrm{clm}_p(f;(\bar{x},\bar{p})) < \infty,$$

then the solution mapping S has a single-valued graphical localization s around $\bar{p}$ for $\bar{x}$ which is calm at $\bar{p}$ with

$$(6) \quad \mathrm{clm}(s;\bar{p}) \leq \frac{\kappa}{1 - \kappa\mu} \mathrm{clm}_p(f;(\bar{p},\bar{x})).$$

If (5) is replaced by the stronger condition

$$(7) \quad \widehat{\mathrm{lip}}_p(f;(\bar{p},\bar{x})) < \infty,$$

then the graphical localization s of S around $\bar{p}$ for $\bar{x}$ is Lipschitz continuous near $\bar{p}$ with

$$(8) \quad \mathrm{lip}(s;\bar{p}) \leq \frac{\kappa}{1 - \kappa\mu} \widehat{\mathrm{lip}}_p(f;(\bar{p},\bar{x})).$$

If $h : X \to Y$ is not only a strict estimator of f, but also a strict first-order approximation of f with respect to x uniformly in p at $(\bar{p},\bar{x})$, then, under (5), we have

$$\mathrm{clm}(s;\bar{p}) \leq \mathrm{lip}(\omega;0)\,\mathrm{clm}_p(f;(\bar{p},\bar{x})),$$

5 Regularity in Infinite Dimensions

and under (7),
$$\text{lip}(s;\bar{p}) \leq \text{lip}(\omega;0)\widehat{\text{lip}}_p(f;(\bar{p},\bar{x})).$$

Proof. Let $\varepsilon > 0$ and choose $\lambda > \kappa$ and $\nu > \mu$ such that

(9) $$\lambda \nu < 1 \quad \text{and} \quad \frac{\lambda}{1-\lambda\nu} \leq \frac{\kappa+\varepsilon}{1-\kappa\mu}.$$

Then there exists a positive scalar α such that for each $y \in B_\alpha(\bar{y})$ the set $(h+F)^{-1}(y) \cap B_\alpha(\bar{x})$ is a singleton, equal to the value $\omega(y)$ of the single-valued localization of $(h+F)^{-1}$, and this localization ω is Lipschitz continuous with Lipschitz constant λ on $B_\alpha(0)$. We adjust α and choose a positive τ to also have, for $e(p,x) = f(p,x) - h(x)$,

(10) $$\sigma(e(p,x'),e(p,x)) \leq \nu\rho(x',x) \quad \text{for all } x',x \in B_\alpha(\bar{x}) \text{ and } p \in B_\tau(\bar{p}).$$

Choose a positive $a \leq \alpha$ satisfying

(11) $$\nu a + \frac{a(1-\lambda\nu)}{\lambda} \leq \alpha$$

and then a positive $r \leq \tau$ such that

(12) $$\sigma(f(p,\bar{x}),f(\bar{p},\bar{x})) \leq \frac{a(1-\lambda\nu)}{\lambda} \quad \text{for all } p \in B_r(\bar{p}).$$

Then for every $x \in B_a(\bar{x})$ and $p \in B_r(\bar{p})$, from (10)–(12) we have

$$\sigma(e(p,x),0) \leq \sigma(e(p,x),e(p,\bar{x})) + \sigma(e(p,\bar{x}),e(\bar{p},\bar{x}))$$
$$\leq \nu\rho(x,\bar{x}) + \sigma(f(p,\bar{x}),f(\bar{p},\bar{x})) \leq \nu a + a(1-\lambda\nu)/\lambda \leq \alpha.$$

Hence, for such x and p, $e(p,x) \in \text{dom } \omega$.

Fix an arbitrary $p \in B_r(\bar{p})$ and consider the mapping

$$\Phi_p : x \mapsto \omega(-e(p,x)) \quad \text{for } x \in B_a(\bar{x}).$$

Observe that for any $x \in B_a(\bar{x})$ having $x = \Phi_p(x)$ implies $x \in S(p) \cap B_a(\bar{x})$, and conversely. Noting that $\bar{x} = \omega(0)$ and using (12) we obtain

$$\rho(\bar{x},\Phi_p(\bar{x})) = \rho(\omega(0),\omega(-e(p,\bar{x}))) \leq \lambda\,\sigma(f(\bar{p},\bar{x}),f(p,\bar{x})) \leq a(1-\lambda\nu).$$

Further, for any $u,v \in B_a(\bar{x})$, using (10) we see that

$$\rho(\Phi_p(u),\Phi_p(v)) = \rho(\omega(-e(u,p)),\omega(-e(v,p)))$$
$$\leq \lambda\,\sigma(e(p,u),e(p,v)) \leq \lambda\nu\rho(u,v).$$

Hence the contraction mapping principle 1A.2 applies, with the λ there taken to be the $\lambda\nu$ here, and it follows that for each $p \in B_r(\bar{p})$ there exists exactly one $s(p)$ in $B_a(\bar{x})$ such that $s(p) \in S(p)$; thus

(13) $$s(p) = \omega(-e(p,s(p))).$$

The function $p \mapsto s(p)$ is therefore a single-valued localization of S around $\bar{p}$ for $\bar{x}$. Moreover, from (13), for each $p', p \in \mathbb{B}_r(\bar{p})$ we have

$$\begin{aligned}
\rho(s(p'),s(p)) &= \rho(\omega(-e(p',s(p'))),\omega(-e(p,s(p)))) \\
&\leq \rho(\omega(-e(p',s(p'))),\omega(-e(p',s(p)))) \\
&\quad + \rho(\omega(-e(p',s(p))),\omega(-e(p,s(p)))) \\
&\leq \lambda\,\sigma(-e(p',s(p')),-e(p',s(p))) \\
&\quad + \lambda\,\sigma(-e(p',s(p)),-e(p,s(p))) \\
&\leq \lambda v \rho(s(p'),s(p)) + \lambda\,\sigma(f(p',s(p)),f(p,s(p))).
\end{aligned}$$

Hence,

$$\rho(s(p'),s(p)) \leq \frac{\lambda}{1-\lambda v}\sigma(f(p',s(p)),f(p,s(p))).$$

Taking into account (9), we obtain (4). In particular, for $p = \bar{p}$, from the continuity of $f(\cdot,\bar{x})$ at $\bar{p}$ we get that s is continuous at $\bar{p}$. Under (5), the estimate (6) directly follows from (4) by passing to zero with ε, and the same for (8) under (7). If h is a strict first-order approximation of f, then μ could be arbitrarily small, and by passing to $\mathrm{lip}(\omega;0)$ with κ and to 0 with μ we obtain from (6) and (8) the last two estimates in the statement. $\square$

Utilizing strict differentiability and ample parameterization we come to the following infinite-dimensional implicit function theorem which parallels 5E.4.

Theorem 5F.5 (using strict differentiability and ample parameterization). *Let X, Y and P be Banach spaces. For $f: P \times X \to Y$ and $F: X \rightrightarrows Y$, consider the generalized equation $f(p,x) + F(x) \ni 0$ with solution mapping S and a pair $(\bar{p},\bar{x})$ with $\bar{x} \in S(\bar{p})$ and suppose that f is strictly differentiable at $(\bar{p},\bar{x})$. If the mapping*

$$h + F \text{ for } h(x) = f(\bar{p},\bar{x}) + D_x f(\bar{p},\bar{x})(x - \bar{x})$$

is strongly metrically regular at $\bar{x}$ for 0, then S has a Lipschitz continuous single-valued localization s around $\bar{p}$ for $\bar{x}$ with

$$\mathrm{lip}(s;\bar{p}) \leq \mathrm{reg}(h+F;\bar{x}|0) \cdot \|D_p f(\bar{p},\bar{x})\|.$$

Furthermore, when f satisfies the ample parameterization condition:

the mapping $D_p f(\bar{p},\bar{x})$ is surjective,

then the converse implication holds as well: the mapping $h + F$ is strongly metrically regular at $\bar{x}$ for 0 provided that S has a Lipschitz continuous single-valued localization around $\bar{p}$ for $\bar{x}$.

5G. The Bartle–Graves Theorem and Extensions

To set the stage, we begin with a Banach space version of the implication (i) $\Rightarrow$ (ii) in the symmetric inverse function theorem 1D.9.

Theorem 5G.1 (inverse function theorem in infinite dimensions). *Let X be a Banach space and consider a function $f : X \to X$ and a point $\bar{x} \in \text{int dom } f$ at which f is strictly (Fréchet) differentiable and the derivative mapping $Df(\bar{x})$ is invertible. Then the inverse mapping f^{-1} has a single-valued graphical localization s around $\bar{y} := f(\bar{x})$ for $\bar{x}$ which is strictly differentiable at $\bar{y}$, and moreover*

$$Ds(\bar{y}) = [Df(\bar{x})]^{-1}.$$

In Section 1F we considered what may happen (in finite dimensions) when the derivative mapping is merely surjective; by adjusting the proof of Theorem 1F.6 one obtains that when the Jacobian $\nabla f(\bar{x})$ has full rank, the inverse f^{-1} has a local selection which is strictly differentiable at $f(\bar{x})$. The claim can be easily extended to Hilbert (and even more general) spaces:

Exercise 5G.2 (differentiable inverse selections). *Let X and Y be Hilbert spaces and let $f : X \to Y$ be a function which is strictly differentiable at $\bar{x}$ and such that the derivative $A := Df(\bar{x})$ is surjective. Then the inverse f^{-1} has a local selection s around $\bar{y} := f(\bar{x})$ for $\bar{x}$ which is strictly differentiable at $\bar{y}$ with derivative $Ds(\bar{y}) = A^*(AA^*)^{-1}$, where A^* is the adjoint of A.*

Guide. Use the argument in the proof of 1F.6 with adjustments to the Hilbert space setting. Another way of proving this result is to consider the function

$$g : (x,u) \mapsto \begin{pmatrix} x + A^*u \\ f(x) \end{pmatrix} \quad \text{for } (x,u) \in X \times Y,$$

which satisfies $g(\bar{x}, 0) = (\bar{x}, \bar{y})$ and whose Jacobian is

$$J = \begin{pmatrix} I & A^* \\ A & 0 \end{pmatrix}.$$

In the Hilbert space context, if A is surjective then the operator J is invertible. Hence, by Theorem 5G.1, the mapping g^{-1} has a single-valued graphical localization $(\xi, \eta) : (v, y) \mapsto (\xi(v, y), \eta(v, y))$ around $(\bar{x}, \bar{y})$ for $(\bar{x}, 0)$. In particular, for some neighborhoods U of $\bar{x}$ and V of $\bar{y}$, the function $s(y) := \xi(\bar{x}, y)$ satisfies $y = f(s(y))$ for $y \in V$. To obtain the formula for the strict derivative, find the inverse of J. $\square$

In the particular case when the function f in 5G.2 is linear, the mapping $A^*(AA^*)^{-1}$ is a continuous linear selection of A^{-1}. A famous result by R. G. Bartle and L. M. Graves [1952] yields that, for arbitrary Banach spaces X and Y, the surjectivity of a mapping $A \in \mathscr{L}(X, Y)$ implies the existence of a continuous local selection of A^{-1}; this selection, however, may not be linear. The original Bartle–Graves theorem is for nonlinear mappings and says the following:

Theorem 5G.3 (Bartle–Graves). *Let X and Y be Banach spaces and let $f : X \to Y$ be a function which is strictly differentiable at $\bar{x}$ and such that the derivative $Df(\bar{x})$ is surjective. Then there is a neighborhood V of $\bar{y} := f(\bar{x})$ along with a continuous function $s : V \to X$ and a constant $\gamma > 0$ such that*

$$(1) \qquad f(s(y)) = y \quad \text{and} \quad \|s(y) - \bar{x}\| \leq \gamma \|y - \bar{y}\| \quad \text{for every } y \in V.$$

In other words, the surjectivity of the strict derivative at $\bar{x}$ implies that f^{-1} has a local selection s which is continuous around $f(\bar{x})$ and calm at $f(\bar{x})$. It is known[3] that, in contrast to the strictly differentiable local selection in 5G.2 for Hilbert spaces, the selection in the Bartle–Graves theorem, even for a bounded linear mapping f, might be not even Lipschitz continuous around $\bar{y}$. For this case we have:

Corollary 5G.4 (inverse selection of a surjective linear mapping in Banach spaces). *For any bounded linear mapping A from X onto Y, there is a continuous (but generally nonlinear) mapping B such that $ABy = y$ for every $y \in Y$.*

Proof. Theorem 5G.3 tells us that A^{-1} has a continuous local selection at 0 for 0. Since A^{-1} is positively homogeneous, this selection is global. □

In this section we develop a generalization of the Bartle–Graves theorem for metrically regular set-valued mappings. First, recall that a mapping $F : Y \rightrightarrows X$ is (sequentially) inner semicontinuous on a set $T \subset Y$ if for every $y \in T$, every $x \in F(y)$ and every sequence of points $y^k \in T$, $y^k \to y$, there exists $x^k \in F(y^k)$ for $k = 1, 2, \ldots$ such that $x^k \to x$ as $k \to \infty$. We also need a basic result which we only state here without proof:

Theorem 5G.5 (Michael's selection theorem). *Let X and Y be Banach spaces and consider a mapping $F : Y \rightrightarrows X$ which is closed-convex-valued and inner semicontinuous on $\operatorname{dom} F \neq \emptyset$. Then F has a continuous selection $s : \operatorname{dom} F \to X$.*

We require a lemma which connects the Aubin property of a mapping with the inner semicontinuity of a truncation of this mapping:

Lemma 5G.6 (inner semicontinuous selection from the Aubin property). *Consider a mapping $S : Y \rightrightarrows X$ and any $(\bar{y}, \bar{x}) \in \operatorname{gph} S$, and suppose that S has the Aubin property at $\bar{y}$ for $\bar{x}$ with constant κ. Suppose, for some $c > 0$, that the sets $S(y) \cap \mathbb{B}_c(\bar{x})$ are convex and closed for all $y \in \mathbb{B}_c(\bar{y})$. Then for any $\alpha > \kappa$ there exists $\beta > 0$ such that the mapping*

$$y \mapsto M_\alpha(y) = \begin{cases} S(y) \cap \mathbb{B}_{\alpha\|y-\bar{y}\|}(\bar{x}) & \text{for } y \in \mathbb{B}_\beta(\bar{y}), \\ \emptyset & \text{otherwise} \end{cases}$$

---

[3] Cf. Deville, Godefroy and Zizler [1993], p. 200.

5 Regularity in Infinite Dimensions

is nonempty-closed-convex-valued and inner semicontinuous on $\mathbb{B}_\beta(\bar{y})$.

Proof. Let a and b be positive numbers such that the balls $\mathbb{B}_a(\bar{x})$ and $\mathbb{B}_b(\bar{y})$ are associated with the Aubin property of S (metric regularity of S^{-1}) with a constant κ. Without loss of generality, let $\max\{a,b\} < c$. Fix $\alpha > \kappa$ and choose β such that

$$0 < \beta \leq \min\left\{\frac{a}{\alpha}, \frac{c}{3\alpha}, b, c\right\}.$$

For such a β the mapping M_α has nonempty closed convex values. It remains to show that M_α is inner semicontinuous on $\mathbb{B}_\beta(\bar{y})$.

Let $(y,x) \in \operatorname{gph} M_\alpha$ and $y^k \to y$, $y^k \in \mathbb{B}_\beta(\bar{y})$. First, let $y = \bar{y}$. Then $M_\alpha(y) = \bar{x}$, and from the Aubin property of S there exists a sequence of points $x^k \in S(y^k)$ such that $\|x^k - \bar{x}\| \leq \kappa\|y^k - \bar{y}\|$. Then $x^k \in M_\alpha(y^k)$, $x^k \to x$ as $k \to \infty$ and we are done in this case.

Now let $y \neq \bar{y}$. The Aubin property of S yields that there exists $\check{x}^k \in S(y^k)$ such that

$$\|\check{x}^k - \bar{x}\| \leq \kappa\|y^k - \bar{y}\|$$

and also there exists $\tilde{x}^k \in S(y^k)$ such that

$$\|\tilde{x}^k - x\| \leq \kappa\|y^k - y\|.$$

Then, the choice of β above yields

$$\|\check{x}^k - \bar{x}\| \leq \kappa\beta \leq \alpha\frac{c}{3\alpha} \leq c$$

and

$$\begin{aligned}\|\tilde{x}^k - \bar{x}\| &\leq \|\tilde{x}^k - x\| + \|x - \bar{x}\| \\ &\leq \kappa\|y^k - y\| + \alpha\|y - \bar{y}\| \\ &\leq 2\kappa\beta + \alpha\beta \leq 3\alpha\beta \leq c.\end{aligned}$$

Let

(2) $$\varepsilon^k = \frac{(\alpha+\kappa)\|y^k - y\|}{(\alpha-\kappa)\|y^k - \bar{y}\| + (\alpha+\kappa)\|y^k - y\|}.$$

Then $0 \leq \varepsilon^k < 1$ and $\varepsilon^k \searrow 0$ as $k \to \infty$. Let $x^k = \varepsilon^k \check{x}^k + (1 - \varepsilon^k)\tilde{x}^k$. Then $x^k \in S(y^k)$. Moreover, we have

$$\begin{aligned}\|x^k - \bar{x}\| &\leq \varepsilon^k\|\check{x}^k - \bar{x}\| + (1-\varepsilon^k)\|\tilde{x}^k - \bar{x}\| \\ &\leq \varepsilon^k \kappa\|y^k - \bar{y}\| + (1-\varepsilon^k)(\|\tilde{x}^k - x\| + \|x - \bar{x}\|) \\ &\leq \varepsilon^k \kappa\|y^k - \bar{y}\| + (1-\varepsilon^k)\kappa\|y^k - y\| + (1-\varepsilon^k)\alpha\|y - \bar{y}\| \\ &\leq \varepsilon^k \kappa\|y^k - \bar{y}\| + (1-\varepsilon^k)\kappa\|y^k - y\| \\ &\quad + (1-\varepsilon^k)\alpha\|y^k - \bar{y}\| + (1-\varepsilon^k)\alpha\|y^k - y\|\end{aligned}$$

$$\leq \alpha\|y^k - \bar{y}\| - \varepsilon^k(\alpha - \kappa)\|y^k - \bar{y}\|$$
$$+ (1 - \varepsilon^k)(\alpha + \kappa)\|y^k - y\| \leq \alpha\|y^k - \bar{y}\|,$$

where in the last inequality we take into account the expression (2) for ε^k. Thus $x^k \in M_\alpha(y^k)$, and since $x^k \to x$, we are done. □

Lemma 5G.6 allows us to apply Michael's selection theorem to the mapping M_α, obtaining the following result:

Theorem 5G.7 (continuous inverse selection from metric regularity). *Consider a mapping $F : X \rightrightarrows Y$ which is metrically regular at $\bar{x}$ for $\bar{y}$. Let, for some $c > 0$, the sets $F^{-1}(y) \cap \mathbb{B}_c(\bar{x})$ be convex and closed for all $y \in \mathbb{B}_c(\bar{y})$. Then for every $\alpha > \mathrm{reg}\,(F;\bar{x}|\bar{y})$ the mapping F^{-1} has a continuous local selection s around $\bar{y}$ for $\bar{x}$ which is calm at $\bar{y}$ with*

(3) $$\mathrm{clm}\,(s;\bar{y}) \leq \alpha.$$

Proof. Choose α such that $\alpha > \mathrm{reg}\,(F;\bar{x}|\bar{y})$, and apply Michael's theorem 5G.5 to the mapping M_α in 5G.6 for $S = F^{-1}$. By the definition of M_α, the continuous local selection obtained in this way is calm with a constant α. □

Note that the continuous local selection s in 5G.7 depends on α and therefore we cannot replace α in (3) with $\mathrm{reg}\,(F;\bar{x}|\bar{y})$.

In the remainder of this section we show that if a mapping F satisfies the assumptions of Theorem 5G.7, then for any function $g : X \to Y$ with $\mathrm{lip}\,(g;\bar{x}) < 1/\mathrm{reg}\,(F;\bar{x}|\bar{y})$, the mapping $(g + F)^{-1}$ has a continuous and calm local selection around $g(\bar{x}) + \bar{y}$ for $\bar{x}$. We will prove this generalization of the Bartle–Graves theorem by repeatedly using an argument similar to the proof of Lemma 5G.6, the idea of which goes back to (modified) Newton's method used to prove the theorems of Lyusternik and Graves and, in fact, to Goursat's proof on his version of the classical inverse function theorem. We put the theorem in the format of the general implicit function theorem paradigm:

Theorem 5G.8 (inverse mapping theorem with continuous and calm local selections). *Consider a mapping $F : X \rightrightarrows Y$ and any $(\bar{x}, \bar{y}) \in \mathrm{gph}\,F$ and suppose that for some $c > 0$ the mapping $\mathbb{B}_c(\bar{y}) \ni y \mapsto F^{-1}(y) \cap \mathbb{B}_c(\bar{x})$ is closed-convex-valued. Let κ and μ be nonnegative constants such that*

$$\mathrm{reg}\,(F;\bar{x}|\bar{y}) \leq \kappa \quad \text{and} \quad \kappa\mu < 1.$$

Then for any function $g : X \to Y$ with $\bar{x} \in \mathrm{int}\,\mathrm{dom}\,g$ and $\mathrm{lip}\,(g;\bar{x}) \leq \mu$ and for every γ with

$$\frac{\kappa}{1 - \kappa\mu} < \gamma$$

the mapping $(g + F)^{-1}$ has a continuous local selection s around $g(\bar{x}) + \bar{y}$ for $\bar{x}$, which moreover is calm at $g(\bar{x}) + \bar{y}$ with

5 Regularity in Infinite Dimensions

(4) $$\text{clm}(s; g(\bar{x}) + \bar{y}) \leq \gamma.$$

Proof. The proof consists of two steps. In the first step, we use induction to obtain a Cauchy sequence of continuous functions $z^0, z^1, \ldots$, such that z^n is a continuous and calm selection of the mapping $y \mapsto F^{-1}(y - g(z^{n-1}(y)))$. Then we show that this sequence has a limit in the space of continuous functions acting from a fixed ball around $\bar{y}$ to the space X and equipped with the supremum norm, and this limit is the selection whose existence is claimed.

Choose κ and μ as in the statement of the theorem and let $\gamma > \kappa/(1 - \kappa\mu)$. Let λ, α and ν be such that $\kappa < \lambda < \alpha < 1/\nu$ and $\nu > \mu$, and also $\lambda/(1 - \alpha\nu) \leq \gamma$. Without loss of generality, we can assume that $g(\bar{x}) = 0$. Let $\mathbb{B}_a(\bar{x})$ and $\mathbb{B}_b(\bar{y})$ be the neighborhoods of $\bar{x}$ and $\bar{y}$, respectively, that are associated with the assumed properties of the mapping F and the function g. Specifically,

(a) For every $y, y' \in \mathbb{B}_b(\bar{y})$ and $x \in F^{-1}(y) \cap \mathbb{B}_a(\bar{x})$ there exists $x' \in F^{-1}(y')$ with

$$\|x' - x\| \leq \lambda \|y' - y\|.$$

(b) For every $y \in \mathbb{B}_b(\bar{y})$ the set $F^{-1}(y) \cap \mathbb{B}_a(\bar{x})$ is nonempty, closed and convex (that is, $\max\{a,b\} \leq c$).

(c) The function g is Lipschitz continuous on $\mathbb{B}_a(\bar{x})$ with a constant ν.

According to 5G.7, there we can find a constant β, $0 < \beta \leq b$, and a continuous function $z^0 : \mathbb{B}_\beta(\bar{y}) \to X$ such that

$$F(z^0(y)) \ni y \text{ and } \|z^0(y) - \bar{x}\| \leq \lambda \|y - \bar{y}\| \text{ for all } y \in \mathbb{B}_\beta(\bar{y}).$$

Choose a positive τ such that

(5) $$\tau \leq (1 - \alpha\nu) \min\left\{a, \frac{a}{2\lambda}, \frac{\beta}{2}\right\}$$

and consider the mapping $y \mapsto M_1(y)$ where

$$M_1(y) = \{x \in F^{-1}(y - g(z^0(y))) \mid \|x - z^0(y)\| \leq \alpha\nu\|z^0(y) - \bar{x}\|\}$$

for $y \in \mathbb{B}_\tau(\bar{y})$ and $M_1(y) = \emptyset$ for $y \notin \mathbb{B}_\tau(\bar{y})$. Clearly, $(\bar{y}, \bar{x}) \in \text{gph } M_1$. Also, for any $y \in \mathbb{B}_\tau(\bar{y})$, we have from the choice of λ and α, using (5), that $z^0(y) \in \mathbb{B}_a(\bar{x})$ and therefore

$$\|y - g(z^0(y)) - \bar{y}\| \leq \tau + \nu\|z^0(y) - \bar{x}\| \leq \tau + \nu\lambda\tau \leq (1 - \alpha\nu)(1 + \nu\lambda)(\beta/2) \leq \beta \leq b.$$

Then from the Aubin property of F^{-1} there exists $x \in F^{-1}(y - g(z^0(y)))$ with

$$\|x - z^0(y)\| \leq \lambda\|g(z^0(y)) - g(\bar{x})\| \leq \alpha\nu\|z^0(y) - \bar{x}\|,$$

which implies $x \in M_1(y)$. Thus M_1 is nonempty-valued. Further, if $(y, x) \in \text{gph } M_1$, using (5) we have that $y \in \mathbb{B}_b(\bar{y})$ and also

$$\|x-\bar{x}\| \leq \|x-z^0(y)\| + \|z^0(y)-\bar{x}\| \leq (1+\alpha v)\lambda\tau \leq (1-(\alpha v)^2)\lambda\frac{a}{2\lambda} \leq \frac{a}{2}.$$

Then, from the property (b) above, since for any $y \in \text{dom } M$ the set $M_1(y)$ is the intersection of a closed ball with a closed convex set, the mapping M_1 is closed-convex-valued on its domain. We will show that this mapping is inner semicontinuous on $B_\tau(\bar{y})$.

Let $y \in B_\tau(\bar{y})$ and $x \in M_1(y)$, and let $y^k \in B_\tau(\bar{y})$, $y^k \to y$ as $k \to \infty$. If $z^0(y) = \bar{x}$, then $M_1(y) = \{\bar{x}\}$ and therefore $x = \bar{x}$. Any $x^k \in M_1(y^k)$ satisfies

$$\|x^k - z^0(y^k)\| \leq \alpha v \|z^0(y^k) - \bar{x}\|.$$

Using the continuity of the function z^0, we see that $x^k \to z^0(y) = \bar{x} = x$; thus M_1 is inner semicontinuous.

Now let $z^0(y) \neq \bar{x}$. Since $z^0(y^k) \in F^{-1}(y^k - g(\bar{x})) \cap B_a(\bar{x})$, the Aubin property of F^{-1} furnishes the existence of $\check{x}^k \in F^{-1}(y^k - g(z^0(y^k)))$ such that

(6) $\quad \|\check{x}^k - z^0(y^k)\| \leq \lambda \|g(z^0(y^k)) - g(\bar{x})\| \leq \lambda v\|z^0(y^k) - \bar{x}\| \leq \alpha v\|z^0(y^k) - \bar{x}\|.$

Then $\check{x}^k \in M_1(y^k)$, and in particular, $\check{x}^k \in B_a(\bar{x})$. Further, the inclusion $x \in F^{-1}(y - g(z^0(y))) \cap B_a(\bar{x})$ combined with the Aubin property of F^{-1} entails the existence of $\tilde{x}^k \in F^{-1}(y^k - g(z^0(y^k)))$ such that

(7) $\quad \|\tilde{x}^k - x\| \leq \lambda(\|y^k - y\| + v\|z^0(y^k) - z^0(y)\|) \to 0 \text{ as } k \to \infty.$

Then $\tilde{x}^k \in B_a(\bar{x})$ for large k. Let

$$\varepsilon^k := \frac{(1+\alpha v)\|z^0(y^k) - z^0(y)\| + \|\tilde{x}^k - x\|}{\alpha v\|z^0(y) - \bar{x}\| - \lambda v\|z^0(y^k) - \bar{x}\|}.$$

Note that, for $k \to \infty$, the numerator in the definition of ε^k goes to 0 because of the continuity of z^0 and (7), while the denominator converges to $(\alpha - \lambda)v\|z^0(y) - \bar{x}\| > 0$; therefore $\varepsilon^k \to 0$ as $k \to \infty$. Let

$$x^k = \varepsilon^k \check{x}^k + (1 - \varepsilon^k)\tilde{x}^k.$$

Since $\tilde{x}^k \to x$ and $\varepsilon^k \to 0$, we get $x^k \to x$ as $k \to \infty$ and also, since $y \mapsto F^{-1}(y) \cap B_a(\bar{x})$ is convex-valued around $(\bar{x}, \bar{y})$, we have $x^k \in F^{-1}(y^k - g(z^0(y^k)))$ for large k. By (6), (7), the assumption that $x \in M_1(y)$, and the choice of ε^k, we have

$$\begin{aligned}
\|x^k - z^0(y^k)\| &\leq \varepsilon^k\|\check{x}^k - z^0(y^k)\| + (1-\varepsilon^k)\|\tilde{x}^k - z^0(y^k)\| \\
&\leq \varepsilon^k \lambda v\|z^0(y^k) - \bar{x}\| + (1-\varepsilon^k)(\|\tilde{x}^k - x\| \\
&\quad + \|x - z^0(y)\| + \|z^0(y) - z^0(y^k)\|) \\
&\leq \varepsilon^k \lambda v\|z^0(y^k) - \bar{x}\| + \|\tilde{x}^k - x\| \\
&\quad + (1-\varepsilon^k)\alpha v\|z^0(y) - \bar{x}\| + \|z^0(y) - z^0(y^k)\| \\
&\leq \alpha v\|z^0(y^k) - \bar{x}\| + \alpha v\|z^0(y^k) - z^0(y)\|
\end{aligned}$$

5 Regularity in Infinite Dimensions

$$+ \|\tilde{x}^k - x\| + \|z^0(y) - z^0(y^k)\|$$
$$- \varepsilon^k \alpha v \|z^0(y) - \bar{x}\| + \varepsilon^k \lambda v \|z^0(y^k) - \bar{x}\|$$
$$\leq \alpha v \|z^0(y^k) - \bar{x}\| + \|\tilde{x}^k - x\| + (1 + \alpha v)\|z^0(y) - z^0(y^k)\|$$
$$- \varepsilon^k(\alpha v \|z^0(y) - \bar{x}\| - \lambda v \|z^0(y^k) - \bar{x}\|)$$
$$= \alpha v \|z^0(y^k) - \bar{x}\|.$$

We obtain that $x^k \in M_1(y^k)$, and since $x^k \to x$, we conclude that the mapping M_1 is inner semicontinuous on its domain $\mathbb{B}_\tau(\bar{y})$. Hence, by Michael's selection theorem 5G.5, there it has a continuous selection $z^1 : \mathbb{B}_\tau(\bar{y}) \to X$; that is, a continuous function z^1 which satisfies

$$z^1(y) \in F^{-1}(y - g(z^0(y))) \text{ and } \|z^1(y) - z^0(y)\| \leq \alpha v \|z^0(y) - \bar{x}\| \text{ for all } y \in \mathbb{B}_\tau(\bar{y}).$$

Then for $y \in \mathbb{B}_\tau(\bar{y})$, by the choice of γ,

$$\|z^1(y) - \bar{x}\| \leq \|z^1(y) - z^0(y)\| + \|z^0(y) - \bar{x}\| \leq (1 + \alpha v)\lambda \|y - \bar{y}\| \leq \gamma \|y - \bar{y}\|.$$

The induction step is parallel to the first step. Let z^0 and z^1 be as above and suppose we have also found functions $z^1, z^2, \ldots, z^n$, such that each z^j, $j = 1, 2, \ldots, n$, is a continuous selection of the mapping $y \mapsto M_j(y)$ where

$$M_j(y) = \left\{ x \in F^{-1}(y - g(z^{j-1}(y))) \mid \|x - z^n(y)\| \leq \alpha v \|z^{j-1}(y) - z^{j-2}(y)\| \right\}$$

for $y \in \mathbb{B}_\tau(\bar{y})$ and $M_j(y) = \emptyset$ for $y \notin \mathbb{B}_\tau(\bar{y})$, where we put $z^{-1}(y) = \bar{x}$ for $y \in \mathbb{B}_\tau(\bar{y})$. Then for $y \in \mathbb{B}_\tau(\bar{y})$ we obtain

$$\|z^j(y) - z^{j-1}(y)\| \leq (\alpha v)^{j-1} \|z^1(y) - z^0(y)\| \leq (\alpha v)^j \|z^0(y) - \bar{x}\|, \quad j = 2, \ldots, n.$$

Therefore,

$$\|z^j(y) - \bar{x}\| \leq \sum_{i=0}^{j} (\alpha v)^i \|z^i(y) - z^{i-1}(y)\|$$
$$\leq \sum_{i=0}^{j} (\alpha v)^i \|z^0(y) - \bar{x}\| \leq \frac{\lambda}{1 - \alpha v} \|y - \bar{y}\| \leq \gamma \|y - \bar{y}\|.$$

Hence, from (5), for $j = 1, 2, \ldots, n$,

(8) $$\|z^j(y) - \bar{x}\| \leq a$$

and also

(9) $$\|y - g(z^j(y)) - \bar{y}\| \leq \tau + v \|z^j(y) - \bar{x}\| \leq \tau + \frac{\lambda v \tau}{1 - \alpha v} \leq \frac{\tau}{1 - \alpha v} \leq \beta \leq b.$$

Consider the mapping $y \mapsto M_{n+1}(y)$ where

$$M_{n+1}(y) = \{x \in F^{-1}(y - g(z^n(y))) \mid \|x - z^n(y)\| \leq \alpha v \|z^n(y) - z^{n-1}(y)\|\}$$

for $y \in B_\tau(\bar{y})$ and $M_{n+1}(y) = \emptyset$ for $y \notin B_\tau(\bar{y})$. As in the first step, we find that M_{n+1} is nonempty-closed-convex-valued. Let $y \in B_\tau(\bar{y})$ and $x \in M_{n+1}(y)$, and let $y^k \in B_\tau(\bar{y})$, $y^k \to y$ as $k \to \infty$. If $z^{n-1}(y) = z^n(y)$, then $M_{n+1}(y) = \{z^n(y)\}$, and consequently $x = z^n(y)$; then from $z^n(y^k) \in F^{-1}(y^k - g(z^{n-1}(y^k))) \cap B_a(\bar{x})$ and $y^k - g(z^{n-1}(y^k)) \in B_b(\bar{y})$, we obtain, using the Aubin property of F^{-1}, that there exists $x^k \in F^{-1}(y^k - g(z^n(y^k)))$ such that

$$\|x^k - z^n(y^k)\| \leq \lambda \|g(z^n(y^k)) - g(z^{n-1}(y^k))\| \leq \alpha v \|z^n(y^k) - z^{n-1}(y^k)\|.$$

Therefore $x^k \in M_{n+1}(y^k)$, $x^k \to z^1(y) = x$ as $k \to \infty$, and hence M_{n+1} is inner semicontinuous for the case considered.

Let $z^n(y) \neq z^{n-1}(y)$. From (8) and (9) for $y = y^k$, since

$$z^n(y^k) \in F^{-1}(y^k - g(z^{n-1}(y^k))) \cap B_a(\bar{x}),$$

the Aubin property of F^{-1} implies the existence of $\check{x}^k \in F^{-1}(y^k - g(z^n(y^k)))$ such that

$$\|\check{x}^k - z^n(y^k)\| \leq \lambda \|g(z^n(y^k)) - g(z^{n-1}(y^k))\| \leq \lambda v \|z^n(y^k) - z^{n-1}(y^k)\|.$$

Similarly, since $x \in F^{-1}(y - g(z^n(y))) \cap B_a(\bar{x})$, there exists $\tilde{x}^k \in F^{-1}(y^k - g(z^n(y^k)))$ such that

$$\begin{aligned}\|\tilde{x}^k - x\| &\leq \lambda(\|y^k - y\| + \|g(z^n(y^k)) - g(z^n(y))\|) \\ &\leq \lambda(\|y^k - y\| + v\|z^n(y^k) - z^n(y)\|) \to 0 \text{ as } k \to \infty.\end{aligned}$$

Put

$$\varepsilon^k := \frac{\alpha v \|z^{n-1}(y) - z^{n-1}(y^k)\| + (1 + \alpha v)\|z^n(y) - z^n(y^k)\| + \|\tilde{x}^k - x\|}{\alpha v \|z^n(y) - z^{n-1}(y)\| - \lambda v \|z^n(y^k) - z^{n-1}(y^k)\|}.$$

Then $\varepsilon^k \to 0$ as $k \to \infty$. Taking

$$x^k = \varepsilon^k \check{x}^k + (1 - \varepsilon^k)\tilde{x}^k,$$

we obtain that $x^k \in F^{-1}(y^k - g(z^n(y^k)))$ for large k. Further, we estimate $\|x^k - z^n(y^k)\|$ in the same way as in the first step, that is,

$$\begin{aligned}\|x^k - z^n(y^k)\| &\leq \varepsilon^k \|\check{x}^k - z^n(y^k)\| + (1 - \varepsilon^k)\|\tilde{x}^k - z^n(y^k)\| \\ &\leq \varepsilon^k \lambda v \|z^n(y^k) - z^{n-1}(y^k)\| \\ &\quad + (1 - \varepsilon^k)(\|\tilde{x}^k - x\| + \|x - z^n(y)\| + \|z^n(y) - z^n(y^k)\|) \\ &\leq \varepsilon^k \lambda v \|z^n(y^k) - z^{n-1}(y^k)\| + \|\tilde{x}^k - x\| \\ &\quad + (1 - \varepsilon^k)\alpha v \|z^n(y) - z^{n-1}(y)\| + \|z^n(y) - z^n(y^k)\| \\ &\leq \alpha v \|z^n(y^k) - z^{n-1}(y^k)\| + \alpha v \|z^n(y^k) - z^n(y)\|\end{aligned}$$

5 Regularity in Infinite Dimensions

$$+\alpha v\|z^{n-1}(y^k) - z^{n-1}(y)\| + \|\tilde{x}^k - x\|$$
$$+\|z^n(y) - z^n(y^k)\| - \varepsilon^k \alpha v\|z^n(y) - z^{n-1}(y)\|$$
$$+\varepsilon^k \lambda v\|z^n(y^k) - z^{n-1}(y^k)\|$$
$$\leq \alpha v\|z^n(y^k) - z^{n-1}(y^k)\| + \|\tilde{x}^k - x\|$$
$$+(1+\alpha v)\|z^n(y) - z^n(y^k)\| + \alpha v\|z^{n-1}(y) - z^{n-1}(y^k)\|$$
$$-\varepsilon^k(\alpha v\|z^n(y) - z^{n-1}(y)\| - \lambda v\|z^n(y^k) - z^{n-1}(y^k)\|)$$
$$= \alpha v\|z^n(y^k) - z^{n-1}(y^k)\|.$$

We conclude that $x^k \in M_{n+1}(y^k)$, and since $x^k \to x$ as $k \to \infty$, the mapping M_{n+1} is inner semicontinuous on $\mathbb{B}_\tau(\bar{y})$. Hence, the mapping M_{n+1} has a continuous selection $z^{n+1} : \mathbb{B}_\tau(\bar{y}) \to X$, that is,

$$z^{n+1}(y) \in F^{-1}(y - g(z^n(y))) \quad \text{and} \quad \|z^{n+1}(y) - z^n(y)\| \leq \alpha v\|z^n(y) - z^{n-1}(y)\|.$$

Thus
$$\|z^{n+1}(y) - z^n(y)\| \leq (\alpha v)^{(n+1)}\|z^0(y) - \bar{x}\|.$$

The induction step is now complete. In consequence, we have an infinite sequence of bounded continuous functions $z^0, \ldots, z^n, \ldots$ such that for all $y \in \mathbb{B}_\tau(\bar{y})$ and for all n,

$$\|z^n(y) - \bar{x}\| \leq \sum_{i=0}^{n} (\alpha v)^i \|z^0(y) - \bar{x}\| \leq \frac{\lambda}{1-\alpha v}\|y - \bar{y}\| \leq \gamma\|y - \bar{y}\|$$

and moreover,

$$\sup_{y \in \mathbb{B}_\tau(\bar{y})} \|z^{n+1}(y) - z^n(y)\| \leq (\alpha v)^n \sup_{y \in \mathbb{B}_\tau(\bar{y})} \|z^0(y) - \bar{x}\| \leq (\alpha v)^n \lambda \tau \quad \text{for } n \geq 1.$$

The sequence $\{z^n\}$ is a Cauchy sequence in the space of functions that are continuous and bounded on $\mathbb{B}_\tau(\bar{y})$ equipped with the supremum norm. Then this sequence has a limit s which is a continuous function on $\mathbb{B}_\tau(\bar{y})$ and satisfies

$$s(y) \in F^{-1}(y - g(s(y)))$$

and
$$\|s(y) - \bar{x}\| \leq \frac{\lambda}{1-\alpha v}\|y - \bar{y}\| \leq \gamma\|y - \bar{y}\| \quad \text{for all } y \in \mathbb{B}_\tau(\bar{y}).$$

Thus, s is a continuous local selection of $(g+F)^{-1}$ which has the calmness property (4). This brings the proof to its end. □

Proof of Theorem 5G.3. Apply 5G.8 with $F = Df(\bar{x})$ and $g(x) = f(x) - Df(\bar{x})x$. Metric regularity of F is equivalent to surjectivity of $Df(\bar{x})$, and F^{-1} is convex-closed-valued. The mapping g has $\text{lip}(g;\bar{x}) = 0$ and finally $F + g = f$. □

Note that Theorem 5G.7 follows from 5G.8 with g the zero function.

We present next an implicit mapping version of Theorem 5G.7.

Theorem 5G.9 (implicit mapping version). *Let X, Y and P be Banach spaces. For $f : P \times X \to Y$ and $F : X \rightrightarrows Y$, consider the generalized equation $f(p,x) + F(x) \ni 0$ with solution mapping*

$$S(p) = \{x \mid f(p,x) + F(x) \ni 0\} \quad \text{having } \bar{x} \in S(\bar{p}).$$

Suppose that F satisfies the conditions in Theorem 5G.7 with $\bar{y} = 0$ and associate constant $\kappa \geq \text{reg}(F; \bar{x} \mid 0)$ and also that f is continuous on a neighborhood of $(\bar{x}, \bar{p})$ and has $\widehat{\text{lip}}_x(f; (\bar{p}, \bar{x})) \leq \mu$, where μ is a nonnegative constant satisfying $\kappa \mu < 1$. Then for every γ satisfying

$$\tag{10} \frac{\kappa}{1 - \kappa \mu} < \gamma$$

there exist neighborhoods U of $\bar{x}$ and Q of $\bar{p}$ along with a continuous function $s : Q \to U$ such that

$$\tag{11} s(p) \in S(p) \quad \text{and} \quad \|s(p) - \bar{x}\| \leq \gamma \|f(p, \bar{x}) - f(\bar{p}, \bar{x})\| \quad \text{for every } p \in Q.$$

Proof. The proof is parallel to the proof of Theorem 5G.7. First we choose γ satisfying (10) and then λ, α and ν such that $\kappa < \lambda < \alpha < \nu^{-1}$ and $\nu > \mu$, and also

$$\tag{12} \frac{\lambda}{1 - \alpha \nu} < \gamma.$$

There are neighborhoods U, V and Q of $\bar{x}$, 0 and $\bar{p}$, respectively, which are associated with the metric regularity of F at $\bar{x}$ for 0 with constant λ and the Lipschitz continuity of f with respect to x with constant ν uniformly in p. By appropriately choosing a sufficiently small radius τ of a ball around $\bar{p}$, we construct an infinite sequence of continuous and bounded functions $z^k : B_\tau(\bar{p}) \to X$, $k = 0, 1, \ldots$, which are uniformly convergent on $B_\tau(\bar{p})$ to a function s satisfying the conditions in (11). The initial z^0 satisfies

$$z^0(p) \in F^{-1}(-f(p, \bar{x})) \quad \text{and} \quad \|z^0(p) - \bar{x}\| \leq \lambda \|f(p, \bar{x}) - f(\bar{p}, \bar{x})\|.$$

For $k = 1, 2, \ldots$, the function z^k is a continuous selection of the mapping

$$M_k : p \mapsto \{x \in F^{-1}(-f(p, z^{k-1}(p))) \mid \|x - z^{k-1}(p)\| \leq \alpha \nu \|z^{k-1}(p) - z^{k-2}(p)\|\}$$

for $p \in B_\tau(\bar{p})$, where $z^{-1}(p) = \bar{x}$. Then for all $p \in B_\tau(\bar{p})$ we obtain

$$z^k(p) \in F^{-1}(-f(p, z^{k-1}(p))) \quad \text{and} \quad \|z^k(p) - z^{k-1}(p)\| \leq (\alpha \nu)^k \|z^0(p) - \bar{x}\|,$$

hence,

$$\tag{13} \|z^k(y) - \bar{x}\| \leq \frac{\lambda}{1 - \alpha \nu} \|f(p, \bar{x}) - f(\bar{p}, \bar{x})\|.$$

The sequence $\{z^k\}$ is a Cauchy sequence of continuous and bounded function, hence it is convergent with respect to the supremum norm. In the limit with $k \to \infty$, taking into account (12) and (13), we obtain a selection s with the desired properties. □

Exercise 5G.10 (specialization for closed sublinear mappings). *Let $F : X \rightrightarrows Y$ have convex and closed graph, let $f : X \to Y$ be strictly differentiable at $\bar{x}$ and let $(\bar{x}, \bar{y}) \in$ gph$(f + F)$. Suppose that*

(14) $$\bar{y} \in \text{int rge}\,(f(\bar{x}) + Df(\bar{x})(x - \bar{x}) + F).$$

Prove that there exist neighborhoods U of $\bar{x}$ and V of $\bar{y}$, a continuous function $s : V \to U$, and a constant γ, such that

$$(f + F)(s(y)) \ni y \quad \text{and} \quad \|s(y) - \bar{x}\| \leq \gamma \|y - \bar{y}\| \quad \text{for every } y \in V.$$

Guide. Apply the Robinson-Ursescu theorem 5B.4. □

Commentary

The equivalence of (a) and (d) in 5A.1 was shown in Theorem 10 on p. 150 of the original treatise of S. Banach [1932]. The statements of this theorem usually include the equivalence of (a) and (b), which is called in Dunford and Schwartz [1958] the "interior mapping principle." Lemma 5A.4 is usually stated for Banach algebras, see, e.g., Theorem 10.7 in Rudin [1991]. Theorem 5A.8 is from Robinson [1972].

The generalization of the Banach open mapping theorem to set-valued mappings with convex closed graphs was obtained independently by Robinson [1976] and Ursescu [1975]; the proof of 5B.3 given here is close to the original proof in Robinson [1976]. A particular case of this result for positively homogeneous mappings was shown earlier by Ng [1973]. The Baire category theorem can be found in Dunford and Schwartz [1958], p. 20. The Robinson–Ursescu theorem is stated in various ways in the literature, see, e.g., Theorem 3.3.1 in Aubin and Ekeland [1984], Theorem 2.2.2 in Aubin and Frankowska [1990], Theorem 2.83 in Bonnans and Shapiro [2000], Theorem 9.48 in Rockafellar and Wets [1998], Theorem 1.3.11 in Zălinescu [2002] and Theorem 4.21 in Mordukhovich [2006].

Sublinear mappings (under the name "convex processes") and their adjoints were introduced by Rockafellar [1967]; see also Rockafellar [1970]. Theorem 5C.9 first appeared in Lewis [1999], see also Lewis [2001]. The norm duality theorem, 5C.10, was originally proved by Borwein [1983], who later gave in Borwein [1986b] a more detailed argument. The statement of the Hahn–Banach theorem 5C.11 is from Dunford and Schwartz [1958], p. 62.

Theorems 5D.1 and 5D.2 are versions of results originally published in Lyusternik [1934] and Graves [1950], with some adjustments to the current setting. Lyusternik apparently viewed his theorem mainly as a stepping stone to obtain the Lagrange multiplier rule for abstract minimization problems, and the title of his paper from 1934 clearly says so. It is also interesting to note that, after the statement of the Lyusternik theorem as 8.10.2 in the functional analysis book by Lyusternik and Sobolev [1965], the authors say that "the proof of this theorem is a modification of the proof of the implicit function theorem, and the [Lyusternik] theorem is a direct generalization of this [implicit function] theorem."

It is quite likely that Graves considered his theorem as an extension of the Banach open mapping theorem for nonlinear mappings. But there is more in its statement and proof; namely, the Graves theorem does not involve differentiation and then, as shown in 5D.3, can be easily extended to become a generalization of the the basic Lemma 5A.4 for nonlinear mappings. This was mentioned already in the historical remarks of Dunford and Schwartz [1958], p. 85. A further generalization in line with the present setting was revealed in Dmitruk, Milyutin and Osmolovskiĭ [1980], where the approximating linear mapping is replaced by a Lipschitz continuous function with a sufficiently small Lipschitz constant. Estimates for the regularity modulus of the kind given in 5D.3 are also present in Ioffe [1979].

In the second part of the last century, when the development of optimality conditions was a key issue, the approach of Lyusternik was recognized for its virtues and

5 Regularity in Infinite Dimensions

extended to great generality. Historical remarks regarding these developments can be found in Ioffe [2000] and Rockafellar and Wets [1998]. The statement in 5D.4 is a slightly modified version of the Lyusternik theorem as given in Section 0.2.4 of Ioffe and Tikhomirov [1974].

Theorem 5E.1 is given as in Dontchev, Lewis and Rockafellar [2003]; earlier results in this vein were obtained by Dontchev and Hager [1993,1994]. The contraction mapping theorem 5E.2 is from Dontchev and Hager [1994]. A related result in abstract metric spaces is given in Arutyunov [2007].

Theorem 5G.3 gives the original form of the Bartle–Graves theorem as contributed in Bartle and Graves [1952]. The particular form 5G.5 of Michael's selection theorem[4] is Lemma 2.1 in Deimling [1992]. Lemma 5G.6 was first given in Borwein and Dontchev [2003], while Theorem 5G.8 is from Dontchev [2004]. These two papers were largely inspired by contacts of the first author of this book with Robert G. Bartle, who was able to read them before he passed away Sept. 18, 2002. Shortly before he died he sent to the first author a letter, where he, among other things, wrote the following:

"Your results are, indeed, an impressive and far-reaching extension of the theorem that Professor Graves and I published over a half-century ago. I was a student in a class of Graves in which he presented the theorem in the case that the parameter domain is the interval $[0,1]$. He expressed the hope that it could be generalized to a more general domain, but said that he didn't see how to do so. By a stroke of luck, I had attended a seminar a few months before given by André Weil, which he titled "On a theorem by Stone." I (mis)understood that he was referring to M. H. Stone, rather than A. H. Stone, and attended. Fortunately, I listened carefully enough to learn about paracompactness and continuous partition of unity[5] (which were totally new to me) and which I found to be useful in extending Graves' proof. So the original theorem was entirely due to Graves; I only provided an extension of his proof, using methods that were not known to him. However, despite the fact that I am merely a 'middleman,' I am pleased that this result has been found to be useful."

In this book we present inverse/implicit function theorems centered around variational analysis, but there are many other theorems that fall into the same category and are designed as tools in other areas of mathematics beyond variational analysis. In particular, we do not discuss in this book the celebrated Nash–Moser theorem, used mainly in geometric analysis and partial differential equations. A rigorous introduction of this theorem and the theory around it would require a lot of space and would tip the balance of topics and ideas away from we want to emphasize. More importantly, we have not been able to identify (as of yet) a specific, sound application of this theorem in variational analysis such as would have justified the inclusion. A rigorous and nice presentation of the Nash–Moser theorem along with

---

[4] The original statement of Michael's selection theorem is for mappings acting from a paracompact space to a Banach space; by a theorem of A. H. Stone every metric space is paracompact and hence every subset of a Banach space is paracompact.

[5] Michael's theorem was not known at that time.

the theory and applications behind it is given in Hamilton [1982]. In the following lines, we only briefly point out a connection to the results in Section 5E.

The Nash–Moser theorem is about mappings acting in *Fréchet* spaces, which are more general than the Banach spaces. Consider a linear (vector) space F equipped with the collection of seminorms $\{\|\cdot\|_n | n \in N\}$ (a seminorm differs from a norm in that the seminorm of a nonzero element could be zero). The topology induced by this (countable) collection of seminorms makes the space F a locally convex topological vector space. If $x = 0$ when $\|x\|_n = 0$ for all n, the space is *Hausdorff*. In a Hausdorff space, one may define a metric based the family of seminorms in the following way:

$$(1) \qquad \rho(x,y) = \sum_{n=1}^{\infty} 2^{-n} \frac{\|x-y\|_n}{1+\|x-y\|_n}.$$

It is not difficult to see that this metric is shift-invariant. A sequence $\{x_k\}$ is said to be Cauchy when $\|x_k - x_j\|_n \to 0$ as k and $j \to \infty$ for all n, or, equivalently, $\rho(x_k, x_j) \to 0$ as $k \to \infty$ and $j \to \infty$. As usual, a space is complete if every Cauchy sequence converges. A Fréchet space is a complete Hausdorff metrizable locally convex topological vector space.

Having two Fréchet spaces F and G, we can now introduce metrics ρ and σ associated with their collections of seminorms as in (1) above, and define Lipschitz continuity and metric regularity accordingly. Then Theorem 5E.3 will apply of course and we can obtain, e.g., a Graves-type theorem in Fréchet spaces, in terms of the metrics ρ and σ, and also an implicit function theorem in Fréchet spaces from the general Theorem 5E.4. To get to the Nash–Moser theorem, however, we have a long way to go, translating the meaning of, e.g., the assumptions Theorem 5E.4, in terms of the metrics ρ and σ for the collections of seminorms and the mappings considered. For that we will need more structure in the spaces, an ordering (grading) of the sequence of seminorms and, moreover, a certain uniform approximation property called the *tameness* condition. For the mappings, the associated tameness property means that certain growth estimates hold. The statement of the Nash–Moser theorem is surprisingly similar to the classical inverse function theorem, but the meaning of the concepts used is much more involved: when a smooth tame mapping f acting between Fréchet spaces has an invertible tame derivative, then f^{-1} has a smooth tame single-valued localization. The rigorous introduction of the tame spaces, mappings and derivatives is beyond the scope of this book; we only note here that extending the Nash–Moser theorem to set-valued mappings, e.g. in the setting of Section 5E, is a challenging avenue for future research.

Chapter 6
Applications in Numerical Variational Analysis

The classical implicit function theorem finds a wide range of applications in numerical analysis. For instance, it helps in deriving error estimates for approximations to differential equations and is often relied on in establishing the convergence of algorithms. Can the generalizations of the classical theory to which we have devoted so much of this book have comparable applications in the numerical treatment of non-classical problems for generalized equations and beyond? In this chapter we provide positive answers in several directions.

We begin with a topic at the core of numerical work, the "conditioning" of a problem and how it extends to concepts like metric regularity. We also explain how the conditioning of a feasibility problem, like solving a system of inequalities, can be understood. Next we take up a general iterative scheme for solving generalized equations under metric regularity, obtaining convergence by means of our earlier basic results. As particular cases, we get various modes of convergence of the age-old procedure known as Newton's method in several guises, and of the much more recently introduced proximal point algorithm. We go a step further with Newton's method by showing that the mapping which assigns to an instance of a parameter the set of all sequences generated by the method obeys, in a Banach space of sequences, the implicit function theorem paradigm in the same pattern as the solution mapping for the underlying generalized equation. Approximations of quadratic optimization problems in Hilbert spaces are then studied. Finally, we apply our methodology to discrete approximations in optimal control.

6A. Radius Theorems and Conditioning

In numerical analysis, a measure of "conditioning" of a problem is typically conceived as a bound on the ratio of the size of solution (output) error to the size of data (input) error. At its simplest, this pattern is seen when solving a linear equation $Ax = y$ for x in terms of y when A is a nonsingular matrix in $\mathbb{R}^{n \times n}$. The data input then is y and the solution output is $A^{-1}y$, but for computational purposes the story cannot just be left at that. Much depends on the extent to which an input error δy leads to an output error δx. The magnitudes of the errors can be measured by the Euclidean norm, say. Then, through linearity, there is the tight bound $|\delta x| \leq |A^{-1}||\delta y|$, in which $|A^{-1}|$ is the corresponding matrix (operator) norm of the mapping $y \mapsto A^{-1}y$ and in fact is the global Lipschitz constant for this mapping. In providing such a bound on the ratio of $|\delta x|$ to $|\delta y|$, $|A^{-1}|$ is called the *absolute condition number* for the problem of solving $Ax = y$. A high value of $|A^{-1}|$ is a warning flag signalling trouble in computing the solution x for a given y.

Another popular conditioning concept concerns relative errors instead of absolute errors. In solving $Ax = y$, the relative error of the input is $|\delta y|/|y|$ (with $y \neq 0$), while the relative error of the output is $|\delta x|/|x|$. It is easy to see that the best bound on the ratio of $|\delta x|/|x|$ to $|\delta y|/|y|$ is the product $|A||A^{-1}|$. Therefore, $|A||A^{-1}|$ is called the *relative condition number* for the problem of solving $Ax = y$. But absolute conditioning will be the chief interest in our present context, for several reasons.

The reciprocal of the absolute condition number $|A^{-1}|$ of a nonsingular matrix A has a geometric interpretation which will serve as an important guide to our developments. It turns out to give an exact bound on how far A can be perturbed to $A + B$ before good behavior breaks down by $A + B$ becoming singular, and thus has more significance for numerical analysis than simply comparing the size of δx to the size of δy. This property of the absolute condition number comes from a classical result about matrices which was stated and proved in Chapter 1 as 1E.9:

$$\inf\left\{ |B| \,\Big|\, A+B \text{ is singular} \right\} = \frac{1}{|A^{-1}|}, \quad \text{for any nonsingular matrix } A.$$

In this sense, $|A^{-1}|^{-1}$ gives the *radius of nonsingularity* around A. As long as B lies within that distance from A, the nonsingularity of $A + B$ is assured. Clearly from this angle as well, a large value of the condition number $|A^{-1}|$ points toward numerical difficulties.

The model provided to us by this example is that of a *radius theorem*, furnishing a bound on how far perturbations of some sort in the specification of a problem can go before some key property is lost. Radius theorems can be investigated not only for solving equations, linear and nonlinear, but also generalized equations, systems of constraints, etc.

We start down that track by stating the version of the cited matrix result that works in infinite dimensions for bounded linear mappings acting in Banach spaces.

6 Applications in Numerical Variational Analysis

Theorem 6A.1 (radius theorem for invertibility of bounded linear mappings). *Let X and Y be Banach spaces and let $A \in \mathscr{L}(X,Y)$ be invertible[1]. Then*

$$\inf_{B \in \mathscr{L}(X,Y)} \left\{ \|B\| \,\Big|\, A + B \text{ is not invertible} \right\} = \frac{1}{\|A^{-1}\|}. \tag{1}$$

Moreover the infimum is the same if restricted to mappings B of rank one.

Proof. The estimation of perturbed inversion in Lemma 5A.4 gives us "$\geq$" in (1). To obtain the opposite inequality and thereby complete the proof, we take any $r > 1/\|A^{-1}\|$ and construct a mapping B of rank one such that $A + B$ is not invertible and $\|B\| < r$. There exists $\hat{x}$ with $\|A\hat{x}\| = 1$ and $\|\hat{x}\| > 1/r$. Choose an $x^* \in X^*$ such that $x^*(\hat{x}) = \|\hat{x}\|$ and $\|x^*\| = 1$. The linear and bounded mapping

$$Bx = -\frac{x^*(x)A\hat{x}}{\|\hat{x}\|} \tag{2}$$

has $\|B\| = 1/\|\hat{x}\|$ and $(A+B)\hat{x} = A\hat{x} - A\hat{x} = 0$. Then $A + B$ is not invertible and hence the infimum in (1) is $\leq r$. It remains to note that B in (2) is of rank one. □

The initial step that can be taken toward generality beyond linear mappings is in the direction of positively homogeneous mappings $H: X \rightrightarrows Y$; here and further on, X and Y are Banach spaces. For such a mapping, ordinary norms can no longer be of help in conditioning, but the outer and inner norms introduced in 4A in finite dimensions and extended in 5A to Banach spaces can come into play:

$$\|H\|^+ = \sup_{\|x\| \leq 1} \sup_{y \in H(x)} \|y\| \quad \text{and} \quad \|H\|^- = \sup_{\|x\| \leq 1} \inf_{y \in H(x)} \|y\|.$$

Their counterparts for the inverse H^{-1} will have a role as well:

$$\|H^{-1}\|^+ = \sup_{\|y\| \leq 1} \sup_{x \in H^{-1}(y)} \|x\| \quad \text{and} \quad \|H^{-1}\|^- = \sup_{\|y\| \leq 1} \inf_{x \in H^{-1}(y)} \|x\|.$$

In thinking of $H^{-1}(y)$ as the set of solutions x to $H(x) \ni y$, it is clear that the outer and inner norms of H^{-1} capture two different aspects of solution behavior, roughly the distance to the farthest solution and the distance to the nearest solution (when multi-valuedness is present). We are able to assert, for instance, that

$$\mathrm{dist}(0, H^{-1}(y)) \leq \|H^{-1}\|^- \|y\| \quad \text{for all } y.$$

From that angle, $\|H^{-1}\|^-$ could be viewed as a sort of *inner* absolute condition number—and in a similar manner, $\|H^{-1}\|^+$ could be viewed as a sort of *outer* absolute condition number. This idea falls a bit short, though, because we only have a comparison between sizes of $\|x\|$ and $\|y\|$, not the size of a shift from x to $x + \delta x$

---

[1] This assumption can be dropped if we identify invertibility of A with $\|A^{-1}\| < \infty$ and adopt the convention $1/\infty = 0$. Similar adjustments can be made in the remaining radius theorems in this section.

caused by a shift from y to $y + \delta y$. Without H being linear, there seems little hope of quantifying that aspect of error, not to speak of relative error. Nonetheless, it will be possible to get radius theorems in which the reciprocals of $\|H^{-1}\|^+$ and $\|H^{-1}\|^-$ are featured.

For $\|H^{-1}\|^+$, we can utilize the inversion estimate for the outer norm in 5A.8. A definition is needed first.

Extended nonsingularity. *A positively homogeneous mapping $H : X \rightrightarrows Y$ is said to be nonsingular if $\|H^{-1}\|^+ < \infty$; it is said to be singular if $\|H^{-1}\|^+ = \infty$.*

As shown in 5A.7, nonsingularity of H in this sense implies that $H^{-1}(0) = \{0\}$; moreover, when $\dim X < \infty$ and $\gph H$ is closed the converse is true as well.

Theorem 6A.2 (radius theorem for nonsingularity of positively homogeneous mappings). *For any $H : X \rightrightarrows Y$ that is positively homogeneous and nonsingular, one has*

$$(3) \qquad \inf_{B \in \mathscr{L}(X,Y)} \left\{ \|B\| \,\Big|\, H + B \text{ is singular} \right\} = \frac{1}{\|H^{-1}\|^+}.$$

Moreover the infimum is the same if restricted to mappings B of rank one.

Proof. The proof is parallel to that of 6A.1. From 5A.8 we get "$\geq$" in (3), and also "$=$" for the case $\|H^{-1}\|^+ = 0$ under the convention $1/0 = \infty$. Let $\|H^{-1}\|^+ > 0$ and consider any $r > 1/\|H^{-1}\|^+$. There exists $(\hat{x}, \hat{y}) \in \gph H$ with $\|\hat{y}\| = 1$ and $\|\hat{x}\| > 1/r$. Let $x^* \in X^*$, $x^*(\hat{x}) = \|\hat{x}\|$ and $\|x^*\| = 1$. The linear and bounded mapping

$$Bx = -\frac{x^*(x)\hat{y}}{\|\hat{x}\|}$$

has $\|B\| = 1/\|\hat{x}\| < r$ and $(H+B)(\hat{x}) = H(\hat{x}) - \hat{y} \ni 0$. Then the nonzero vector $\hat{x}$ belongs to $(H+B)^{-1}(0)$, hence $\|(H+B)^{-1}\|^+ = \infty$, i.e., $H+B$ is singular. The infimum in (3) must therefore be less than r. Appealing to the choice of r we conclude that the infimum in (3) cannot be more than $1/\|H^{-1}\|^+$, and we are done. □

To develop a radius theorem about $\|H^{-1}\|^-$, we have to look more narrowly at *sublinear* mappings, which are characterized by having graphs that are not just cones, as corresponds to positive homogeneity, but *convex* cones. For such a mapping H, if its graph is also closed, we have an inversion estimate for the inner norm in 5C.9. Furthermore, we know from 5C.2 that the surjectivity of H is equivalent to having $\|H^{-1}\|^- < \infty$. We also have available the notion of the adjoint mapping as introduced in Section 5C: the upper adjoint of $H : X \rightrightarrows Y$ is the sublinear mapping $H^{*+} : Y^* \rightrightarrows X^*$ defined by

$$(y^*, x^*) \in \gph H^{*+} \iff \langle x^*, x \rangle \leq \langle y^*, y \rangle \text{ for all } (x,y) \in \gph H.$$

Recall too, from 5C.13, that for a sublinear mapping H with closed graph,

$$(4) \qquad \|(H^{*+})^{-1}\|^+ = \|H^{-1}\|^-,$$

and also, from 5C.14,

(5) $$(H+B)^{*+} = H^{*+} + B^* \quad \text{for any } B \in \mathscr{L}(X,Y).$$

Theorem 6A.3 (radius theorem for surjectivity of sublinear mappings). *For any $H : X \rightrightarrows Y$ that is sublinear, surjective, and with closed graph,*

$$\inf_{B \in \mathscr{L}(X,Y)} \left\{ \|B\| \,\Big|\, H+B \text{ is not surjective} \right\} = \frac{1}{\|H^{-1}\|^-}.$$

Moreover the infimum is the same if restricted to B of rank one.

Proof. For any $B \in \mathscr{L}(X,Y)$, the mapping $H+B$ is sublinear with closed graph, so that $(H+B)^{*+} = H^{*+} + B^*$ by (5). By the definition of the adjoint, $H+B$ is surjective if and only if $H^{*+} + B^*$ is nonsingular. It follows that

(6) $$\begin{aligned}\inf_{B \in \mathscr{L}(X,Y)} & \left\{ \|B\| \,\Big|\, H+B \text{ is not surjective} \right\} \\ &= \inf_{B \in \mathscr{L}(X,Y)} \left\{ \|B^*\| \,\Big|\, H^{*+} + B^* \text{ is singular} \right\}.\end{aligned}$$

The right side of (6) can be identified through Theorem 6A.2 with

(7) $$\inf_{C \in \mathscr{L}(Y^*, X^*)} \left\{ \|C\| \,\Big|\, H^{*+} + C \text{ is singular} \right\} = \frac{1}{\|(H^{*+})^{-1}\|^+}$$

by the observation that any $C \in \mathscr{L}(Y^*, X^*)$ of rank one has the form B^* for some $B \in \mathscr{L}(X,Y)$ of rank one. It remains to apply the relation in (4). In consequence of that, the left side of (7) is $1/\|H^{-1}\|^-$, and we get the desired equality. □

In the case of H being a bounded linear mapping $A : X \to Y$, Theorems 6A.2 and 6A.3 both furnish results which complement Theorem 6A.1, since nonsingularity just comes down to A^{-1} being single-valued on rge A, while surjectivity corresponds only to dom A^{-1} being all of Y, and neither of those properties automatically entails the other. When $X = Y = \mathbb{R}^n$, of course, all three theorems reduce to the matrix result recalled at the beginning of this section.

The surjectivity result in 6A.3 offers more than an extended insight into equation solving, however. It can be applied also to systems of inequalities. This is true even in infinite dimensions, but we are not yet prepared to speak of inequality constraints in that framework, so we limit the following illustration to solving $Ax \leq y$ in the case of a matrix $A \in \mathbb{R}^{m \times n}$. It will be convenient to say that

$Ax \leq y$ is *universally solvable* if it has a solution $x \in \mathbb{R}^n$ for every $y \in \mathbb{R}^m$.

We adopt for $y = (y_1, \ldots, y_m) \in \mathbb{R}^m$ the maximum norm $|y|_\infty = \max_{1 \leq k \leq m} |y_k|$ but equip $\mathbb{R}^n$ with any norm. The associated operator norm for linear mappings acting from $\mathbb{R}^n$ to $\mathbb{R}^m$ is denoted by $|\cdot|_\infty$. Also, we use the notation for $y = (y_1, \ldots, y_m)$ that

$$y^+ = (y_1^+, \ldots, y_m^+), \text{ where } y_k^+ = \max\{0, y_k\}.$$

Example 6A.4 (radius of universal solvability for systems of linear inequalities). Suppose for a matrix $A \in \mathbb{R}^{m \times n}$ that $Ax \leq y$ is universally solvable. Then

$$\inf_{B \in \mathbb{R}^{m \times n}} \left\{ |B|_\infty \,\big|\, (A+B)x \leq y \text{ is not universally solvable} \right\} = \frac{1}{\sup_{|x| \leq 1} |[Ax]^+|_\infty}.$$

Detail. We apply Theorem 6A.3 to the mapping $F : x \mapsto \{y \,|\, Ax \leq y\}$. Then $|F^{-1}|^- = \sup_{|x| \leq 1} \inf_{y \geq Ax} |y|_\infty$, where the infimum equals $|[Ax]^+|_\infty$. □

More will be said about constraint systems in Section 6B.

Since surjectivity of a sublinear mapping is also equivalent to its metric regularity at 0 for 0, we could restate Theorem 6A.3 in terms of metric regularity as well. Such a result is actually true for any strictly differentiable function. Specifically, Corollary 5D.5 says that, for a function $f : X \to Y$ which is strictly differentiable at $\bar{x}$, one has

$$\operatorname{reg}(f; \bar{x} | \bar{y}) = \operatorname{reg} Df(\bar{x}) \text{ for } \bar{y} = f(\bar{x}).$$

Since any linear mapping is sublinear, this equality combined with 6A.3 gives us yet another radius result.

Corollary 6A.5 (radius theorem for metric regularity of strictly differentiable functions). *Let* $f : X \to Y$ *be strictly differentiable at* $\bar{x}$, *let* $\bar{y} := f(\bar{x})$, *and let* $Df(\bar{x})$ *be surjective. Then*

$$\inf_{B \in \mathscr{L}(X,Y)} \left\{ \|B\| \,\Big|\, f+B \text{ is not metrically regular at } \bar{x} \text{ for } \bar{y} + B\bar{x} \right\} = \frac{1}{\|Df(\bar{x})^{-1}\|^-}.$$

It should not escape attention here that in 6A.5 we are not focused any more on the origins of X and Y but on a general pair $(\bar{x}, \bar{y})$ in the graph of f. This allows us to return to "conditioning" from a different perspective, if we are willing to think of such a property in a local sense only.

Suppose that a y near to $\bar{y}$ is perturbed to $y + \delta y$. The solution set $f^{-1}(y)$ to the problem of solving $f(x) \ni y$ is thereby shifted to $f^{-1}(y + \delta y)$, and we have an interest in understanding the "error" vectors δx such that $x + \delta x \in f^{-1}(y + \delta y)$. Since anyway x need not be the only element of $f^{-1}(y)$, it is appropriate to quantify the shift by looking for the smallest size of δx, or in other words at $\operatorname{dist}(x, f^{-1}(y + \delta y))$ and how it compares to $\|\delta y\|$. This ratio, in its limit as (x,y) goes to $(\bar{x}, \bar{y})$ and $\|\delta y\|$ goes to 0, is precisely $\operatorname{reg}(f; \bar{x} | \bar{y})$.

In this sense, $\operatorname{reg}(f; \bar{x} | \bar{y})$ can be deemed the absolute condition number *locally* with respect to $\bar{x}$ and $\bar{y}$ for the problem of solving $f(x) \ni y$ for x in terms of y. We then have a *local, nonlinear* analog of Theorem 6A.1, tying a condition number to a radius. It provides something more even for linear f, of course, since in contrast to Theorem 6A.1, it imposes no requirement of invertibility.

Corollary 6A.5 can be stated in a more general form, which we give here as an exercise:

6 Applications in Numerical Variational Analysis   317

Exercise 6A.6. Let $F : X \rightrightarrows Y$ with $(\bar{x},\bar{y}) \in \text{gph}\, F$ being a pair at which gph F is locally closed. Let F be metrically regular at $\bar{x}$ for $\bar{y}$, and let $f : X \to Y$ satisfy $\bar{x} \in \text{int dom}\, f$ and $\text{lip}(f;\bar{x}) = 0$. Then

$$\inf_{B \in \mathscr{L}(X,Y)} \left\{ \|B\| \,\Big|\, F+B \text{ is not metrically regular at } \bar{x} \text{ for } \bar{y}+B\bar{x} \right\}$$
$$= \inf_{B \in \mathscr{L}(X,Y)} \left\{ \|B\| \,\Big|\, F+f+B \text{ is not metrically regular at } \bar{x} \text{ for } \bar{y}+f(\bar{x})+B\bar{x} \right\}.$$

Guide. Observe that, by the Banach space version of 3F.4 (which follows from 5E.1), the mapping $F+B$ is metrically regular at $\bar{y}+B\bar{x}$ if and only if the mapping $F+f+B$ is metrically regular at $\bar{y}+f(\bar{x})+B\bar{x}$. □

We will show next that, in *finite* dimensions at least, the radius result in 6A.5 is valid when f is replaced by *any set-valued mapping* F whose graph is locally closed around the reference pair $(\bar{x},\bar{y})$.

Theorem 6A.7 (radius theorem for metric regularity). *Let X and Y be finite-dimensional normed linear spaces, and for $F : X \rightrightarrows Y$ and $\bar{y} \in F(\bar{x})$ let gph F be locally closed at $(\bar{x},\bar{y})$. Suppose F is metrically regular at $\bar{x}$ for $\bar{y}$. Then*

$$(8) \quad \inf_{B \in \mathscr{L}(X,Y)} \left\{ \|B\| \,\Big|\, F+B \text{ is not metrically regular at } \bar{x} \text{ for } \bar{y}+B\bar{x} \right\} = \frac{1}{\text{reg}(F;\bar{x}|\bar{y})}.$$

Moreover, the infimum is unchanged if taken with respect to linear mappings of rank 1, but also remains unchanged when the class of perturbations B is enlarged to all locally Lipschitz continuous functions g, with $\|B\|$ replaced by the Lipschitz modulus $\text{lip}(g;\bar{x})$ of g at $\bar{x}$.

Proof. The general perturbation inequality derived in Theorem 5E.1, see 5E.7, produces the estimate

$$\inf_{g:X \to Y} \left\{ \text{lip}(g;\bar{x}) \,\Big|\, F+g \text{ is not metrically regular at } \bar{x} \text{ for } \bar{y}+g(\bar{x}) \right\} \geq \frac{1}{\text{reg}(F;\bar{x}|\bar{y})},$$

which becomes the equality (8) in the case when $\text{reg}(F;\bar{x}|\bar{y}) = 0$ under the convention $1/0 = \infty$. To confirm the opposite inequality when $\text{reg}(F;\bar{x}|\bar{y}) > 0$, we apply Theorem 4B.9, according to which

$$(9) \quad \text{reg}(F;\bar{x}|\bar{y}) = \limsup_{\substack{(x,y) \to (\bar{x},\bar{y}) \\ (x,y) \in \text{gph}\, F}} \|\tilde{D}F(x|y)^{-1}\|^-,$$

where $\tilde{D}F(x|y)$ is the convexified graphical derivative of F at x for y. Take a sequence of positive real numbers $\varepsilon_k \searrow 0$. Then for k sufficiently large, say $k > \bar{k}$, by (9) there exists $(x_k, y_k) \in \text{gph}\, F$ with $(x_k, y_k) \to (\bar{x},\bar{y})$ and

$$\text{reg}(F;\bar{x}|\bar{y}) + \varepsilon_k \geq \|\tilde{D}F(x_k|y_k)^{-1}\|^- \geq \text{reg}(F;\bar{x}|\bar{y}) - \varepsilon_k > 0.$$

Let $H_k := \tilde{D}F(x_k|y_k)$ and $S_k := H_k^{*+}$; then norm duality gives us $\|H_k^{-1}\|^- = \|S_k^{-1}\|^+$, see 5C.13.

For each $k > \bar{k}$ choose a positive real r_k satisfying $\|S_k^{-1}\|^+ - \varepsilon_k < 1/r_k < \|S_k^{-1}\|^+$. From the last inequality there must exist $(\hat{y}_k, \hat{x}_k) \in \text{gph } S_k$ with $\|\hat{x}_k\| = 1$ and $\|S_k^{-1}\|^+ \geq \|\hat{y}_k\| > 1/r_k$. Pick $y_k^* \in Y$ with $\langle \hat{y}_k, y_k^* \rangle = \|\hat{y}_k\|$ and $\|y_k^*\| = 1$, and define the rank-one mapping $\hat{G}_k \in \mathscr{L}(Y,X)$ by

$$\hat{G}_k(y) := -\frac{\langle y, y_k^* \rangle}{\|\hat{y}_k\|} \hat{x}_k.$$

Then $\hat{G}_k(\hat{y}_k) = -\hat{x}_k$ and hence $(S_k + \hat{G}_k)(\hat{y}_k) = S_k(\hat{y}_k) + \hat{G}_k(\hat{y}_k) = S_k(\hat{y}_k) - \hat{x}_k \ni 0$. Therefore, $\hat{y}_k \in (S_k + \hat{G}_k)^{-1}(0)$, and since $\hat{y}_k \neq 0$ and S_k is positively homogeneous with closed graph, we have by Proposition 5A.7, formula 5A(10), that

(10) $$\|(S_k + \hat{G}_k)^{-1}\|^+ = \infty.$$

Note that $\|\hat{G}_k\| = \|\hat{x}_k\|/\|\hat{y}_k\| = 1/\|\hat{y}_k\| < r_k$.

Since the sequences $\hat{y}_k, \hat{x}_k$ and y_k^* are bounded (and the spaces are *finite-dimensional*), we can extract from them subsequences converging respectively to $\hat{y}, \hat{x}$ and y^*. The limits then satisfy $\|\hat{y}\| = \text{reg}(F;\bar{x}|\bar{y})$, $\|\hat{x}\| = 1$ and $\|y^*\| = 1$. Define the rank-one mapping $\hat{G} \in \mathscr{L}(Y,X)$ by

$$\hat{G}(y) := -\frac{\langle y, y^* \rangle}{\|\hat{y}\|} \hat{x}.$$

Then we have $\|\hat{G}\| = \text{reg}(F;\bar{x}|\bar{y})^{-1}$ and $\|\hat{G}_k - \hat{G}\| \to 0$.

Let $B := (\hat{G})^*$ and suppose $F + B$ is metrically regular at $\bar{x}$ for $\bar{y} + B\bar{x}$. Theorem 4B.9 yields that there is a finite positive constant c such that for $k > \bar{k}$ sufficiently large, we have

$$c > \|\tilde{D}(F+B)(x_k|y_k + Bx_k)^{-1}\|^-.$$

Through 4B.10, this gives us

$$c > \|(\tilde{D}F(x_k|y_k) + B)^{-1}\|^- = \|(H_k + B)^{-1}\|^-.$$

Since B is linear, we have $B^* = ((\hat{G})^*)^* = \hat{G}$, and since $H_k + B$ is sublinear, it follows further by 5C.14 that

(11) $$c > \|([H_k + B]^{*+})^{-1}\|^+ = \|(H_k^{*+} + B^*)^{-1}\|^+ = \|(S_k + \hat{G})^{-1}\|^+.$$

Take $k > \bar{k}$ sufficiently large such that $\|\hat{G} - \hat{G}_k\| \leq 1/(2c)$. Setting $P_k := S_k + \hat{G}$ and $Q_k := \hat{G}_k - \hat{G}$, we have that

$$[\|P_k^{-1}\|^+]^{-1} \geq 1/c > 1/(2c) \geq \|Q_k\|.$$

By using the inversion estimate for the outer norm in Theorem 5C.9, we have

6 Applications in Numerical Variational Analysis

$$\|(S_k + \hat{G}_k)^{-1}\|^+ = \|(P_k + Q_k)^{-1}\|^+ \leq \left([\|P_k^{-1}\|^+]^{-1} - \|Q_k\|\right)^{-1} \leq 2c < \infty.$$

This contradicts (10). Hence, $F + B$ is not metrically regular at $\bar{x}$ for $\bar{y} + B\bar{x}$. Noting that $\|B\| = \|\hat{G}\| = 1/\operatorname{reg}(F;\bar{x}|\bar{y})$ and that B is of rank one, we are finished. □

In a pattern just like the one laid out after Corollary 6A.5, it is appropriate to consider $\operatorname{reg}(F;\bar{x}|\bar{y})$ as the *local* absolute condition number with respect to $\bar{x}$ and $\bar{y}$ for the problem of solving $F(x) \ni y$ for x in terms of y. An even grander extension of the fact in 6A.1, that the reciprocal of the absolute condition number gives the radius of perturbation for preserving an associated property, is thereby achieved.

Based on Theorem 6A.7, it is now easy to obtain a parallel radius result for strong metric regularity.

Theorem 6A.8 (radius theorem for strong metric regularity). *For finite-dimensional normed linear spaces X and Y, let $F : X \rightrightarrows Y$ have $\bar{y} \in F(\bar{x})$. Suppose that F is strongly metrically regular at $\bar{x}$ for $\bar{y}$. Then*

$$(12) \quad \inf_{B \in \mathscr{L}(X,Y)} \left\{ \|B\| \,\Big|\, F + B \text{ is not strongly regular at } \bar{x} \text{ for } \bar{y} + B\bar{x} \right\} = \frac{1}{\operatorname{reg}(F;\bar{x}|\bar{y})}.$$

Moreover, the infimum is unchanged if taken with respect to linear mappings of rank 1, but also remains unchanged when the class of perturbations B is enlarged to the class of locally Lipschitz continuous functions g with $\|B\|$ replaced by the Lipschitz modulus $\operatorname{lip}(g;\bar{x})$.

Proof. Theorem 5F.1 reveals that "$\geq$" holds in (12) when the linear perturbation is replaced by a Lipschitz perturbation, and moreover that (12) is satisfied in the limit case $\operatorname{reg}(F;\bar{x}|\bar{y}) = 0$ under the convention $1/0 = \infty$. The inequality becomes an equality with the observation that the assumed strong metric regularity of F implies that F has locally closed graph at $(\bar{x},\bar{y})$ and is metrically regular at $\bar{x}$ for $\bar{y}$. Hence the infimum in (12) is not greater than the infimum in (8). □

Next comes a radius theorem for strong subregularity to go along with the ones for metric regularity and strong metric regularity.

Theorem 6A.9 (radius theorem for strong metric subregularity). *Let X and Y be finite-dimensional normed linear spaces X and Y, and for $F : X \rightrightarrows Y$ and $\bar{y} \in F(\bar{x})$ let $\operatorname{gph} F$ be locally closed at $(\bar{x},\bar{y})$. Suppose that F is strongly metrically subregular at $\bar{x}$ for $\bar{y}$. Then*

$$\inf_{B \in \mathscr{L}(X,Y)} \left\{ \|B\| \,\Big|\, F + B \text{ is not strongly subregular at } \bar{x} \text{ for } \bar{y} + B\bar{x} \right\} = \frac{1}{\operatorname{subreg}(F;\bar{x}|\bar{y})}.$$

Moreover, the infimum is unchanged if taken with respect to mappings B of rank 1, but also remains unchanged when the class of perturbations is enlarged to the class of functions $g : X \to Y$ that are calm at $\bar{x}$ and continuous around $\bar{x}$, with $\|B\|$ replaced by the calmness modulus $\operatorname{clm}(g;\bar{x})$.

Proof. From the equivalence of the strong subregularity of a mapping F at $\bar{x}$ for $\bar{y}$ with the nonsingularity of its graphical derivative $DF(\bar{x}|\bar{y})$, as shown in Theorem 4C.1, we have

$$
(13) \quad \inf_{B \in \mathscr{L}(X,Y)} \left\{ \|B\| \,\Big|\, F+B \text{ is not strongly subregular at } \bar{x} \text{ for } \bar{y}+B\bar{x} \right\}
$$
$$
= \inf_{B \in \mathscr{L}(X,Y)} \left\{ \|B\| \,\Big|\, D(F+B)(\bar{x}|\bar{y}+B\bar{x}) \text{ is singular} \right\}.
$$

We know from the sum rule for graphical differentiation (4A.1) that $D(F+B)(\bar{x}|\bar{y}+B\bar{x}) = DF(\bar{x}|\bar{y}) + B$, hence

$$
(14) \quad \inf_{B \in \mathscr{L}(X,Y)} \left\{ \|B\| \,\Big|\, D(F+B)(\bar{x}|\bar{y}+B\bar{x}) \text{ is singular} \right\}
$$
$$
= \inf_{B \in \mathscr{L}(X,Y)} \left\{ \|B\| \,\Big|\, DF(\bar{x}|\bar{y})+B \text{ is singular} \right\}.
$$

Since $DF(\bar{x}|\bar{y}) + B$ is positively homogeneous, 6A.2 translates to

$$
(15) \quad \inf_{B \in \mathscr{L}(X,Y)} \left\{ \|B\| \,\Big|\, DF(\bar{x}|\bar{y})+B \text{ is singular} \right\} = \frac{1}{\|DF(\bar{x}|\bar{y})^{-1}\|^+},
$$

including the case $\|DF(\bar{x}|\bar{y})^{-1}\|^+ = 0$ with the convention $1/0 = \infty$. Theorem 4C.1 tells us also that $\|DF(\bar{x}|\bar{y})^{-1}\|^+ = \text{subreg}(F;\bar{x}|\bar{y})$ and then, putting together (13), (14) and (15), we get the desired equality. $\square$

As with the preceding results, the modulus $\text{subreg}(F;\bar{x}|\bar{y})$ can be regarded as a sort of local absolute condition number. But in this case only the ratio of $\text{dist}(\bar{x}, F^{-1}(\bar{y}+\delta y))$ to $\|\delta y\|$ is considered in its limsup as δy goes to 0, not the limsup of all the ratios $\text{dist}(x, F^{-1}(y+\delta y))/\|\delta y\|$ with $(x,y) \in \text{gph}\, F$ tending to $(\bar{x},\bar{y})$, which gives $\text{reg}(F;\bar{x}|\bar{y})$. Specifically, with $\text{reg}(F;\bar{x}|\bar{y})$ appropriately termed the absolute condition number for F locally with respect to $\bar{x}$ and $\bar{y}$, $\text{subreg}(F;\bar{x}|\bar{y})$ is the corresponding *subcondition number*.

The radius-type theorems above could be rewritten in terms of the associated equivalent properties of the inverse mappings. For example, Theorem 6A.7 could be restated in terms of perturbations B of a mapping F whose inverse has the Aubin property.

6B. Constraints and Feasibility

Universal solvability of systems of linear inequalities in finite dimensions was already featured in Example 6A.4 as an application of one of our radius theorems, but now we will go more deeply into the subject of constraint systems and their solvability. We take as our focus problems of the very general type

6 Applications in Numerical Variational Analysis

(1) $$\text{find } x \text{ such that } F(x) \ni 0$$

for a set-valued mapping $F : X \rightrightarrows Y$ from one Banach space to another. Of course, the set of all solutions is just $F^{-1}(0)$, but we are thinking of F as representing a kind of constraint system and are concerned with whether the set $F^{-1}(0)$ might shift from nonempty to empty under some sort of perturbation. Mostly, we will study the case where F has convex graph.

Feasibility. *Problem (1) will be called feasible if $F^{-1}(0) \neq \emptyset$, i.e., $0 \in \text{rge } F$, and strictly feasible if $0 \in \text{int rge } F$.*

Two examples will point the way toward progress. Recall that any closed, convex cone $K \subset Y$ with nonempty interior induces a partial ordering "$\leq_K$" under the rule that $y_0 \leq_K y_1$ means $y_1 - y_0 \in K$. Correspondingly, $y_0 <_K y_1$ means $y_1 - y_0 \in \text{int } K$.

Example 6B.1 (convex constraint systems). Let $C \subset X$ be a closed convex set, let $K \subset Y$ be a closed convex cone, and let $A : C \to Y$ be a continuous and convex mapping with respect to the partial ordering in Y induced by K; that is,

$$A((1-\theta)x_0 + \theta x_1) \leq_K (1-\theta)A(x_0) + \theta A(x_1) \text{ for } x_0, x_1 \in C \text{ when } 0 < \theta < 1.$$

Define the mapping $F : X \rightrightarrows Y$ by

$$F(x) = \begin{cases} A(x) + K & \text{if } x \in C, \\ \emptyset & \text{if } x \notin C. \end{cases}$$

Then F has closed, convex graph, and feasibility of solving $F(x) \ni 0$ for x refers to

$$\exists \bar{x} \in C \text{ such that } A(\bar{x}) \leq_K 0.$$

On the other hand, as long as $\text{int } K \neq \emptyset$, strict feasibility refers to

$$\exists \bar{x} \in C \text{ such that } A(\bar{x}) <_K 0.$$

Detail. To see this, note that, when $\text{int } K \neq \emptyset$, we have $K = \text{cl int } K$, so that the convex set $\text{rge } F = A(C) + K$ is the closure of the open set $O := A(C) + \text{int } K$. Also, O is convex. It follows then that $\text{int rge } F = O$. □

Example 6B.2 (linear-conic constraint systems). Consider the convex constraint system in Example 6B.1 under the additional assumptions that A is linear and C is a cone, so that the condition $\bar{x} \in C$ can be written equivalently as $\bar{x} \geq_C 0$. Then F is sublinear, and its adjoint $F^{*+} : Y^* \rightrightarrows X^*$ is given in terms of the adjoint A^* of A and the dual cones $K^+ = -K^*$ and $C^+ = -C^*$ (where * denotes polar) by

$$F^{*+}(y^*) = \begin{cases} A^*(y^*) - C^+ & \text{if } y^* \in K^+, \\ \emptyset & \text{if } y^* \notin K^+, \end{cases}$$

so that $F^{*+}(y^*) \ni x^*$ if and only if $y^* \geq_{K^+} 0$ and $A^*(y^*) \geq_{C^+} x^*$.

Detail. In this case the graph of F is clearly a convex cone, and that means F is sublinear. The claims about the adjoint of F follow by elementary calculation. □

Along the lines of the analysis in 6A, in dealing with the feasibility problem (1) we will be interested in perturbations in which F is replaced by $F + B$ for some $B \in \mathscr{L}(X,Y)$, and at the same time, the zero on the right is replaced by some other $b \in Y$. Such a double perturbation, the magnitude of which can be quantified by the norm

(2) $$\|(B,b)\| = \max\{\,\|B\|, \|b\|\,\},$$

transforms the condition $F(x) \ni 0$ to $(F+B)(x) \ni b$ and the solution set $F^{-1}(0)$ to $(F+B)^{-1}(b)$, creating infeasibility if $(F+B)^{-1}(b) = \emptyset$, i.e., if $b \notin \mathrm{rge}\,(F+B)$. We want to understand how large $\|(B,b)\|$ can be before this happens.

Distance to infeasibility. For $F : X \rightrightarrows Y$ with convex graph and $0 \in \mathrm{rge}\,F$, the *distance to infeasibility* of the system $F(x) \ni 0$ is defined to be the value

(3) $$\inf_{B \in \mathscr{L}(X,Y),\, b \in Y} \Big\{\, \|(B,b)\| \,\Big|\, b \notin \mathrm{rge}\,(F+B) \Big\}.$$

Surprisingly, perhaps, it turns out that there would be no difference if feasibility were replaced by strict feasibility in this definition. In the next pair of lemmas, it is assumed that $F : X \rightrightarrows Y$ has convex graph and $0 \in \mathrm{rge}\,F$.

Lemma 6B.3 (distance to infeasibility versus distance to strict infeasibility). *The distance to infeasibility is the same as the distance to strict infeasibility, namely the value*

(4) $$\inf_{B \in \mathscr{L}(X,Y),\, b \in Y} \Big\{\, \|(B,b)\| \,\Big|\, b \notin \mathrm{int}\,\mathrm{rge}\,(F+B) \Big\}.$$

Proof. Let S_1 denote the set of (B,b) over which the infimum is taken in (3) and let S_2 be the corresponding set in (4). Obviously $S_1 \subset S_2$, so the first infimum cannot be less than the second. We must show that it also cannot be greater. This amounts to demonstrating that for any $(B,b) \in S_2$ and any $\varepsilon > 0$ we can find $(B',b') \in S_1$ such that $\|(B',b')\| \leq \|(B,b)\| + \varepsilon$. In fact, we can get this with $B' = B$ simply by noting that when $b \notin \mathrm{int}\,\mathrm{rge}\,(F+B)$ there must exist $b' \in Y$ with $b' \notin \mathrm{rge}\,(F+B)$ and $\|b' - b\| \leq \varepsilon$. □

By utilizing the Robinson–Ursescu theorem 5B.4, we can see furthermore that the distance to infeasibility is actually the same as the distance to metric nonregularity:

Lemma 6B.4 (distance to infeasibility equals radius of metric regularity). *The distance to infeasibility in problem (1) coincides with the value*

(5) $$\inf_{(B,b) \in \mathscr{L}(X,Y) \times Y} \Big\{\, \|(b,B)\| \,\Big|\, F+B \text{ is not metrically regular at any } \bar{x} \text{ for } b \Big\}.$$

Proof. In view of the equivalence of infeasibility with strict feasibility in 6B.3, the Robinson–Ursescu theorem 5B.4 just says that problem (1) is feasible if and only if F is metrically regular at $\bar{x}$ for 0 for any $\bar{x} \in F^{-1}(0)$, hence (5). □

In order to estimate the distance to infeasibility in terms of the modulus of metric regularity, we pass from F to a special mapping $\bar{F}$ constructed as a "homogenization" of F. We will then be able to apply to $\bar{F}$ the result on distance to metric nonregularity of sublinear mappings given in 6A.3.

We use the *horizon mapping* F^∞ associated with F, the graph of F^∞ in $X \times Y$ being the recession cone of gph F in the sense of convex analysis:

$$(x', y') \in \text{gph } F^\infty \iff \text{gph } F + (x', y') \subset \text{gph } F.$$

Homogenization. For $F : X \rightrightarrows Y$ and $0 \in \text{rge } F$, the *homogenization* of the constraint system $F(x) \ni 0$ in (1) is the system $\bar{F}(x,t) \ni 0$, where $\bar{F} : X \times \mathbb{R} \rightrightarrows Y$ is defined by

$$\bar{F}(x,t) = \begin{cases} tF(t^{-1}x) & \text{if } t > 0, \\ F^\infty(x) & \text{if } t = 0, \\ \emptyset & \text{if } t < 0. \end{cases}$$

The solution sets to the two systems are related by

$$x \in F^{-1}(0) \iff (x, 1) \in \bar{F}^{-1}(0).$$

Note that if F is positively homogeneous with closed graph, then $tF(t^{-1}x) = F(x) = F^\infty(x)$ for all $t > 0$, so that we simply have $\bar{F}(x,t) = F(x)$ for $t \geq 0$, but $\bar{F}(x,t) = \emptyset$ for $t < 0$.

In what follows, we adopt the norm

(6) $$\|(x,t)\| = \|x\| + |t| \text{ for } (x,t) \in X \times \mathbb{R}.$$

We are now ready to state and prove a result which gives a quantitative expression for the magnitude of the distance to infeasibility:

Theorem 6B.5 (distance to infeasibility for the homogenized mapping). *Let $F : X \rightrightarrows Y$ have closed, convex graph and let $0 \in \text{rge } F$. Then in the homogenized system $\bar{F}(t,x) \ni 0$ the mapping $\bar{F}$ is sublinear with closed graph, and*

(7) $$0 \in \text{int rge } F \iff 0 \in \text{int rge } \bar{F} \iff \bar{F} \text{ is surjective.}$$

Furthermore, for the given constraint system $F(x) \ni 0$ one has

(8) $$\text{distance to infeasibility } = 1/\text{reg}(\bar{F}; 0, 0 | 0).$$

Proof. The definition of $\bar{F}$ corresponds to gph $\bar{F}$ being the closed convex cone in $X \times \mathbb{R} \times Y$ that is generated by $\{(x, 1, y) \mid (x, y) \in \text{gph } F\}$. Hence $\bar{F}$ is sublinear, and also, rge $\bar{F}$ is a convex cone. We have $(\text{rge } F) = F(X) = \bar{F}(X, 1)$. So it is obvious that if $0 \in \text{int rge } F$, then $0 \in \text{int rge } \bar{F}$. Since $\text{rge } \bar{F}$ is a convex cone, the latter is equivalent to having $\text{rge } \bar{F} = Y$, i.e., surjectivity.

Conversely now, suppose $\bar{F}$ is surjective. Theorem 5B.4 (Robinson–Ursescu) informs us that in this case, $0 \in \text{int } \bar{F}(W)$ for every neighborhood W of the origin in $\mathbb{R} \times X$. It must be verified, however, that $0 \in \text{int rge } F$. In terms of $C(t) = \bar{F}(\mathbb{B}, t) \subset Y$, it will suffice to show that $0 \in \text{int } C(t)$ for some $t > 0$. Note that the sublinearity of $\bar{F}$ implies that

(9) $\qquad C((1-\theta)t_0 + \theta t_1) \supset (1-\theta)C(t_0) + \theta C(t_1) \quad \text{for } 0 < \theta < 1$.

Our assumption that $0 \in \text{rge } F$ ensures having $F^{-1}(0) \neq \emptyset$. Choose $\tau \in (0, \infty)$ small enough that $1/(2\tau) > d(0, F^{-1}(0))$. Then

(10) $\qquad 0 \in C(t) \quad \text{for all } t \in [0, 2\tau]$,

whereas, because $[-2\tau, 2\tau] \times \mathbb{B}$ is a neighborhood W of the origin in $\mathbb{R} \times X$, we have

(11) $\qquad 0 \in \text{int } \bar{F}(\mathbb{B}, [-2\tau, 2\tau]) = \text{int} \bigcup_{0 \leq t \leq 2\tau} C(t)$.

We will use this to show that actually $0 \in \text{int } C(\tau)$. For $y^* \in Y^*$ define

$$\sigma(y^*, t) := \sup_{y \in C(t)} \langle y, y^* \rangle, \qquad \lambda(t) := \inf_{\|y^*\|=1} \sigma(y^*, t).$$

The property in (9) makes $\sigma(y^*, t)$ concave in t, and the same then follows for $\lambda(t)$. As long as $0 \leq t \leq 2\tau$, we have $\sigma(y^*, t) \geq 0$ and $\lambda(t) \geq 0$ by (10). On the other hand, the union in (11) includes some ball around the origin. Therefore,

(12) $\qquad \exists \varepsilon > 0$ such that $\sup_{0 \leq t \leq 2\tau} \sigma(y^*, t) \geq \varepsilon$ for all $y^* \in Y^*$ with $\|y^*\| = 1$.

We argue next that $\lambda(\tau) > 0$. If not, then since λ is a nonnegative concave function on $[0, 2\tau]$, we would have to have $\lambda(t) = 0$ for all $t \in [0, 2\tau]$. Supposing that to be the case, choose $\delta \in (0, \varepsilon/2)$ and, in the light of the definition of $\lambda(\tau)$, an element $\hat{y}^*$ with $\sigma(\hat{y}^*, \tau) < \delta$. The nonnegativity and concavity of $\sigma(\hat{y}^*, \cdot)$ on $[0, 2\tau]$ imply then that $\sigma(\hat{y}^*, t) \leq (\delta/\tau)t$ when $\tau \leq t \leq 2\tau$ and $\sigma(\hat{y}^*, t) \leq 2\delta - (\delta/\tau)t$ when $0 \leq t \leq \tau$. But that gives us $\sigma(\hat{y}^*, t) \leq 2\delta < \varepsilon$ for all $t \in [0, 2\tau]$, in contradiction to the property of ε in (12). Therefore, $\lambda(\tau) > 0$, as claimed.

We have $\sigma(y^*, \tau) \geq \lambda(\tau)$ when $\|y^*\| = 1$, and hence by positive homogeneity $\sigma(y^*, \tau) \geq \lambda(\tau)\|y^*\|$ for all $y^* \in Y^*$. In this inequality, $\sigma(\cdot, \tau)$ is the support function of the convex set $C(\tau)$, or equivalently of $\text{cl } C(\tau)$, whereas $\lambda(\tau)\|\cdot\|$ is the support function of $\lambda(\tau)\mathbb{B}$. It follows therefore that $\text{cl } C(\tau) \supset \lambda(\tau)\mathbb{B}$, so that at least $0 \in \text{int cl } C(\tau)$.

Now, remembering that $C(\tau) = \tau F(\tau^{-1}\mathbb{B})$, we obtain $0 \in \text{int cl } F(\tau^{-1}\mathbb{B})$. Consider the mapping

6 Applications in Numerical Variational Analysis   325

$$\tilde{F}(x) = \begin{cases} F(x) & \text{if } x \in \tau^{-1}\mathbb{B}, \\ \emptyset & \text{otherwise.} \end{cases}$$

Clearly rge $\tilde{F} \subset$ rge F. Applying Theorem 5B.1 to the mapping $\tilde{F}$ gives us

$$0 \in \text{int cl rge } \tilde{F} = \text{int rge } \tilde{F} \subset \text{int rge } F.$$

This completes the proof of (7).

Let us turn now to (8). The first thing to observe is that every $\bar{B} \in \mathscr{L}(X \times \mathbb{R}, Y)$ can be identified with a pair $(B,b) \in \mathscr{L}(X,Y) \times Y$ under the formula $\bar{B}(x,t) = B(x) - tb$. Moreover, under this identification we get $\|\bar{B}\|$ equal to the expression in (2), due to the choice of norm in (6). The next thing to observe is that

$$(\bar{F} + \bar{B})(x,t) = \begin{cases} t(F+B)(t^{-1}x) - tb & \text{if } t > 0, \\ (F+B)^{\infty}(x) & \text{if } t = 0, \\ \emptyset & \text{if } t < 0, \end{cases}$$

so that $\bar{F} + \bar{B}$ gives the homogenization of the perturbed system $(F+B)(x) \ni b$. Therefore, on the basis of what has so far been proved, we have

$$b \in \text{int rge}(F+B) \iff \bar{F} + \bar{B} \text{ is surjective.}$$

Hence, through Lemma 6B.4, the distance to infeasibility for the system $F(x) \ni 0$ is the infimum of $\|\bar{B}\|$ over all $\bar{B} \in \mathscr{L}(X \times \mathbb{R}, Y)$ such that $\bar{F} + \bar{B}$ is not surjective. Theorem 6A.3 then furnishes the conclusion in (8).   □

Passing to the adjoint mapping, we can obtain a "dual" formula for the distance to infeasibility:

Corollary 6B.6 (distance to infeasibility for closed convex processes). *Let* $F: X \rightrightarrows Y$ *have closed, convex graph, and let* $0 \in \text{rge } F$. *Define the convex, positively homogeneous function* $h: X^* \times Y^* \to (-\infty, \infty]$ *by*

$$h(x^*, y^*) = \sup_{x,y} \{ \langle x, x^* \rangle - \langle y, y^* \rangle \mid y \in F(x) \}.$$

Then for the system $F(x) \ni 0$,

(13) $\qquad$ distance to infeasibility $= \inf_{\|y^*\|=1,\, x^*} \max \{ \|x^*\|, h(x^*, y^*) \}.$

Proof. By Theorem 6B.5, the distance to infeasibility is $1/\text{reg}(\bar{F}; 0, 0 | 0)$. On the other hand, $\text{reg}(\bar{F}; 0, 0|0) = \|(\bar{F}^{*+})^{-1}\|^+$ for the adjoint mapping $\bar{F}^{*+} : Y^* \rightrightarrows X^* \times \mathbb{R}$. By definition, $(x^*, s) \in \bar{F}^{*+}(y^*)$ if and only if $(x^*, s, -y^*)$ belongs to the polar cone (gph $\bar{F}$)*. Because gph $\bar{F}$ is the closed convex cone generated by $\{(x, 1, y) \mid (x, y) \in$ gph $F\}$, this condition is the same as

$$s + \langle x, x^* \rangle - \langle y, y^* \rangle \leq 0 \text{ for all } (x,y) \in \text{gph } F$$

and can be expressed as $s + h(x^*, y^*) \leq 0$. Hence

(14) $$\|(\bar{F}^{*+})^{-1}\|^+ = \sup\left\{\|y^*\| \,\middle|\, \|(x^*,s)\| \leq 1,\ s+h(x^*,y^*) \leq 0\right\},$$

where the norm on $X^* \times \mathbb{R}$ dual to the one in (6) is $\|(x^*,s)\| = \max\{\|x^*\|,|s|\}$. The distance to infeasibility, being the reciprocal of the quantity in (14), can be expressed therefore (through the positive homogeneity of h) as

(15) $$\inf_{\|y^*\|=1,\, x^*,s}\left\{\max\{\|x^*\|,|s|\} \,\middle|\, s+h(x^*,y^*) \leq 0\right\}.$$

(In converting from (14) to an infimum restricted to $\|y^*\| = 1$ in (15), we need to be cautious about the possibility that there might be no elements $(x^*,s,y^*) \in \text{gph}(\bar{F}^{*+})^{-1}$ with $y^* \neq 0$, in which case the infimum in (15) is ∞. But then the expression in (10) is 0, so the statement remains correct under the convention $1/\infty = 0$.) Observe next that, in the infimum in (15), s will be taken to be as near to 0 as possible while maintaining $-s \geq h(x^*,y^*)$. Thus, $|s|$ will be the max of 0 and $h(x^*,y^*)$, and $\max\{\|x\|,|s|\}$ will be the max of these two quantities and $\|x\|$ —but then the 0 is superfluous, and we end up with (15) equaling the expression on the right side of (3). □

We can now present our result for homogeneous systems:

Corollary 6B.7 (distance to infeasibility for sublinear mappings). *Let $F : X \rightrightarrows Y$ be sublinear with closed graph and let $0 \in \text{rge}\,F$. Then for the inclusion $F(x) \ni 0$,*

$$\text{distance to infeasibility} = \inf_{\|y^*\|=1} d\big(0, F^{*+}(y^*)\big).$$

Proof. In this case the function h in 6B.6 has $h(x^*,y^*) = 0$ when $x^* \in F^{*+}(y^*)$, but $h(x^*,y^*) = \infty$ otherwise. □

In particular, for a linear-conic constraint system of type $x \geq_C 0,\ A(x) \leq_K 0$, with respect to a continuous linear mapping $A : X \to Y$ and closed, convex cones $C \subset X$ and $K \subset Y$, we obtain

$$\text{distance to infeasibility} = \inf_{y^* \in K^+,\, \|y^*\|=1} d(A^*(y^*), C^+).$$

6C. Iterative Processes for Generalized Equations

Our occupation with numerical matters turns even more serious in this section, where we consider computational methods for solving generalized equations. The problem is to

(1) $$\text{find } x \text{ such that } f(x) + F(x) \ni 0,$$

where $f : X \to Y$ is a continuous function and $F : X \rightrightarrows Y$ is a set-valued mapping with closed graph; both X and Y are Banach spaces. As we already know, the model of a generalized equation covers huge territory. The classical case of nonlinear equations corresponds to having $F = 0$, whereas by taking $F \equiv -K$ for a fixed set K one gets various constraint systems. When F is the normal cone mapping N_C associated with a closed, convex set $C \subset X$, and $Y = X^*$, we have a variational inequality.

With the aim of approximating a solution to the generalized equation (1), we consider the following general iterative process: *Choose a sequence of functions $A_k : X \times X \to Y$ and an initial point x_0, and generate a sequence $\{x_k\}_{k=0}^{\infty}$ iteratively by taking x_{k+1} to be a solution to the auxiliary generalized equation*

$$(2) \qquad A_k(x_{k+1}, x_k) + F(x_{k+1}) \ni 0, \quad \text{for } k = 0, 1, \ldots.$$

Our goal is to specify conditions on the sequence of functions A_k in relation to the function f under which the process (2) is convergent, at least in a certain sense. We don't take on the task of explaining how the subproblems in (2) might themselves be solved. That is a separate issue, but of course those subproblems ought to be chosen to be simpler and easier to solve, depending on the form of F. Our concern here lies only with the process defined by (2) and what our earlier results are able to say about it.

Specific choices of the sequence of mappings A_k in the general iterative process (2) lead to known computational methods for solving (1). Under the assumption that f is differentiable, if we take $A_k(x, u) = f(u) + Df(u)(x - u)$ for all k, the iteration (2) becomes the following version of *Newton's method* for solving the generalized equation (1):

$$(3) \qquad f(x_k) + Df(x_k)(x_{k+1} - x_k) + F(x_{k+1}) \ni 0, \quad \text{for } k = 0, 1, \ldots.$$

This approach uses "partial linearization," in which we linearize f at the current point but leave F intact. It reduces to the standard version of Newton's method for solving the nonlinear equation $f(x) = 0$ when F is the zero mapping. We used this method to prove the classical inverse function theorem 1A.1.

In the case when (1) represents the optimality systems for a nonlinear programming problem, the iteration (3) becomes the popular sequential quadratic programming (SQP) algorithm for optimization. We will briefly describe the SQP algorithm later in the section.

Although one might imagine that a "true" Newton-type method for (1) ought to involve some kind of approximation to F as well as f, such an extension runs into technical difficulties, in particular for infinite-dimensional variational problems.

If we choose $A_k(x, u) = \lambda_k(x - u) + f(x)$ in (2) for some sequence of positive numbers λ_k, we obtain the basic form of the *proximal point method*:

$$(4) \qquad \lambda_k(x_{k+1} - x_k) + f(x_{k+1}) + F(x_{k+1}) \ni 0, \quad \text{for } k = 0, 1, \ldots.$$

At the end of the section we will provide more details about this method, putting it in the perspective of monotone mappings and optimization problems.

Our main result, which follows, concerns convergence of the iterative process (2) under the assumption of metric regularity of the mapping $f + F$.

Theorem 6C.1 (convergence under metric regularity). *Let $\bar{x}$ be a solution to (1), let the mapping $f + F$ be metrically regular at $\bar{x}$ for 0 and let $\kappa > \operatorname{reg}(f + F; \bar{x}|0)$. Consider a sequence of mappings $A_k : X \times X \to Y$ with the following property: there exist sequences of nonnegative numbers $\{\varepsilon_k\}$ and $\{\mu_k\}$ satisfying*

(5) $$\sup_k \kappa \varepsilon_k < 1 \quad \text{and} \quad \sup_k \frac{\kappa \mu_k}{1 - \kappa \varepsilon_k} < 1,$$

and a neighborhood U of $\bar{x}$ so that

(6) $$\|f(x) - A_k(x, u) - [f(x') - A_k(x', u)]\| \leq \varepsilon_k \|x - x'\| \quad \text{for every } x, x', u \in U,$$

(7) $$\|A_k(\bar{x}, u) - f(\bar{x})\| \leq \mu_k \|u - \bar{x}\| \quad \text{for every } u \in U,$$

for all $k = 0, 1, \ldots$.

Then there is a neighborhood O of $\bar{x}$ such that, for any starting point $x_0 \in O$ and any sequence $\delta_k \searrow 0$ satisfying

(8) $$\gamma_k := \frac{\kappa \mu_k + \delta_k}{1 - \kappa \varepsilon_k} < 1 \text{ for } k = 0, 1, \ldots,$$

there exists a sequence $\{x_k\}$ generated by the procedure (2) which converges to $\bar{x}$ with

(9) $$\|x_{k+1} - \bar{x}\| \leq \gamma_k \|x_k - \bar{x}\| \quad \text{for all } k = 0, 1, \ldots.$$

Proof. Let constants $a > 0$ and $b > 0$ be such that the mapping $f + F$ is metrically regular at $\bar{x}$ for 0 with constant κ and neighborhoods $\mathbb{B}_a(\bar{x})$ and $\mathbb{B}_b(0)$. Make a smaller if necessary so that $\mathbb{B}_a(\bar{x}) \subset U$ and

(10) $$(\varepsilon_k + \mu_k)a \leq b \quad \text{for all } k.$$

From the second inequality in (5) there exists a sequence $\delta_k \searrow 0$ satisfying (8); choose such a sequence and determine γ_k from (8). Pick $x_0 \in \mathbb{B}_a(\bar{x})$. If $x_0 = \bar{x}$ then take $x_k = \bar{x}$ for all k and there is nothing more to prove. If not, consider the function $x \mapsto g_0(x) := f(x) - A_0(x, x_0)$. For any $x \in \mathbb{B}_a(\bar{x})$, using (6), (7) and (10), and noting that $A_k(\bar{x}, \bar{x}) = f(\bar{x})$ from (7), we have

(11) $$\begin{aligned} \|g_0(x)\| &= \|f(x) - A_0(x, x_0) - f(\bar{x}) + A_0(\bar{x}, \bar{x})\| \\ &\leq \|f(x) - A_0(x, x_0) - f(\bar{x}) + A_0(\bar{x}, x_0)\| + \|A_0(\bar{x}, x_0) - A_0(\bar{x}, \bar{x})\| \\ &\leq \varepsilon_0 \|x - \bar{x}\| + \mu_0 \|x_0 - \bar{x}\| \leq \varepsilon_0 a + \mu_0 a \leq b. \end{aligned}$$

We will demonstrate that the mapping $\Phi_0 : x \mapsto (f+F)^{-1}(g_0(x))$ satisfies the assumptions of the contraction mapping principle for set-valued mappings (Theorem 5E.2).

By virtue of the metric regularity of $f+F$, the form of g_0, the fact that $-f(\bar{x}) = -A(\bar{x},\bar{x}) \in F(\bar{x})$, and (7), we have

$$d(\bar{x}, \Phi_0(\bar{x})) = d(\bar{x}, (f+F)^{-1}(g_0(\bar{x}))) \leq \kappa d(g_0(\bar{x}), (f+F)(\bar{x}))$$
$$= \kappa d(-A_0(\bar{x},x_0), F(\bar{x})) \leq \kappa \|A_0(\bar{x},x_0) - A(\bar{x},\bar{x})\|$$
$$\leq \kappa \mu_0 \|x_0 - \bar{x}\| < \gamma_0 \|x_0 - \bar{x}\|(1 - \kappa \varepsilon_0),$$

where γ_0 is defined in (8) and hence, $\gamma_0 \|x_0 - \bar{x}\| \leq a$. Let $u, v \in \mathbb{B}_{\gamma_0 \|x_0 - \bar{x}\|}(\bar{x})$. Invoking again the metric regularity of $f+F$ as well as the estimate (11), for any $u, v \in \mathbb{B}_{\gamma_0 \|x_0 - \bar{x}\|}(\bar{x})$ we obtain

$$e(\Phi_0(u) \cap \mathbb{B}_{\gamma_0 \|x_0 - \bar{x}\|}(\bar{x}), \Phi_0(v)) \leq e(\Phi_0(u) \cap \mathbb{B}_a(\bar{x}), \Phi_0(v))$$
$$= \sup\{d(x, (f+F)^{-1}(g_0(v))) \mid x \in (f+F)^{-1}(g_0(u)) \cap \mathbb{B}_a(\bar{x})\}$$
$$\leq \sup\{\kappa d(g_0(v), f(x) + F(x)) \mid x \in (f+F)^{-1}(g_0(u)) \cap \mathbb{B}_a(\bar{x})\}$$
$$\leq \kappa \|f(v) - A_0(v,x_0) - [f(u) - A_0(u,x_0)]\| \leq \kappa \varepsilon_0 \|u - v\|.$$

Hence, by the contraction mapping principle 5E.2 there exists a fixed point $x_1 \in \Phi_0(x_1) \cap \mathbb{B}_{\gamma_0 \|x_0 - \bar{x}\|}(\bar{x})$. This translates to $g_0(x_1) = f(x_1) - A_0(x_1, x_0) \in (f+F)(x_1)$, meaning that x_1 is obtained from x_0 by iteration (2) and satisfies (9) for $k = 0$.

The induction step is now clear. If $x_k \in \mathbb{B}_a(\bar{x})$ and $x_k \neq \bar{x}$, by defining $g_k(x) = f(x) - A_k(x, x_k)$, we obtain as in (11) that $\|g_k(x)\| \leq b$ for all $x \in \mathbb{B}_{a_k}(\bar{x})$. Then Theorem 5E.2 applies to $\Phi_k : x \mapsto (f+F)^{-1}(g_k(x))$ on the ball $\mathbb{B}_{\gamma_k \|x_k - \bar{x}\|}(\bar{x})$ and yields the existence of an iterate x_{k+1} satisfying (9). The condition in (8) ensures that the sequence $\{x_k\}$ is convergent and its limit is $\bar{x}$. □

It is a standard concept in numerical analysis that a sequence $\{x_k\}$ is *linearly convergent* to $\bar{x}$ when

$$\limsup_{k \to \infty} \frac{\|x_{k+1} - \bar{x}\|}{\|x_k - \bar{x}\|} < 1.$$

Thus, the sequence $\{x_k\}$ whose existence is claimed in 6C.1, is linearly convergent to $\bar{x}$. If the stronger condition

$$\lim_{k \to \infty} \frac{\|x_{k+1} - \bar{x}\|}{\|x_k - \bar{x}\|} = 0$$

holds, then x_k is said to be *superlinearly convergent* to $\bar{x}$. If an iterative method, like (2), produces a sequence which is linearly (superlinearly) convergent, one says that the method itself is linearly (superlinearly) convergent.

Corollary 6C.2 (superlinear convergence). *Under the conditions of Theorem 6C.1, assume in addition, that the sequence $\{\mu_k\}$ can be chosen to satisfy*

(12) $$\lim_{k\to\infty} \mu_k = 0.$$

Then the sequence of iterates x_k according to (2), whose existence is claimed in 6C.1, is superlinearly convergent to $\bar{x}$.

Proof. In this case the sequence $\{\gamma_k\}$ in (8) converges to zero and then the claimed mode of convergence follows from (9). □

Exercise 6C.3 (convergence under strong metric subregularity). *Let $\bar{x}$ be a solution of (1). Let $f + F$ be strongly subregular at $\bar{x}$ for 0, and consider the iteration process (2) under the conditions (5), (6) and (7) in 6C.1, where $\kappa > \text{subreg}(f + F; \bar{x}|0)$. Prove there exists a neighborhood O of $\bar{x}$ such that, if a sequence $\{x_k\}$ is generated by (2) and has all its elements $x_k \in O$, then $\{x_k\}$ is linearly convergent to $\bar{x}$. Under (12), this sequence is superlinearly convergent.*

Guide. Utilizing the strong metric subregularity of $f + F$ and in particular from formula 3I(5), show the existence of a positive constant a such that

(13) $$\|x - \bar{x}\| \leq \kappa d(0, (f + F)(x)) \quad \text{for all } x \in \mathbb{B}_a(\bar{x}).$$

Let $x_k \in \mathbb{B}_a(\bar{x})$ for $k = 0, 1, \ldots$ be generated by (2), and let $g_k(x) = f(x) - A_k(x, x_k)$. From (6) and (7),

$$\|g_k(x_{k+1})\| \leq \|f(x_{k+1}) - A_k(x_{k+1}, x_k) - [f(\bar{x}) - A_k(\bar{x}, x_k)]\|$$
$$+ \|A_k(\bar{x}, x_k) - f(\bar{x})\| \leq \varepsilon_k \|x_{k+1} - \bar{x}\| + \mu_k \|x_k - \bar{x}\|.$$

Since $g_k(x_{k+1}) \in (f + F)(x_{k+1})$, (13) for $x = x_{k+1}$ gives us

$$\|x_{k+1} - \bar{x}\| \leq \kappa \varepsilon_k \|x_{k+1} - \bar{x}\| + \kappa \mu_k \|x_k - \bar{x}\|,$$

which leads to the desired conclusion. □

Note that 6C.1 claims the existence of a convergent sequence which of course may be not unique, whereas 6C.3 doesn't guarantee the existence of any sequence generated by (2), inasmuch as strong subregularity doesn't guarantee local solvability of (2). We do obtain existence and uniqueness of a sequence generated by (2) when we assume strong metric regularity of the mapping in (1).

Exercise 6C.4 (convergence under strong metric regularity). *Under the conditions of Theorem 6C.1, assume in addition that the mapping $f + F$ is strongly metrically regular at $\bar{x}$ for 0. Then there exists a neighborhood O of $\bar{x}$ such that, for any $x_0 \in O$, there is a unique sequence $\{x_k\}$ generated by the iterative process (2). This sequence is linearly convergent to $\bar{x}$. If (12) holds, the sequence is superlinearly convergent.*

Guide. This could be verified in several ways, one of which is to repeat the proof of 6C.1 using the standard contraction mapping principle, 1A.2, instead of 5E.2. □

We will see next what the assumptions in 6C.1 mean in the specific cases of Newton's method (3) and proximal point method (4).

6 Applications in Numerical Variational Analysis   331

Corollary 6C.5 (convergence of Newton's method). *Consider Newton's method (3) as a specific case of the iteration (2) under the assumptions that $\bar{x}$ is a solution to (1) and the function f is continuously differentiable near $\bar{x}$. Then we have:*

(i) *if $f + F$ is metrically regular at $\bar{x}$ for 0, then there exists a neighborhood O of $\bar{x}$ such that, for any $x_0 \in O$, there is a sequence $\{x_k\}$ generated by the method starting at x_0 which is linearly convergent to $\bar{x}$;*

(ii) *if $f + F$ is strongly metrically subregular at $\bar{x}$ for 0, then there exists a neighborhood O of $\bar{x}$ such that any sequence $\{x_k\}$ generated by the method which is contained in O is linearly convergent to $\bar{x}$;*

(iii) *if $f + F$ is strongly metrically regular at $\bar{x}$ for 0, then for the neighborhood O in (i) and any $x_0 \in O$ the method generates a unique sequence $\{x_k\}$. This sequence, according to (i), is linearly convergent to $\bar{x}$.*

Proof. In this case the sequence of mapping is $A(x,u) = f(u) + Df(u)(x-u)$, the same for all k. It is straightforward to check that the assumed smoothness of f near $\bar{x}$ implies both $\widehat{\text{lip}}_x(f - A; (\bar{x}, \bar{x})) = 0$ and $\text{clm}(A(\bar{x}, \cdot); \bar{x}) = 0$. Thus, we can choose constant sequences, $\varepsilon_k = \varepsilon$ and $\mu_k = \mu$, for sufficiently small ε and μ, obtaining through 6C.1–4 the claimed modes of convergence. □

Under a stronger condition on the function f, Newton's method (3) converges *quadratically*. The precise result is as follows.

Theorem 6C.6 (quadratic convergence of Newton's method). *Consider Newton's method (3) for a function f which is continuously differentiable near $\bar{x}$ and such that $\text{lip}(Df; \bar{x}) < \infty$. Assume that the mapping $f + F$ is metrically regular at $\bar{x}$ for 0. Then for any γ satisfying*

(14) $$\gamma > \frac{1}{2}\text{reg}(f + F; \bar{x}|0) \cdot \text{lip}(Df; \bar{x})$$

there exists a neighborhood O of $\bar{x}$ such that, for any $x_0 \in O$, there is a sequence $\{x_k\}$ generated by the method which is quadratically convergent to $\bar{x}$ in the sense that

(15) $$\|x_{k+1} - \bar{x}\| \leq \gamma \|x_k - \bar{x}\|^2 \ \text{ for } k = 0, 1, \ldots.$$

Proof. The proof follows the fixed point argument in the proof of 6C.1, but with some modifications that require attention. Choose γ as in (14) and let

(16) $$\kappa > \text{reg}(f + F; \bar{x}|0) \ \text{ and } \ \mu > \text{lip}(Df; \bar{x}) \ \text{ be such that } \gamma > \kappa\mu/2.$$

Further, choose $a > 0$ and $b > 0$ so that $f + F$ is metrically regular at $\bar{x}$ for 0 with constant κ and neighborhoods $I\!B_a(\bar{x})$ and $I\!B_b(0)$. Make $a > 0$ smaller if necessary so that

(17) $$\|Df(x') - Df(x)\| \leq \mu \|x' - x\| \ \text{ for } x', x \in I\!B_a(\bar{x})$$

and also, taking into account (16),

(18) $$\frac{5}{2}\mu a^2 \leq b, \quad \kappa\mu a < 1, \quad \frac{1}{2}\kappa\mu < \gamma(1-\kappa\mu a) \quad \text{and} \quad \gamma a < 1.$$

Here and in the following section we use an estimate for smooth functions obtained by elementary calculus. From the standard equality

$$f(u) - f(v) = \int_0^1 Df(v + t(u-v))(u-v)\,dt,$$

we have through (17) that, for all $u, v \in I\!B_a(\bar{x})$,

$$\|f(u) - f(v) - Df(v)(u-v)\|$$
$$= \|\int_0^1 D_x f(v + t(u-v))(u-v)\,dt - Df(v)(u-v)\| \leq \mu \int_0^1 t\,dt \|u-v\|^2.$$

This yields

(19) $$\|f(u) - f(v) - Df(v)(u-v)\| \leq \frac{1}{2}\mu\|u-v\|^2.$$

Fix $w \in I\!B_a(\bar{x})$ and consider the function

$$x \mapsto g(w,x) = f(\bar{x}) + Df(\bar{x})(x-\bar{x}) - f(w) - Df(w)(x-w).$$

Using (19), and then the first condition in (18), we get

(20) $$\begin{aligned}\|g(w,x)\| &\leq \|f(\bar{x}) + Df(\bar{x})(x-\bar{x}) - f(x)\| \\ &\quad + \|f(x) - f(w) - Df(w)(x-w)\| \\ &\leq \tfrac{1}{2}\mu\|x-\bar{x}\|^2 + \tfrac{1}{2}\mu\|x-w\|^2 \leq \tfrac{1}{2}\mu a^2 + \tfrac{1}{2}\mu 4a^2 \leq b.\end{aligned}$$

Pick $x_0 \in I\!B_a(\bar{x})$, $x_0 \neq \bar{x}$, and consider the mapping $\Phi_0(x) = (f+F)^{-1}(g(x_0,x))$. Noting that $0 \in (f+F)(\bar{x})$ and using the metric regularity of $f+F$ together with (20), and also (18) and (19), we obtain

$$\begin{aligned}d(\bar{x}, \Phi_0(\bar{x})) &= d(\bar{x}, (f+F)^{-1}(g(x_0,\bar{x}))) \leq \kappa d(g(x_0,\bar{x}), f(\bar{x}) + F(\bar{x})) \\ &\leq \kappa\|f(\bar{x}) - f(x_0) - Df(x_0)(\bar{x} - x_0)\| \\ &\leq \frac{1}{2}\kappa\mu\|x_0 - \bar{x}\|^2 < r_0(1 - \kappa\mu a),\end{aligned}$$

where $r_0 = \gamma\|x_0 - \bar{x}\|^2 \leq a$. Moreover, for any $u, v \in I\!B_{r_0}(\bar{x})$,

$$\begin{aligned}e(\Phi_0(u) \cap I\!B_{r_0}(\bar{x}), \Phi_0(v)) &\leq \kappa\|g(x_0,u) - g(x_0,v)\| \\ &\leq \kappa\|(Df(\bar{x}) - Df(x_0))(u-v)\| \leq \kappa\mu a\|u-v\|.\end{aligned}$$

Hence, by 5E.2 there exists $x_1 \in \Phi(x_1) \cap I\!B_{r_0}(\bar{x})$, which translates to having x_1 obtained from x_0 as a first iterate of Newton's method (3) and satisfying (15) for $k = 0$. The induction step is completely analogous, giving us a sequence $\{x_k\}$ which satisfies (15). Since $\gamma a < 1$ as required in (18), this sequence is convergent. □

In the following section we will present a more elaborate parameterized version of 6C.6 under strong metric regularity, in which case, the sequence $\{x_k\}$, whose existence is claimed in 6C.6, is unique in O.

Corollary 6C.7 (convergence of proximal point method). *Consider the proximal point method* (4) *as a specific case of the iteration* (2) *under the assumptions that $\bar{x}$ is a solution to* (1) *and the function f is continuous at $\bar{x}$.*
  (i) *If $f + F$ is metrically regular at $\bar{x}$ for 0 and*

$$\sup_k \lambda_k < \frac{1}{2\operatorname{reg}(f+F;\bar{x}|0)}, \tag{21}$$

then there exists a neighborhood O of $\bar{x}$ such that for any $x_0 \in O$ there is a sequence $\{x_k\}$ generated by the method starting at x_0 which is linearly convergent to $\bar{x}$.
  (ii) *If $f + F$ is strongly metrically subregular at $\bar{x}$ for 0 and*

$$\sup_k \lambda_k < \frac{1}{2\operatorname{subreg}(f+F;\bar{x}|0)}, \tag{22}$$

then there exists a neighborhood O of $\bar{x}$ such that any sequence $\{x_k\}$ generated by the method which is contained in O is linearly convergent to $\bar{x}$.
  (iii) *If $f+F$ is strongly metrically regular at $\bar{x}$ for 0, then for the neighborhood O in* (i) *and any $x_0 \in O$ the method generates a unique sequence $\{x_k\}$. This sequence, according to* (i), *is linearly convergent to $\bar{x}$.*

If the sequence of numbers λ_k in (4) *is chosen such that $\lim_{k \to \infty} \lambda_k = 0$, the convergence claimed in* (i), (ii) *and* (iii) *is superlinear.*

Proof. With $A_k(x,u) = \lambda_k(x-u) + f(x)$, in (6)–(8) we can take $\varepsilon_k = \mu_k = \lambda_k$ for all $k = 0, 1, \dots$. Then, with a particular choice of κ, from (21) or (22) we get (6) and also (7)(8) for any neighborhood U of $\bar{x}$. If $\lambda_k \to 0$, then (12) holds, implying superlinear convergence. □

Let us go back to Newton's method in the general form (3) and apply it to the nonlinear programming problem considered in sections 2A and 2G:

$$\text{minimize } g_0(x) \text{ over all } x \text{ satisfying } g_i(x) \begin{cases} \leq 0 & \text{for } i \in [1,s], \\ = 0 & \text{for } i \in [s+1,m], \end{cases} \tag{23}$$

where the functions $g_i : \mathbb{R}^n \to \mathbb{R}$ are twice continuously differentiable. In terms of the Lagrangian function

$$L(x,y) = g_0(x) + y_1 g_1(x) + \cdots + y_m g_m(x),$$

the first-order optimality (Karush–Kuhn–Tucker) condition takes the form

$$\begin{cases} \nabla_x L(x,y) = 0, \\ g(x) \in N_{\mathbb{R}^s_+ \times \mathbb{R}^{m-s}}(y), \end{cases} \tag{24}$$

where we denote by $g(x)$ the vector with components $g_1(x),\ldots,g_m(x)$. Let $\bar{x}$ be a local minimum for (23) satisfying the constraint qualification, and let $\bar{y}$ be an associated Lagrange multiplier vector. As applied to the variational inequality (24), Newton's method (3) consists in generating a sequence $\{(x_k,y_k)\}$ starting from a point (x_0,y_0), close enough to $(\bar{x},\bar{y})$, according to the iteration

(25) $\begin{cases} \nabla_x L(x_k,y_k) + \nabla^2_{xx} L(x_k,y_k)(x_{k+1} - x_k) + \nabla g(x_k)^\mathsf{T}(y_{k+1} - y_k) = 0, \\ g(x_k) + \nabla g(x_k)(x_{k+1} - x_k) \in N_{\mathbb{R}^s_+ \times \mathbb{R}^{m-s}}(y_{k+1}). \end{cases}$

Theorem 2G.9 (with suppressed dependence on the parameter p in its statement) provides conditions under which the variational inequality (24) is strongly metrically regular at the reference point: linear independence of the gradients of the active constraints and a strong form of the second-order sufficient optimality condition; we recall these conditions below in Example 6C.8. Under these conditions, we can find (x_{k+1}, y_{k+1}) which satisfies (25) by solving the quadratic programming problem

(26)
$$\text{minimize } \left[\frac{1}{2}\langle x - x_k, \nabla^2_{xx} L(x_k, y_k)(x - x_k)\rangle \right.$$
$$\left. + \langle \nabla_x L(x_k, y_k) - \nabla g(x_k)^\mathsf{T} y_k, (x - x_k)\rangle\right]$$
$$\text{subject to } g_i(x_k) + \nabla g_i(x_k)(x - x_k) \begin{cases} \leq 0 & \text{for } i \in [1,s], \\ = 0 & \text{for } i \in [s+1,m]. \end{cases}$$

Thus, in the circumstances of (23) under strong metric regularity of the mapping in (24), Newton's method (3) comes down to sequentially solving quadratic programs of the form (26). This specific application of Newton's method is therefore called the *sequential quadratic programming* (SQP) method.

We summarize the conclusions obtained so far about the SQP method as an illustration of the power of the theory developed in this section.

Example 6C.8 (quadratic convergence of SQP). Consider the nonlinear programming problem (23) with the associated Karush–Kuhn–Tucker condition (24) and let $\bar{x}$ be a solution with an associated Lagrange multiplier vector $\bar{y}$. In the notation

$$I = \{i \in [1,m] \mid g_i(\bar{x}) = 0\} \supset \{s+1,\ldots,m\},$$
$$I_0 = \{i \in [1,s] \mid g_i(\bar{x}) = 0 \text{ and } \bar{y}_i = 0\} \subset I$$

and

$$M^+ = \{w \in \mathbb{R}^n \mid w \perp \nabla_x g_i(\bar{x}) \text{ for all } i \in I \setminus I_0\},$$
$$M^- = \{w \in \mathbb{R}^n \mid w \perp \nabla_x g_i(\bar{x}) \text{ for all } i \in I\},$$

suppose that the following conditions are both fulfilled:

(a) the gradients $\nabla_x g_i(\bar{x})$ for $i \in I$ are linearly independent,
(b) $\langle w, \nabla^2_{xx} L(\bar{x},\bar{y})w\rangle > 0$ for every nonzero $w \in M^+$ with $\nabla^2_{xx} L(\bar{p},\bar{x},\bar{y})w \perp M^-$.

Then there exists a neighborhood O of $(\bar{x},\bar{y})$ such that, for any starting point $(x_0,y_0) \in O$, the SQP method (26) generates a unique sequence which converges quadratically to $(\bar{x},\bar{y})$.

There are various numerical issues related to implementation of the SQP method that have been investigated in the last several decades, and various enhancements are available as commercial software, but we shall not go into this further.

Lastly, we will discuss a bit more the proximal point method in the context of monotone mappings. First, note that the iterative process (4) can be equally well written as

(27) $\qquad x_{k+1} \in [I + \lambda_k^{-1} T]^{-1}(x_k) \quad$ for $i = 1,2,\ldots,$ where $T = f + F$.

It has been extensively studied under the additional assumption that X is a Hilbert space (e.g., consider $I\!R^n$ under the Euclidean norm) and T is a *maximal monotone* mapping from X to X. Monotonicity, which we considered in 2F only for functions from $I\!R^n$ to $I\!R^n$, refers in the case of a potentially set-valued mapping T to the property of having

(28) $\qquad \langle y' - y, x' - x \rangle \geq 0 \quad$ for all $(x,y),(x',y') \in \mathrm{gph}\, T$.

It is called maximal when no more points can be added to gph T without running into a violation of (28). (A localized monotonicity for set-valued mappings was introduced at the end of 3G, but again only in finite dimensions.)

The following fact about maximal monotone mappings, recalled here without its proof, underlies much of the literature on the proximal point method in basic form and indicates its fixed-point motivation.

Theorem 6C.9 (resolvents of maximal monotone mappings). *Let X be a Hilbert space, and let $T : X \rightrightarrows X$ be maximal monotone. Then for any $c > 0$ the mapping $P_c = (I + cT)^{-1}$ is single-valued with all of X as its domain and moreover is nonexpansive; in other words, it is globally Lipschitz continuous from X into X with Lipschitz constant 1. The fixed points of P_c are the points $\bar{x}$ such that $T(\bar{x}) \ni 0$ (if any), and they form a closed, convex set.*

According to this, x_{k+1} always exists and is uniquely determined from x_k in the proximal point iterations (4) as expressed in (27), when $f + F$ is maximal monotone. Here is an important example of that circumstance, which we again state without bringing out its proof:

Theorem 6C.10 (maximal monotonicity in a variational inequality). *Let X be a Hilbert space, and let $F = N_C$ for a nonempty, closed, convex set in C. Let $f : C \to X$ be continuous and monotone. Then $f + F$ is maximal monotone.*

The "proximal point" terminology comes out of this framework through an application to optimization, as now explained. For a real-valued function g on a Hilbert space X with derivative $Dg(x)$, we denote by $\nabla g(x)$, as in the case of $X = I\!R^n$, the unique element of X such that $Dg(x)w = \langle \nabla g(x), w \rangle$ for all $w \in X$.

Example 6C.11 (connections with minimization). Let X be a Hilbert space, let C be a nonempty, closed, convex subset of X, and let $h : X \to \mathbb{R}$ be convex and continuously (Fréchet) differentiable. Let $f(x) = \nabla h(x)$. Then f is continuous and monotone, and the variational inequality

$$f(x) + N_C(x) \ni 0,$$

as an instance of the generalized equation (1), describes the points x (if any) which minimize h over C. In comparison, in the iterations for this case of the proximal point method in the basic form (4), the point x_{k+1} determined from x_k is the unique minimizer of $h(x) + (\lambda_k/2)\|x - x_k\|^2$ over C.

Detail. This invokes the gradient monotonicity property associated with convexity in 2F.3(a) (which is equally valid in infinite dimensions), along with the optimality condition in 2A.6. The addition of the quadratic expression $(\lambda_k/2)\|x - x_k\|^2$ to h creates a function h_k which is strongly convex with constant λ_k and thus attains its minimum, moreover uniquely. □

The expression $(\lambda_k/2)\|x - x_k\|^2$ in 6C.11 is called a *proximal term* because it helps to keep x near to the current point x_k. Its effect is to stabilize the procedure while inducing technically desirable properties like strong convexity in place of plain convexity. It's from this that the algorithm got its name.

Instead of adding a quadratic term to h, the strategy in Example 6C.11 could be generalized to adding a term $r_k(x - x_k)$ for some other convex function r_k having its minimum at the origin, and adjusting the algorithm accordingly.

Exercise 6C.12. *Prove Theorem 6C.1 by using the result stated in Exercise 5E.5.*

6D. An Implicit Function Theorem for Newton's Iteration

In this section we get back to the parameterized generalized equation

(1) $\qquad f(p,x) + F(x) \ni 0$, or equivalently $-f(p,x) \in F(x)$,

for a function $f : P \times X \to Y$ and a (generally set-valued) mapping F, where $p \in P$ is a parameter and P, X and Y are Banach spaces. We assume that the function f is Fréchet differentiable with respect to x and continuous together with its partial derivative $D_x f(p,x)$ everywhere, and that the mapping F has closed graph[2]. Associated with the generalized equation (1) as usual is its *solution mapping*

(2) $\qquad S : p \mapsto \{x \,|\, f(p,x) + F(x) \ni 0\}$ for $p \in P$.

---

[2] Here, as well as in the preceding section, we could of course use local versions of these assumptions since our analysis is local.

6 Applications in Numerical Variational Analysis

We will focus on the version of Newton's method treated in the previous section, the only difference being that now we utilize the partial derivative of the function f with respect to x:

(3) $\quad f(p,x_k) + D_x f(p,x_k)(x_{k+1} - x_k) + F(x_{k+1}) \ni 0, \quad \text{for } k = 0, 1, \ldots,$

with a given starting point x_0. We will consider the method (3) more broadly by reconceiving Newton's iteration as an inclusion, the solution of which gives a *whole sequence* instead of just an element in X. Let $l_\infty(X)$ be the Banach space consisting of all infinite sequences $\xi = \{x_1, x_2, \ldots, x_k, \ldots\}$ with elements $x_k \in X$, $k = 1, 2, \ldots$, equipped with the supremum norm

$$\|\xi\|_\infty = \sup_{k \geq 1} \|x_k\|.$$

Define a mapping $\Xi : X \times P \rightrightarrows l_\infty(X)$ as

(4) $\quad \Xi : (u,p) \mapsto \left\{ \xi \in l_\infty(X) \;\middle|\; \bigcap_{k=0}^\infty (f(p,x_k) + D_x f(p,x_k)(x_{k+1} - x_k) + F(x_{k+1})) \ni 0 \text{ with } x_0 = u \right\},$

whose value for a given (u,p) is the set of all sequences $\{x_k\}_{k=1}^\infty$ generated by Newton's iteration (3) for p that start from u. If $\bar{x}$ is a solution to (1) for $\bar{p}$, then the constant sequence $\bar{\xi} = \{\bar{x}, \bar{x}, \ldots, \bar{x}, \ldots\}$ satisfies $\bar{\xi} \in \Xi(\bar{x}, \bar{p})$.

Our first result reveals uniform quadratic convergence under strong metric regularity.

Theorem 6D.1 (uniform convergence of Newton's iteration). *For the generalized equation (1) with solution mapping S in (2), let $\bar{x} \in S(\bar{p})$. Let*

$$\widehat{\operatorname{lip}}_p(f;(\bar{p},\bar{x})) + \widehat{\operatorname{lip}}_x(D_x f;(\bar{p},\bar{x})) < \infty,$$

and let the mapping

(5) $\quad G(x) = f(\bar{p},\bar{x}) + D_x f(\bar{p},\bar{x})(x - \bar{x}) + F(x) \quad \text{for which } G(\bar{x}) \ni 0$

be strongly metrically regular at $\bar{x}$ for 0 with associated Lipschitz continuous single-valued localization σ around 0 for $\bar{x}$ of the inverse G^{-1}. Then for every

(6) $\quad \gamma > \dfrac{1}{2} \operatorname{lip}(\sigma;0) \cdot \widehat{\operatorname{lip}}_x(D_x f;(\bar{p},\bar{x}))$

there exist neighborhoods Q of $\bar{p}$ and U of $\bar{x}$ such that, for every $p \in Q$ and $u \in U$, there is exactly one sequence $\xi(u,p)$ with components $x_1, \ldots, x_k, \ldots$ all belonging to U and generated by Newton's iteration (3) starting from u for the value p of the parameter. This sequence converges to the value $s(p)$ of the Lipschitz continuous localization s of the solution mapping S around $\bar{p}$ for $\bar{x}$ whose existence is claimed in Theorem 5F.4. Moreover the convergence is quadratic with constant γ, that is,

(7) $$\|x_{k+1} - s(p)\| \leq \gamma \|x_k - s(p)\|^2 \quad \text{for } k = 0, 1, \ldots.$$

Thus, the mapping Ξ in (4) has a single-valued graphical localization ξ around $(\bar{x}, \bar{p})$ for $\bar{\xi}$. In addition, for u close to $\bar{x}$ and p close to $\bar{p}$ the value $\xi(u, p)$ of this localization is a sequence which converges quadratically to the associated solution $s(p)$ for p as described in (7).

Proof. Choose γ as in (6) and then $\kappa > \text{lip}(\sigma; 0)$ and $\mu > \widehat{\text{lip}}_x(D_x f; (\bar{p}, \bar{x}))$ such that $\kappa\mu < 2\gamma$. Next, choose $\varepsilon > 0$ so that $\kappa\varepsilon < 1$ and furthermore

(8) $$\frac{\kappa\mu}{2(1 - \kappa\varepsilon)} \leq \gamma.$$

The assumed strong regularity of the mapping G in (6) at $\bar{x}$ for 0 and the choice of κ guarantee the existence of positive constants α' and b' such that the mapping $y \mapsto \sigma(y) = G^{-1}(y) \cap B_{\alpha'}(\bar{x})$ is a Lipschitz continuous function on $B_{b'}(0)$ with Lipschitz constant κ. Along with the mapping G consider the parameterized mapping

$$x \mapsto G_{p,w}(x) = f(p, w) + D_x f(p, w)(x - w) + F(x).$$

Note that $G_{p,w}(x) = G(x) + r(p, w; x)$, where the function

$$r(p, w; x) = f(p, w) + D_x f(p, w)(x - w) - f(\bar{p}, \bar{x}) - D_x f(\bar{p}, \bar{x})(x - \bar{x})$$

is affine, and hence Lipschitz continuous, with Lipschitz constant

$$\eta(p, w) = \|D_x f(p, w) - D_x f(\bar{p}, \bar{x})\|.$$

Now, let κ' be such that $\kappa > \kappa' > \text{lip}(\sigma; 0)$, and let $\chi > 0$ satisfy

$$\chi\kappa' < 1 \quad \text{and} \quad \frac{\kappa'}{1 - \chi\kappa'} < \kappa.$$

Applying 5F.3, which is a special case of 5F.1, and taking into account that $r(\bar{p}, \bar{x}; \bar{x}) = 0$, we obtain the existence of positive constants $\alpha \leq \alpha'$ and $b \leq b'$ such that, for p and w satisfying $\eta(p, w) \leq \chi$, the mapping $y \mapsto G_{p,w}^{-1}(y) \cap B_\alpha(\bar{x})$ is a Lipschitz continuous function on $B_b(0)$ with Lipschitz constant κ. We denote this function by $\Theta(p, w; \cdot)$.

Since $D_x f$ is continuous, there are positive constants c and a such that $\eta(p, w) \leq \chi$ as long as $p \in B_c(\bar{p})$ and $w \in B_a(\bar{x})$. Make a and c smaller if necessary so that $a \leq \alpha$ and moreover

(9) $$\|D_x f(p, x) - D_x f(p, x')\| \leq \mu \|x - x'\| \quad \text{for } x, x' \in B_a(\bar{x}) \text{ and } p \in B_c(\bar{p}).$$

By Theorem 5F.4, we can further adjust a and c so that the truncation $S(p) \cap B_a(\bar{x})$ of the solution mapping S in (2) is a function s which is Lipschitz continuous on

$I\!B_c(\bar{p})$ with Lipschitz constant some $\lambda > \mathrm{lip}(\sigma;0) \cdot \widehat{\mathrm{lip}}_p(f;(\bar{p},\bar{x}))$. Next, take a even smaller if necessary so that

(10) $\qquad \dfrac{27}{8}\mu a^2 \leq b, \quad \dfrac{3}{2}\mu a \leq \varepsilon, \quad \dfrac{1}{2}\kappa\mu a < 1 - \kappa\varepsilon \text{ and } \dfrac{9}{2}\gamma a \leq 1.$

The first and the third inequality in (10) allow us to choose $\delta > 0$ satisfying

(11) $\qquad \delta + \dfrac{1}{8}\mu a^2 \leq b \quad \text{and} \quad \kappa\delta + \dfrac{1}{2}\kappa\mu a^2 \leq a(1-\kappa\varepsilon).$

Then make c even smaller if necessary so that

(12) $\qquad \|s(p) - \bar{x}\| \leq a/2 \quad \text{and} \quad \|f(p,\bar{x}) - f(\bar{p},\bar{x})\| \leq \delta \text{ for } p \in I\!B_c(\bar{p}).$

Summarizing to this point, we have determined constants a, b and c such that, for each $p \in I\!B_c(\bar{p})$ and $w \in I\!B_a(\bar{x})$, the function $\Theta(p,w,\cdot)$ is Lipschitz continuous on $I\!B_b(0)$ with constant κ, and also, the conditions (9)–(12) are satisfied.

From 6C(19) applied now to the function $f(p,x)$ we have through (9) that, for all $u,v \in I\!B_a(\bar{x})$ and $p \in I\!B_c(\bar{p})$,

(13) $\qquad \|f(p,u) - f(p,v) - D_x f(p,v)(u-v)\| \leq \dfrac{1}{2}\mu \|u-v\|^2.$

Fix $p \in I\!B_c(\bar{p})$ and $w \in I\!B_a(\bar{x})$, and consider the function

(14) $\qquad \begin{aligned} x \mapsto g(p,w;x) = &-f(p,w) - D_x f(p,w)(x-w) \\ &+ f(p,s(p)) + D_x f(p,s(p))(x-s(p)). \end{aligned}$

Recall that here $s(p) = S(p) \cap I\!B_{a/2}(\bar{x})$ for all $p \in I\!B_c(\bar{p})$. For any $x \in I\!B_a(\bar{x})$, we have using (9) and (13) that

$$\begin{aligned} \|g(p,w;x)\| \leq & \|f(p,s(p)) - f(p,w) - D_x f(p,w)(s(p)-w)\| \\ & + \|(D_x f(p,w) - D_x f(p,s(p)))(x-s(p))\| \\ \leq & \tfrac{1}{2}\mu\|w - s(p)\|^2 + \mu\|w - s(p)\|\|x - s(p)\| \leq \tfrac{27}{8}\mu a^2. \end{aligned}$$

Then, from the first inequality in (10),

(15) $\qquad \|g(p,w;x)\| \leq b.$

By way of (12), (13) and the first inequality in (11), we come to

(16) $\qquad \begin{aligned} &\|f(\bar{p},\bar{x}) - f(p,s(p)) - D_x f(p,s(p))(\bar{x}-s(p))\| \\ &\leq \|f(\bar{p},\bar{x}) - f(p,\bar{x})\| + \|f(p,\bar{x}) - f(p,s(p)) - D_x f(p,s(p))(\bar{x}-s(p))\| \\ &\leq \delta + \tfrac{1}{2}\mu\|s(p) - \bar{x}\|^2 \leq \delta + \tfrac{1}{8}\mu a^2 \leq b. \end{aligned}$

Hence, remembering that $p \in I\!B_c(\bar{p})$ and $s(p) \in I\!B_a(\bar{x})$, we see that both $g(p,w;x)$ and $f(\bar{p},\bar{x}) - f(p,s(p)) - D_x f(p,s(p))(\bar{x}-s(p))$ are in the domain of $\Theta(p,s(p);\cdot)$ where this function is Lipschitz continuous with Lipschitz constant κ.

We now choose $p \in I\!B_c(\bar{p})$ and $u \in I\!B_a(\bar{x})$, and construct a sequence $\xi(u,p)$ generated by Newton's iteration (3) starting from u for the value p of the parameter, whose existence, uniqueness and quadratic convergence is claimed in the statement of the theorem.

If $u = s(p)$ there is nothing to prove, so assume $u \neq s(p)$. Our first step is to show that, for the function g defined in (14), the mapping

$$\Phi_0 : x \mapsto \Theta(p, s(p); g(p, u; x))$$

has a unique fixed point in $I\!B_a(\bar{x})$. Utilizing the equality

$$\bar{x} = \Theta(p, s(p); -f(\bar{p}, \bar{x}) + f(p, s(p)) + D_x f(p, s(p))(\bar{x} - s(p))),$$

plus (15), (16) and the Lipschitz continuity of $\Theta(p, s(p); \cdot)$ in $I\!B_b(0)$ with constant κ, and then the second inequality in (12), (13) and the second inequality in (11), we get

(17)
$$\begin{aligned}
\|\bar{x} - \Phi_0(\bar{x})\| &= \|\Theta(p, s(p); -f(\bar{p}, \bar{x}) + f(p, s(p)) + D_x f(p, s(p))(\bar{x} - s(p))) \\
&\quad - \Theta(p, s(p); g(p, u; \bar{x}))\| \\
&\leq \kappa\| -f(\bar{p}, \bar{x}) + f(p, s(p)) + D_x f(p, s(p))(\bar{x} - s(p)) \\
&\quad - [-f(p, u) - D_x f(p, u)(\bar{x} - u) \\
&\quad + f(p, s(p)) + D_x f(p, s(p))(\bar{x} - s(p))]\| \\
&= \kappa\| -f(\bar{p}, \bar{x}) + f(p, u) + D_x f(p, u)(\bar{x} - u)\| \\
&\leq \kappa\| -f(\bar{p}, \bar{x}) + f(p, \bar{x})\| + \kappa\|f(p, u) - f(p, \bar{x}) - D_x f(p, u)(u - \bar{x})\| \\
&\leq \kappa\delta + \tfrac{1}{2}\kappa\mu\|u - \bar{x}\|^2 \leq \kappa\delta + \tfrac{1}{2}\kappa\mu a^2 \leq a(1 - \kappa\varepsilon).
\end{aligned}$$

Further, for any $v, v' \in I\!B_a(\bar{x})$, by (15), the Lipschitz continuity of $\Theta(p, s(p); \cdot)$, (9), and the second inequality in (10), we obtain

(18)
$$\begin{aligned}
\|\Phi_0(v) - \Phi_0(v')\| &= \|\Theta(p, s(p); g(p, u; v)) - \Theta(p, s(p); g(p, u; v'))\| \\
&\leq \kappa\|g(p, u; v) - g(p, u; v')\| = \kappa\|(-D_x f(p, u) + D_x f(p, s(p)))(v - v')\| \\
&\leq \kappa\mu\|u - s(p)\|\|v - v'\| \leq \tfrac{3}{2}a\kappa\mu\|v - v'\| \leq \kappa\varepsilon\|v - v'\|.
\end{aligned}$$

Hence, by 1A.2, there is a fixed point $x_1 \in \Phi_0(x_1) \cap I\!B_a(\bar{x})$. This translates to $g(p, u; x_1) \in G_{p, s(p)}(x_1)$ or, equivalently,

$$0 \in f(p, u) + D_x f(p, u)(x_1 - u) + F(x_1).$$

This means that x_1 is obtained by Newton's iteration (3) from u for p, and there is no more than just one such iterate in $I\!B_a(\bar{x})$.

Now we will demonstrate that x_1 satisfies a tighter estimate. Let

$$\omega_0 = \gamma\|u - s(p)\|^2.$$

Then $\omega_0 > 0$ and, by the last inequality in (10), $\omega_0 \leq \gamma(a + a/2)^2 \leq a/2$. We apply again the basic contraction mapping principle 1A.2 to the mapping Φ_0 but now on

$\mathbb{B}_{\omega_0}(s(p))$. Noting that $s(p) = \Theta(p, s(p); 0)$ and using (8), (13) and (15), we have

(19)
$$\begin{aligned}\|s(p) - \Phi_0(s(p))\| &= \|\Theta(p, s(p); 0) - \Theta(p, s(p); g(p, u; s(p)))\| \\ &\leq \kappa \|g(p, u; s(p))\| = \kappa \| - f(p, u) - D_x f(p, u)(s(p) - u) + f(p, s(p))\| \\ &\leq \tfrac{1}{2}\kappa\mu \|u - s(p)\|^2 \leq \gamma(1 - \kappa\varepsilon)\|u - s(p)\|^2 = \omega_0(1 - \kappa\varepsilon).\end{aligned}$$

Since $\mathbb{B}_{\omega_0}(s(p)) \subset \mathbb{B}_a(\bar{x})$, we immediately get from (18) that

(20) $$\|\Phi_0(v) - \Phi_0(v')\| \leq \kappa\varepsilon \|v - v'\| \quad \text{for any } v, v' \in \mathbb{B}_{\omega_0}(s(p)).$$

Thus, the contraction mapping principle applied to the function Φ_0 on the ball $\mathbb{B}_{\omega_1}(s(p))$ yields the existence of x'_1 in this ball such that $x'_1 = \Phi_0(x'_1)$. But the fixed point x'_1 of Φ_0 in $\mathbb{B}_{\omega_0}(s(p))$ must then coincide with the unique fixed point x_1 of Φ_0 in the larger set $\mathbb{B}_a(\bar{x})$. Hence the fixed point x_1 of Φ_0 on $\mathbb{B}_a(\bar{x})$ satisfies

$$\|x_1 - s(p)\| \leq \gamma \|u - s(p)\|^2,$$

which means that (7) holds for $k = 0$ with $x_0 = u$.

The induction step is now clear: if the claim holds for $k = 1, 2, \ldots, n$, then by defining $\Phi_n : x \mapsto \Theta(p, s(p); g(p, x_n; x))$ and replacing u by x_n in (17) and (18), we obtain that Φ_n has a unique fixed point x_{n+1} in $\mathbb{B}_a(\bar{x})$. This tells us that $g(p, x_n; x_{n+1}) \in G_{p, s(p)}(x_{n+1})$ and in consequence that x_{n+1} is the unique Newton's iterate from x_n for p which is in $\mathbb{B}_a(\bar{x})$.

Next, by employing again the contraction mapping principle as in (19) and (20) to Φ_n, but now on the ball $\mathbb{B}_{\omega_n}(s(p))$ for $\omega_n = \gamma \|x_n - s(p)\|^2$, we obtain that x_{n+1} is at distance ω_n from $s(p)$. Invoking the first inequality in (12) and then the last one in (10) we have

$$\theta := \gamma \|x_0 - s(p)\| \leq \gamma(\|x_0 - \bar{x}\| + \|s(p) - \bar{x}\|) \leq \gamma\left(a + \frac{a}{2}\right) < 1.$$

Therefore

(21) $$\|x_k - s(p)\| \leq \theta^{2^k - 1} \|x_0 - s(p)\|$$

so the sequence $\{x_k\}$ is convergent to $s(p)$ and moreover the convergence is quadratic as described in (7). This completes the proof of the theorem. □

We can go a step further and arrive at an implicit function theorem for Newton's iteration which is strikingly similar to the implicit function theorem 5F.4.

Theorem 6D.2 (implicit function theorem for Newton's iteration). *In addition to the assumptions of Theorem 6D.1, suppose that*

$$\text{lip}(D_x f; (\bar{p}, \bar{x})) < \infty.$$

Then the single-valued localization ξ of the mapping Ξ in (4) around $(\bar{x}, \bar{p})$ for $\bar{\xi}$ described in Theorem 6D.1 is Lipschitz continuous near $(\bar{x}, \bar{p})$, moreover with

(22) $\widehat{\text{lip}}_u(\xi;(\bar{x},\bar{p})) = 0$ and $\widehat{\text{lip}}_p(\xi;(\bar{x},\bar{p})) \leq \text{lip}(\sigma;0)\cdot\widehat{\text{lip}}_p(f;(\bar{p},\bar{x}))$.

Proof. First, recall some notation and facts established in Theorem 6D.1 and its proof. We know that for any $\kappa > \text{lip}(\sigma;0)$, there exist positive constants a, α, b and c such that $a \leq \alpha$ and, for every $p \in \mathbb{B}_c(\bar{p})$ and $w \in \mathbb{B}_a(\bar{x})$, the mapping $y \mapsto G_{p,w}^{-1}(y) \cap \mathbb{B}_\alpha(\bar{x})$ is a function, with values $\Theta(p,w;y)$, which is Lipschitz continuous on $\mathbb{B}_b(0)$ with Lipschitz constant κ; moreover, the truncation $S(p) \cap \mathbb{B}_a(\bar{x})$ of the solution mapping in (2) is a Lipschitz continuous function on $\mathbb{B}_c(\bar{p})$ and its values are in $\mathbb{B}_{a/2}(\bar{x})$; also, for any starting point $u \in \mathbb{B}_a(\bar{x})$ and any $p \in \mathbb{B}_c(\bar{p})$, there is a unique sequence $\xi(u,p)$ starting from u and generated by Newton's method (3) for p whose components are contained in $\mathbb{B}_a(\bar{x})$, with this sequence being quadratically convergent to $s(p)$ as described in (7).

Our starting observation is that, for any positive $a' \leq a$, by adjusting the size of the constant c and taking as a starting point $u \in \mathbb{B}_{a'}(\bar{x})$, we can arrange that, for any $p \in \mathbb{B}_c(\bar{p})$, all elements x_k of the sequence $\xi(u,p)$ are actually in $\mathbb{B}_{a'}(\bar{x})$. Indeed, by taking $\delta > 0$ to satisfy (11) with a replaced by a' and then choosing c so that (12) holds for the new δ and for a', then all requirements for a will hold for a' as well and hence all Newton's iterates x_k will be at distance a' from $\bar{x}$.

Choose
$$\eta > \text{lip}(D_x f;(\bar{p},\bar{x})) \quad \text{and} \quad \nu > \widehat{\text{lip}}_p(f;(\bar{p},\bar{x})).$$

Pick a positive constant $d \leq a/2$ and make c smaller if necessary, so that for every $p, p' \in \mathbb{B}_c(\bar{p})$ and every $w, w' \in \mathbb{B}_d(\bar{x})$, we have

(23) $\quad \|D_x f(p',w') - D_x f(p,w)\| \leq \eta(\|p'-p\| + \|w'-w\|),$

(24) $\quad \|f(p',w) - f(p,w)\| \leq \nu\|p'-p\|,$

and, in addition, for every $x \in \mathbb{B}_d(\bar{x})$, every $p, p' \in \mathbb{B}_c(\bar{p})$ and every $w, w' \in \mathbb{B}_d(\bar{x})$, we have

(25) $\quad \|f(p',w') + D_x f(p',w')(x-w') - f(p,w) - D_x f(p,w)(x-w)\| \leq b.$

Choose a positive τ such that $\tau\kappa < 1/3$. Make d and c smaller if necessary so that

(26) $\quad 3\eta(d+c) < \tau.$

Since
$$\frac{\kappa\tau}{1-\kappa\tau} < \frac{1}{2}$$
we can take c still smaller in order to have

(27) $\quad \dfrac{\kappa\tau(2d) + 3\kappa(\tau+\nu)(2c)}{1-\kappa\tau} \leq d.$

Let $p, p' \in \mathbb{B}_c(\bar{p})$, $u, u' \in \mathbb{B}_d(\bar{x})$, $(p,u) \neq (p',u')$. In accordance with Theorem 6D.1 and the observation above, let $\xi(p,u) = (x_1, \ldots, x_k, \ldots)$ be the unique sequence generated by Newton's iteration (3) starting from u whose components x_k are all in $\mathbb{B}_d(\bar{x})$ and hence in $\mathbb{B}_{a/2}(\bar{x})$. For this sequence, denoting $x_0 = u$, we know that for all $k \geq 0$

(28) $\quad x_{k+1} = \Theta(p, x_k; 0) := (f(p, x_k) + D_x f(p, x_k)(\cdot - x_k) + F(\cdot))^{-1}(0) \cap \mathbb{B}_\alpha(\bar{x})$.

Let
$$\gamma_0 = \frac{\kappa \tau \|u - u'\| + \kappa(\tau + v)\|p - p'\|}{1 - \kappa \tau}.$$

By using (27) we get that $\gamma_0 \leq d$ and then $\mathbb{B}_{\gamma_0}(x_1) \subset \mathbb{B}_a(\bar{x})$. Consider the function
$$\Phi_0 : x \mapsto \Theta(p, u; -f(p', u') - D_x f(p', u')(x - u') + f(p, u) + D_x f(p, u)(x - u)).$$

Employing (25) and then the Lipschitz continuity of $\Theta(p, u; \cdot)$ on $\mathbb{B}_b(0)$, and applying (13), (23), (24), (26) and (28), we obtain

(29)
$$\begin{aligned}
\|x_1 - \Phi_0(x_1)\| &= \|\Theta(p, u; 0) - \\
&\quad \Theta(p, u; -f(p', u') - D_x f(p', u')(x_1 - u') + f(p, u) + D_x f(p, u)(x_1 - u))\| \\
&\leq \kappa \|f(p', u) - f(p', u') - D_x f(p', u')(u - u') \\
&\quad - D_x f(p', u')(x_1 - u) - f(p', u) + f(p, u) + D_x f(p, u)(x_1 - u)\| \\
&\leq \kappa \|f(p', u) - f(p', u') - D_x f(p', u')(u - u')\| \\
&\quad + \kappa \|(D_x f(p, u) - D_x f(p', u'))(x_1 - u)\| + \kappa \| - f(p', u) + f(p, u)\| \\
&\leq \tfrac{1}{2} \kappa \eta \|u - u'\|^2 + \kappa \eta \|u - u'\| \|x_1 - u\| \\
&\quad + \kappa \eta \|p - p'\| \|x_1 - u\| + \kappa v \|p - p'\| \\
&\leq 3 \kappa \eta d \|u - u'\| + \kappa (2 \eta d + v) \|p - p'\| \\
&\leq \kappa \tau \|u - u'\| + \kappa (\tau + v) \|p - p'\| = \gamma_0 (1 - \kappa \tau).
\end{aligned}$$

For $v, v' \in \mathbb{B}_{\gamma_0}(x_1)$, we have by way of (23), (24) and (26) that

(30) $\quad \|\Phi_0(v) - \Phi_0(v')\| \leq \kappa \|(-D_x f(p', u') + D_x f(p, u))(v - v')\|$
$\leq 2 \kappa \eta (d + c) \|v - v'\| \leq \kappa \tau \|v - v'\|.$

Hence, by the contraction mapping principle 1A.2, there is a unique x_1' in $\mathbb{B}_{\gamma_0}(x_1)$ such that
$$x_1' = \Theta(p, u; -f(p', u') - D_x f(p', u')(x_1' - u') + f(p, u) + D_x f(p, u)(x_1' - u)).$$

But then
$$f(p', u') + Df(p', u')(x_1' - u') + F(x_1') \ni 0,$$
that is, x_1' is the unique Newton's iterate from u' for p' which satisfies
$$\|x_1' - x_1\| \leq \gamma_0.$$

Since $\gamma_0 \leq d$, we obtain that $x_1' \in \mathbb{B}_a(\bar{x})$ and then x_1' is the unique Newton's iteration from u' for p' which is in $\mathbb{B}_a(\bar{x})$.

By induction, we construct a sequence $\xi' = \{x'_1, x'_2, \ldots, x'_k, \ldots\} \in \Xi(p', u')$ such that the distance from x'_k to the corresponding component x_k of ξ satisfies the estimate

$$(31) \quad \|x'_k - x_k\| \leq \gamma_{k-1} := \frac{\kappa\tau\|x_{k-1} - x'_{k-1}\| + \kappa(\tau+\nu)\|p - p'\|}{1 - \kappa\tau} \quad \text{for } k = 2, 3, \ldots.$$

Suppose that for some $n > 1$ we have found $x'_2, x'_3, \ldots, x'_n$ with this property. First, observe that

$$\gamma_k \leq \left[\frac{\kappa\tau}{1-\kappa\tau}\right]^{k+1} \|u - u'\| + \frac{\kappa(\tau+\nu)}{1-\kappa\tau}\|p - p'\| \sum_{i=0}^{k} \left[\frac{\kappa\tau}{1-\kappa\tau}\right]^i,$$

from which we get the estimate that, for all $k = 0, 1, \ldots, n-1$,

$$(32) \quad \gamma_k \leq \frac{\kappa\tau}{1-\kappa\tau}\|u - u'\| + \frac{\kappa(\tau+\nu)}{1-2\kappa\tau}\|p - p'\|.$$

In particular, we obtain through (27) that $\gamma_k \leq d$ for all k, and consequently $x'_k \in \mathbb{B}_d(x_k) \subset \mathbb{B}_a(\bar{x})$.

To show that x'_{n+1} is a Newton's iterate from x'_n for p', we proceed in the same way as in obtaining x'_1 from u' for p'. Consider the function

$$\Phi_k : x \mapsto \Theta(p, x_k; -f(p', x'_k) - D_x f(p', x'_k)(x - x'_k) + f(p, x_k) + D_x f(p, x_k)(x - x_k)).$$

By replacing Φ_0 by Φ_k, u by x_k, u' by x'_k, and x_1 by x_{k+1} in (29) and (30), we get

$$\|x_{k+1} - \Phi_k(x_{k+1})\| \leq \kappa\tau\|x_k - x'_k\| + \kappa(\tau+\nu)\|p - p'\| = \gamma_k(1-\kappa\tau)$$

and

$$\|\Phi_k(v) - \Phi_k(v')\| \leq \kappa\tau\|v - v'\| \quad \text{for any } v, v' \in \mathbb{B}_{\gamma_k}(x_{k+1}).$$

Then, by the contraction mapping principle 1A.2 there is a unique x'_{k+1} in $\mathbb{B}_{\gamma_k}(x_{k+1})$ with $x'_{k+1} = \Phi_k(x'_{k+1})$, which gives us

$$f(p', x'_k) + D_x f(p', x'_k)(x'_{k+1} - x'_k) + F(x'_{k+1}) \ni 0.$$

Moreover, since $\gamma_k \leq d$, we have that $x'_{k+1} \in \mathbb{B}_a(\bar{x})$.

We have constructed a sequence $x'_1, \ldots, x'_k, \ldots$, generated by Newton's iteration for p' starting from u', and whose components are in $\mathbb{B}_a(\bar{x})$. According to Theorem 6D.1, this sequence must be the value $\xi(u', p')$ of the single-valued localization ξ whose value $\xi(u, p)$ is the sequence $x_1, \ldots, x_k, \ldots$. Taking into account (31) and (32), we come to the estimate

$$\|\xi(u, p) - \xi(u', p')\|_\infty \leq O(\tau)\|u - u'\| + (\kappa\nu + O(\tau))\|p - p'\|.$$

Since τ can be chosen arbitrarily small, this yields (22). $\square$

6 Applications in Numerical Variational Analysis

As in the case of the classical implicit function theorem, the inverse function version of Theorem 6D.2 turns into an "if and only if" result.

Consider the generalized equation (1) with $f(p,x) = g(x) - p$ whose solution mapping $S = (g+F)^{-1}$, and let $\bar{x} \in S(0)$. In order to apply 6D.2 suppose that g is differentiable near $\bar{x}$ with $\text{lip}(Dg;\bar{x}) < \infty$. The corresponding Newton's iteration mapping in (3) then has the form

$$(33) \quad \Upsilon : (u,p) \mapsto \left\{ \xi \in l_\infty(X) \;\middle|\; \bigcap_{k=0}^\infty (g(x_k) + Dg(x_k)(x_{k+1} - x_k) + F(x_{k+1})) \ni p \text{ with } x_0 = u \right\}.$$

Theorem 6D.3 (inverse function theorem for Newton's iteration). *The mapping $g + F$ is strongly regular at $\bar{x}$ for 0 if and only if the mapping Υ in (33) has a Lipschitz continuous single-valued localization ξ around $(\bar{x}, 0)$ for $\bar{\xi}$ with*

$$(34) \quad \widehat{\text{lip}}_u(\xi;(\bar{x},0)) < 1$$

and is such that, for each (u,p) close to $(\bar{x}, 0)$, the sequence $\xi(u,p)$ is convergent. Moreover, in this case

$$(35) \quad \widehat{\text{lip}}_p(\xi;(\bar{x},0)) = \text{reg}(g+F;\bar{x}|0).$$

Proof. The "only if" part follows from the combination of 6D.1 and 6D.2. Noting that the Lipschitz modulus of the single-valued localization σ in 6D.1 equals the regularity modulus of $g + F$, from (22) we get

$$(36) \quad \widehat{\text{lip}}_p(\xi;(\bar{x},0)) \leq \text{lip}(\sigma;0) = \text{reg}(g+F;\bar{x}|0).$$

To prove the "if" part, choose $\kappa > \widehat{\text{lip}}_p(\xi;(\bar{x},0))$, a positive $\varepsilon < 1$ and corresponding neighborhoods U of $\bar{x}$ and Q of 0 such that the sequence $\xi(u,p)$ is the only element of $\Upsilon(u,p)$ whose components $x_1, \ldots, x_k, \ldots$ are in U, and moreover the function ξ acting from $X \times Y$ to $l_\infty(X)$ is Lipschitz continuous with Lipschitz constants κ in $p \in Q$ uniformly in $u \in U$ and still more, from (34), ξ is Lipschitz continuous with Lipschitz constants ε in $u \in U$ uniformly in $p \in Q$. From the assumed local closedness of gph F it is possible to make Q and U smaller if necessary so that for any $p \in Q$ and any sequence with components $v_k \in U$ convergent to v and satisfying

$$(37) \quad g(v_k) + Dg(v_k)(v_{k+1} - v_k) + F(v_{k+1}) \ni p \text{ for all } k = 1, 2, \ldots,$$

one has $g(v) + F(v) \ni p$.

Let $p, p' \in Q$ and let $x \in (g+F)^{-1}(p) \cap U$. The constant sequence all elements of which are equal x, namely, $\chi = (x, x, \ldots, x, \ldots)$, is obviously convergent to the solution x of the inclusion $g(x) + F(x) \ni p$. Then $\chi \in \Upsilon(x,p)$ and all its components are in U, hence $\chi = \xi(x,p)$. By assumption,

(38) $$\|\chi - \xi(x,p')\|_\infty \leq \kappa\|p-p'\|,$$

and moreover $\xi(x,p') = \{x'_1, \ldots, x'_k, \ldots\}$ is convergent. By definition, $\xi(x,p')$ satisfies

$$g(x'_k) + Dg(x'_k)(x'_{k+1} - x'_k) + F(x'_{k+1}) \ni p' \quad \text{for all } k = 1, 2, \ldots.$$

From the property described in (37) we obtain that the sequence $\xi(x,p')$ is convergent to a solution $x' \in (g+F)^{-1}(p') \cap U$. Hence, using (38), we have

$$\|x - x'\| \leq \|x - x'_k\| + \|x'_k - x'\|$$
$$\leq \|\chi - \xi(x,p')\|_\infty + \|x'_k - x'\| \leq \kappa\|p-p'\| + \|x'_k - x'\|.$$

Since $x'_k \to x'$ as $k \to \infty$, we conclude by passing to the limit in this last inequality that

(39) $$\|x - x'\| \leq \kappa\|p-p'\|.$$

This means that the mapping $g+F$ is metrically regular at $\bar{x}$ for 0. We will demonstrate that the mapping $(g+F)^{-1}$ has a single-valued localization around 0 for $\bar{x}$. We know that $\text{dom}\,(g+F)^{-1}$ contains a neighborhood of 0. Assume that for any neighborhoods U of $\bar{x}$ and Q of 0 there exist $p \in Q$ and $w, w' \in U$ such that $w \neq w'$ and both w and w' are in $(g+F)^{-1}(p)$. Then the constant sequences $\{w, w, \ldots, w, \ldots\} \in \Upsilon(w,p)$ and $\{w', w', \ldots, w', \ldots\} \in \Upsilon(w',p)$ and all their components are in U, hence $\{w, w, \ldots, w, \ldots\} = \xi(w,p)$ and $\{w', w', \ldots, w', \ldots\} = \xi(w',p)$. In the beginning of the proof we have chosen the neighborhoods U and Q such that for a fixed $p \in V$ the mapping $u \mapsto \xi(u,p)$ is a Lipschitz continuous function from X to $l_\infty(X)$ with Lipschitz constant $\varepsilon < 1$, and hence this condition holds for all of its components. This yields

$$\|w - w'\| \leq \varepsilon\|w - w'\| < \|w - w'\|,$$

which is absurd. Hence, $(g+F)^{-1}$ has a single-valued localization s around 0 for $\bar{x}$. But then from (39) this localization is Lipschitz continuous around 0 with $\text{lip}\,(s;0) \leq \kappa$. The Banach space versions of Theorems 2B.10 and 3G.1 say that $\text{lip}\,(\sigma;0) = \text{lip}\,(s;0) = \text{reg}\,(g+F;\bar{x}|0)$ and hence $\text{lip}\,(\sigma;0) \leq \kappa$. Since κ could be arbitrarily close to $\widehat{\text{lip}}_p(\xi;(\bar{x},0))$, we get the inequality opposite to (36), and hence the equality (35) holds. □

As an illustration of possible applications of the results in Theorems 6D.1 and 6D.2 in studying complexity of Newton's iteration, we will produce an estimate for the number of iterations needed to achieve a particular accuracy of the method, which is the same for all values of the parameter p in some neighborhood of the reference point $\bar{p}$. Given an accuracy measure ρ, suppose that Newton's method (3) is to be terminated at the k-th step if

(40) $$d(0, f(p, x_k) + F(x_k)) \leq \rho.$$

Also suppose that the constant μ and the constants a and c are chosen to satisfy (9). For $p \in B_c(\bar{p})$ consider the unique sequence $\{x_k\}$ generated by (3) for p, all elements of which are in $B_a(\bar{x})$. Since x_k is a Newton's iterate from x_{k-1}, we have that

$$f(p, x_k) - f(p, x_{k-1}) - D_x f(p, x_{k-1})(x_k - x_{k-1}) \in f(p, x_k) + F(x_k).$$

Using (13), we have

(41)
$$\begin{aligned} d(0, f(p, x_k) + F(x_k)) \\ \leq \|f(p, x_k) - f(p, x_{k-1}) - D_x f(p, x_{k-1})(x_k - x_{k-1})\| \\ \leq \tfrac{1}{2}\mu \|x_k - x_{k-1}\|^2. \end{aligned}$$

Let k_ρ be the first iteration at which (40) holds; then for $k < k_\rho$ from (41) we obtain

(42)
$$\rho < \frac{1}{2}\mu \|x_k - x_{k-1}\|^2.$$

Further, utilizing (21) we get

$$\|x_k - x_{k-1}\| \leq \|x_k - s(p)\| + \|x_{k-1} - s(p)\| \leq \theta^{2^k-2}(1+\theta)(\|x_0 - \bar{x}\| + \|s(p) - \bar{x}\|),$$

and from the choice of x_0 and the first inequality in (12) we have

$$\|x_k - x_{k-1}\| \leq \theta^{2^k-2}(1+\theta)\frac{3a}{2}.$$

But then, taking into account (42), we obtain

$$\rho < \frac{1}{2}\mu \theta^{2^{k+1}}\frac{9a^2(1+\theta)^2}{4\theta^4}.$$

Therefore k_ρ satisfies

$$k_\rho \leq \log_2\left(\log_\theta\left(\frac{8\theta^4 \rho}{9a^2\mu(1+\theta)^2}\right)\right) - 1.$$

Thus, we have obtained an upper bound of the number of iterations needed to achieve a particular accuracy, which, most importantly, is *the same for all values of the parameter p in some neighborhood of the reference value $\bar{p}$*. This tells us, for example, that, under the assumptions of Theorem 6D.2, small changes of parameters in a problem don't affect the performance of Newton's method as applied to this problem.

Exercise 6D.4 (using ample parameterization). *In addition to the conditions in 6D.2, let f be strictly differentiable at $(\bar{p}, \bar{x})$ and the derivative $D_p f(\bar{p}, \bar{x})$ be surjective. Show that the metric regularity of the mapping G in (5) is not only sufficient,* but also necessary for the existence of a single-valued localization of the mapping Ξ in (4) whose values are convergent, as in the statement of 6D.1.

6E. Galerkin's Method for Quadratic Minimization

The topic of this section is likewise a traditional scheme in numerical analysis and its properties of convergence, again placed in a broader setting than the classical one. The problem at which this scheme will be directed is quadratic optimization in a Hilbert space setting:

(1) $$\text{minimize } \frac{1}{2}\langle x, Ax\rangle - \langle v, x\rangle \text{ over } x \in C,$$

where C is a nonempty, closed and convex set in a Hilbert space X, and $v \in X$ is a parameter. Here $\langle \cdot, \cdot \rangle$ denotes the inner product in X; the associated norm is $\|x\| = \sqrt{\langle x, x\rangle}$. We take $A : X \to X$ to be a linear and bounded mapping, entailing dom $A = X$; furthermore, we take A to be self-adjoint, $\langle x, Ay\rangle = \langle y, Ax\rangle$ for all $x, y \in X$ and require that

(2) $$\langle x, Ax\rangle \geq \mu \|x\|^2 \text{ for all } x \in C - C, \text{ for a constant } \mu > 0.$$

This property of A, sometimes called *coercivity* (a term which can have conflicting manifestations), corresponds to A being strongly monotone relative to C in the sense defined in 2F, as well as to the quadratic function in (1) being strongly convex relative to C. For $X = \mathbb{R}^n$, (2) is equivalent to positive definiteness of A relative to the subspace generated by $C - C$. For any Hilbert space X in which that subspace is dense, it entails A being invertible with $\|A^{-1}\| \leq \mu^{-1}$.

In the usual framework for Galerkin's method, C would be all of X, so the targeted problem would be unconstrained. The idea is to consider an increasing sequence of finite-dimensional subspaces X_k of X, and by iteratively minimizing over X_k, to get a solution point $\hat{x}_k$, generate a sequence which, in the limit, solves the problem for X.

This approach has proven valuable in circumstances where X is a standard function space and the special functions making up the subspaces X_k are familiar tools of approximation, such as trigonometric expansions. Here, we will work more generally with convex sets C_k furnishing "inner approximations" to C, with the eventual possibility of taking $C_k = C \cap X_k$ for a subspace X_k.

In Section 2G with $X = \mathbb{R}^n$, we looked at a problem like (1) in which the function was not necessarily quadratic, and we studied the dependence of its solution on the parameter v. Before proceeding with anything else, we must update to our Hilbert space context with a quadratic function the particular facts from that development which will be called upon.

6 Applications in Numerical Variational Analysis

Theorem 6E.1 (optimality and its characterization). *For problem (1) under condition (2), there exists for each v a unique solution x. The solution mapping $S : v \to x$ is thus single-valued with $\mathrm{dom}\, S = X$. Moreover, this mapping S is Lipschitz continuous with constant μ^{-1}, and it is characterized by a variational inequality:*

$$(3) \qquad x = S(v) \iff -v + Ax + N_C(x) \ni 0.$$

Proof. The existence of a solution x for a fixed v comes from the fact that, for each sufficiently large $\alpha \in \mathbb{R}$ the set C_α of $x \in C$ for which the function being minimized in (1) has value $\leq \alpha$ is nonempty, convex, closed and bounded, with the bound coming from (2). Such a subset of X is weakly compact; the intersection of the C_α that are nonempty is therefore nonempty. That intersection is comprised of all possible solutions x. The uniqueness of such x follows however from the strong convexity of the function in question. The characterization of x in (3) is proved exactly as in the case of $X = \mathbb{R}^n$ in 2A.6. The Lipschitz property of S comes out of the same argument that was used in the second half of the proof of 2F.9, utilizing the strong monotonicity of A. □

As an important consequence of Theorem 6E.1, we get a Hilbert space version of the projection result in 1D.5 for convex sets in $\mathbb{R}^n$.

Corollary 6E.2 (projections onto convex sets). *For a nonempty, closed, convex set C in a Hilbert space X, there exists for each $v \in X$ a unique nearest point x of C, called the projection of v on C and denoted by $P_C(v)$. The projection mapping $P_C : X \to C$ is Lipschitz continuous with constant 1.*

Proof. Take $A = I$ in (1), noting that (2) holds then with $\mu = 1$. Problem (1) is equivalent then to minimizing $\|x - v\|$ over $x \in C$, because the expression being minimized differs from $\frac{1}{2}\|x - v\|^2$ only by the constant term $\frac{1}{2}\|v\|^2$. □

In Galerkin's method, when we get to it, there will be need of comparing solutions to (1) with solutions to other problems for the same v but sets different from C. In effect, we have to be able to handle the choice of C as another sort of parameter. For a start, consider just two different sets, D_1 and D_2. How might solutions to the versions of (1) with D_1 and D_2 in place of C, but with fixed v, relate to each other? To get anywhere with this we require a *joint* strong monotonicity condition extending (2):

$$(4) \quad \langle x, Ax \rangle \geq \mu \|x\|^2 \text{ for all } x \in D_i - D_j \text{ and } i, j \in \{1,2\}, i \neq j, \text{ where } \mu > 0.$$

Obviously (4) holds without any fuss over different sets if we simply have A strongly monotone with constant μ on all of X.

Proposition 6E.3 (solution estimation for varying sets). *Consider any nonempty, closed, convex sets D_1 and D_2 in X satisfying (4). If x_1 and x_2 are the solutions of problem (1) with constraint sets D_1 and D_2, respectively, in place of C, then*

$$(5) \quad \mu \|x_1 - x_2\|^2 \leq \langle Ax_1 - v, u_1 - x_2 \rangle + \langle Ax_2 - v, u_2 - x_1 \rangle \text{ for all } u_1 \in D_1, u_2 \in D_2.$$

Proof. From (4) we have

(6) $$\mu \|x_1 - x_2\|^2 \leq \langle A(x_1 - x_2), x_1 - x_2 \rangle,$$

whereas for any $u_1 \in D_1$ and $u_2 \in D_2$, (3) gives us

(7) $$0 \leq \langle Ax_1 - v, u_1 - x_1 \rangle, \qquad 0 \leq \langle Ax_2 - v, u_2 - x_2 \rangle.$$

Adding the inequalities in (7) to the one in (6) and rearranging the sum, we obtain

$$\mu \|x_1 - x_2\|^2 \leq \langle A(x_1 - x_2), x_1 - x_2 \rangle + \langle Ax_1 - v, u_1 - x_1 \rangle + \langle Ax_2 - v, u_2 - x_2 \rangle$$
$$= \langle Ax_1 - v, u_1 - x_2 \rangle + \langle Ax_2 - v, u_2 - x_1 \rangle,$$

as claimed in (5). □

Having this background at our disposal, we are ready to make progress with our generalized version of Galerkin's method. We consider along with C a sequence of sets $C_k \subset X$ for $k = 1, 2, \ldots$ which, like C, are nonempty, closed and convex. We suppose that

(8) $$C_k \subset C_{k+1} \subset \cdots \subset C, \quad \text{with } \mathrm{cl}\,[C_1 \cup C_2 \cup \cdots] = C,$$

and let

(9) $$S_k = \text{the solution mapping for (1) with } C_k \text{ in place of } C,$$

as provided by Theorem 6E.1 through the observation that (2) carries over to any subset of C. By generalized Galerkin's sequence associated with (8) for a given v, we will mean the sequence of solutions $\hat{x}_k = S_k(v), k = 1, 2, \ldots$.

Theorem 6E.4 (general rate of convergence). *Let S be the solution mapping to (1) as provided by Theorem 6E.1 under condition (2), and let $\{C_k\}$ be a sequence of nonempty, closed, convex sets satisfying (8). Then for any v the associated Galerkin's sequence $\{\hat{x}_k\} = \{S_k(v)\}$ converges to $\hat{x} = S(v)$. In fact, there is a constant c such that*

(10) $$\|\hat{x}_k - \hat{x}\| \leq c\, d(\hat{x}, C_k)^{1/2} \quad \text{for all } k.$$

Proof. On the basis of (8), we have $\mathrm{dist}(\hat{x}, C_k) \to 0$. The sequence of projections $\bar{x}_k = P_{C_k}(\hat{x})$ with $\|\bar{x}_k - \hat{x}\| = d(\hat{x}, C_k)$, whose existence is guaranteed by 6E.2, converges then to $\hat{x}$. From 6E.3 applied to $D_1 = C_k$ and $D_2 = C$, with $\hat{x}_k$ and $\hat{x}$ in the place of the x_1 and x_2 there, and on the other hand $u_1 = \bar{x}_k$ and $u_2 = \hat{x}_k$, we get $\mu \|\hat{x}_k - \hat{x}\|^2 \leq \langle A\hat{x}_k - v, \bar{x}_k - \hat{x} \rangle$ and therefore

(11) $$\mu \|\hat{x}_k - \hat{x}\|^2 \leq \langle A(\hat{x}_k - \hat{x}) + A\hat{x} - v, \bar{x}_k - \hat{x} \rangle$$
$$\leq (\|A\| \|\hat{x}_k - \hat{x}\| + \|A\| \|\hat{x}\| + \|v\|)\, \mathrm{dist}(\hat{x}, C_k).$$

This quadratic inequality in $d_k = \|\hat{x}_k - \hat{x}\|$ implies that the sequence $\{d_k\}$ is bounded, say by b. Putting this b in place of $\|\hat{x}_k - \hat{x}\|$ on the right side of (11), we get a bound of the form in (10). □

Is the square root describing the rate of convergence through the estimate in (10) exact? The following example shows that this is indeed the case, and no improvement is possible, in general.

Example 6E.5 (counterexample to improving the general estimate). Consider problem (1) in the case of $X = \mathbb{R}^2$, $C = \{(x_1, x_2) \mid x_2 \leq 0\}$ (lower half-plane), $v = (0,1)$ and $A = I$, so that the issue revolves around projecting v on C and the solution is $\hat{x} = (0,0)$. For each $k = 1, 2, \ldots$ let $a_k = (1/k, 0)$ and let C_k consist of the points $x \in C$ such that $\langle x - a_k, v - a_k \rangle \leq 0$. Then the projection $\hat{x}_k$ of v on C_k is a_k, and

$$|\hat{x}_k - \hat{x}| = 1/k, \qquad d(\hat{x}, C_k) = \frac{1}{k\sqrt{1+k^2}}.$$

In this case the ratio $|\hat{x}_k - \hat{x}|/d(\hat{x}, C_k)^p$ is unbounded in k for any $p > 1/2$.

Detail. The fact that the projection of v on C_k is a_k comes from the observation that $v - a_k \in N_{C_k}(a_k)$. A similar observation confirms that the specified $\bar{x}_k$ is the projection of $\hat{x}$ on C_k. The ratio $|\hat{x}_k - \hat{x}|/d(\hat{x}, C_k)^p$ can be calculated as $k^{2p-1}(1 + 1/(k^2)^{p/2}$, and from that the conclusion is clear that it is bounded with respect to k if and only if $2 - (1/p) \leq 0$, or in other words, $p \leq 1/2$. □

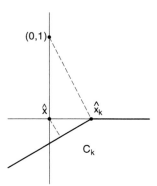

Fig. 6.1 Illustration to Example 6E.5.

There is, nevertheless, an important case in which the exponent $1/2$ in (10) can be replaced by 1. This case is featured in the following result:

Theorem 6E.6 (improved rate of convergence for subspaces). *Under the conditions of Theorem 6E.4, if the sets C and C_k are subspaces of X, then there is a constant c such that*

(12) $$\|\hat{x}_k - \hat{x}\| \leq c d(\hat{x}, C_k) \quad \text{for all } k.$$

Proof. In this situation the variational inequality in (3) reduces to the requirement that $Ax - v \perp C$. We then have $A\hat{x} - v \in C^\perp \subset C_k^\perp$ and $A\hat{x}_k - v \in C_k^\perp$, so that $A(\hat{x}_k - \hat{x}) \in C_k^\perp$. Consider now an arbitrary $x \in C_k$, noting that since $\hat{x}_k \in C_k$ we also have $\hat{x}_k - x \in C_k$. We calculate from (2) that

$$\begin{aligned} \mu \|\hat{x}_k - \hat{x}\|^2 &\leq \langle A(\hat{x}_k - \hat{x}), \hat{x}_k - \hat{x} \rangle \\ &= \langle A(\hat{x}_k - \hat{x}), \hat{x}_k - x \rangle + \langle A(\hat{x}_k - \hat{x}), x - \hat{x} \rangle \\ &= \langle A(\hat{x}_k - \hat{x}), x - \hat{x} \rangle \leq \|A\| \|\hat{x}_k - \hat{x}\| \|x - \hat{x}\|. \end{aligned}$$

This gives us the estimate (12). □

The result in 6E.6 corresponds to the classical Galerkin method, at least if C is all of X. We can combine it with the one in 6E.4 as follows.

Corollary 6E.7 (application to intersections with subspaces). *Let S be the solution mapping to (1) as provided by Theorem 6E.1 under condition (2). Let $\{X_k\}$ be an increasing sequence of closed subspaces of X such that (8) holds for the sets $C_k = C \cap X_k$. Then for any v the associated Galerkin's sequence $\{\hat{x}_k\} = \{S_k(v)\}$ converges to $\hat{x} = S(v)$ at the rate indicated in (10), but if C itself is a subspace, it converges at the rate indicated in (12).*

The closure condition in (8), in the case of $C_k = C \cap X_k$, says that $\text{dist}(x, C \cap X_k) \to 0$ as $k \to \infty$ for every $x \in C$. When the Hilbert space X is separable we may choose the subspaces X_k by taking a countable dense subset $x_1, x_2, \ldots$ of X and letting X_k be the span of $x_1, \ldots, x_k$. Because the subspaces are finite-dimensional, Galerkin's method in this case can be viewed as a discretization scheme. The property that $\text{dist}(x, C \cap X_k) \to 0$ as $k \to \infty$ for every $x \in C$ is called the *consistency* of the discretization scheme. In the following section we will look at the discretization of a specific variational problem.

6F. Approximations in Optimal Control

For an example which illustrates how the theory of solution mappings can be applied in infinite dimensions with an eye toward numerical approximations, we turn to a basic problem in optimal control, the so-called linear-quadratic regulator problem. That problem takes the form:

(1) $\quad$ minimize $\displaystyle\int_0^1 \left(\frac{1}{2}[x(t)^\mathsf{T} Q x(t) + u(t)^\mathsf{T} R u(t)] + s(t)^\mathsf{T} x(t) - r(t)^\mathsf{T} u(t) \right) dt$

subject to

(2) $\quad \dot{x}(t) = Ax(t) + Bu(t) + p(t) \quad \text{for a.e. } t \in [0,1], \qquad x(0) = a,$

6 Applications in Numerical Variational Analysis

and the constraint that

(3) $$u(t) \in U \quad \text{for a.e. } t \in [0, 1].$$

This concerns the *control system* governed by (2) in which $x(t) \in \mathbb{R}^n$ is the *state* at time t and $u(t)$ is the *control* exercised at time t. The choice of the control function $u : [0, 1] \to \mathbb{R}^m$ yields from the initial state $a \in \mathbb{R}^n$ and the dynamical equation in (2) a corresponding *state trajectory* $x : [0, 1] \to \mathbb{R}^n$ with derivative $\dot{x}$. The matrices A, B, Q and R have dimensions fitting these circumstances, with Q and R symmetric and positive semidefinite so as to ensure (as will be seen below) that the function being minimized in (1) is convex.

The set $U \subset \mathbb{R}^m$ from which the values of the control have to be selected from in (3) is nonempty, convex and compact[3]. We also assume that the matrix R is positive definite relative to U; in other words, there exists $\mu > 0$ such that

(4) $$u^\mathsf{T} R u \geq \mu |u|^2 \quad \text{for all } u \in U - U.$$

Any control function $u : [0, 1] \to U$ is required to be measurable ("a.e." refers as usual to "almost everywhere" with respect to Lebesgue measure), and since it takes values in the bounded set U for a.e. $t \in [0, 1]$, it is essentially bounded. But we take the set of *feasible control* functions to be a larger subset of $L^2(\mathbb{R}^m, [0, 1])$ functions, in which space the inner product and the norm are

$$\langle u, v \rangle = \int_0^1 u(t)^\mathsf{T} v(t)\, dt, \qquad \|u\|_2 = \sqrt{\langle u, u \rangle}.$$

We follow that Hilbert space pattern throughout, assuming that the function r in (1) belongs to $L^2(\mathbb{R}^m, [0, 1])$ while p and s belong to $L^2(\mathbb{R}^n, [0, 1])$. This is a convenient compromise which will put us in the framework of quadratic optimization in 6E.

There are two ways of looking at problem (1). We can think of it in terms of minimizing over function pairs (u, x) constrained by both (2) and (3), or we can regard x as a "dependent variable" produced from u through (2) and standard facts about differential equations, so as to think of the minimization revolving only around the choice of u. For any u satisfying (3) (and therefore essentially bounded), there is a unique state trajectory x specified by (2) in the sense of x being an absolutely continuous function of t and therefore differentiable a.e. Due to the assumption that $p \in L^2(\mathbb{R}^n, [0, 1])$, the derivative $\dot{x}$ can then be interpreted as an element of $L^2(\mathbb{R}^n, [0, 1])$ as well. Indeed, x is given by the Cauchy formula

(5) $$x(t) = e^{At} a + \int_0^t e^{A(t-\tau)} (B u(\tau) + p(\tau))\, d\tau \quad \text{for all } t \in [0, 1].$$

In particular, we can view it as belonging to the Banach space $C(\mathbb{R}^n, [0, 1])$ of continuous functions from $[0, 1]$ to $\mathbb{R}^n$ equipped with the norm

---

[3] We do not really need U to be bounded, but this assumption simplifies the analysis.

$$\|x\|_\infty = \max_{0 \le t \le 1} |x(t)|.$$

The relation between u and x can be cast in a frame of inputs and outputs. Define the mapping $T : L^2(\mathbb{R}^n, [0,1]) \to L^2(\mathbb{R}^n, [0,1])$ as

(6) $$(Tw)(t) = \int_0^t e^{A(t-\tau)} w(\tau) d\tau \quad \text{for a.e. } t \in [0,1],$$

and, on the other hand, let $W : L^2(\mathbb{R}^n, [0,1]) \to L^2(\mathbb{R}^n, [0,1])$ be the mapping defined by

(7) $$\text{for } p \in L^2(\mathbb{R}^n, [0,1]),\ W(p) \text{ is the solution to } \dot{W} = AW + p,\ W(0) = a.$$

Finally, with a slight abuse of notation, denote by B the mapping from $L^2(\mathbb{R}^m, [0,1])$ to $L^2(\mathbb{R}^n, [0,1])$ associated with the matrix B, that is $(Bu)(t) = Bu(t)$; later we do the same for the mappings Q and R. Then the formula for x in (5) comes out as

(8) $$x = (TB)(u) + W(p),$$

where u is the input, x is the output, and p is a parameter. Note that in this case we are treating x as an element of $L^2(\mathbb{R}^n, [0,1])$ instead of $C(\mathbb{R}^n, [0,1])$. This makes no real difference but will aid in the analysis.

Exercise 6F.1 (adjoint in the Cauchy formula). *Prove that the mapping T defined by (6) is linear and bounded. Also show that the adjoint (dual) mapping T^*, satisfying $\langle x, Tu \rangle = \langle T^*x, u \rangle$, is given by*

$$(T^*x)(t) = \int_t^1 e^{A^\mathsf{T}(\tau-t)} x(\tau) d\tau \quad \text{for a.e. } t \in [0,1].$$

Also show $(TB)^* = B^* T^*$, where B^* is the mapping $L^2(\mathbb{R}^n, [0,1])$ to $L^2(\mathbb{R}^m, [0,1])$ associated with the transposed matrix B^T; that is

$$((TB)^*x)(t) = \int_t^1 B^\mathsf{T} e^{A^\mathsf{T}(\tau-t)} x(\tau) d\tau \quad \text{for a.e. } t \in [0,1].$$

Guide. Apply the rule for changing the order of integration

$$\int_0^1 x(t)^\mathsf{T} \int_0^t e^{A(t-\tau)} w(\tau) d\tau dt = \int_0^1 \int_\tau^1 x(t)^\mathsf{T} e^{A(t-\tau)} w(\tau) dt d\tau,$$

and interpret what it says. □

To shorten notation in what follows, we will just write L^2 for both $L^2(\mathbb{R}^m, [0,1])$ and $L^2(\mathbb{R}^n, [0,1])$, leaving it to the reader to keep in mind which elements lie in $\mathbb{R}^m$ and which lie in $\mathbb{R}^n$.

The change of variables $z = x - w$ with $w = W(p)$ as in (7) gives the following reformulation of (1)–(3), where the parameter p is transferred to the problem of

6 Applications in Numerical Variational Analysis

minimizing the objective function

$$(9) \quad \int_0^1 \left(\frac{1}{2} [z(t)^\mathsf{T} Q z(t) + u(t)^\mathsf{T} R u(t)] + (s(t) + QW(p)(t))^\mathsf{T} z(t) - r(t)^\mathsf{T} u(t) \right) dt$$

subject to

$$(10) \quad \dot{z} = Az + Bu, \quad z(0) = 0, \quad u(t) \in U \text{ for a.e. } t \in [0,1].$$

In (9) we have dropped the constant terms that do not affect the solution. Noting that $z = (TB)(u)$ and utilizing the adjoint T^* of the mapping T, let

$$(11) \quad V(y) = -r + B^* T^*(s + QW(p)) \quad \text{for } y = (p,s,r),$$

and define the self-adjoint bounded linear mapping $\mathscr{A} : L^2 \to L^2$ by

$$(12) \quad \mathscr{A} = B^* T^* Q T B + R.$$

Here, as for the mapping B, we regard Q and R as linear bounded mappings acting between L^2 spaces: for $(Ru)(t) = Ru(t)$, and so forth. Let

$$(13) \quad C = \{ u \in L^2 \,|\, u(t) \in U \text{ for a.e. } t \in [0,1] \}.$$

With this notation, problem (9)–(10) can be written in the form treated in 6E:

$$(14) \quad \text{minimize } \frac{1}{2} \langle u, \mathscr{A} u \rangle + \langle V(y), u \rangle \text{ subject to } u \in C.$$

Exercise 6F.2 (coercivity in control). Prove that the set C in (13) is a closed and convex subset of L^2 and that the mapping $\mathscr{A} \in \mathscr{L}(L^2, L^2)$ in (12) satisfies the condition

$$\langle u, \mathscr{A} u \rangle \geq \mu \|u\|_2^2 \quad \text{for all } u \in C - C,$$

where μ is the constant in (4).

Applying Theorem 6E.1 in the presence of 6F.2, we obtain a necessary and sufficient condition for the optimality of u in problem (14), namely the variational inequality

$$(15) \quad V(y) + \mathscr{A} u + N_C(u) \ni 0.$$

For (15), or equivalently for (14) or (1)–(3), we arrive then at the following result of implicit-function type:

Theorem 6F.3 (implicit function theorem for optimal control in L^2). *Under (4), the solution mapping S which goes from parameter elements $y = (p,s,r)$ to pairs (u,x) solving (1)–(3) is single-valued and globally Lipschitz continuous from the space $L^2(\mathbb{R}^n \times \mathbb{R}^n \times \mathbb{R}^m, [0,1])$ to the space $L^2(\mathbb{R}^m, [0,1]) \times C(\mathbb{R}^n, [0,1])$.*

Proof. Because V in (11) is an affine function of $y = (p,s,r)$, we obtain from 6E.1 that for each y problem (14) has a unique solution $u(y)$ and, moreover, the function $y \mapsto u(y)$ is globally Lipschitz continuous in the respective norms. The value $u(y)$ is the unique optimal control in problem (1)–(3) for y. Taking norms in the Cauchy formula (5), we see further that for any $y = (p,s,r)$ and $y' = (p',s',r')$, if x and x' are the corresponding solutions of (2) for $u(y)$, p, and $u(y')$, p', then, for some constants c_1 and c_2, we get

$$|x(t) - x'(t)| \leq c_1 \int_0^t (|B||u(y)(\tau) - u(y')(\tau)| + |p(\tau) - p'(\tau)|)d\tau$$
$$\leq c_2(\|u(y) - u(y')\|_2 + \|p - p'\|_2).$$

Taking the supremum on the left and having in mind that $y \mapsto u(y)$ is Lipschitz continuous, we obtain that the optimal trajectory mapping $y \mapsto x(y)$ is Lipschitz continuous from the L^2 space of y to $C(\mathbb{R}^n, [0,1])$. Putting these facts together, we confirm the claim in the theorem. $\square$

The optimal control u whose existence and uniqueness for a given y is asserted in 6F.3 is actually, as an element of L^2, an equivalence class of functions differing from each other only on sets of measure zero in $[0,1]$. Thus, having specified an optimal control function u, we may change its values $u(t)$ on a t-set of measure zero without altering the value of the expression being minimized or affecting optimality. We will go on to show now that one can pick a particular function from the equivalence class which has better continuity properties with respect to both time and the parameter dependence.

For a given control u and parameter $y = (p,s,r)$, let

$$\psi = T^*(Qx + s),$$

where x solves (8). Then, through the Cauchy formula and 6F.1, ψ is given by

$$\psi(t) = \int_t^1 e^{A^T(\tau - t)} (Qx(\tau) + s(\tau)) d\tau.$$

Hence, ψ is a continuous function which is differentiable almost everywhere in $[0,1]$ and its derivative $\dot\psi$ is in L^2. Further, taking into account (6) and (8), $\dot\psi$ satisfies

(16) $\quad \dot\psi(t) = -A^T \psi(t) - Qx(t) - s(t) \quad$ for a.e. $t \in [0,1], \quad \psi(1) = 0,$

where x is the solution of (2) for the given u. The function ψ is called the *adjoint* or *dual* trajectory associated with a given control u and its corresponding state trajectory x, and (16) is called the *adjoint equation*. Bearing in mind the particular form of $V(y)$ in (11) and that, by definition,

$$\psi = T^*(Qx + s) = T^*[QTBu + s + QW(p)],$$

we can re-express the variational inequality (15) in terms of ψ as

(17) $\langle -r + Ru + B^*\psi, v - u\rangle \geq 0$ for all $v \in C$,

where B^* stands for the linear mapping associated with the transpose of the matrix B. The boundary value problem combining (2) and (16), coupled with the variational inequality (17), fully characterizes the solution to problem (1)–(3).

We need next a standard fact from Lebesgue integration. For a function φ on $[0,1]$, a point $\hat{t} \in (0,1)$ is said to be a *Lebesgue point* of φ when

$$\lim_{\varepsilon \to 0} \frac{1}{2\varepsilon} \int_{\hat{t}-\varepsilon}^{\hat{t}+\varepsilon} \varphi(\tau)d\tau = \varphi(t).$$

It is known that when φ is integrable on $[0,1]$, its set of Lebesgue points is of full measure 1.

Now, let u be the optimal control for a particular parameter value y, and let x and ψ be the associated optimal trajectory and adjoint trajectory, respectively. Let $\hat{t}$ be a Lebesgue point of both u and r (the set of such $\hat{t}$ is of full measure). Pick any $w \in U$, and for $0 < \varepsilon < \min\{\hat{t}, 1-\hat{t}\}$ consider the function

$$\hat{u}_\varepsilon(t) = \begin{cases} w & \text{for } t \in (\hat{t}-\varepsilon, \hat{t}+\varepsilon), \\ u(t) & \text{otherwise.} \end{cases}$$

Then for every sufficiently small ε the function $\hat{u}_\varepsilon$ is a feasible control, i.e., belongs to the set C in (13), and from (17) we obtain

$$\int_{\hat{t}-\varepsilon}^{\hat{t}+\varepsilon} (-r(\tau) + Ru(\tau) + B^T \psi(\tau))^T (w - u(\tau))d\tau \geq 0.$$

Since $\hat{t}$ is a Lebesgue point of the function under the integral (we know that ψ is continuous and hence its set of Lebesgue points is the entire interval $[0,1]$), we can pass to zero with ε and by taking into account that $\hat{t}$ is an arbitrary point from a set of full measure in $[0,1]$ and that w can be any element of U, come to the following *pointwise* variational inequality which is required to hold for a.e. $t \in [0,1]$:

(18) $(-r(t) + Ru(t) + B^T \psi(t))^T (w - u(t)) \geq 0$ for every $w \in U$.

As is easily seen, (18) implies (17) as well, and hence these two variational inequalities are equivalent.

Summarizing, we can now say that a feasible control u is the solution of (1)–(3) for a given $y = (p, s, r)$ with corresponding optimal trajectory x and adjoint trajectory ψ if and only if the triple (u, x, ψ) solves the following boundary value problem coupled with a pointwise variational inequality:

(19a) $\begin{cases} \dot{x}(t) = Ax(t) + Bu(t) + p(t), & x(0) = a, \\ \dot{\psi}(t) = -A^T\psi(t) - Qx(t) - s(t), & \psi(1) = 0, \end{cases}$

(19b) $r(t) \in Ru(t) + B^T\psi(t) + N_U(u(t))$ for a.e. $t \in [0,1]$.

That is, for an optimal control u and associated optimal state and adjoint trajectories x and ψ, there exists a set of full measure in $[0,1]$ such that (19b) holds for every t in this set. Under an additional condition on the function r we obtain the following result:

Theorem 6F.4 (implicit function theorem for continuous optimal controls). *Let the parameter $y = (p,s,r)$ in (1)–(3) be such that the function r is Lipschitz continuous on $[0,1]$. Then, from the equivalence class of optimal control functions for this y, there exists an optimal control $u(y)$ for which (19b) holds for all $t \in [0,1]$ and which is Lipschitz continuous with respect to t on $[0,1]$. Moreover, the solution mapping $y \mapsto u(y)$ is Lipschitz continuous from the space $L^2(\mathbb{R}^n \times \mathbb{R}^n, [0,1]) \times C(\mathbb{R}^m, [0,1])$ to the space $C(\mathbb{R}^m, [0,1])$.*

Proof. It is clear that the adjoint trajectory ψ is Lipschitz continuous in t on $[0,1]$ for any feasible control; indeed, it is the solution of the linear differential equation (16), the right side of which is a function in L^2. Let x and ψ be the optimal state and adjoint trajectories and let u be a function satisfying (19b) for all $t \in \sigma$ where σ is a set of full measure in $[0,1]$. For $t \notin \sigma$ we define $u(t)$ to be the unique solution of the following strongly monotone variational inequality in $\mathbb{R}^n$:

$$(20) \qquad q(t) \in Ru + N_U(u), \text{ where } q(t) = r(t) - B^\mathsf{T} \psi(t).$$

Then this u is within the equivalence class of optimal controls, and, moreover, the vector $u(t)$ satisfies (20) for all $t \in [0,1]$. Noting that q is a Lipschitz continuous function in t on $[0,1]$, we get from 2F.10 that for each fixed $t \in [0,1]$ the solution mapping of (20) is Lipschitz continuous with respect to $q(t)$. Since the composition of Lipschitz continuous functions is Lipschitz continuous, the particular optimal control function u which satisfies (19b) for all $t \in [0,1]$ is Lipschitz continuous in t on $[0,1]$.

We already know from 6F.3 that the optimal trajectory mapping $y \mapsto x(y)$ is Lipschitz continuous into $C(\mathbb{R}^n, [0,1])$. By the same argument, the associated adjoint mapping $y \mapsto \psi(y)$ is Lipschitz continuous from L^2 to $C(\mathbb{R}^n, [0,1])$. But then, according to 2F.10 again, for every y, y' with r, r' Lipschitz continuous and every $t \in [0,1]$, with the optimal control values $u(y)(t)$ and $u(y')(t)$ at t being the unique solutions of (20), we have

$$|u(y)(t) - u(y')(t)| \leq \mu^{-1} \left(|r(t) - r'(t)| + |B||\psi(y)(t) - \psi(y')(t)| \right).$$

This holds for every $t \in [0,1]$, so by invoking the maximum norm we get the desired result. □

We focus next on the issue of solving problem (1)–(3) numerically. By this we mean determining the optimal control function u. This is a matter of recovering a function on $[0,1]$ which is only specified implicitly, in this case by a variational problem. Aside from very special cases, it means producing numerically an acceptable approximation of the desired function u. For simplicity, let us assume that $y = (p,s,r) = 0$.

6 Applications in Numerical Variational Analysis

For such an approximation we may focus on a finite-dimensional space of functions on $[0,1]$ within L^2. Suppose that the interval $[0,1]$ is divided into N pieces $[t_i, t_{i+1}]$ by equally spaced nodes t_i, $i = 0, 1, \ldots, N$, with $t_0 = 0$ and $t_N = 1$, the fixed mesh size being $h = t_i - t_{i-1} = 1/N$. To approximate the optimal control function u that is Lipschitz continuous on $[0,1]$ according to 6F.4, we will employ *piecewise constant* functions across the grid $\{t_i\}$ that are continuous from the right at each t_i, $i = 0, 1, \ldots, N-1$ and from the left at $t_N = 1$. Specifically, for a given N, we consider the subset of L^2 given by

$$\left\{ u \,\Big|\, u(t) = u(t_i) \text{ for } t \in [t_i, t_{i+1}),\ i = 0, 1, \ldots, N-2,\ u(t) = u(t_{N-1}) \text{ for } t \in [t_{N-1}, t_N] \right\}.$$

In order to fully discretize problem (1)–(3) and transform it into a finite-dimensional optimization problem, we also need to use finite-dimensional approximations of the operators of integration and differentiation involved.

Rather than (1)–(3), we now invoke a discretization of the optimality system (19ab). For solving the differential equations in (19a) we use the simplest Euler scheme over the mesh $\{t_i\}$. The Euler scheme applied to (19a), forward for the state equation and backward for the adjoint equation, combined with restricting the functional variational inequality (19b) to the nodes of the scheme, results in the following discrete-time boundary value problem coupled with a finite-dimensional variational inequality:

$$(21) \quad \begin{cases} x_{i+1} = (I + hA)x_i + hBu_i, & x_0 = a, \\ \psi_i = (I + hA^\mathsf{T})\psi_{i+1} + hQx_{i+1}, & \psi_N = 0, \\ 0 \in Ru_i + B^\mathsf{T}\psi_i + N_U(u_i) & \end{cases} \quad \text{for } i = 0, 1, \ldots, N-1.$$

There are various numerical techniques for solving problems of this form; here we shall not discuss this issue.

We will now derive an estimate for the error in approximating the solution of problem (1)–(3) by use of discretization (21) of the optimality system (19ab).

We suppose that for each given N we can solve (21) exactly, obtaining vectors $u_i^N \in U$, $i = 0, \ldots, N-1$, and $x_i^N \in \mathbb{R}^n$, $\psi_i^N \in \mathbb{R}^n$, $i = 0, \ldots, N$. For a given N, the solution (u^N, x^N, ψ^N) of (21) is identified with a function on $[0,1]$, where x^N and ψ^N are the piecewise linear and continuous interpolations across the grid $\{t_i\}$ over $[0,1]$ of $(a, x_1^N, \ldots, x_N^N)$ and $(\psi_0^N, \psi_1^N, \ldots, \psi_{N-1}^N, 0)$, respectively, and u^N is the piecewise constant interpolation of $(u_0^N, u_1^N, \ldots, u_{N-1}^N)$ which is continuous from the right across the grid points $t_i = ih$, $i = 0, 1, \ldots, N-1$ and from the left at $t_N = 1$. The functions x^N and ψ^N are piecewise differentiable and their derivatives $\dot{x}^N$ and $\dot{\psi}^N$ are piecewise constant functions which are assumed to have the same continuity properties in t as the control u^N. Thus, (u^N, x^N, ψ^N) is a function defined in the whole interval $[0,1]$, and it belongs to L^2.

Theorem 6F.5 (error estimate for discrete approximation). *Consider problem (1)–(3) with $r = 0$, $s = 0$ and $p = 0$ under condition (4) and let, according to 6F.4, (u, x, ψ) be the solution of the equivalent optimality system (19ab) for all $t \in [0,1]$,* with u Lipschitz continuous in t on $[0,1]$. Consider also the discretization (21) and, for $N = 1,2,\ldots$ and mesh size $h = 1/N$, denote by (u^N, x^N, ψ^N) its solution extended by interpolation to the interval $[0,1]$ in the manner described above. Then the following estimate holds:

(22) $$\|u^N - u\|_\infty + \|x^N - x\|_\infty + \|\psi^N - \psi\|_\infty = O(h).$$

Proof. For $t \in [t_i, t_{i+1})$, $i = 0, 1, \ldots, N-1$, let

$$\begin{aligned} p^N(t) &= A(x^N(t_i) - x^N(t)), \\ s^N(t) &= A^\mathsf{T}(\psi^N(t_{i+1}) - \psi^N(t)) + Q(x^N(t_{i+1}) - x^N(t)), \\ r^N(t) &= -B^\mathsf{T}(\psi^N(t_i) - \psi^N(t)). \end{aligned}$$

By virtue of the control u^N being piecewise constant and

$$\dot{x}^N(t) = \frac{x^N(t_{i+1}) - x^N(t_i)}{h}, \quad \dot{\psi}^N(t) = \frac{\psi^N(t_{i+1}) - \psi^N(t_i)}{h} \quad \text{for } t \in [t_i, t_{i+1}),$$

for $i = 0, 1, \ldots, N-1$, the discretized optimality system (21) can be written as follows: for all $t \in [t_i, t_{i+1})$, $i = 0, 1, \ldots, N-1$, and $t = 1$,

(23) $$\begin{cases} \dot{x}^N(t) = Ax^N(t) + Bu^N(t) + p^N(t), & x^N(0) = a, \\ \dot{\psi}^N(t) = -A^\mathsf{T}\psi^N(t) - Qx^N(t) - s^N(t), & \psi^N(1) = 0, \\ r^N(t) \in Ru^N(t) + B^\mathsf{T}\psi^N(t) + N_U(u^N(t)). \end{cases}$$

Observe that system (23) has the same form as (19ab) with a particular choice of the parameters. Specifically, (u^N, x^N, ψ^N) is the solution of (19ab) for the parameter value $y^N := (p^N, s^N, r^N)$, while (u, x, ψ) is the solution of (19ab) for $y = (p, s, r) = (0, 0, 0)$. Then, by the implicit function theorem 6F.4, the solution mapping of (19) is Lipschitz continuous in the maximum norms, so there exists a constant c such that

(24) $$\|u^N - u\|_\infty + \|x^N - x\|_\infty + \|\psi^N - \psi\|_\infty \leq c\|y^N\|_\infty.$$

To finish the proof, we need to show that

(25) $$\|y^N\|_\infty = \max\{\|p^N\|_\infty, \|s^N\|_\infty, \|r^N\|_\infty\} = O(h).$$

For that purpose we employ the following standard result in the theory of difference equations which we state here without proof:

Lemma 6F.6 (discrete Gronwall lemma). *Consider reals α_i, $i = 0, \ldots, N$, which satisfy*

$$0 \leq \alpha_0 \leq a \quad \text{and} \quad 0 \leq \alpha_{i+1} \leq a + b\sum_{j=0}^{i} \alpha_j \quad \text{for } i = 0, \ldots, N.$$

Then $0 \leq \alpha_i \leq a(1+b)^i$ for $i = 0, \ldots, N$. Similarly, if

$$0 \leq \alpha_N \leq a \text{ and } 0 \leq \alpha_{i+1} \leq a + b \sum_{j=i+1}^{N} \alpha_j \text{ for } i = 0, \ldots, N,$$

then $0 \leq \alpha_i \leq a(1+b)^{N-i}$ for $i = 0, \ldots, N$.

Continuing on this basis with the proof of (25), we observe that x^N is piecewise linear across the grid $\{t_i\}$; clearly

$$|x^N(t) - x^N(t_i)| \leq |x^N(t_{i+1}) - x^N(t_i)| \text{ for } t \in [t_i, t_{i+1}], i = 0, 1, \ldots, N-1.$$

Then, since all u_i are from the compact set U, from the first equation in (21) we get

$$|p^N(t)| \leq h(c_1 |x^N(t_i)| + c_2) \text{ for } t \in [t_i, t_{i+1}], i = 0, 1, \ldots, N-1$$

with some constants c_1, c_2 independent of N. On the other hand, the first equation in (21) can be written equivalently as

$$x^N(t_{i+1}) = a + \sum_{j=1}^{i} h(A x^N(t_j) + B u^N(t_j)),$$

and then, by taking norms and applying the direct part of discrete Gronwall lemma 6F.6, we obtain that $\sup_{0 \leq i \leq N} |x^N(t_i)|$ is bounded by a constant which does not depend on N. This gives us error of order $O(h)$ for p^N in the maximum norm. By repeating this argument for the discrete adjoint equation (the second equation in (21)), but now applying the backward part of 6F.7, we get the same order of magnitude for s^N and r^N. This proves (25) and hence also (22). □

Note that the order of the discretization error is $O(h)$, which is sharp for the Euler scheme. Using higher-order schemes may improve the order of approximation, but this may require better continuity properties of the optimal control.

In the proof of 6F.5 we used the combination of the implicit function theorem 6F.4 for the variational system involved and the estimate (25) for the *residual* $y^N = (p^N, s^N, r^N)$ of the approximation scheme. The convergence to zero of the residual comes out of the approximation scheme and the continuity properties of the solution of the original problem with respect to time t; in numerical analysis this is called the *consistency* of the problem and its approximation. The property emerging from the implicit function theorem 6F.4, that is, the Lipschitz continuity of the solution with respect to the residual, is sometimes called *stability*. Theorem 6F.5 furnishes an illustration of a well-known paradigm in numerical analysis: *stability plus consistency yields convergence*.

Having the analysis of the linear-quadratic problem as a basis, we could proceed to more general nonlinear and nonconvex optimal control problems and obtain convergence of approximations and error estimates by applying more advanced implicit function theorems using, e.g., linearization of the associated nonlinear optimality systems. However, this would involve more sophisticated techniques which go beyond the scope of this book, so here is where we stop.

Commentary

Theorem 6A.2 is from Dontchev, Lewis and Rockafellar [2003], where it is supplied with a direct proof. Theorem 6A.3 was first shown by Lewis [1999]; see also Lewis [2001]. Theorem 6A.7 was initially proved in Dontchev, Lewis and Rockafellar [2003] by using the characterization of the metric regularity of a mapping in terms of the nonsingularity of its coderivative (see Section 4H) and applying the radius theorem for nonsingularity in 6A.2. The proof given here is from Dontchev, Quincampoix and Zlateva [2006]. For extensions to infinite-dimensional spaces see Ioffe [2003a,b]. Theorems 6A.8 and 6A.9 are from Dontchev and Rockafellar [2004].

The material in Section 6B is basically from Dontchev, Lewis and Rockafellar [2003]. The results in Sections 6C and 6D have roots in several papers; see Rockafellar [1976a,b], Robinson [1994], Dontchev [2000], Aragón Artacho, Dontchev, and Geoffroy [2007] and Dontchev and Rockafellar [2009b].

Most of the results in Section 6E can be found in basic texts on variational methods; for a recent such book see Attouch, Buttazzo and Michaille [2006]. Section 6F presents a very simplified version of a result in Dontchev [1996]; for advanced studies in this area see Malanowski [2001] and Veliov [2006].

References

Alt, W. [1990], The LagrangeNewton method for infinite-dimensional optimization problems, Numerical Functional Analysis and Optimization, 11, 201–224.

Apostol, T. [1962], *Calculus*, Blaisdell Publ. Co., Waltham, MA.

Aragón Artacho, F. J., A. L. Dontchev and M. H. Geoffroy [2007], Convergence of the proximal point method for metrically regular mappings, ESAIM Proceedings, 17, 1–8.

Arutyunov, A. [2007], Covering mappings in metric spaces and fixed points, Doklady Mathematics, 76, 665–668.

Attouch, H., G. Buttazzo and G. Michaille [2006], *Variational analysis in Sobolev and BV spaces, Applications to PDEs and optimization,* MPS/SIAM Series on Optimization, 6, Society for Industrial and Applied Mathematics, Philadelphia, PA; Mathematical Programming Society, Philadelphia, PA.

Aubin, J-P. [1981], Contingent derivatives of set-valued maps and existence of solutions to nonlinear inclusions and differential inclusions, in *Mathematical analysis and applications, Part A,* edited by L. Nachbin, Advances in Mathematics: Supplementary Studies, 7A, pp. 159–229, Academic Press, New York–London.

Aubin, J-P. [1984], Lipschitz behavior of solutions to convex minimization problems, Mathematics of Operations Research, 9, 87–111.

Aubin, J-P. and I. Ekeland [1984], *Applied nonlinear analysis,* John Wiley & Sons, Inc., New York.

Aubin, J-P. and H. Frankowska [1987], On inverse function theorems for set-valued maps, Journal de Mathématiques Pures et Appliquées, 66, 71–89.

Aubin, J-P. and H. Frankowska [1990], *Set-valued analysis,* Systems and Control: Foundations & Applications, 2, Birkhäuser, Boston, MA.

Banach, S. [1922], Sur les opérations dans les ensembles abstraits et leur application aux équations intégrales, Fundamenta Mathematicae, 3, 133–181.

Banach, S. [1932], *Théorie des Operations Linéaires*, Monografje Matematyczne, Warszawa, English translation by North Holland, Amsterdam, 1987.

Bank, B., J. Guddat, D. Klatte, B. Kummer and K. Tammer [1983], *Nonlinear parametric optimization,* Birkhäuser, Basel–Boston, MA.

Bartle, R. G. and L. M. Graves [1952], Mappings between function spaces, Transactions of the American Mathematical Society, 72, 400–413.

References

Bartle, R. G. and D. R. Sherbert [1992], *Introduction to real analysis*, Second edition, John Wiley, New York.

Bessis, D. N., Yu. S. Ledyaev and R. B. Vinter [2001], Dualization of the Euler and Hamiltonian inclusions, Nonlinear Analysis, 43, 861–882.

Bonnans, J. F. and A. Shapiro [2000], *Perturbation analysis of optimization problems*, Springer Series in Operations Research, Springer, New York.

Borwein, J. M. [1983], Adjoint process duality, Mathematics of Operations Research, 8, 403–434.

Borwein, J. M. [1986a], Stability and regular points of inequality systems, Journal of Optimization Theory and Applications, 48, 9–52.

Borwein, J. M. [1986b], Norm duality for convex processes and applications, Journal of Optimization Theory and Applications, 48, 53–64.

Borwein, J. M. and A. L. Dontchev [2003], On the Bartle-Graves theorem, Proceedings of the American Mathematical Society, 131, 2553–2560.

Borwein, J. M. and A. S. Lewis [2006], *Convex analysis and nonlinear optimization. Theory and examples*, Second edition, CMS Books in Mathematics/Ouvrages de Mathématiques de la SMC, 3. Springer, New York.

Borwein, J. M. and Q. J. Zhu [2005], *Techniques of variational analysis*, CMS Books in Mathematics/Ouvrages de Mathématiques de la SMC, 20, Springer, New York.

Cauchy, A-L. [1831], Résumé d'un mémoir sur la mécanique céleste et sur un nouveau calcul appelé calcul des limites, in *Oeuvres Complétes d'Augustun Cauchy*, volume 12, pp. 48–112, Gauthier-Villars, Paris 1916. Edited by L'Académie des Sciences. The part pertaining to series expansions was read at a meeting of the Academie de Turin, 11 October, 1831.

Chipman, J. S. [1997], "Proofs" and proofs of the Eckart-Young theorem, in *Stochastic processes and functional analysis*, 71–83, Lecture Notes in Pure and Applied Mathematics, 186, Dekker, New York.

Clarke, F. H. [1976], On the inverse function theorem, Pacific Journal of Mathematics, 64, 97–102.

Clarke, F. H. [1983], *Optimization and nonsmooth analysis*, Canadian Mathematical Society Series of Monographs and Advanced Texts. A Wiley-Interscience Publication. John Wiley & Sons, Inc., New York.

Cottle, R. W., J.-S. Pang and R. E. Stone [1992], *The linear complementarity problem*, Academic Press, Inc., Boston, MA.

Courant, R. [1988], *Differential and integral calculus*, Vol. II. Translated from the German by E. J. McShane. Reprint of the 1936 original. Wiley Classics Library. A Wiley-Interscience Publication. John Wiley & Sons, Inc., New York.

Deimling, K. [1992], *Multivalued differential equations*, Walter de Gruyter, Berlin.

Deville, R., G. Godefroy and V. Zizler [1993], *Smoothness and renormings in Banach spaces*, Pitman Monographs and Surveys in Pure and Applied Mathematics, 64. Longman Scientific & Technical, Harlow; copublished in the United States with John Wiley & Sons, Inc., New York.

Deutsch, F. [2001], *Best approximation in inner product spaces*, CMS Books in Mathematics/Ouvrages de Mathématiques de la SMC, 7. Springer, New York.

Dieudonné, J. [1969], *Foundations of modern analysis*, Enlarged and corrected printing. Pure and Applied Mathematics, Vol. 10-I, Academic Press, New York.

References

Dini, U. [1877/78], *Analisi infinitesimale*, Lezioni dettate nella R. Universitá di Pisa.

Dmitruk, A. V., A. A. Milyutin and N. P. Osmolovskiĭ [1980], Lyusternik's theorem and the theory of extremum, Uspekhi Mat. Nauk, 35, no. 6 (216), 11–46 (Russian).

Dontchev, A. L. [1995a], Characterizations of Lipschitz stability in optimization, in *Recent developments in well-posed variational problems*, 95–115, Kluwer, Dordrecht.

Dontchev, A. L. [1995b], Implicit function theorems for generalized equations, Mathematical Programming, 70, Ser. A, 91–106.

Dontchev, A. L. [1996], An a priori estimate for discrete approximations in nonlinear optimal control, SIAM Journal of Control and Optimization, 34, 1315–1328.

Dontchev, A. L. [2000], Lipschitzian stability of Newton's method for variational inclusions, in *System modelling and optimization (Cambridge, 1999)*, 119–147, Kluwer Acad. Publ., Boston, MA.

Dontchev, A. L. [2004], A local selection theorem for metrically regular mappings. Journal of Convex Analysis, 11, 81–94.

Dontchev, A. L. and W. W. Hager [1993], Lipschitzian stability in nonlinear control and optimization, SIAM Journal of Control and Optimization, 31, 569–603.

Dontchev, A. L. and W. W. Hager [1994], An inverse mapping theorem for set-valued maps, Proceedings of the American Mathematical Society, 121, 481–489.

Dontchev, A. L. and R. T. Rockafellar [1996], Characterizations of strong regularity for variational inequalities over polyhedral convex sets, SIAM Journal on Optimization, 6, 1087–1105.

Dontchev A. L. and R. T. Rockafellar [1998], Characterizations of Lipschitzian stability in nonlinear programming, in *Mathematical programming with data perturbations*, 65–82, Lecture Notes in Pure and Appl. Math., 195, Dekker, New York.

Dontchev, A. L. and R. T. Rockafellar [2001], Ample parameterization of variational inclusions, SIAM Journal on Optimization, 12, 170–187.

Dontchev, A. L. and R. T. Rockafellar [2004], Regularity and conditioning of solution mappings in variational analysis, Set-Valued Analysis, 12, 79–109.

Dontchev, A. L. and R. T. Rockafellar [2009a], Robinson's implicit function theorem and its extensions, Mathematical Programming, 117, 129–147.

Dontchev, A. L. and R. T. Rockafellar [2009b], Newton's method for generalized equations: a sequential implicit function theorem, to appear in Mathematical Programming.

Dontchev, A. L. and T. Zolezzi [1993], *Well-posed optimization problems*, Lecture Notes in Mathematics 1543, Springer, Berlin.

Dontchev, A. L., A. S. Lewis and R. T. Rockafellar [2003], The radius of metric regularity, Transactions of the American Mathematical Society, 355, 493–517.

Dontchev, A. L., M. Quincampoix and N. Zlateva [2006], Aubin criterion for metric regularity, Journal of Convex Analysis, 13, 281–297.

Dunford, N. and J. T. Schwartz [1958], *Linear Operators I. General Theory*, with the assistance of W. G. Bade and R. G. Bartle. Pure and Applied Mathematics, Vol. 7 Interscience, Publishers, Inc., New York; Interscience Publishers, Ltd., London.

Eckart, C. and G. Young [1936], The approximation of one matrix by another of lower rank, Psychometrica, 1, 211–218.

Ekeland, I. [1990], The ε-variational principle revisited, with notes by S. Terracini, in *Methods of nonconvex analysis (Varenna 1989)*, 1–15, Lecture Notes in Mathematics, 1446, Springer, Berlin.

Facchinei, F. and J-S. Pang [2003], *Finite-dimensional variational inequalities and complementarity problems*, Springer, New York.

Fiacco, A. V. and G. P. McCormick [1968], *Nonlinear programming: Sequential unconstrained minimization techniques*, John Wiley & Sons, Inc., New York–London–Sydney.

Fitzpatrick, P. [2006], *Advanced calculus*, Thomson Brooks/Cole.

Golub, G. H. and C. F. Van Loan [1996], *Matrix computations*, The John Hopkins University Press, Baltimore, MD, 3d edition.

Goursat, Ed. [1903], Sur la théorie des fonctions implicites, Bulletin de la Société Mathématiques de France, 31, 184–192.

Goursat, Ed. [1904], *A course in mathematical analysis*, English translation by E. R. Hedrick, Ginn Co., Boston.

Graves, L. M. [1950], Some mapping theorems, Duke Mathematical Journal, 17, 111–114.

Halkin, H. [1974], Implicit functions and optimization problems without continuous differentiability of the data, SIAM Journal on Control, 12, 229–236.

Hamilton, R. S. [1982], The inverse function theorem of Nash and Moser, Bulletin of the American Mathematical Society (N.S.), 7, 65–222.

Hausdorff, F. [1927], *Mengenlehre*, Walter de Gruyter and Co., Berlin.

Hedrick, E. R. and W. D. A. Westfall [1916], Sur l'existence des fonctions implicites, Bulletin de la Société Mathématiques de France, 44, 1–13.

Hildebrand, H. and L. M. Graves [1927], Implicit functions and their differentials in general analysis, Transactions of the American Mathematical Society, 29, 127–153.

Hoffman, A. J. [1952], On approximate solutions of systems of linear inequalities, Journal of Research of the National Bureau of Standards, 49, 263–265.

Hurwicz, L. and M. K. Richter [2003], Implicit functions and diffeomorphisms without C^1, Advances in mathematical economics, 5, 65–96.

Ioffe, A. D. [1979], Regular points of Lipschitz functions, Transactions of the American Mathematical Society, 251, 61–69.

Ioffe, A. D. [1981], Nonsmooth analysis: differential calculus of nondifferentiable mappings, Transactions of the American Mathematical Society, 266, 1–56.

Ioffe, A. D. [1984], Approximate subdifferentials and applications, I. The finite-dimensional theory, Transactions of the American Mathematical Society, 281, 389–416.

Ioffe, A. D. [2000], Metric regularity and subdifferential calculus, Uspekhi Mat. Nauk, 55, no. 3 (333), 103–162 (Russian), English translation in Russian Mathematical Surveys, 55, 501–558.

Ioffe, A. D. [2003a], On robustness of the regularity property of maps, Control and Cybernetics, 32, 543–554.

Ioffe, A. D. [2003b], On stability estimates for the regularity property of maps, in H. Brezis, K.C. Chang, S.J. Li, and P. Rabinowitz, editors, *Topological methods, variational methods and their applications*, pp. 133–142, World Science Publishing, NJ.

References

Ioffe, A. D. and V. M. Tikhomirov [1974], *Theory of extremal problems*, Nauka, Moscow (Russian).

Jongen, H. Th., T. Möbert, J. Rückmann and K. Tammer [1987], On inertia and Schur complement in optimization, Linear Algebra and Applications, 95, 97–109.

Josephy, N. H. [1979], Newton's method for generalized equations and the PIES energy model, Ph.D. Dissertation, Department of Industrial Engineering, University of Wisconsin-Madison.

Kahan, W. [1966], Numerical linear algebra, Canadian Mathematics Bulletin, 9, 757–801.

Kantorovich, L. V. and G. P. Akilov [1964], *Functional analysis in normed spaces*, Macmillan, New York. Translation by D. E. Brown from Russian original, Fizmatgiz, Moscow, 1959.

Kenderov, P. [1975], Semi-continuity of set-valued monotone mappings, Fundamenta Mathematicae, 88, 61–69.

King, A. J. and R. T. Rockafellar [1992], Sensitivity analysis for nonsmooth generalized equations, Mathematical Programming, Ser. A, 55(2), 193–212.

Klatte, D. and B. Kummer [2002], *Nonsmooth equations in optimization. Regularity, calculus, methods and applications*. Kluwer, Dordrecht.

Kojima, M. [1980], Strongly stable stationary solutions in nonlinear programs, in *Analysis and computation of fixed points (Madison, Wisconsin, 1979)*, pp. 93–138, Academic Press, New York–London.

Krantz, S. G. and H. R. Parks [2002], *The implicit function theorem. History, theory, and applications*, Birkhäuser, Boston.

Kruger, A. [1982], On characterizing the covering property of nonsmooth operators, in *Proceedings of the School on Theory of Operators in Function Spaces, Minsk*, 94–95 (Russian).

Kummer, B. [1991], An implicit-function theorem for $C^{0,1}$-equations and parametric $C^{1,1}$-optimization, Journal of Mathematical Analysis and Applications, 158, 35–46.

Kuratowski, K. [1933], *Topologie*, I & II, Panstwowe Wydawnictwo Naukowe, Warszawa.

Lamson, K. W. [1920], A general implicit function theorem with an application to problems of relative minima, American Journal of Mathematics, 42, 243–256.

Leach, E. B. [1961], A note on inverse function theorems, Proceedings of the American Mathematical Society, 12, 694–697.

Ledyaev, Yu. S. and Q. J. Zhu [1999], Implicit multifunction theorems, Set-Valued Analysis, 7, 209–238.

Levitin, E. S. [1992], *Perturbation theory in mathematical programming and its applications*, Nauka, Moscow (Russian).

Levy, A. B. [1996], Implicit multifunction theorems for the sensitivity analysis of variational conditions, Mathematical Programming, Ser. A, 74(3), 333–350.

Levy, A. B. and R. A. Poliquin [1997], Characterizing the single-valuedness of multifunctions. Set-Valued Analysis, 5, 351–364.

Lewis, A. S. [1999], Ill-conditioned convex processes and conic linear systems, Mathematics of Operations Research, 24, 829–834.

Lewis A. S. [2001], Ill-conditioned inclusions, Set-Valued Analysis, 9, 375–381.

Li, Wu [1994], Sharp Lipschitz constants for basic optimal solutions and basic feasible solutions of linear programs, SIAM Journal on Control and Optimization, 32, 140–153.

Lyusternik, L. A. [1934], On the conditional extrema of functionals, Mat. Sbornik, 41, 390–401 (Russian).

Lyusternik, L. A. and V. I. Sobolev [1965], *Elements of functional analysis,* Nauka, Moscow (Russian).

Malanowski, K. [2001], Stability and sensitivity analysis for optimal control problems with control-state constraints, Dissertationes Mathematicae, 394, 55pp.

Mordukhovich, B. S. [1984], Nonsmooth analysis with nonconvex generalized differentials and conjugate mappings, Doklady Akad. Nauk BSSR, 28, 976–979, (Russian).

Mordukhovich, B. S. [2006], *Variational analysis and generalized differentiation, I. Basic theory,* Springer, Berlin.

Nadler, Sam B., Jr. [1969], Multi-valued contraction mappings, Pacific Journal of Mathematics, 30, 475–488.

Ng, Kung Fu [1973], An open mapping theorem, Proceedings of Cambridge Philosophical Society, 74, 61–66.

Nijenhuis, A. [1974], Strong derivatives and inverse mappings, The American Mathematical Monthly, 81, 969–980.

Noble, B. and J. W. Daniel [1977], *Applied linear algebra,* Second edition, Prentice-Hall, Inc., Englewood Cliffs, N.J.

Páles, Z. [1997], Inverse and implicit function theorems for nonsmooth maps in Banach spaces, Journal of Mathematical Analysis and Applications, 209, 202–220.

Pompeiu, D. [1905], Fonctions de variable complexes, Annales de la Faculté des Sciences de l´ Université de Toulouse, 7, 265–345.

Ralph, D. [1993], A new proof of Robinson's homeomorphism theorem for PL-normal maps, Linear Algebra and Applications, 178, 249–260.

Robinson, S. M. [1972], Normed convex processes, Transactions of the American Mathematical Society, 174, 127–140.

Robinson, S. M. [1976], Regularity and stability for convex multivalued functions, Mathematics of Operations Research, 1, 130–143.

Robinson, S. M. [1980], Strongly regular generalized equations, Mathematics of Operations Research, 5, 43–62.

Robinson, S. M. [1981], Some continuity properties of polyhedral multifunctions, Mathematical Programming Study, 14, 206–214.

Robinson, S. M. [1984], Local structure of feasible sets in nonlinear programming, Part II: Nondegeneracy, Mathematical Programming Study, 22, 217–230.

Robinson, S. M. [1991], An implicit-function theorem for a class of nonsmooth functions, Mathematics of Operations Research, 16, 292–309.

Robinson, S. M. [1992], Normal maps induced by linear transformations, Mathematics of Operations Research, 17, 691–714.

Robinson, S. M. [1994], Newton's method for a class of nonsmooth functions, Set-valued Analysis, 2, 291–305.

References

Robinson, S. M. [2007], Solution continuity in monotone affine variational inequalities, SIAM Journal in Optimization, 18, 1046–1060.

Rockafellar, R. T. [1967], Monotone processes of convex and concave type, Memoirs of the American Mathematical Society, 77, Providence, RI.

Rockafellar, R. T. [1970], *Convex analysis,* Princeton University Press, Princeton, NJ.

Rockafellar, R. T. [1976a], Monotone operators and the proximal point algorithm, SIAM Journal on Control and Optimization, 14, 877–898.

Rockafellar, R. T. [1976b], Augmented Lagrangians and applications of the proximal point algorithm in convex programming, Mathematics of Operations Research, 1, 97–116.

Rockafellar, R. T. [1985], Lipschitzian properties of multifunctions, Nonlinear Analysis, 9, 867–885.

Rockafellar, R. T. [1989], Proto-differentiability of set-valued mappings and its applications in optimization, in *Analyse non linéaire (Perpignan 1987),* Ann. Inst. H. Poincaré, Anal. Non Linéaire, 6, suppl., 448–482.

Rockafellar, R. T. and R. J-B. Wets [1998], *Variational Analysis,* Springer, Berlin.

Rudin, W. [1991], *Functional analysis,* Second edition, McGraw-Hill, Inc., New York.

Schirotzek, W. [2007], *Nonsmooth analysis,* Springer, Berlin.

Scholtes, S. [1994], *Introduction to piecewise differentiable equations,* Preprint 53, Institut für Statistik und Mathematische Wirtschaftstheorie, Universität Karlsruhe, Karlsruhe.

Spanier, E. H. [1966], *Algebraic topology,* McGraw-Hill, New York.

Stewart, G. W. and J. G. Sun [1990], *Matrix perturbation theory,* Academic Press, Boston.

Schwartz, L. [1967], *Analyse mathématique,* I, Hermann, Paris.

Ursescu, C. [1975], Multifunctions with convex closed graph, Czechoslovak Mathematical Journal, 25 (100), 438–441.

Veliov, V. M. [2006], Approximations with error estimates for optimal control problems for linear systems, in *Large-scale scientific computing,* 263–270, Lecture Notes in Computer Science, 3743, Springer, Berlin.

Walkup, D. W. and R. J-B. Wets [1969], A Lipschitzian characterization of convex polyhedra, Proceedings of the American Mathematical Society, 23, 167–173.

Zălinescu, C. [2002], *Convex analysis in general vector spaces,* World Scientific Publishing, NJ.

Notation

2C(4): formula (4) in Section 2C
R: the real numbers
N: the natural numbers
$\mathcal{N}$: the collection of all subsets N of N such that $N \setminus N$ is finite
$\mathcal{N}^\sharp$: the collection of all infinite subsets of N
$\{x^k\}$: a sequence with elements x^k
$\varepsilon_k \searrow 0$: a sequence of positive numbers ε_k tending to 0
$\limsup_k C^k$: outer limit
$\liminf_k C^k$: inner limit
$|x|$: Euclidean norm
$\|x\|$: any norm
$\langle x, y \rangle$: canonical inner product, bilinear form
$|H|^+$: outer norm
$|H|^-$: inner norm
$B_a(x)$: closed ball with center x and radius r
B: closed unit ball
cl C: closure
int C: interior
core C: core
rc C: recession cone
P_C: projection mapping
$T_C(x)$: tangent cone
$N_C(x)$: normal cone
K^*: polar to cone K, mapping adjoint to K, space dual to K
$K_C(x, v)$: critical cone
$A^\top$: transposition
rank A: rank
ker A: kernel
det A: determinant
$d_C(x), d(x, C)$: distance from x to C
$e(C, D)$: the excess of C beyond D
$h(C, D)$: Pompeiu-Hausdorff distance
dom F: domain
rge F: range
gph F: graph
$\nabla f(x)$: Jacobian
$Df(x)$: derivative

$\mathscr{C}^k$: the space of k-times continuously differentiable functions
$DF(x|y)$: graphical derivative
$D^*F(x|y)$: coderivative
$\mathrm{clm}(f;x), \mathrm{clm}(S;y|x)$: calmness modulus
$\mathrm{lip}(f;x), \mathrm{lip}(S;y|x)$: Lipschitz modulus
$\widehat{\mathrm{clm}}_p(f;(p,x))$: partial calmness modulus
$\widehat{\mathrm{lip}}_p(f;(p,x))$: partial Lipschitz modulus
$\mathrm{reg}(F;x|y)$: regularity modulus
$\mathrm{subreg}(F;x|y)$: subregularity modulus

Index

adjoint
  equation, 356
  upper and lower, 269
ample parameterization, 85
Aubin property, 159
  partial, 167

calmness
  isolated, 186
  partial, 25
Cauchy formula, 353
complementarity problem, 64
cone, 62
  critical, 98
  normal, 62
  polar, 64
  recession, 267
  tangent, 65
constraint qualification, 70
contraction mapping principle, 15
  for set-valued mappings, 284
convergence
  linear, 329
  of iterations under metric regularity, 328
  Painlevé–Kuratowski, 135
  Pompeiu–Hausdorff, 140
  quadratic, 331
  set, 134
  superlinear, 329
convex programming, 73

derivative
  convexified graphical, 213
  Fréchet, 275
  graphical, 199
  one-sided directional, 89
  strict graphical, 239

  strict partial, 34
discretization, 359
distance, 28
  Pompeiu–Hausdorff, 138
  to infeasibility, 322

Ekeland variational principle, 207
estimator, 38
  partial, 45
excess, 138

first-order approximation, 36
  partial, 45
function
  calm, 22
  convex, 66
  Lipschitz continuous, 5
  monotone, 106
  piecewise smooth, 93
  positively homogeneous, 88
  semidifferentiable, 89
  strictly differentiable, 31
  upper semicontinuous, 4

Galerkin method, 348
generalized equation, 62
generalized Jacobian, 238

homogenization, 323

implicit function theorem
  classical (Dini), 17
  for generalized equations, 79
  for local minima, 125
  for Newton iteration, 341
  for optimal control, 355
  for stationary points, 123

Goursat, 20
Robinson, 75
utilizing semiderivatives, 92
with first-order approximations, 79
with strong metric regularity, 180
with strong metric regularity in metric spaces, 294
implicit mapping theorem
for a constraint system, 221
with graphical derivatives, 211
with metric regularity, 172
with metric regularity in metric spaces, 286
with strong metric subregularity, 192
inner and outer limits, 134
inverse function theorem
beyond differentiability, 38
Clarke, 238
classical, 10
for local diffeomorphism, 48
Kummer, 240
symmetric, 24
with strong metric regularity in metric spaces, 292
with strong metric regularity, 179
inverse mapping theorem
with continuous and calm local selections, 300
with metric regularity, 169
with metric regularity in metric spaces, 280
with strong metric subregularity, 190

Karush–Kuhn–Tucker conditions, 72

Lagrange multiplier rule, 70
lemma
critical face, 229
discrete Gronwall, 360
Hoffman, 150
reduction, 98

Mangasarian–Fromovitz constraint qualification, 176
mapping
adjoint, 252
calm, 182
feasible set, 145
horizon, 323
inner semicontinuous, 142
linear, 5
Lipschitz continuous, 148
locally monotone, 181
maximal monotone, 335
optimal set, 145
optimal value, 145

outer Lipschitz continuous, 154
outer semicontinuous, 142
Painlevé–Kuratowski continuous, 142
polyhedral, 155
polyhedral convex, 150
Pompeiu–Hausdorff continuous, 142
positively homogeneous, 200
stationary point, 115
sublinear, 265
with closed convex graph, 259
metric regularity, 164
coderivative criterion, 246
derivative criterion, 205
of sublinear mappings, 265
strong, 179
metric subregularity, 182
derivative criterion for strong, 218
strong, 186
modulus
calmness, 22
Lipschitz, 26
metric regularity, 164
metric subregularity, 183
partial calmness, 25
partial uniform Lipschitz, 34

Nash equilibrium, 73
necessary condition for optimality, 69
Newton method, 11
uniform convergence, 337
for generalized equations, 327
nonlinear programming, 72
norm
duality, 270
outer and inner, 202
operator, 6

openness, 56
linear, 166
optimal control, 352
optimization, 67

parametric robustness, 88
projection, 28
proximal point method, 327

saddle point, 73
second-order optimality, 113
selection, 49
seminorm, 22
set
adsorbing, 259
convex, 27
locally closed, 169

Index 375

polyhedral convex, 97
space
   dual, 252
   metric, 251
SQP method, 334

theorem
   Baire category, 260
   Banach open mapping, 253
   Bartle–Graves, 298
   Brouwer fixed point, 52
   Brouwer invariance of domain, 47
   correction function, 19
   Graves, 276
   Hahn–Banach, 270
   Hildebrand–Graves, 58
   Lyusternik, 275
   Michael selection, 298
   Minkowski–Weyl, 97
   Nadler, 291
   Nash–Moser, 310
   radius for metric regularity, 317
   radius for strong metric regularity, 319
   radius for strong metric subregularity, 319
   Robinson–Ursescu, 263
two-person zero-sum game, 73

variational inequality, 62
   for a Nash equilibrium, 73
   affine polyhedral, 100
   Lagrangian, 71
   monotone, 110

Lightning Source UK Ltd.
Milton Keynes UK
27 January 2011
166489UK00007B/5/P